Why Cities Need Large Parks

Why Cities Need Large Parks

LARGE PARKS IN LARGE CITIES

Edited by Richard Murray

Routledge
Taylor & Francis Group

LONDON AND NEW YORK

First published 2022
by Routledge
605 Third Avenue, New York, NY 10158

and by Routledge
2 Park Square, Milton Park, Abingdon, Oxon, OX14 4RN

Routledge is an imprint of the Taylor & Francis Group, an informa business

Library of Congress Cataloging-in-Publication Data
A catalog record for this title has been requested

ISBN: 978-1-032-07297-5 (hbk)
ISBN: 978-1-032-07293-7 (pbk)
ISBN: 978-1-003-20637-8 (ebk)

DOI: 10.4324/9781003206378

Typeset in Adobe Garamond Pro and News Gothic
by Adobe Systems Incorporated

Publisher's Note
This book has been prepared from camera-ready copy provided by the editor.

For the realization of this book the generous contributions from the following foundations, organizations, and individuals have been invaluable:

Formas—a Swedish research council for sustainable development

Frances Boylston

King Gustaf VI Adolf's Foundation for Swedish Culture

The Committee for the Gustavian Park—from the donations of Stig Ramel and Greger Carpelan

The Royal Patriotic Society

The Royal Swedish Academy of Letters, History, and Antiquities

Åke Wiberg Foundation

Richard Murray, editor of this book, has been an environmental activist since the 1960s. He is President of the Swedish Ekoparken Association (Förbundet för Ekoparken) and is the Co-chair of Large Urban Parks Committee of the World Urban Parks Association. He has a PhD in political economy from the University of Stockholm.

Contents

Visions of Urban Green

Quality of Life

Ecosystem Services

Social Cohesion

City Meets Nature

Heritage and Identity

Design for All

The accelerating growth of cities and the expansion of urban areas around the world present not only many opportunities but also difficult challenges for cities. Stockholm is no exception. Responding to these challenges, WWF Sweden initiated Ekoparken WWF, a project aiming to protect a series of connected royal parks from being exploited.

Despite the parks' locations in the heart of the Stockholm metropolitan area, an astonishing amount of biodiversity was discerned. For many people life in cities has severed their connection to nature: the Ekoparken project would bring Stockholmers closer to nature. Recognizing this, the WWF committed itself to the project. In addition, these parks serve to conserve a heritage landscape of great historic significance, as well as enhancing and preserving sites for recreation. Last but not least, the Ekoparken has a dampening effect on storm water flooding and heat waves, factors important in the context of a future marked by more extreme weather and climate.

On May 19th, 1995, Swedish King Carl XVI Gustaf inaugurated the Royal National City Park at a ceremony in front of Ulriksdal Palace, located in one of the royal parks and home to the headquarters of WWF Sweden. This was the result of a collaboration between WWF and a group of dedicated NGOs. The King released three white doves to symbolize three strategic aims for the park: Protect, Care and Present. The white dove for protection had difficulties taking off – an omen and a call for the conservation of urban green spaces in expanding cities all around the world.

WWF wants to secure a future in which people live in harmony with nature. To that end, the WWF works with business, NGOs, and government to protect the natural environment of the land and the sea, including nature within and close to cities. The conference Large Parks in Large Cities, held in Stockholm in 2015 together with Ekoparken Association, was an important moment in stressing the importance of having nature close to cities. The book you are holding in your hands is based on the many accounts presented at the conference about the necessity of large urban parks. An important outcome of the conference was the collaboration of WWF Sweden and Ekoparken Association with the World Urban Parks organization. It is clear that urban green space is an issue of paramount importance across the globe, not least in the fast-growing megacities of Asia, Africa and Latin America.

This book provides some important answers to the question of Why Cities Need Large Parks. Read it: I am sure it will inspire you to care for and share the green treasures of your cities!

Håkan Wirtén
Former Secretary General, WWF Sweden

Demand for large urban parks emerged at the height of the First Industrial Revolution in the mid-1800s, when many citizens were experiencing poor quality of life and unhealthy conditions. Large urban parks represented new ideas of accessible public spaces, often established on land previously owned by aristocracy, royalty or the army. They represented new ideas on how city life could be improved and how large green spaces could enhance urban citizens' physical and psychological well-being (e.g. Birkenhead Park in Liverpool, Bois de Boulogne in Paris, Tiergarten in Berlin and Central Park in New York City).

The multifunctionality of large urban parks is at their core. They are places for citizens to relax or play sports; venues for cultural or recreational events; and very often contain buildings and facilities of great historical and cultural value, for example the Güell Park in Barcelona by the master Gaudí, and the Meiji Shrine in Yoyogi park in Tokyo.

In today's world we add increasingly important benefits to the list. Large urban parks are habitats for biodiversity and also serve functions of climate change adaptation. The notion that cities may have rich biodiversity is today recognized by the Convention of Biological Diversity, and large urban parks are often habitats for an astonishingly rich fauna and flora (e.g. Cape Town, Singapore, Mexico City). For people living in cities, this biodiversity may represent high cultural, recreational and aesthetic values, but is also important for other aspects of health and well-being, for example by reducing the urban heat island effect, air pollution and risks of flooding.

At a time when we are seriously reconsidering how we live in cities and our urban quality of life, while also grappling with serious challenges of climate change, the authors of this book detail the much-needed evidence, pathways and vision for a future of more liveable, resilient cities where large urban parks are at the core.

Thomas Elmqvist, Professor in Natural Resource Management
at Stockholm Resilience Centre, Stockholm University

RICHARD MURRAY

Why Cities Need Large Parks

This book has grown from a conference—Large Parks in Large Cities—held in Stockholm in September 2015, and organized by Ekoparken Association and WWF Sweden. Four Royal Academies (Science, Forestry and Agriculture, Letters, History and Antiquities and Fine Arts), Stockholm University, and Swedish Agricultural University participated in the organizational committee. People from twenty-six countries from six continents participated. Their accounts of important parks around the world and the conference's keynote speeches form the core of this book, with a few parks added in order to create an even more worldwide perspective. Since then, I have worked to form this book to reflect the understanding emanating from the conference of how important large parks and green infrastructure in cities are. The conference and the book constitute a concious effort to illuminate the value of large parks and urban green and to help ward off attacks on green spaces in cities all over the world.

The book spans all inhabited continents with examples from thirty major cities. Together with the nine survey articles ca. 300 urban and peri-urban parks and nature reserves are mentioned and discussed. No other such overview exists to date.[1] This forms a basis for further inquiry regarding green infrastructure in large cities world wide. Just to mention one issue: to find out how and to what extent green infrastructure is integrated into master planning for large cities.

It is my hope that this book will help park people, managers, NGOs and city planners to develop the green city. Several international organizations actively promote learning, such as World Urban Parks, Europarc and The International Union for the Conservation of Nature. And Central Park Conservancy has set up The Institute for Urban Parks for education and peer-to-peer exhanges, building on its long experience as a large urban and public park.

In the early 1990s most cities in the Western world did not grow, some even shrank. This had been going on for more than a decade. Thus a window of opportunity to set aside land for parks was open. This happened in Stockholm, and also in other places. In 1995,

LEFT: Djurgårdsbrunnskanalen, a canal dug in the early 19th century for the beautification of Djurgården, part of the Royal National City Park in Stockholm, Sweden.

Minister for Environment Görel Thurdin with Henrik Waldenström, WWF, arguing for the preservation of Ekoparken, on top of a ski-jump from the 1920s in the park. The parliament enacted, in complete unison, legislation worked out by Thurdin and her ministry in 1994.

Fedenatur, an organization for peri-urban parks in Southern Europe was started. Many cities can tell storeis of land that has been purchased in boom times not being used for development but instead being ear-marked as nature reserves when economic recession has set in. More remarkable is the early foresight to set aside large tracts of land in fast growing cities. For example, New York created Central Park in the 1850s, Colonel William Light, set aside a green belt of several blocks around Adelaide in 1837 and Charles V of Spain ordered the forest of Chapulte-pec to become the property in perpetuity of Mexico City as a place of recreation for the inhabitants of the city—which he did in 1530!

Since around the year 2000 cities all over the world have been grow-ing fast. It has been projected that within the next thirty years city population will increase by 50 percent. A doctrine is that resilient cities should be built compact. But cities still grow faster in area than in population. This is a trend since at least 1800 and the trend continues up to this day. In Europe and Japan population densities have shrunk from about 80 persons per hectare in 1990 to 65 in 2000. And in developing countries from 175 to 135. Land-rich developed countries, such as Australia and the USA, have a much lower city population density, but have still decreased from about 25 to 20 persons per hectare.

Given this trend, there ought to be more room for parks, large and small, but that seems not to be the case. Real estate capital all around the world, looking for profitable investments, shows a great interest in centrally located green areas. Local politicians find using green spots to be the least complicated way to solve the need for new schools, day care centers, hospitals, parking places, roads and other public amenities. In addition, budgets for managing green areas have diminished over a long period of time. Peter Clark summarizes, in his survey: "In recent times, however, green space in its many varieties has been under attack."

Why do cities, especially large cities, need large parks? My commitment stems from the campaign I was brought into, in 1993, to save a collection of royal parks in the center of the Stockholm region. Twenty-two NGOs had formed Ekoparken Association, and I was asked to chair it. People had begun to call attention to a massive development threatening these parks around 1990. Henrik Waldenström, an ornithologist who had noted the rich bird life in these urban areas, adjacent to the very city center, got support from the Royal Court to prepare a report, *Ekoparken—A Nature and Culture Reserve in a Big City*, arguing for the preservation of this large tract of land, right in the middle of Stockholm, capital of Sweden. At the same time Peter Schantz, a scholar in the field of human movement biology, focused on that part of the area that includes several English landscape parks and a Baroque park. To that end, he and a group of academics, in 1991, founded the Committee for the Gustavian Park. Both contacted parliamentarians who wrote propositions. Görel Thurdin, Minister for the Environment, ordered an investigation and on the basis of that proposed legislation in 1994. A law went into force on the 1ˢᵗ of January 1995 and reads:

> *Ulriksdal–Haga–Brunnsviken–Djurgården is a national urban park. New development, new buildings and other measures shall only be permissible in national urban parks if they can be undertaken without encroaching on park landscapes or the natural environment and without detriment to any other natural and cultural assets of the historical landscape.*

The Royal National City Park of Stockholm is the first and today the sole national urban park in Sweden. The protected area comprises 27 square kilometers of land, lakes and sea in the midst of the 2 million inhabitants of the conurbation of Greater Stockholm. A government commission suggested that two other national urban parks be formed but city councilors in those cities objected.

Finland, however, picked up the idea of National Urban Parks very quickly. Legislation opened up for National Urban Parks to be instituted in 2000. The Finns made a slight change of emphasis, explicitly including historic city centers "important for an understanding of national history

Given this trend, there ought to be more room for parks, large and small, but that seems not to be the case.

JUKKA-PEKKA FLANDER

JUKKA-PEKKA FLANDER

or of that of the city itself." Biodiversity, recreation and social values are as important in the Finnish parks as in the Swedish one. In Finland it was realized that even the not very large cities contain important treasures of biodiversity as well as of a rich cultural heritage. Establishing a National Urban Park is also seen as a strategic means to limit urban sprawl.

Thus far, ten cities have up to now applied for and been designated National Urban Park status. The first one was established in 2001, in Hämeenlinna (in Swedish: Tavastehus). It covers 10 square kilometers, in a city of 70,000 inhabitants. In 2002 Pori (in Swedish: Björneborg) National Urban Park was established. It covers also 10 square kilometers, in a city with a population around 100,000. The National Urban Park in Pori includes the Kokemäki River, which has the largest river mouth wetlands in the Nordic countries—a fantastic bird sanctuary. The tree clad boulevards connect the inner city with the Pori Forest, an island used as a recreational area, inhabited by flying squirrels (*Pteromys volans*). The two largest National Urban Parks are in Hanko (in Swedish: Hangö), with 63 square kilometers of which land comprises 3 and sea 60 square kilometers, and in Kuopio, with 72 square kilometers (land 2 and sea and lakes 70 square kilometers).

Prior to the decision to establish a National Urban Park, a dialogue must take place, involving the Ministry of Environment, citizens and organizations of the city concerned. This process of participation has meant that broad segments of the city population, business, civic organizations and local politicians, eventually endorse and embrace the park with enthusiasm.

Large urban parks are seldom shaped in virgin land, more often taking up unbuildable and degraded areas and transforming them. Central Park, Adelaide and Chapultepec are notable exceptions as is Kings Park in Perth, Australia. Graham Fairclough points out that most large urban parks have had a cultural origin. Is that still so? Contemporary park projects that Julia Czerniak presents are often efforts to turn marginalized land—swamps, rocky hills, dumps, decaying railway yards and former factory sites—into green, recreational areas. In the harsh competition for land, residential and business developments usually win. Parks may be created out of what is left. Around the world, there are many projects—many more than those exemplified in this book—

TOP LEFT: Birger Jarl (the statesman who led the consolidation of Sweden, ca.1210–1266) had the castle in Hämeenlinna built in the 13th century, when Finland became a part of Sweden.
BOTTOM LEFT: Will Helsinki apply for a NUP? Mustikkamaa (Blueberry land) is a recreational island close to the city center of Helsinki, the capital of Finland.

that do just that; revitalizing derelict land. This is done in several US cities, and also in Latin America, Africa and Asia. Doing this has multiple purposes. The most obvious one is to rehabilitate nature from previous human abuse. But aside from recreation opportunities and touristic fame, it also gives cities fresh air, fresh water, flood protection, lower temperatures, biodiversity—and reconnects urbanites with nature. These are examples of what, nowadays, is termed 'nature-based solutions'. Estimating monetary benefits from urban ecosystem services shows that such projects are profitable, with cost savings from better health, energy savings, reduced flood damages, fresh water supply and other amenities estimated to amount to many billions of dollars.

There are very little comparable data on land use in cities—what proportion of the city area is green? Greater London National Park City claims that 49.5 percent are green/blue spaces in that city, evidencing this with a very detailed and interesting mapping. The United Nations Sustainable Development Goal no. 11 targets the universal access to safe, inclusive and accessible, green and public spaces. Only fragmentary statistics in order to monitor progress on this goal have as yet been produced. Measurement of land use is still a research topic. An investigation of the availability of urban green space in European cities shows that the share of city population that live within 500 meters from an urban green space two hectares or more varies tremendously—between less than 20 percent to more than 90 percent. Stockholm, with its Royal National City Park is close to the top. Other measures have to be taken into account, for example square meters of urban greenspace per resident, since if a lot of people live very close to a park it may be over-crowded. A lot more needs to be done to improve measurements—type of urban green, quality, connectivity, availability etc.—in order to understand what the requirements are for a healthy life. Richard Forman argues, based on available research, that large parks and small parks and their connectivity have different values in relation to recreation, urban heat, flood reduction and biodiversity. His and others' research would benefit from comparable data.

Typically, availability of green space is very unevenly distributed in cities, favoring the better off. The mechanisms explaining why more greenery leads to better health and longer lives are dissected by Peter Schantz. From time to time, park projects have been initiated to rectify the uneven distribution. The People's Park Movement in the early 20th century, important for the formation of several of the parks presented in this book, is ample evidence.

Cities grow, cities spread. A large-scale example to control urban sprawl

Typically, the availability of green space is very unevenly distributed in cities, favoring the better off.

is Midpeninsula Regional Open Space District south of San Francisco, established in 1972, now totaling 250 square kilometers. Teresa Pastor writes about other peri-urban parks that have been used to control urban sprawl, whilst at the same time providing urbanites with recreation opportunities. Projections of global urban area growth based on historical time series data show an increase in urban land by up to 67 percent in the years 2013–2050, very much concentrated to Africa and Asia. Anna Persson shows that city expansion becomes a real problem for biodiversity, because cities are often surrounded by areas with high biodiversity and threatened species. Tijuca forest outside of Rio de Janeiro is an example. Many of the growing cities are close to protected areas and biodiversity hotspots. Urban Emanuelsson describes different ways in which the city can meet countryside and nature and how this impacts recreational and other values. Richard Forman stresses the need for forward looking urban-region plans for every city.

Patricia O'Donnell reminds the reader that urban parks not only function as inclusive recreational assets and meeting places but also serve as important platforms for public assembly and free speech. Peter Clark, in accounting for the historic development of urban parks, points out the close connection between democratic developments and demands for public space. Graham Fairclough recalls, that the international community, having ratified the Florence and Faro charters, has agreed that "the actors in the heritage process are all of us, not only heritage experts or specialists, not only owners and governments, but everyone, whether individually or collectively." Thus, it is a democratic human right to have a say on landscape and parks.

The many large parks and green infrastructure systems that are presented in this volume illustrate the diverse uses, multiple values, and many benefits of large urban parks. The presentations are grouped according to themes but that is only part of the story, since all of them touch upon many different values of urban parks. Richard Forman, in his introductory chapter, provides the reader with analytical tools to understand and assess the characteristics and merits of the various parks presented in the book.

I am deeply grateful and want to thank the 45 persons (and others behind them) around the globe that have contributed to make my idea become a real book.

[1] Steenbergen and Reh (2011) is the closest. Their book details the history of eleven parks/park systems in Europe and United States from a landscape architecture point of view. Large Parks by J. Czerniak and G. Hargreaves (eds.) should also be mentioned.

RICHARD T. T. FORMAN

Values of Large-versus-Small Urban Greenspaces and their Arrangement

A large urban park glistens like an emerald in the city. Everyone has been in the place and treasures it. Yet the city also has numerous small parks and greenspaces scattered across diverse neighborhoods. Residents walk to and meet neighbors in these small spots. Indeed, parks also affect the air, water, and biodiversity of a city. What values do the big and little parks provide? Does their spatial arrangement matter? Which is more valuable, the large one or the many small ones?

Here I evaluate, apparently for the first time, the effect of urban park size, using four key values or benefits, i.e., effects on: air cooling, biodiversity, flood reduction, and recreation. These and other values or ecosystem services are also compared in selected large and medium-size urban greenspaces globally.

Furthermore, most likely the arrangement of greenspaces strongly affects their collective or cumulative value for a city. Greenspaces may be near the center or at the edge of a city. How does that affect their values or roles? And how are parks arranged relative to each other, and does that affect their values or functions? Indeed, arrangement is a key to converting a collection of parks or greenspaces into a park system with viable linkages.

At the heart of the urban park-size question is the relative value of one large park versus that of the same area subdivided into many small parks.

I consider these topics in the following order: (1) large, medium, and small parks; (2) arrangements of greenspaces; and (3) values in major urban greenspaces globally. While the administrative concept of a park may include both greenspace and built area, such as the impressive Royal National City Park of Stockholm, only vegetation-dominated or pure greenspace parts of parks are included in this analysis. Although greenspace as a more inclusive concept, includes parks and other unbuilt areas, such as nature reserves, sports fields, brownfields, and community gardens (allotments), for convenience the terms park and greenspace will be used interchangeably here.

LARGE, MEDIUM, AND SMALL PARKS

Many combinations of large, medium, and small greenspaces exist in cities. Some may provide considerable cumulative value, such as: (1) a few (or several) medium-size parks; (2) many small parks; (3) one large and many small; and (4) a few medium and many small parks.

My analysis assumes that the surroundings are all built, the shape of parks is square or rounded (Forman 1995, 2014), and no major green corridors are present. In reality, few cities have one large park, while some have a few or several medium-size parks (in shorthand, FM or SM parks). Almost all cities have many small parks. Green corridors of varied size may be common.

Clearly one large park is better than one small park. At the heart of the urban park-size question though is the relative value of one large park versus that of the same area subdivided into many small parks (OL or MS). The essential background for this question lies in early studies of biodiversity in large-to-small woodlands (SLOSS, single large or several small) in rural and remote landscapes (Simberloff and Abele 1976; Forman *et al.* 1976). Small woodlands are widely spread out and together they normally contain many habitats and common species. In contrast, a large woodland usually contains uncommon or rare species, especially in the woodland interior away from the edge.

Simple spatial models using limited but useful empirical data, plus several well-documented ecological patterns or principles, are used to evaluate the park-size question. (1) Air cooling, (2) biodiversity, (3) flood reduction, and (4) recreation are evaluated, while indirect evidence is provided for (5) uptake of CO_2 and other air pollutants.

AIR COOLING

Measurements of 42 greenspaces in Berlin found that the amount of cooling (i.e., temperature in a park compared with temperature in the surrounding built area) correlated with greenspace size (Figure 1a). On average a small greenspace (<30 ha) was approximately 1°C cooler, a medium one (30–500 ha) 3°C cooler (though with much variability), and a large greenspace (>500 ha) 5°C cooler.

Furthermore, a cooling effect extends outward from a greenspace. In Berlin, the rough average distances of cooling from a park boundary were (Figure 1b): small park, 150 m; medium, 250 m; medium-large, 500 m; and large (extrapolated), 750 m. Combining the cooling patterns from Figures 1a and 1b suggests that theoretically an entire city area could be cooled in the heat of summer by greenspaces. Thus, in the example of Figure 1c, the approximate average amount of cooling is: 5°C for the large park; < 2.5°C for its 750-m-wide surrounding area (a curvilinear decrease from 5° at park edge to 0° at 750 m out); 1°C for the small parks; and <0.5°C for the space around them (curvilinear decrease from 1° to 0°). Large parks cool a city much more than do small parks.

Tiergarten, a medium-large park with planted vegetation in the central area of Berlin, Germany.

Thus, a large greenspace cools a lot, and cooling extends far outwards, whereas a small one cools a little, and cooling extends over a small area (Chang *et al.* 2007). In the Berlin case (Figure 1c), the urban area (white space) cooled around one large park equals the area cooled around approximately 70 small parks. Furthermore, the area cooled is almost all cooled to a lower temperature in the one-large greenspace case, compared with the many-small case. These results therefore may be diagrammed as OL>>MS, one large is much better than many small greenspaces for air cooling.

BIODIVERSITY

To simplify, let us consider biodiversity to be mainly the number of species present, and secondarily the number of habitats or natural communities present. Since the small parks are distributed over a wider area than the large one, they are likely to include many habitats/communities, and hence many species. On the other hand, a large park is more likely to contain a high hill or large wetland or waterbody, which also supports many habitats and species. Note that species number or diversity does not indicate species composition, i.e., what species are present. A botanical study of 10 large urban parks ($\geq$400 ha each) within and at the edge of major cities in the Boston-to-Washington megalopolis recorded 1391 vascular plant species, 65 percent being native species (Loeb 2006). Less than 1 percent of the species are in all ten parks, and less than 2.5 percent in nine or more parks. No distinct park flora exists. Based on the plant diversity and composition, no correlation exists with park size or spatial distribution of the large parks.

To compare biodiversity in large versus small parks, let us first eliminate the preceding variables and assume that all parks are covered by similar semi-natural vegetation. Several ecological processes determine the one-large-versus-many-small-parks outcome for biodiversity (Figure 2).

- Common generalist species that thrive in many environmental conditions are abundant in both cases.
- A 'species rain' from a large natural-area source outside the city provides a continuous immigration of species to all urban parks, with more species entering the nearby, and fewer the more-distant, small parks.
- This 'species rain' includes specialist native species, which do poorly

RIGHT: In summer, cool shady urban areas attract people. Lisbon, Portugal.

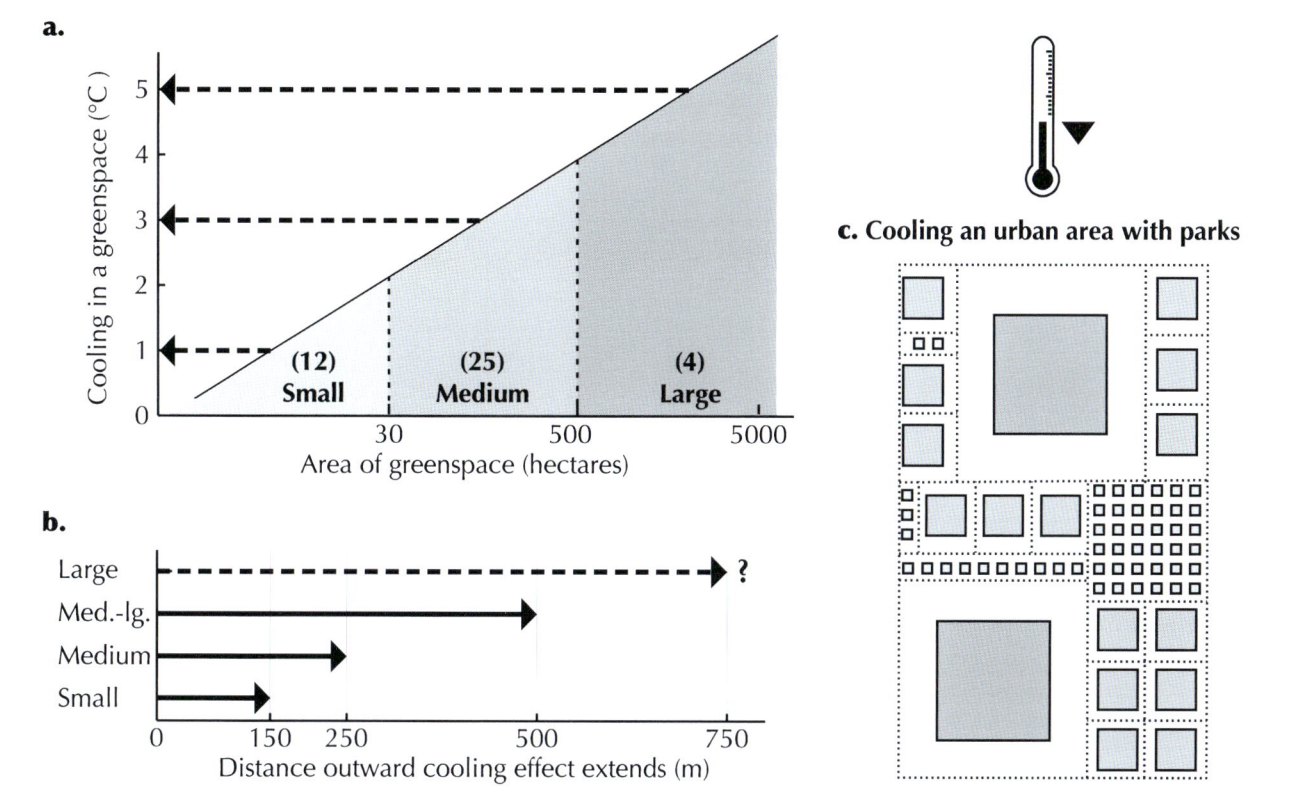

a.

Cooling in a greenspace (°C)

c. Cooling an urban area with parks

b.

Distance outward cooling effect extends (m)

Figure 1. Air cooling in and around small, medium, and large urban greenspaces. (a) and (b) based on Berlin (von Stulpnagel et al. 1990). (a) Cooling = approximate average temperature decrease compared with surrounding urban area (number of greenspaces indicated in parentheses). (b) Based on moderate wind (2-4 m/sec), and averaging limited measurements of upwind and downwind data. (c) Schematic diagram based on the results in (a) and (b) (Forman 2014). Rough estimate for large greenspaces extrapolated from the distances in (b).

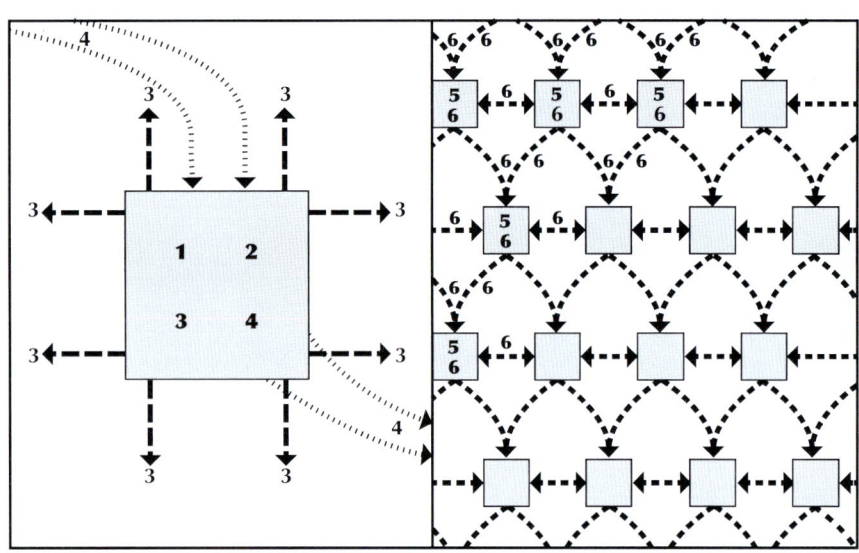

TACO I. MATTHEWS

1 Large populations of common generalist species
2 Uncommon specialist species briefly surviving
3 Source of "species rain" in surrounding areas
4 Rest stop for birds migrating across urban area
5 Dispersed sites with numerous common species
6 Stepping stones for widespread species movement

Figure 2. Ecological processes affecting the biodiversity of large and small urban greenspaces. Arrows indicate relative distance that effects extend outward from a greenspace. The large urban greenspace on the left and the small ones on the right have the same total area.

GUSTAVO PEDRO

in the small parks, but which on average survive for a period in a large park. This process maintains a relatively species-rich large urban greenspace of moderate conservation value.

• Small parks serve as stepping stones especially for widespread movement of common generalist species across the city.

• A large urban greenspace is the source of a 'species rain' for the local area surrounding it.

• Finally, birds migrating across the city are most likely to stop en route in a large greenspace for their needed food and rest.

Combining these patterns indicates that biodiversity is likely to be higher in one large patch than in many small ones (OL>MS).

Rather than having all semi-natural vegetation, a somewhat more realistic model has small greenspaces overwhelmingly covered by mowed grass, or by grass, benches, walkways, and scattered trees. Typically, a large greenspace is mainly covered by grass-bench-walkway-tree habitat, but also maintains some semi-natural vegetation present (Kowarik and Korner 2005; Konijnendijk *et al.* 2005). With this spatial pattern, the large greenspace containing semi-natural vegetation is likely to be much richer in biodiversity than the small ones (OL>>MS).

FLOOD REDUCTION

Generally, floodwaters rapidly enter a city from upriver or upslope. Greenspaces may then accelerate the lowering or drying up of flood-waters in three ways (Todd and Mays 2005; Welty 2009; Forman 2019): evapotranspiration[1], surface-runoff interception, and infiltration into the soil. These three hydrologic processes are evaluated in large versus small parks, using a simple spatial model analogous to that in Figure 2.

An elevated rate of evapotranspiration on the windy side of a large semi-natural greenspace extends farther into it than on a protected side. This creates a wider edge effect (Forman 1995) distinct from the green-space interior (Figure 3). Solar radiation on the sunny side causes the same effect. Wind and solar radiation are typically the prime drivers of evapotranspiration, which pumps water out of the soil into the air. On the other hand, wind and sun effects penetrate throughout small parks. Consequently, the elevated rate of evapotranspiration occurs over the entire surface of small parks, but only in the peripheral portion of the typical large greenspace. Therefore, for flood reduction by evapo-

---

BOTTOM LEFT: Squirrel monkey in neighborhood close to Tijuca National Park, Rio de Janeiro, Brazil.

transpiration, many small greenspaces are better than one large greenspace (MS>OL).

Water evaporated and air cooling by park vegetation extends outward (Figure 3), ameliorating the adjacent microclimate (see Figure 1). Microclimatic amelioration extends farthest downwind from a large greenspace.

Surface water runoff from upper to lower areas of a city, encounters parks en route that may intercept or slow the flow. The total length of upslope park edges is a measure of the parks' capacity to intercept (stop or slow) water runoff. Compared with the large park in the model or diagram, the small parks have four times the amount of upslope edge, and hence surface runoff interception (Figure 3).

Runoff interception can be further increased by diagonal pipes extending a short distance outward and upslope that funnel water from the surrounding built area to a greenspace (Figure 3). Relative to a single large park, the small parks offer many more opportunities over a larger area to capture such water runoff from the surroundings. Therefore, combined with upslope park-edge length, for water runoff interception MS>>OL.

Infiltration of water into the soil is high in the interior of semi-natural vegetation, and typically lower in the more intensively altered edge areas where surface runoff is characteristically greater (Figure 3). Therefore, infiltration is considerably greater in one large than in many small parks (OL>>MS). This difference may be minimized, even reversed, by having all rainwater falling on small parks funneled to a bioretention basin(s) (pond or wetland) in each park (Davis *et al.* 2009; Hurley and Forman 2011). Adding short-distance diagonal pipes from the upslope surroundings connected to the bioretention basin(s) in a park would produce greater water infiltration into soil in many small rather than one large greenspace (MS>OL).

Summarizing the value of greenspaces for flood reduction, water infiltration into soil would be greater in one large versus many small greenspaces, unless mitigation with diagonal pipes and bioretention basins were widespread and reversed the pattern. However, for both evapotranspiration and runoff interception, clearly many small greenspaces provide more hydrologic benefit than one large greenspace. In short, for reducing floodwater, MS>>OL; overall, many small are much better than one large park.

BOTTOM RIGHT: Storm water from the Tony Ireland Park entering the Murrumbidgee River, Wagga Wagga, Australia.

1 Flood-reduction by
  infiltration/evapotranspiration
2 Microclimate amelioration of
  adjacent areas
3 Flood reduction by intercepted
  runoff & evapotranspiration
4 Small microclimate
  amelioration

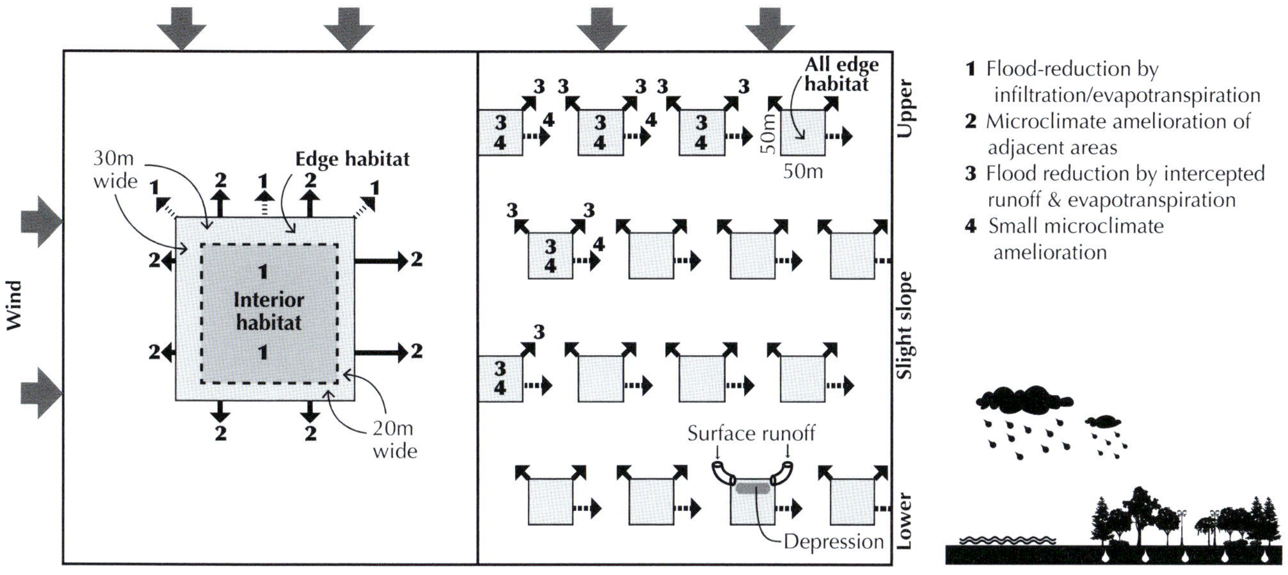

Figure 3. Flood reduction by evapotranspiration, runoff interception, and soil infiltration in one large versus several small greenspaces. The large patch has both edge and interior habitat; the small patches only edge habitat. Diagram, bottom row, illustrates two short pipes draining surface runoff into an elongated oval detention/bioretention basin, which could be added to any greenspace.

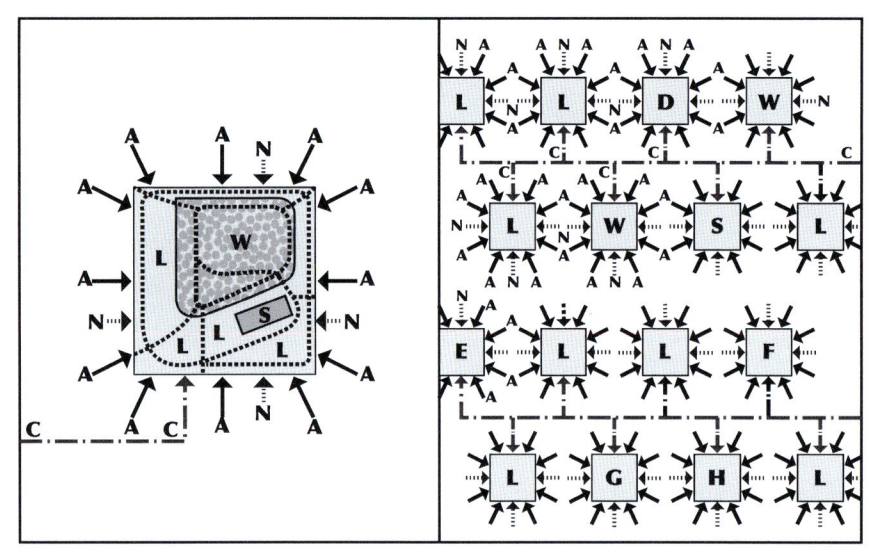

W   Woods or other
        semi-natural area
S   Sports/games area
L   Lawn + other open areas
------ Walking trail network
A   Access distance of most
        park users
C   City park maintenance access
N   Neighbors care for park,
        especially when city
        maintenance budgets drop
D, E, F, G, H Other major park uses

Figure 4. Recreational resources, users, and maintenance in large versus small urban parks.

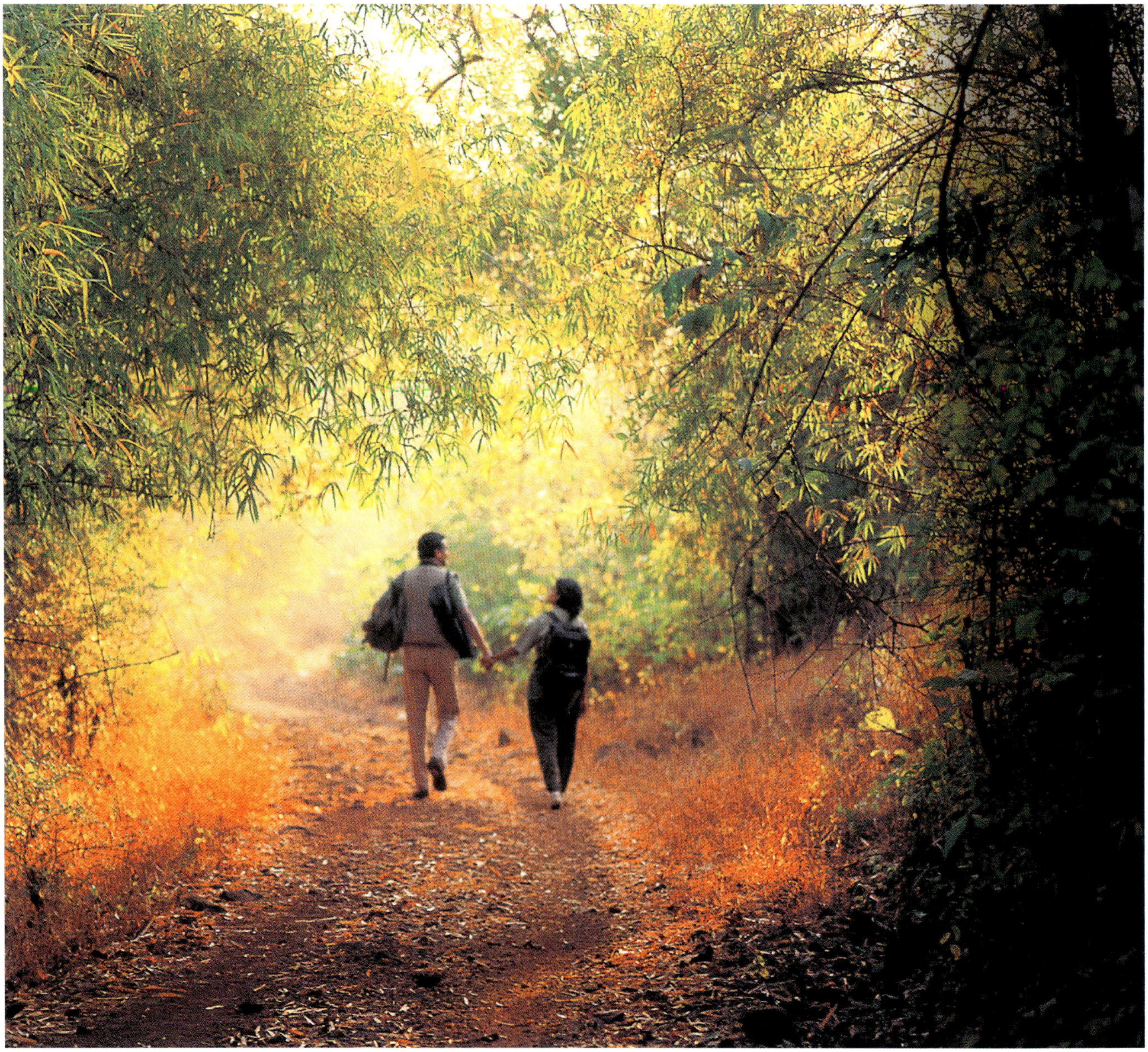

RECREATION

Typically, mowed grass (or grass-bench-walkway-scattered-tree habitat) almost entirely covers small parks, and is more abundant than the usual mowed-grass and semi-natural vegetation of a large park. These patterns are used to compare large-versus-small greenspaces for recreation or leisure value. Several recreational values are important in the evaluation, as follows (Figure 4):

- A trail system in a large park facilitates park walking and jogging exercise, whereas trail networks in small parks are limited or absent.
- A semi-natural area in the large park provides educational, health, and inspirational benefits, commonly limited or missing in small parks.
- Many small parks typically provide opportunities for more people to participate in a diversity of sports and games than does a single large park.
- Many small parks serve people in more widely dispersed neighbor-hoods, and on average residences are closer to small parks than to one large park.
- Small parks tend to differ markedly from one another, and together usually provide a wider range of specialized resources or facilities than does a single large park (Forman 2008).
- Dog-walking in small parks generally causes little ecological degrad-ation by disturbing wildlife and scent-marking wildlife habitats, unlike the considerable impact produced in the semi-natural habitat of a large urban park (Forman 2019).
- Small parks surrounded by numerous diverse neighborhoods are likely to have on average more care or stewardship by neighbors (especially when government park budgets are low), than is the case for a large park.

In effect for recreation, although large parks provide some unique benefits, overall, many small parks are much better than one large park, MS>>OL.

CO$_2$ AND OTHER AIR POLLUTANTS

A study of the uptake of air pollutants by white ash trees (*Fraxinus americana*, 28-cm-diameter) in different Chicago locations provides indirect insight into large-versus-small park values (McPherson 1994). Ash trees in varied-size parks absorbed more CO$_2$ than did trees in house

BOTTOM LEFT: Dawn recreation at a forest trail in Sanjay Gandhi park, Mumbai, India.

yards and street-sides (50, 43, 35 kg/yr, respectively). Park trees also exceeded yard and street trees in the direct uptake of pollutants affecting the health of plants, animals, and people: PM_{10} particulate matter (1.00, 0.80, 0.67 kg/yr); O_3 ozone (0.35, 0.30, 0.23); CO carbon monoxide (0.02, 0.01, 0.01); SO_2 sulfur dioxide (0.23, 0.19, 0.15); and NO_2 nitrogen dioxide (0.25, 0.20, 0.16). These results probably result from better growth (due to more soil moisture and organic matter on average) of park trees than yard and street trees.

It seems likely that the Chicago park trees are more representative of those in large parks, while the yard and street trees better represent those in small parks. Thus, the data suggest that a large park cleans the urban air better than do many small parks, i.e., more absorption of CO_2 and uptake of unhealthful air pollutants (OL>MS).

One large park is much better for air cooling and biodiversity, whereas many small parks are much better for flood reduction and recreation.

HEALTH-RELATED MORTALITY

Similarly, a Philadelphia study linking health-related mortality of surrounding residents to the shape of greenspaces (Wang and Tassinary 2009) provides indirect evidence relative to large-versus-small park values. Health statistics were compared for residents in census tracts (population 2500–8000) with greenspaces >84 square meters per person varying from rounded/squarish to complex (e.g., convoluted boundary, relatively distinct sections, connected nodes). Cause-specific mortality (related to heart disease, chronic lower respiratory diseases or neoplasms) and all-cause mortality were determined for 369 census tracts. Census tracts with more complex-shaped greenspaces showed a lower risk of both cause-specific and all-cause mortality. The authors suggest that this may be linked to more access points for residents provided by complex-shaped greenspaces.

Complex-shaped greenspaces are more similar to a number of small greenspaces than to a single one. Thus, the results suggest that many small greenspaces reduce surrounding health-related mortality better than does one large greenspace.

SUMMARIZING

Summarizing the one large park versus many small parks (OL versus MS) question highlights the four prime values or roles of greenspaces.

One large better
    For air cooling: OL>>MS
    For biodiversity: OL>>MS
    For CO2 and other air pollutants: tentatively OL>MS

Many small better
    For flood reduction: MS>>OL
    For recreation: MS>>OL
    For health-related mortality: tentatively MS>OL

In short, both large and small greenspaces provide value, but different values. One large park is much better for air cooling and biodiversity, whereas many small parks are much better for flood reduction and recreation.

It would be useful to extend this type of analysis to include medium-size green spaces. As suggested at the beginning of this section, a few (or several) medium-size parks, or a few medium parks and many small ones, may offer significant benefits for a city. Furthermore, the arrangement of greenspaces seems extremely important in providing value.

ARRANGEMENT OF GREENSPACES

For this topic, I consider: (1) a system versus a collection of parks; (2) arrangements of large, medium, and small parks; and (3) park arrangements relative to city center and surroundings.

A SYSTEM VERSUS A COLLECTION OF PARKS

Providing routes for effective movement of people and/or species among parks changes a group of parks into a park system, with consequent benefits to both nature and people (Forman 2008, 2014). Unlike the small parks, a large park normally has a considerable network of trails for walking. However, the more a collection of small parks is converted into a park system with flows/movements among them, the better this is for recreation, as well as species movement. For example, tree lines with shade between urban parks facilitate summer walkers between them, even in a loop through small parks.

The "emerald network" model of large interconnected greenspaces may be considered an optimal system. For instance, a plan for the San Diego (California) urban region is primarily composed of large greenspaces (often brown in this dry climate, and with some containing reservoirs) interconnected by green corridors or greenways. In 1997 about one-third of the planned network was protected land, and today over two-thirds is protected—a remarkable success story. The corridors linking large greenspaces help create a network with loops or alternative routes for movement through the network. A similar emerald network emerged in a 2004 plan for the Greater Barcelona urban region (Forman 2004). These are effectively urban-region greenspace systems.

Biological core areas

Biological linkages

---- Study area boundary

Conservation plan for an urban region where the arrangement of greenspaces is a key. "Emerald network" of large greenspaces (green) connected by green corridors (light brown) for San Diego County, California, United States. [Based on New York Times, page 1, February 16, 1997 Martin Thelander.]

Connecting parks with a continuous corridor is normally considered best for movement between parks (Bennett 2003, Forman 2014). Yet a row of stepping stones facilitates some movement and is often more feasible in an urban area (Figure 5b). A cluster of stepping stones between parks is still better because it offers alternative routes (Forman 1995). Even a broken corridor, or peninsulas (lobes) of nearby patches projecting toward each other, facilitates inter-park movement.

The internal design or arrangement of habitats or land uses within parks may also affect movement between them. Consider a system where all parks are heterogeneous (Gilbert 1991) and similar to one another, versus a system where each park is relatively homogeneous and quite different from other parks (see the first and third illustrations in Figure 5a) (Forman 2008). Which system is better? The heterogeneous similar parks all have a diversity of about the same land uses packed together. In contrast, the homogeneous different parks are each specialized, e.g., as relatively unique flagship parks. As a collection of parks, the first case is monotonous while the second case is highly diverse. Considerable inter-park movement of species and people would occur in the first case (a system), but not in the second case (a collection of parks).

Yet the heterogeneous similar parks each support most uses of a local neighborhood, such as a playground, flower garden, tiny woods, picnic spot, and tiny ballfield. This system of parks serves the common needs of neighbors around parks throughout the city. In contrast, the collection of homogeneous different parks (e.g., composed of specialized flagship parks) serves the specialized needs of the city's population, such as Olympic swimmers, certain handicapped residents, remnant old-growth woods for birdwatchers, and the highlights of historic sites. Most users of specialized flagship parks come from well beyond the local neighborhood.

Park maintenance is also relevant to deciding between similar heterogeneous parks or specialized ones, because government budgets fluctuate. Both large parks and specialized flagship parks are mainly maintained by government. On the other hand, small heterogeneous parks primarily serving the local neighborhood usually receive some care or stewardship from neighbors during low-budget periods (Forman 2008).

Movements of people and species between similar parks are greater than between dissimilar specialized ones (Figure 5a). Furthermore, movements between parks may be greater if the sides of nearby parks facing each other are of the same habitat or land use (Figure 5b) (Forman 2014). Thus, birds fly between the same type of woods, butterflies between butterfly gardens (Giuliano 2005), and people between flower

RICHARD MURRAY

Small parks, even private, have a value and may be well kept by residents. Kista, Stockholm.

gardens or ballfields that face one another in separate nearby parks. In essence, the spatial arrangement of land uses, both within parks and among parks, determines whether the city has a collection or a linked system of parks.

ARRANGEMENTS OF LARGE, MEDIUM, AND SMALL PARKS

Some spatial arrangements of parks tell much about movements and other interactions between large, medium, and small patches, almost irrespective of their surroundings. For instance, a small park near a large one normally has a net movement from the large to the small park (Forman 2014). Two parks of a similar type close together have much movement between them, and may affect the microclimate of one or both. A large park with several or many small ones close-by probably has more species entering, and sustains them longer, than an isolated large park (Figure 5c) (Opdam 1991). Movements in and out of a large park are more frequent in the directions of nearby small parks. A cluster of small parks provides alternative routes for movement between two medium parks. Two or a few small parks clustered closely together are perceived by some species, such as certain owls, as analogous to a single large park.

Park shape is also relevant to functioning and movement (Forman 1995). A corner of a square or rectangular park serves as a short projection or lobe providing good outward visibility. Also, the corner may be a common site of arriving and departing movements from or to another park. A long projection from a park serves as a funnel channeling movement from the park outward toward another park in that direction. The long lobe projecting from a park also serves as a 'drift fence', which catches species moving across the built area, and channels them to the main park area. An elongated park catches both people and species moving in the surroundings, and may funnel them to the ends of the park. Many other inter-park interactions are doubtless predictable almost irrespective of context.

An interesting example is Portland (Oregon, USA). Over half a century it has added dozens of new smallish parks, and increased the size of some others, so that nearly every resident has convenient access to one or more local parks (Ozawa 2004). The parks are also stepping stones for wildlife movement across the city. In fact, Portland has a huge green-wedge park (with elk, deer, bears, eagles, and salmon) which serves as a source providing a 'species rain' for many of the smaller parks (Houck and Cody 2000; Forman 2008).

Over half a century Portland has added dozens of new smallish parks, and increased the size of some others, so that nearly every resident has convenient access to one or more local parks.

PARK ARRANGEMENTS RELATIVE TO CITY CENTER AND SURROUNDINGS

The values of a park depend on whether it is in the central portion of a city or near the edge. Consider the common types of greenspaces within a city versus those on the edge of a city.

One list identified the common types of medium-size or large greenspaces within cities as (Forman 2008): (1) large urban wood-lawn park; (2) large semi-natural park; (3) lower floodplain-delta area; (4) railway yard; and (5) zoo. The small greenspaces listed are: (1) small urban wood-lawn park; (2) small semi-natural area; (3) historic/cultural site; (4) vacant plot; (5) school yard; (6) cemetery; (7) street side planted spot; and (8) green roof.

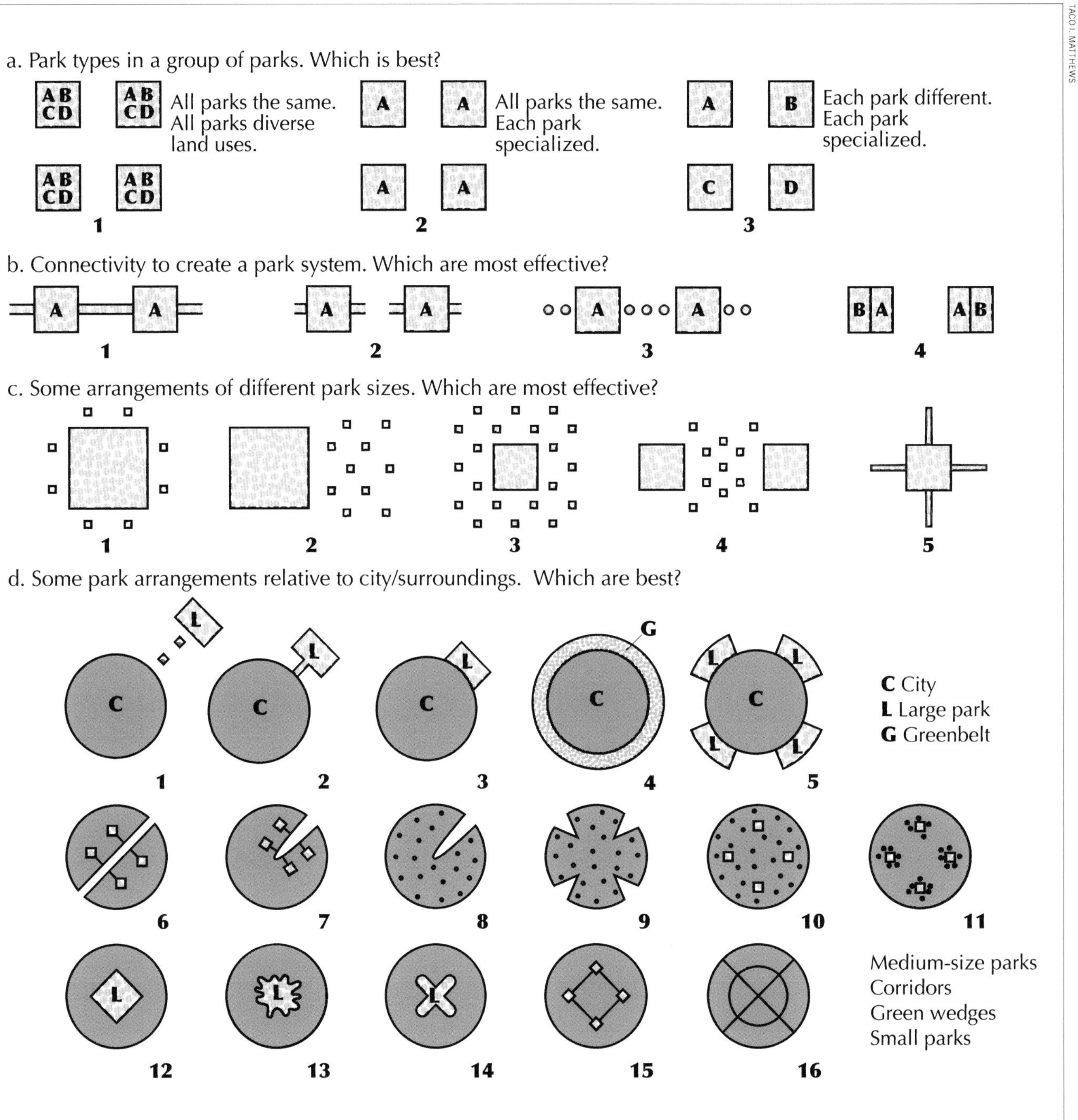

a. Park types in a group of parks. Which is best?

All parks the same. All parks diverse land uses.

All parks the same. Each park specialized.

Each park different. Each park specialized.

b. Connectivity to create a park system. Which are most effective?

c. Some arrangements of different park sizes. Which are most effective?

d. Some park arrangements relative to city/surroundings. Which are best?

C City
L Large park
G Greenbelt

Medium-size parks
Corridors
Green wedges
Small parks

Figure 5. Some promising park types, connections, and arrangements for urban areas.

TACO I. MATTHEWS

Large greenspaces in the edge area of a city are more diverse (Forman 2014): (1) heterogeneous suburban park; (2) large semi-natural or natural area; (3) golf course; (4) botanical garden: (5) market-gardening (truck-farming) area; (6) nursery plants area; (7) lake or pond area; (8) wetland; and (9) fairground or race-track site. Other large edge features are often mainly greenspace, but sometimes much covered with buildings and roads: (10) business office area; (11) brownfield (with chemical soil pollution); (12) industrial area; (13) shopping center area; and (14) municipal-use space. Ten additional types are listed as medium-size greenspaces.

Thus, the types of greenspace differ considerably relative to location in the interior or near the edge of a city. Large and medium greenspaces within the city are the least variable, or most predictable. All probably (Konijnendijk *et al.*, 2005): cool the downwind built area; reduce flooding; attract migrating birds; and are a local species source for the surrounding area.

The highly diverse greenspace types near the city edge play both positive and negative roles, including providing cool clean air, dusty polluted air, protection of a local water body, spread of invasive species, nature appreciation, and much more. Assuming that all cities are peppered with small parks, here are some of the arrangements of large, medium, and/or small greenspaces (some illustrated in Figure 5d) that may be compared:

- Greenbelt (e.g., London), number 4 in Figure 5d.
- Ring of medium and/or large greenspaces around the edge (e.g., Seoul) (5).
- Green wedge(s) projecting part way or all the way to city center (Stockholm and Valparaiso, Chile) (7 or 8).
- Few mediums connected to a major green corridor bisecting the city (Leipzig, Germany) (6).
- Few mediums connected to a major green wedge into the city (7).
- One central large surrounded by many smalls (New York) (12 or 14 if you add many smalls).
- One medium and many smalls in each major section of the city (10).
- One large at, or close to, the edge (3).
- One large greenspace farther out from, but connected by corridor or stepping stones to, the urban edge (1 or 2).

Other interesting park arrangements exist in and around cities (Figure 5d). The relative value of such combinations and arrangements of urban greenspaces should be evaluated.

VALUES IN MAJOR URBAN GREENSPACES GLOBALLY

A comparison of medium-size and large urban parks/greenspaces in different locations highlights their major values or roles (Table 1) (Forman 2014): (1) large greenspaces adjoining or close-by nine different cities; (2) medium-size greenspaces adjoining or close-by nine other cities; and (3) medium greenspaces inside a third set of nine cities. The greenspaces were selected based on the author's familiarity or in some cases the literature.

Overall, the large and medium greenspaces close-by or adjoining a city provide the highest total number of major values (16 and 28 respectively), and medium greenspaces within a city the fewest (8) (Table 1). Specialized or uncommon values are mainly limited to the greenspaces near city edges. A single large greenspace near the edge provides 4.1 major values on average, a medium one near the edge 4.3, and a medium greenspace inside the city an average of 2.7 values. Recreation is the most important value of both large and medium-sized urban greenspaces. This is followed in abundance by tourism, social/cultural value, rich biodiversity, flood reduction, water supply, and wood products. In effect, close-by large or medium-sized greenspaces provide numerous values, whereas a medium-sized one inside a city focuses almost entirely on recreation, tourism, and/or social/cultural value.

PLAN AHEAD

Early city greenspaces served mainly to grow food for urban residents. But for more than three centuries city parks have commonly presented the savanna image of low, mainly grassy cover with scattered trees and clusters of trees. Walkways are abundant to facilitate residents' leisure or recreation throughout the green space. Such pleasing tidy parks highlight Paris, New York, Tokyo, Melbourne, Ottawa, and many other cities in industrialized nations today.

Yet think of Rio de Janeiro, Lagos, Mexico City, Sao Paulo, and other cities where squatters arrive from rural areas with no money (Benton-Short and Short 2008; Giusti de Perez and Perez 2008). The newcomers need to promptly find or construct a shelter, preferably near potential jobs. Well-located greenspaces with few buildings are prime locations attracting informal squatter settlers. This dramatically changes a greenspace. With billions more people, all urban, expected in the few decades ahead, half are projected to join the urban poor (UN Population Division 2017). Worldwide, tidy pleasant parks and squatter settlers are on a collision course. Which will increasingly win? What is the solution?

For the next billion people arriving within 14 years, the most suitable

TABLE 1
Major values/roles of large and medium-size greenspaces at edge or inside major cities.

LARGE URBAN EDGE GREENSPACES (>100 km²)

Brasilia	Parque Brasilia	Wetland protection, water supply, informal squatter settlers
Santiago	Andes slopes	Water supply, recreation, flood reduction, tourism
Paris	Fontainebleau	Traditional weekend escape for royalty; recreation, hunting, wood products
Cape Town	Table Mountain	Recreation, aesthetics, tourism, rich biodiversity
Canberra	Forest	Water supply, recreation, wood products
Hong Kong	Forested hills	Recreation, flood reduction, biodiversity, wood products
Caracas	Rainforest slopes	Water supply, recreation, flood reduction, mudslide reduction, wood products
Sapporo	Forest on volcano	Recreation, tourism, flood reduction, mudslide reduction, wood products
Ottawa	Greenbelt in south	Recreation, farming, wetland/ponds protection

MEDIUM-LARGE URBAN EDGE GREENSPACES (10–100 km²)

Barcelona	Collserola	Recreation, aesthetics, carbon storage
Mumbai	Sanjay Gandhi	Protect leopards, indigenous communities, water supply
Kyoto	Forested slopes	Social/cultural value, flood reduction, mudslide reduction
Moscow	Elk Island	Government uses, nature protection, recreation
Miami	Loxahatchee	Wetland protection, biodiversity
Rio de Janeiro	Tijuca	Recreation, tourism, biodiversity, flood reduction, landslide reduction, protected jaguars until 1980s
Berlin	City Forest	Recreation, water supply, mining reclamation, rich wildlife biodiversity, wood products, climate & noise amelioration
Stuttgart	Hillslopes (1990s)	Vegetation cover with cool-air drainage at night that both cools and cleans the urban air
Mexico City	Ajusco (1990s)	Water supply, biodiversity, hunting, squatter settlements, weekend houses for wealthy, drug gangs, police, anti-gov. groups, military

MEDIUM-LARGE GREENSPACES INSIDE CITIES (1–10 km²)

Edinburgh	Holyrood	Recreation, social/cultural value, tourism
San Francisco	Golden Gate	Recreation, botanical garden
Mexico City	Chapultepec	Recreation, social/cultural value, tourism
Tokyo	Imperial Palace	Royalty area
Berlin	Tiergarten	Recreation, social/cultural value, tourism
New York	Central Park	Recreation, social/cultural value
Barcelona	Montjuic	Recreation, social/cultural value, tourism, botanical garden
Buenos Aires	Reserva ecológica	Original wetland, then solid waste dump, now biodiversity nature reserve
Madrid	Retiro	Recreation, social/cultural value, tourism

places appear to be in existing outer suburbs, sprawl areas, satellite cities, and towns and villages in farmland (Forman and Wu 2016a). Nearly all are just outside a big city, where large and medium parks seem poised to house squatter settlements. Urban region plans for cities and their surrounding areas, drawn up expeditiously, approved, and followed, seem to be the only solution to minimize this park destruction and provide viable communities (Forman and Wu 2016b).

Today city growth follows four main patterns: bulges from the city; growth of satellite cities; strip/ribbon development along transportation corridors; and sprawl of low-density development (Forman 2008, 2014). Strip development and sprawl widely degrade the land. In contrast, bulges and satellite-city growth, if consistent with a good urban-region plan, can provide sustainable community patterns, and also sustain large and medium-size greenspaces adjoining or close-by a city.

Alternatively, internal urbanization, or infill, specifically targets development to greenspaces such as community gardens (allotments) within the city. Infill seems better than outside sprawl. Yet infill is desirable only up to the point that it does not disrupt the connectivity tying urban greenspaces together into a system. In addition, infill should not disrupt the convenient access for all city residents to a local park.

Climate change is especially apt to change the parks of coastal cities (Keenan and Weisz 2017). Therefore, targeting today's parks for future flood abatement and saltwater wetlands may make good sense. As development moves inland, the existing large and medium parks just outside coastal cities could become the iconic parks of the future cities. Still, major disturbances happen in greenspaces. Consider fire in Valparaiso, tsunami, flooding, and Berlin's Tiergarten Park trees cut for firewood in wartime. "Change, nature's mighty law" (Robert Burns) operates in cities.

In addition to such natural and societal forces, endless fluctuations in park maintenance budgets keep changing parks. During high-budget times, new structures often appear in parks; in low-budget phases, the places degrade. However, during tough times, a dedicated neighboring community may step in to help care for and sustain the park. Planning for future change, natural or human-caused and rapid or slow, in large as well as small parks is a sure sign of wisdom in cities.

RIGHT: Leopard paw prints along a wet patch above Kanheri Caves, Sanjay Gandhi park in Mumbai, India.

SUMMARY AND CONCLUSION

Parks or greenspaces, as highlights of an urban area providing many unique benefits, vary in both size and arrangement. What is the relative value of one large urban park versus many small ones? How does park arrangement affect the overall value, specifically the location of a large or medium-size park relative to other parks, the city center, and city edge? How can a collection of parks become a park system? These questions are addressed mainly based on ecological principles and the literature. Park size is evaluated using four key values: air cooling, biodiversity, flood reduction, and recreation.

For air cooling, one large park is much better than many small ones. For biodiversity, one large park is also much better. For flood reduction, many small parks are much better than one large park. For recreation, many small parks are also much better. Thus, one large park and many small parks provide different yet important values. An optimum design or arrangement probably maintains one large park (greenspace) surrounded by many small parks in each major section of an urban area.

A park system with active interactions among parks can be enhanced by the arrangement of parks, types of parks, features between parks, and arrangement of land uses within parks. Larger and medium-size greenspaces within a city are relatively similar, mainly providing for recreation, socio-cultural value, and tourism. Large ones at the urban edge of cities differ markedly and provide many additional values for society.

In short, small parks throughout an urban area, plus one or a few large or medium-size parks, are both enormously valuable for a city. Urban region plans for every city, especially highlighting parks, are needed to address big changes ahead: outward city growth; infill disrupting connectivity; major disturbances; climate change; and greenspaces targeted by informal squatter settlers.

[1] The sum of evaporation and plant transpiration.

LEFT: Informal squatter settlers rebuilding and colonizing after major fire in a large urban greenspace. Green wedge of Valparaiso, Chile. Luis Alvarez kindly showed this rare sight to the author.

Visions of Urban Green

In ancient Persia, gardens were traditionally likened to wine that does not intoxicate. The meaning of the old Persian word "pardis," from which the word "paradise" is derived, is a "walled in garden"—a true paradise! Kings and nobles could afford such gardens around their palaces and houses, but for the common man the countryside was the great refuge for many centuries. Now that more than half of the world's population lives in cities—and more and more will do so, in ever larger cities—the need to take refuge in nature has to be satisfied within cities themselves.

PETER CLARK

Large Urban Parks and Urban Green Space: A Historical Perspective

It is impossible to understand the history of large urban parks without seeing their evolution in the context of the wider development of green space in cities. In this chapter I want first to explore the broad pattern of urban green space from the pre-modern period to the present day, before going on to the describe the historic trajectory of large urban parks. In the process I hope to shed light not only on the factors influencing the advent of large urban parks but also to say something about the actors shaping their development and about the people using them. I will be talking mainly about European cities but, as we will discover, green space developments have a global outreach.

Clearly, however important large urban parks are in ecological, economic, social and cultural terms—what James Corner vividly calls "the great outdoor nature theatres of the city" (Czerniak and Hargreaves 2007, p.11)—they represent only a proportion of urban green space in a city and often for urban residents one of the less proximate.

A 2009 survey of European cities found that the proportion of green space to total area mostly ranged between 10 and 20 per cent, though a few cities having much higher or lower levels. Figures for other parts of the world are more disparate, from 25 per cent for Seoul and 29 percent for Guangzhou down to 5 per cent for Kuala Lumpur and 1.5 per cent for Hong Kong (Baycan-Levent et al. 2009, p. 207–8; Clark 2017, p. 192).

For European cities recent studies have revealed the extraordinary variety of urban green space, embracing not only large and small parks but sport and recreation grounds, institutional gardens, private domestic gardens, allotment gardens, urban farms, tree-lined boulevards, garden suburbs, green belts, parkways and roadside verges, green infrastructure, and the many species of wasteland or vacant ground. Overlapping often with wasteland, blue space is also another important constituent of urban open space in many cities (Clark ed. 2006; Brantz and Dümpelmann 2011; Elliot 2016; Conway 1991; Clark et al. eds. 2017).

We should not think of urban green space as being a modern construction. Into the 19th century most town dwellers, even in the biggest

LEFT: Skyscrapers in Mumbai, India, viewed from Sanjay Gandhi National Park.

cities, had walking access to fields, woods and other green spaces on the outskirts, and within cities there were many areas of untended waste or vacant ground, just as orchards and commercial gardens remained common in parts of Victorian cities, despite accelerating urbanization and industrialization.

Again, we find designed green space, especially private or institutional gardens, from early times. Imperial parks and gardens appear in ancient China, often linked to capital cities, and royal households in medieval and Mughal India had famous urban water gardens, as did cities like Isfahan in Iran under Shah 'Abbas. By the early modern period private gardens of landowners, high officials and merchants became common in Chinese cities like Suzhou and Hangzhou, in Japan's Edo (Tokyo), and in European cities such as London, Paris and Vienna. European monasteries and Asian temples and shrines often had extensive gardens attached (Clark 2017, p.193).

The 17th and 18th centuries saw a proliferation of other green spaces in major cities. In Europe and Japan, we find pleasure gardens to entertain better off visitors. In Paris the old fortifications were converted into boulevards, while botanical gardens, the first at Pisa in 1543, multiplied across Europe, Asia, and the Americas, encouraged by mounting global trade, scientific interest, and colonization. Not least important, large urban parks start to make an appearance in leading cities, as we will see shortly (Clark 2017, p.194).

However, the big take-off in designed green space in cities was during the 19th century, as accelerating urban and industrial growth absorbed many traditional open spaces, as fears about disease and infection increased, and middle-class pressure for social segregation and the social disciplining of the lower orders became influential. Thus, while some large parks originated earlier, the main wave of medium and smaller civic parks occurred in the 19th and early 20th centuries. In South London the major Battersea Park, opened in 1858 and constructed by the Metropolitan Board of Works, exemplified the surge of new civic parks across Britain. Paris saw a cascade of park creation in the years 1851–1870 with 33 new spaces covering over 1,000 hectares. Many of the early ornamental parks, with their flowerbeds armed with regiments of exotic plants, were top down creations for the bourgeoisie to display themselves, for walking or riding, for combating city pollution—the green lungs effect—and as places where the lower orders could be instructed in civilized manners. Wealthy developers often gave land and sometimes money to councils to create parks, in order to inflate property values on their nearby estates. Municipal parks became markers of civic pride and identity—and late 19th century cities across Europe and beyond competed with one another for international prominence in park provision. The diffusion of civic parks outside Europe was often spread via official missions by city politicians, through visits by architects and landscape gardeners, and by colonial administrators. But the European model was adapted on the ground. In British colonies in Asia and Africa green spaces were often segregated on racial lines with parks limited to European districts (Reeder 2006, p. 32 et seq.; Pinol 2017, pp. 27–29; Clark 2017, pp. 195–196).

By the 1880s Western city park policy had to contend with changing patterns of demand, including growing popular pressure for recreational space—for sports and other informal leisure activity. As a result, many small recreation grounds and pocket parks proliferated after 1890. In addition, specialist sports grounds like tennis clubs and golf courses multiplied: London had 100 golf courses by 1914 and others sprang up in Europe, the United States and Asia (Clark 2017, pp. 195–196; Mack and Parscher 2017, pp. 88–89).

As well as municipal parks and sports grounds, the 19th and early 20th centuries saw a proliferation of other types of designed green space. Middle-class villas, often in new city suburbs, increasingly had their own extensive private gardens screened by walls and extensive tree cover from public gaze. Allotment gardens spread in many European cities before the First World War, their advance promoted largely through voluntary activity, often by women activists, concerned both to make ordinary people aware of nature and to help feed their families. In Stockholm, for instance, the leading proponent and allotment activist was Anna Lindhagen. Another new international development was the diffusion of garden suburbs, with their leafy streets and modest private gardens, modelled on the ideas of Ebenezer Howard and the Garden City movement, which by the inter-war period were a feature of cities across Europe, in the United States, the Middle East and Japan (Clark 2017, pp. 195–196; Mack and Spencer 2018, pp. 88–89).

After the Second World War, European cities used park creation as an instrument of urban renewal. There was a large-scale expansion of London green space in the 1960s linked to extensive Modernist housing development. In Paris during the 1980s and 1990s many new parks were created on former industrial estates, as Paris sought to burnish its credentials as a green city (Garside 2006, p. 70 et seq.; Hannikainen 2017 p. 39 et seq.; Pinol 2017, p. 30 et seq.).

In recent times, however, green space in its many varieties has been under attack. British newspapers claim that "Britain's urban green space

In recent times, however, green space in its many varieties has been under attack.

is at risk of rapid decline" and speak of "the crisis in British parks" (Owen 2014; Duxbury 2016). From the 1980s growing financial problems in cities, especially in Britain and the United States, have led municipal policymakers into making drastic changes—privatizing park services, slashing staffing levels and scaling back park regulation, relying in some countries on outside donors and commercial support. In London in 1997 80 percent of councils planned to cut back on park expenditure and many small parks were left with no park keeper. In the United States many small parks have closed and park systems in New York, Chicago and elsewhere rely heavily on business sponsors' external funding. Other kinds of urban green space under pressure include small sports grounds, while many private domestic gardens are being eroded by infilling, and the new planning dogma of the sustainable 'compact city' (Hannikainen 2017, pp.47–48; Taylor 2009, p. 353 et seq.; Ojala *et al.* 2017, pp. 73–82).

Even so, not all urban green spaces are disappearing. In big US cities allotments or community gardens have multiplied in recent decades, often on waste land and frequently worked by minorities and women. In Philadelphia the Pennsylvania Horticultural Society helped establish over 500 community gardens during the 1990s. Still many of these are on temporary sites which are frequently taken over by developers. In Europe too allotment gardens are often under threat despite strong demand. In recent years there has been a dramatic increase of golf courses. In Europe the number of courses in and around major German cities nearly trebled between 1987 and 2005. In China the number of courses has risen sharply since the 1980s, with around 600 by 2015, mostly close to rapidly expanding big cities, sometimes linked to gated communities for the new elite. Japan has even more—2,400 courses in 2009. In China, with accelerating urbanization since the 1980s, there has been an important new development of parks, including small neighborhood parks (Clark 2017, pp. 197–201).

Several points are evident from this short survey of urban green space trends, which are relevant to large urban parks. Firstly, we can say that while there has been an important development of urban green space in cities, especially from the 19th century, and that this has an increasingly global coverage, the process has varied widely on the ground and in European and North American cities green space is increasingly under pressure, if not attack. Secondly it is evident that different types of urban green space have different trajectories, for instance some developing earlier or later than others. Finally, we have touched on some of the factors contributing to the creation of designed green space including

St. James's Park and the Mall, London, United Kingdom, after 1745,
painting by Joseph Nickolls.

urbanization and the growth of big cities, the advent of urban planning
and civic park policies, changing concepts of nature and leisure, competition and emulation between cities—never to be forgotten and crucial for
the dissemination of ideas, along with globalization and colonization. As
we shall see, many of these factors influenced the history of large urban parks.

Turning now to the development of large urban parks, it is likely that
under the medieval Chinese emperors some private aristocratic and
imperial gardens were open to the public at specific times, but the first
public gardens to be permanently open, usually royal parks, appear in
European capital cities during the 17th and 18th centuries. Among the
earliest in Europe was St James' Park in London (34 hectares) opened
by king Charles II to the public in the 1660s; the contiguous Hyde Park,
another royal park, was opened some years before, and together with
nearby Green Park and Kensington Gardens cover 300 hectares. In
the compact mid-17th century city of London with a population of less
than half a million, a city easily walkable from one side to another,

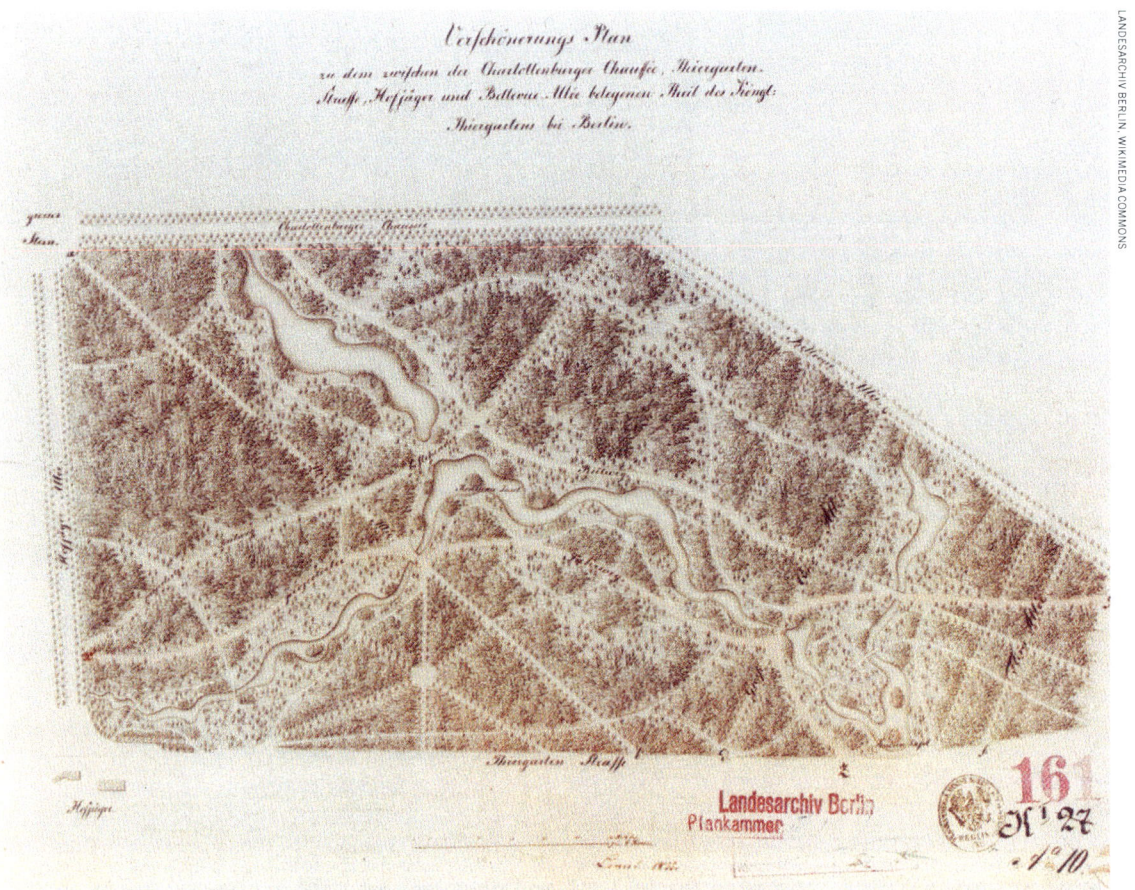

St James' park clearly represented a large urban park, even before the later additions. London's Royal Parks have always been under state control—handed over by the Crown to the government in 1851. As a result, Royal Parks policy has been rather different from that of London's municipal parks: willing to shut off parts of the park for elite activity; a reluctance of administrators to bow to populist pressure for sports grounds; until recently at least, less under financial pressure. At the same time, usage has been mixed. In the 18th century the upper classes used the parks mainly on weekdays when the lower orders were at work: they came just on Sundays and Mondays. Victorian users included the unemployed and homeless who dossed down there at night, as well as those engaged in illicit sexual activity, again mostly after dusk (Clark 2017, p. 193; for

TOP: A park plan for Tiergarten, Berlin, Germany, by Peter Joseph Lenné, drawn by Gerhard Koeber 1835.
RIGHT: Bois de Boulogne, France, Plan de Paris 1921, édition L. Guilmain à Paris.

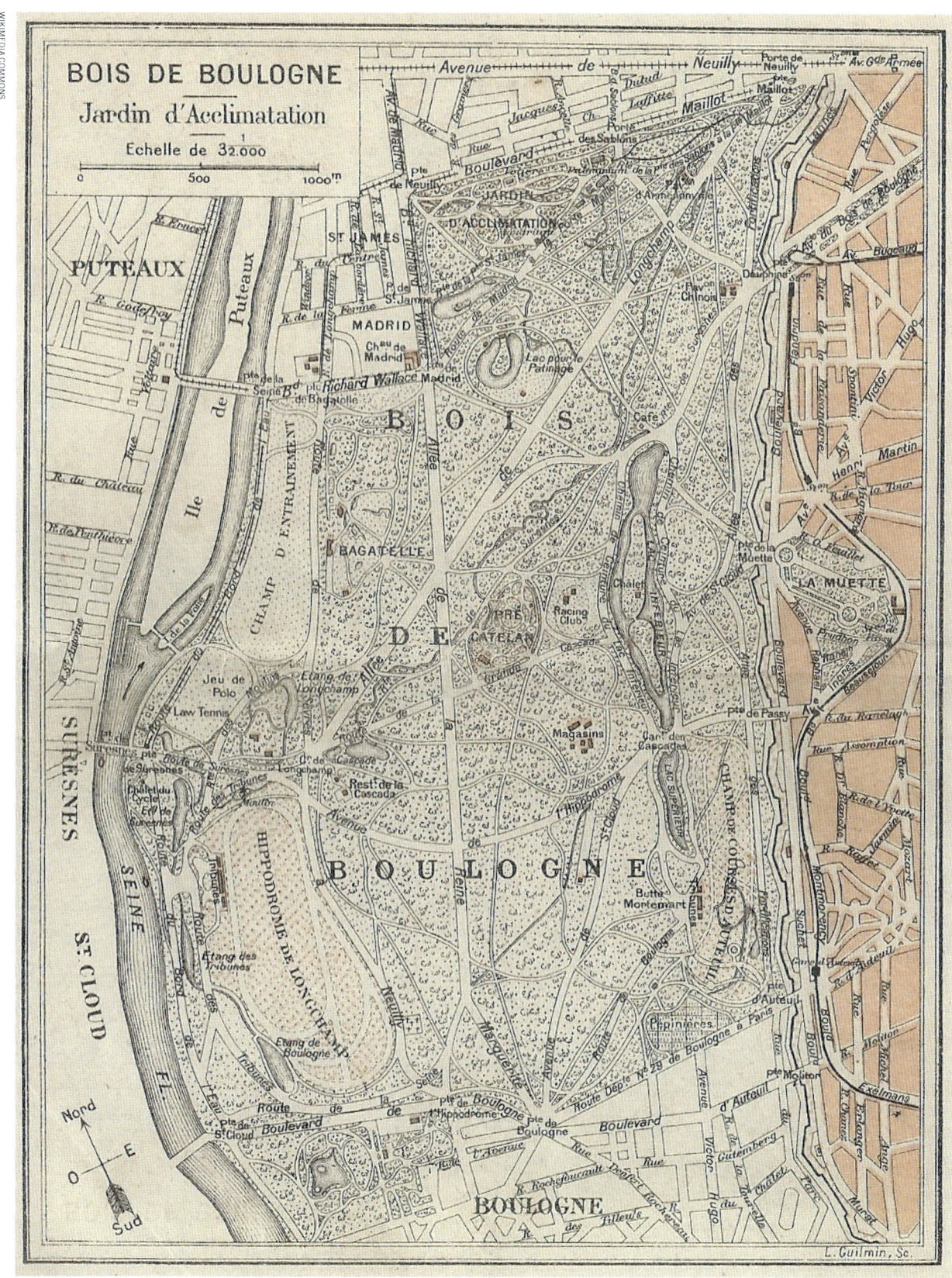

WIKIMEDIA COMMONS

PHOTOCHROME PRINT BY THE DETROIT PHOTOGRAPHIC CO. WIKIMEDIA COMMONS

London's Royal Parks in the modern period see Hannikainen 2014).

Other royal parks opened to the 'public' in the early modern period included the Haagsche Bosch, a hunting park close to the Dutch royal capital of The Hague. From the end of the 16th century first the elite and then the wider citizenry were granted access to it. In Berlin the Tiergarten was thrown open in the 1740s by the king of Prussia, and Stockholm citizens were granted entry (for a fee) to the Djurgården in the 1760s; this now forms part of the Royal National City Park (discussed elsewhere in this volume) (Clark 2017, pp. 193–194; Brantz 2017, pp. 142–143).

The opening of royal parks was not just a European trend. In Istanbul, the largest Middle Eastern city at the time, several imperial parks by the Bosphorus were opened to the public in the 18th century and visited by a variety of users—not just the better off, including women, but by middling groups. Thus, even before the onset of large-scale urbanization, the creation of new large urban parks probably reflected new concepts of urban nature and cultural and leisure activity in big cities, as well as improved living standards for the better off (Hamadeh 2007, pp. 272–312).

Accelerating city growth and industrialization with all its problems spawned new pressures from the 1840s. Yet royal or government sponsorship of large urban parks remained important in Europe. In Paris, Emperor Napoleon III opened the former royal hunting ground, the Bois de Boulogne, to the public in 1852, and it was laid out by Adolphe Alphand with curving walks and carriageways. In the later 19th century it was filled with boaters, picnickers, strollers and runners, its lakeside area teeming with Parisian life at the weekends: but it also became notorious for prostitution and other illicit activity (Hargreaves 2007, pp. 128–139).

In the US the creation of large urban parks followed on closely their development in Europe, spurred on by urbanization, property development, new ideas of urban nature, sanitary concerns and inter-city rivalry —every up and coming city wanted its big park. The iconic Central Park in New York, opened finally in 1873, was heavily influenced by European park ideas. Judiciously combining European large park and municipal park models in a distinctive romantic style, Olmsted's design (replicated in the many other parks he designed or influenced across North America) owed much to the idea of highly regulated, passive recreation, then in vogue in Europe, and to a preference for polite elite

The big take-off in designed green space in cities was during the 19th century.

TOP LEFT: Sunday stroll in Bois de Boulogne, Paris, France, Henri Evenepoel 1899, La Boverie.
BOTTOM LEFT: Lower end of mall, Central Park, New York, United States, 1901.

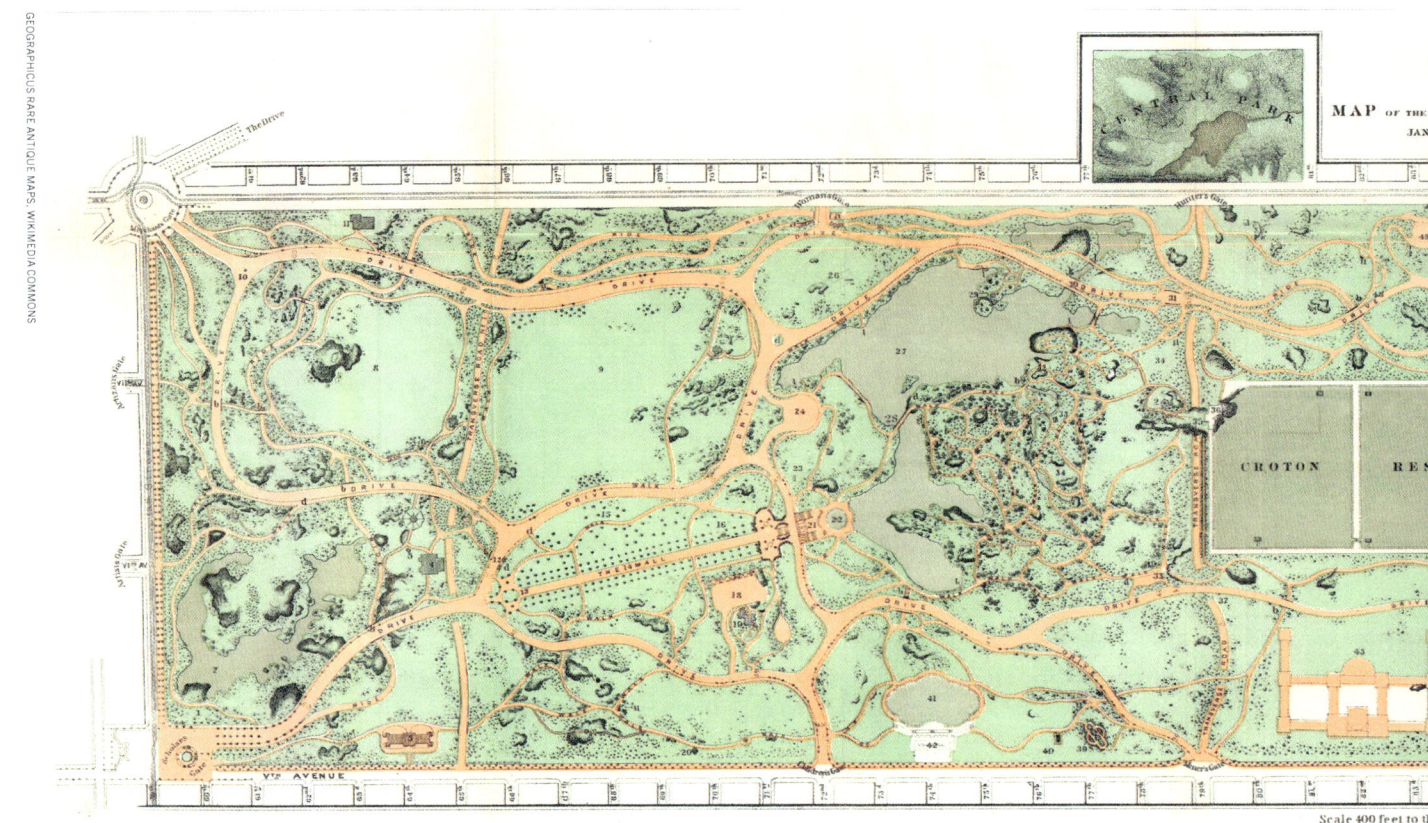

Calvert Vaux and Fredrick Law Olmsted map of Central Park, New York, United States, 1870.

users. But both priorities came under acute pressure from the 1880s if not before with the impact of new recreational activities and popular demands. We see similar pressures in other North American large parks like Franklin Park in Boston, Golden Gate Park in San Franscisco and Stanley Park, Vancouver, where park life was transformed by the early 20th century with the installation of attractions such as tennis and baseball, animal rides, boating, golf courses and zoos. Unlike some of the European parks most of the new large urban parks in North America were not government parks but municipal parks. This has become particularly important since the 1970s and 1980s, as municipal budgets have declined with major implications for park expenditure. However, the iconic large urban parks have been able to exploit their fame to access donor support and commercial sponsorship, as in the case of Central Park or Millennium Park Chicago (Taylor 2009, p. 247 et passim and p. 348 et passim; von Hoffman 1988, pp. 339–350).

AL PARK

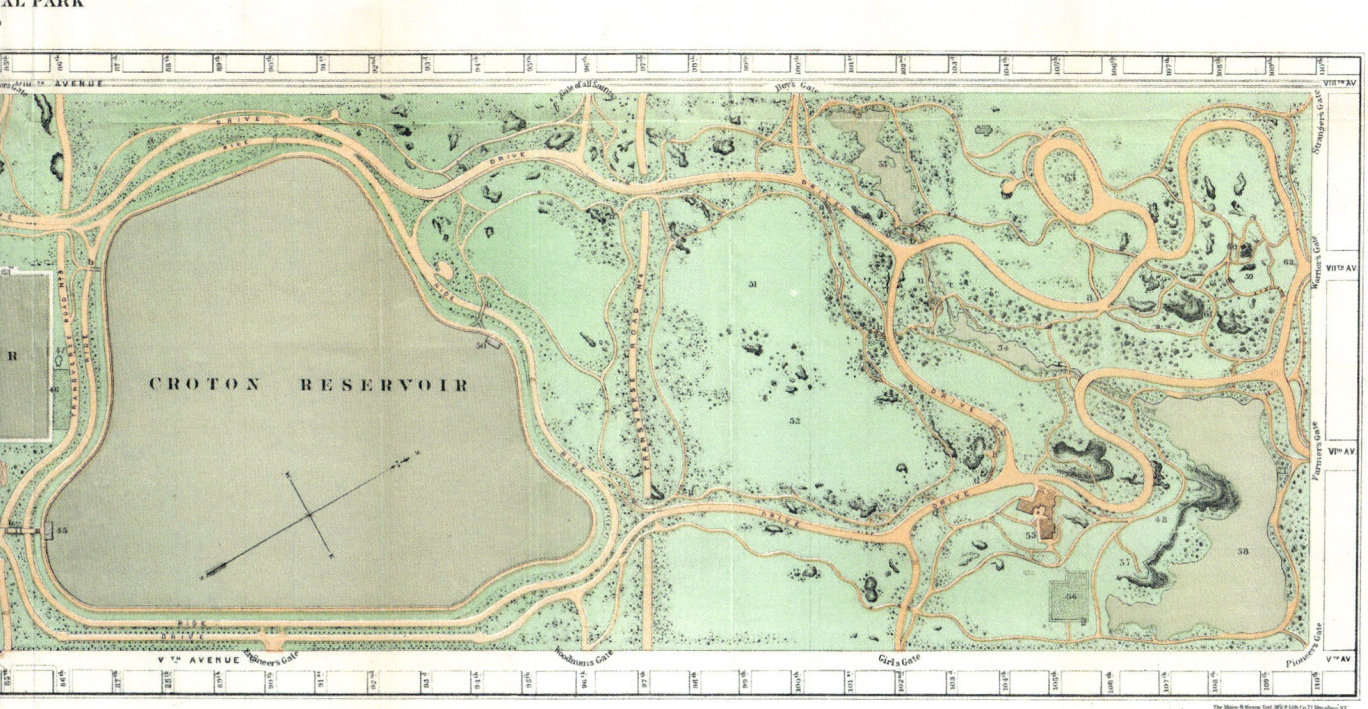

CROTON RESERVOIR

Elsewhere, it would seem a variety of bodies have been the primary initiators of large urban parks. In the Netherlands the powerful City of Amsterdam, faced by mass unemployment in the 1930s, sought to mobilize the jobless to make the Amsterdamse Bos a public park of 1000 hectare. Between 1934 and 1940, the forestation project provided work for more than 20,000 unemployed people. The designers borrowed English and German models to create a park of natural landscape, recreation and relaxation (Hargreaves 2007, pp. 147–150; https://en.wikipedia.org/wiki/Amsterdamse_Bos).

In Beijing the imperial Temple of Heaven Park was opened to the public in 1918 after the end of the Qing empire. Covering 273 hectares the site includes extensive historic pine woods and remains a government asset. Some of it consists of playgrounds, exercise and game areas. These facilities are used by families bringing children to play. Some of the open spaces and side buildings host choral shows, ethnic dances, and other

MARJOLIJN POKORNY

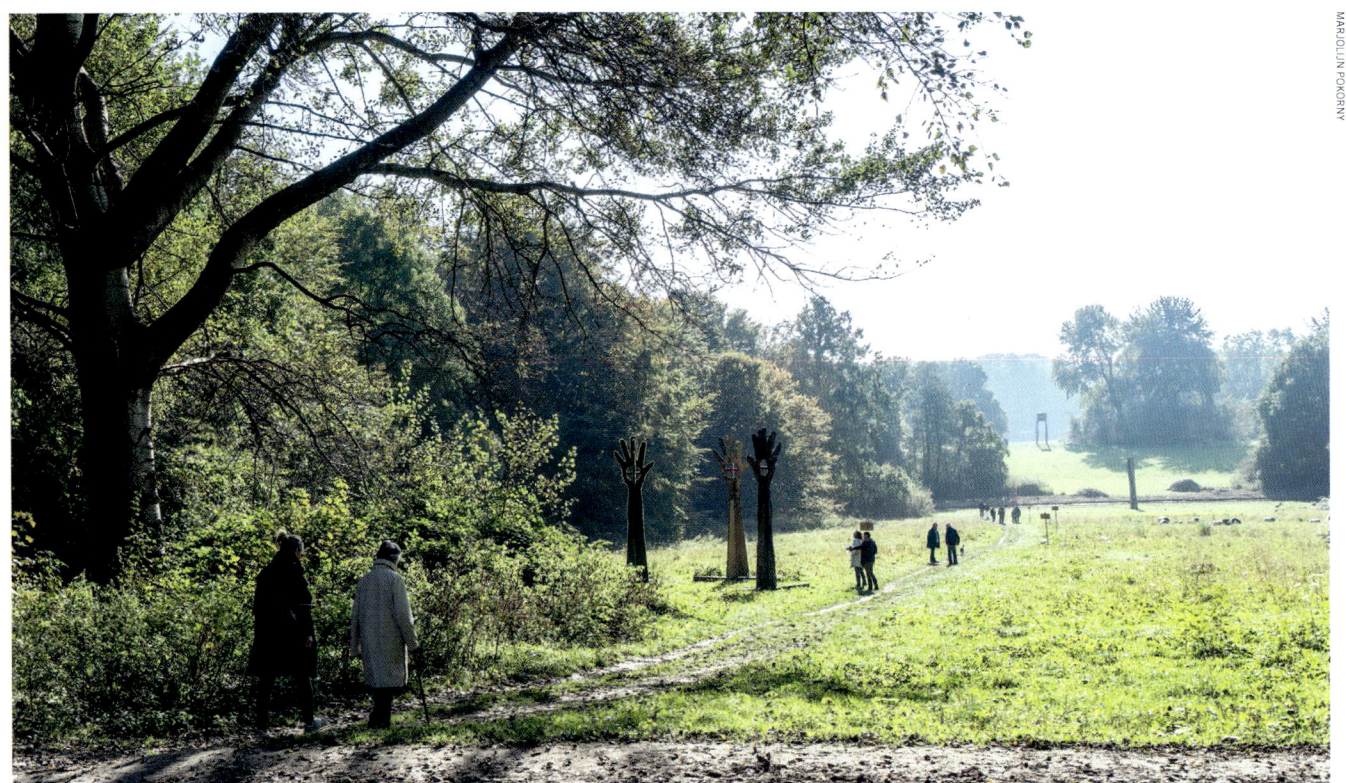

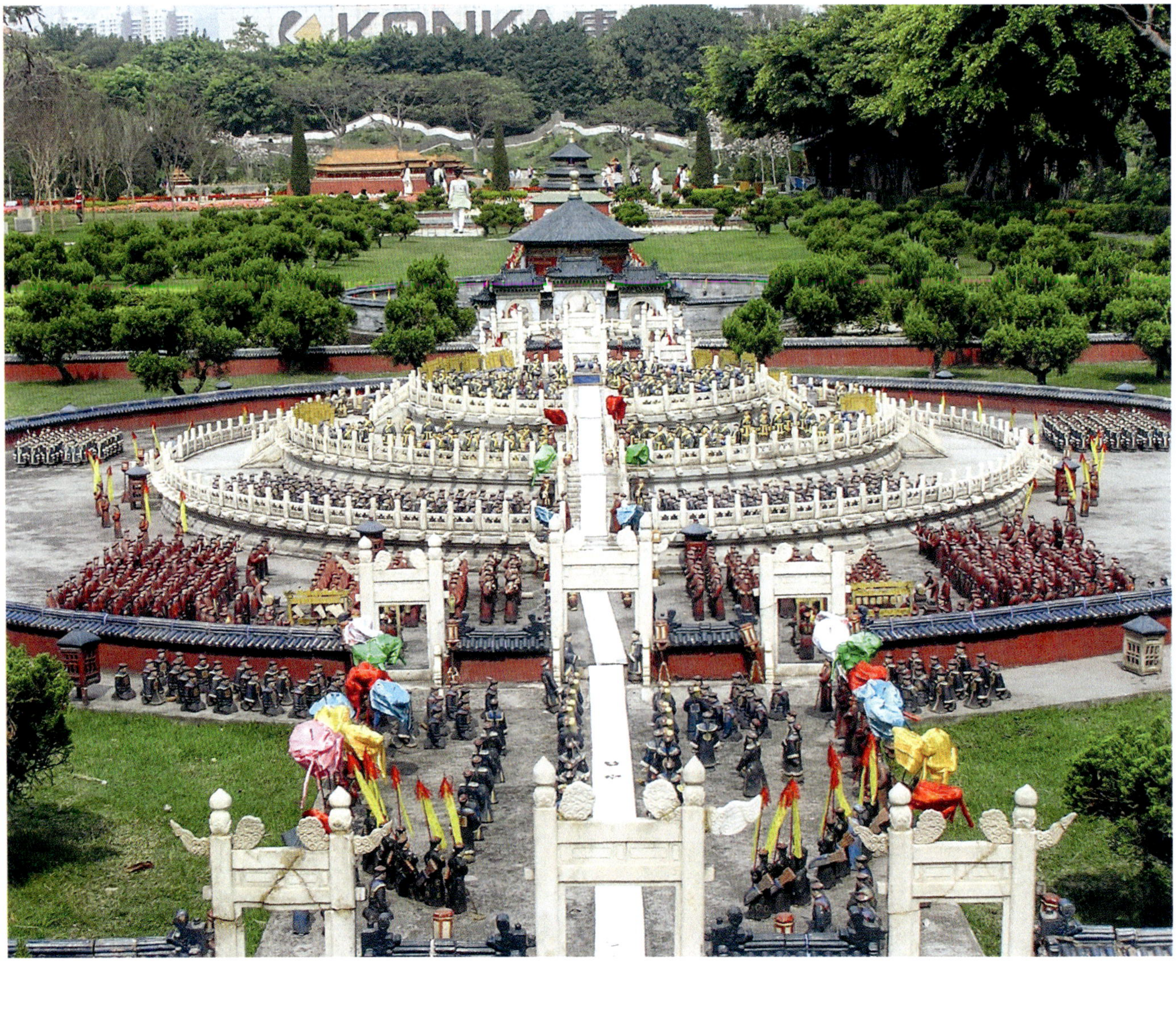

DMPENDSE WIKIMEDIA COMMONS

presentations. Visitors are strictly regulated (http://whc.unesco.org/en/list/881; https://en.wikipedia.org/wiki/Temple_of_Heaven).

In Mumbai the extensive Sanjay Gandhi National Park was first, before independence, established by British colonial officials and then extended by the National Forest Department (to 104 square kilometers) in the 1960s. Surrounded on three sides by an ever-expanding metropolis, the park provides conventional natural relief against pollution and the social problems of urbanization, as well as offering important opportunities for leisure and cultural activities—there are major archaeological features. The park is home to a number of endangered species of flora and fauna. The forest area of the park houses over 1000 plant species, 251 species of migratory, land and water birds, 50,000 species of insects and 40 species of mammals (https://en.wikipedia.org/wiki/Sanjay_Gandhi_National_Park).

In recent decades large urban parks, like urban green spaces in general, have experienced serious financial, development and management challenges. Nonetheless, despite these problems, there is evidence that the large park momentum continues. Not only have existing parks found new sources of revenue (as in the US) but new large parks have been created.

One obvious example is the Royal National City Park in Stockholm, established by the Swedish parliament in 1994 to ward off the threat of development in part of the area. In a similar fashion Helsinki is also acquiring a national urban park, based around the older Central Park. In London the Lee Valley Park (40 square kilometers) stretching from metropolitan London into Hertfordshire and Essex is managed by a special Lee Valley Regional Park Authority (authorized by Parliament) and combines previously derelict land, conventional urban green space, heritage sites, country parks, nature reserves and lakes as well as important sports facilities. Use ranges from passive recreation to horse-riding, team sports, and golf. The list of recent new parks also includes Landschaftspark Duisburg-Nord, built on industrial wasteland in the Ruhr Valley, Germany, and the Fresh Kills Staten Island park, started in 2008 and constructed on a huge landfill site and advertised as "from trash to treasure." In the pipeline are new large urban parks being planned or developed in the US, in Canada, Spain and elsewhere (Schantz 2006; https://en.wikipedia.org/wiki/Lee_Valley_Park; Hargreaves 2007,

---

TOP LEFT: Amsterdamse Bos, The Netherlands.
BOTTOM LEFT: A model version of the Temple of Heaven in Splendid China amusement park, Shenzhen, China.

p. 162 et seq.; Czerniak 2007, p. 224; http://www.fieldoperations.net/project-details/project/freshkills-park.html).

What are the key factors affecting the resilience, even dynamic, of large urban parks on the contemporary scene? Government support in a number of cases has shielded them from development pressures; their iconic heritage character has enabled them to attract external funds; park policy has diversified to attract a wide variety of users—some parks have allowed minorities and alternative or illicit groups to appropriate areas of parks (Czerniak and Hargreaves 2007, pp. 51, 133, 208; Orvell and Meikle 2009, p. 339 et seq.). Large urban parks have likewise benefitted from renewed inter-city competition in a globalizing world, where large areas of green space are seen as crucial for a city's international identity and capacity to attract affluent residents, businessmen and tourists, as much as art museums, fancy concert-halls and public libraries. Ecological research arguing that large urban parks have a higher level and quality of biodiversity than smaller green spaces may also be a significant asset in helping to justify the scale and cost of large urban parks in an age of public austerity—at least in the West.

This brings me back to the vital relationship of large urban parks and urban green space in general. Firstly, one of the great abiding strengths of large urban parks is that they often constitute and bring together in their different layers a great diversity of urban green spaces. They frequently include as well as conventional park aspects—flower beds, tree-lined walks and so on—important areas of previous wasteland (recent ones regularly built on polluted brownfield sites), areas for sport and recreation, sometimes golf courses. Again, they often incorporate extensive blue space, roadside verges and institutional gardens as well as urban forest.

Following on from this we can say that large urban parks attract a wider range of users than we find in other types of green space: not only children and young people, sportspeople, and local residents but also tourists, women, minorities, clubs, and even 'alternative' groups. Again, in terms of people playing a key role in their development we need to reiterate in many cases the wide range of actors: the special role of government authorities, as well as of city councils, planners, architects and landscape designers, park boards, business sponsors, property developers, contractors, and the media.

However, these Big Daddies of urban green space also have a crucial

Nonetheless, despite these problems, there is evidence that the large park momentum continues.

RIGHT: Lee Valley Park, United Kingdom: Walthamstow Marshes.

RAIMOND SPEKKING WIKIMEDIA COMMONS

ZAIRON WIKIMEDIA COMMONS

function because of their extent, their diversity, their eco-system services impacting on and interacting with other specialist types of green space across the city. They may serve as a large hub for green space activity with birds, insects, and other fauna and flora spreading out from them to other green spaces such as private domestic gardens and wasteland. At the same time, since large urban parks are often not compact entities, other forms, types of urban open space—smaller parks, gardens, blue space etc.—may provide important connecting corridors between different sections of the park.

In conclusion then, the importance of large urban parks as a distinct category of urban green space is evident, but we still have much to learn about this important landscape phenomenon in the context of the history of urban green space on a European and global scale.

TOP AND BOTTOM LEFT: Landschaftspark Duisburg Nord, Germany.

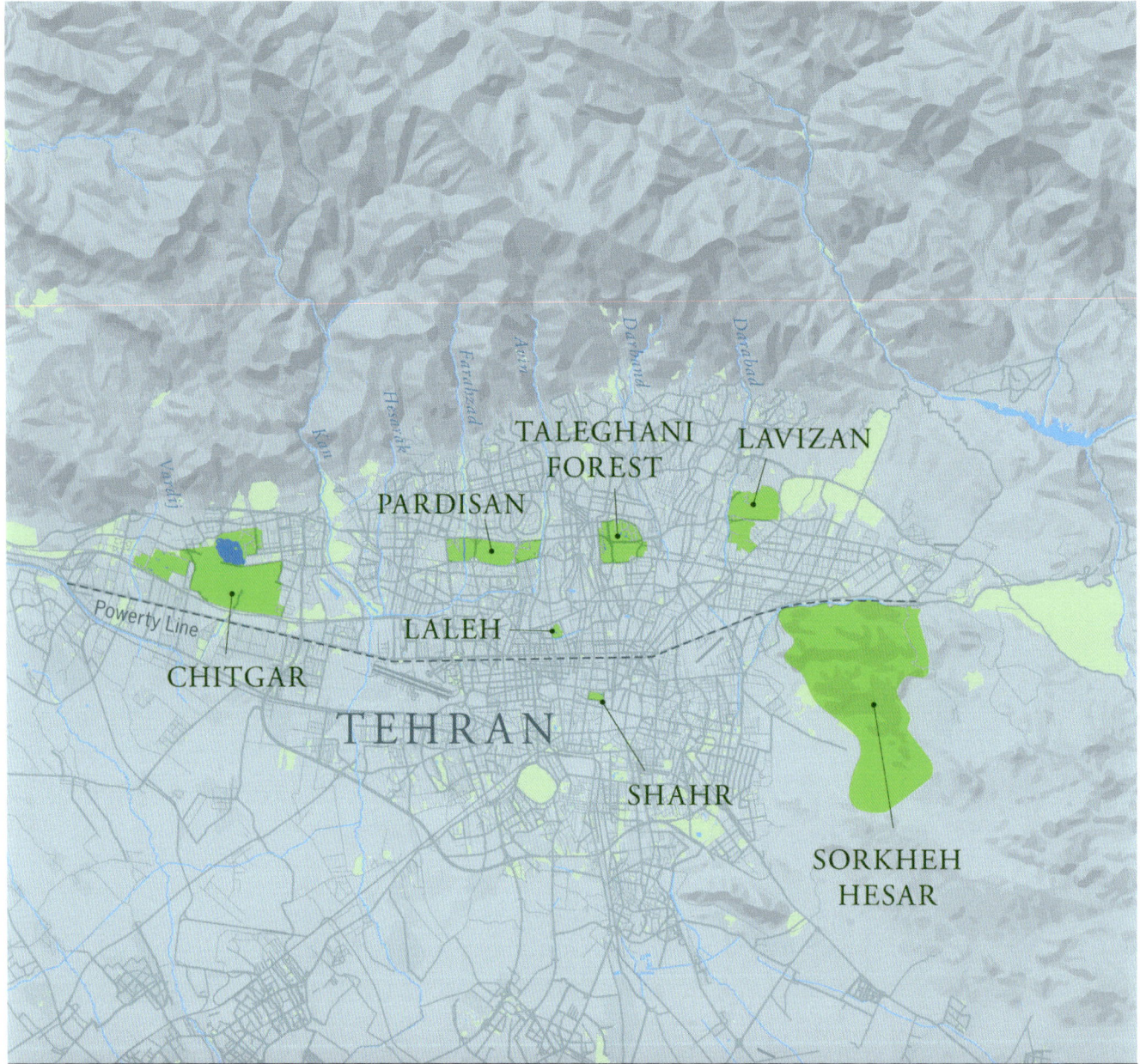

LARGE PARKS IN TEHRAN, IRAN

Park Name	Area (Hectare)	Year of construction	Location	Park Name	Area (Hectare)	Year of construction	Location
Park e Shahr	26	1959	North	Jamshidieh Park	69	1977	North
Saei Park	12	1963	North	Sorkheh Hesar	9.380	1980	South
Laleh Park	35	1966	North	Taleghani Park	31	1981	North
Mellat Park	34	1967	North	Almahdi Park	6.5	1986	South
Daneshjoo Park	3.2	1967	South	Artists Park	6	1990	North
Lavizan Forest Park	1,000	1967	North	Goftogu Park	14.5	2003	North
Chitgar Park	950	1968	North	Nahj ol-Balagha Park	35	2007	North
Shafaq Park	1.6	1969	North	Ab-o-Atash Park	2.4	2009	North
Niavaran Park	6.2	1969	North	Velayat Park	60	2011	North
Tehran Besat Park	53	1974	South	Javanmardan Park	20	2012	North
Pardisan Park	275	1975	North	Tehran Jurassic Park	3	2013	North

HOOSHMAND ALIZADEH AND KAYOUMARS HABIBI
Urban Park Traditions in Iran and Tehran

It was in the 6th century B.C. that Cyrus the Great (559–530) established the garden at the Palace in Pasargadae and so began the history of the garden (Bāg) in Iran, particularly in the Iranian central plateau. King Cyrus created a fertile garden out of barren land, bringing symmetry and order out of chaos and, duplicating the divine paradise on earth, constituted a powerful statement of authority, fertility, and legitimacy. The Persian word "Bāg" refers to all cultivated trees and shrubs enclosed behind walls. Based on the mystical thought of Zoroastrianism, a garden is perceived as a symbolic site with streams dividing it into four gardens. It is likely that later Persian gardens were influenced by descriptions of the archetypal Garden of Paradise in the Koran, which stresses the garden's green color, shade, fruits, fountains of running water, and cool pavilions where the inhabitants may drink a wine that does not intoxicate. The purpose of garden creation was also for interaction between man and nature.

The city of Tehran has experienced a rapid transformation from an enclosed city, with a population of 15,000 in 1794, to an ever-growing megalopolis, with a current population of around 12 million. Historically, the city was known for its central location on the intersection of the Silk Road and the Spice Road. Its site is a fertile alluvial plain encircled by Alborz Mountains, like a fortified castle with ever-flowing rivers, thus abundant water and mild weather. Numerous gardens from the era of the Qajar dynasty (1794–1925) have been identified within the enclosure, as well as outside it, along river-valleys and on the Alborz mountainside. The British Orientalist Edward G. Browne (1862–1926) described the gardens of northern Tehran as places for walking and picnics, the most popular ways of passing leisure time. These gardens linked the mountainside to the city fabric. Some, like Negarestan (now a museum), Atabak (now the Russian Embassy) and Masoudieh (currently the Department of Education), have survived. But, unfortunately, many have been demolished in the course of time, due to the process of city growth.

Stronger ties with Europe in the period of Nasser al-Din Shah (1848–1896) paved the way for cultural exchange and borrowing of other forms of art and garden design. As a result, the significance of garden-planning increased and the rich competed in planting a larger variety of flowers and plants. Main features of the art of garden-planning during this period included: changes in the use of water; pruning trees; arranging plants in geometrical arrays; installing statues; arranging lawns and plants before the edifices; and using curved lines in designing green spaces.

In 1925, Reza Shah came to power, and an era of modernization started. At that time Tehran lacked modern urban parks; there were private gardens but no public parks. There had been, in the mid-19th century, only one large park open to the public, namely, the Lālehzār garden. Covered with wild tulips during spring, Lālehzār Garden was a kind of paradise on earth. With modernization there was the creation of new spaces, known as Bagh-e Melli (national gardens), created in large cities such as Tehran, Tabriz and Mashhad. The design of the new gardens was simple: one or two axial lines, a center and some parcels of green spaces between the pathways.

King Cyrus created a fertile garden out of barren land, bringing symmetry and order out of chaos and, duplicating the divine paradise on earth, constituted a powerful statement of authority, fertility, and legitimacy.

Reza Shah's main preoccupation was to modernize railways, judiciary and educational systems, traditional attire and religious customs etc., paying little attention to parks and green spaces. It was only after the 1953 Iranian coup d'état that Iranian city planning came to be more influenced by western ideas.[1] In 1959 a public park was built upon the old quarter of Sanglaj. It was named Park e Shahr, literally the City Park, and includes 26 hectares. The City Park was created with the intention of creating a quasi-natural environment in the city of Tehran that would be inviting for walks. The design of the park features a mix of European influence with a dominant north-south promenade that displays a spatial resemblance to the Royal Qajar garden, the Golestan Palace, within the old part of the city.

It was only after Tehran's first master plan in 1968 that parks became an issue in Tehran's city planning. City growth, high rise buildings, traffic congestion and related environmental problems called for more parks. There was also a growing sense among the Iranian design-elite that "the technological progress made possible by Western industrialism was no longer adequate—that it had somehow caused people to lose contact with their surroundings and with others" (John-Alder 2016). Jahangir Sedaghatfar (an expatriate Iranian architect who played the main role in the formation of the Pardisan Park in 1975) said that the vision of a modern city was mainly driven by commerce, technology, and the leisure pursuits of an aspiring middle class rather than by a shared sense

TOP RIGHT: Red line (added by authors) shows the boundary of the old quarter of Sanglaj in 1920, which was demolished to give way to Park e Shahr in 1959.
BOTTOM RIGHT: Park e Shahr (City Park) in Tehran.

MITTELHOLZER, W.: PERSIENFLUG, 1926/WIKIMEDIA COMMONS

VLADIMIR LYSENKO WIKIMEDIA COMMONS

Tehran from Ab-o-Atash park on a clear day May 14 2016.

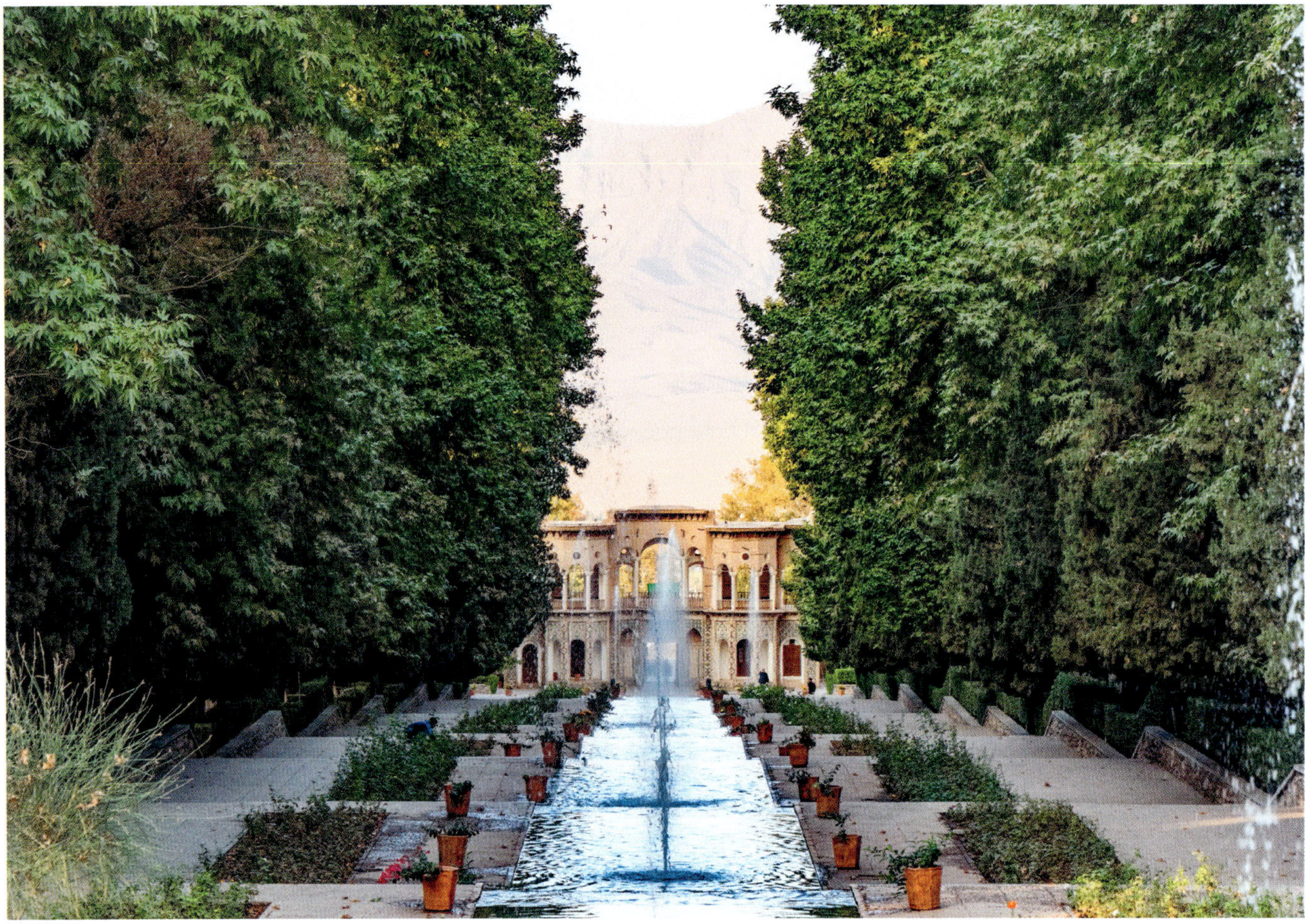

IMEDUARD SHUTTERSTOCK

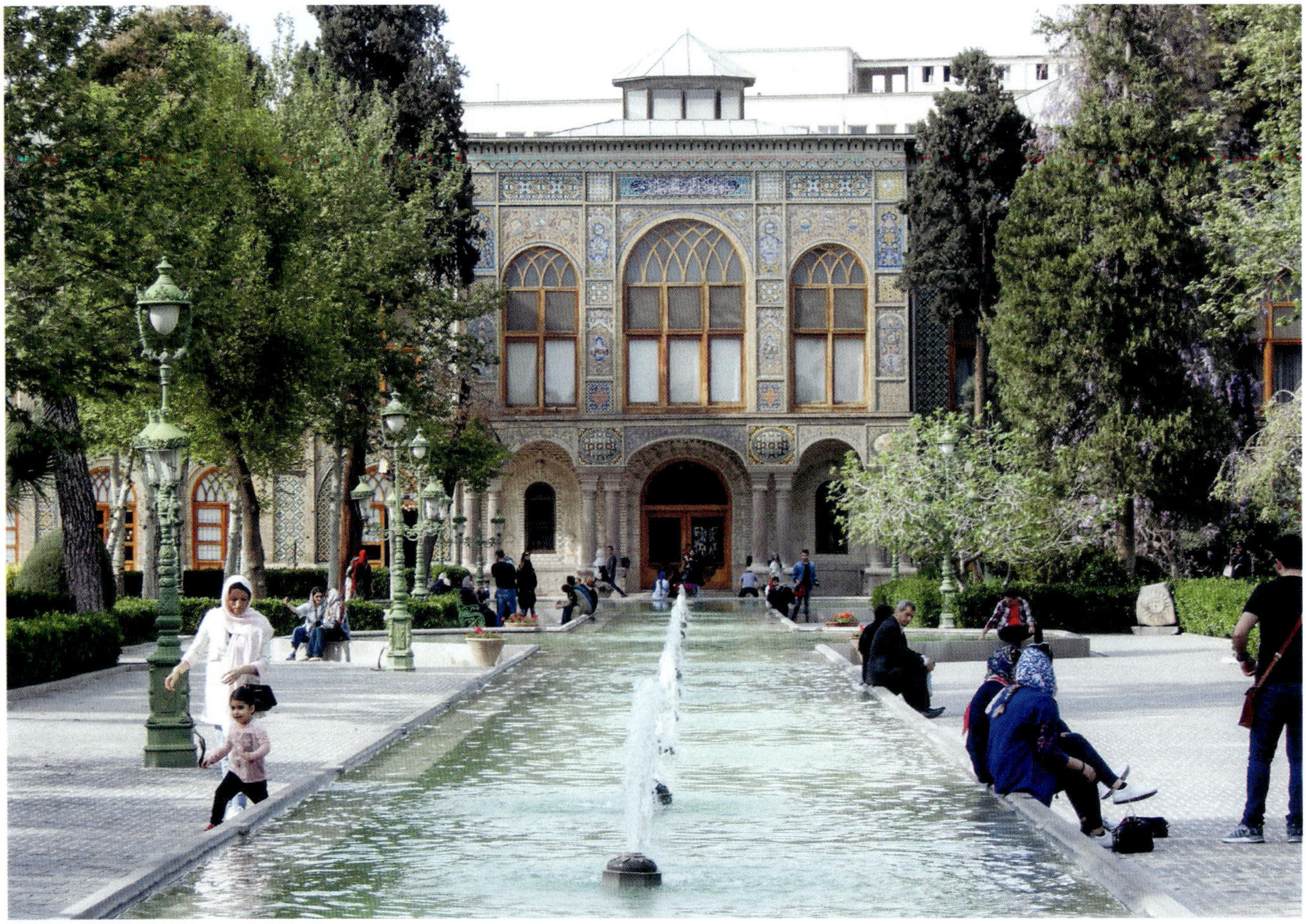

GÖRAN JOHNSON

of place (Sedaghatfar 1971). The traditional Iranian garden construction gradually sank into oblivion. Modern public parks were created, overwhelmingly in the rich neighborhoods in the north, towards the slopes of Alborz Mountain, where ever-flowing rivers and high precipitation create favorable conditions. These include the Gheytarieh, Niavaran, Jamshidieh Stone Garden, Pardisan and Shafaq parks. As a result, more than 80 percent of the public park space has been located in the northern part of the city, even though only 50 percent of the city's residents live there. The height difference between the northernmost and southernmost areas of the city is 1,000 meter. The flat desert areas in the south are inhabited by people with lower incomes.

The aim of the first master plan of Tehran in 1968 was "to bring new order to the irregular urban expansion of the city and respond to the growing number of rural migrants" and to respond to the issue of the congested city center "by proposing a series of centers to reorient growth and reorder social structures" based on a linear decentralization, stretching the city westward (Mashayekhi, 2018). This modernist idea of planning sought to create the 'ideal city' using three key elements: the neighborhood unit, green spaces and the super-highway. The plan conceptualized 10 regional centers, separated by large green areas and linked by a network of highways and rapid public transportation. Five strategic large parks were located along the main corridor of city development westward. One of these is Pardisan Park in the northwest. Another was the AbassAbad Park of which nothing is left today.

Pardisan Park is a forest park covering more than 270 hectares, located in the northwest of Tehran . It is bordered on three sides by expressways, as well as by the Nahj Ol-Balagheh linear park to the west along the Ponak River valley, and the Milad recreational complex to the east. Altogether this comprises almost 500 hectares of green space. Due to its strategic location in the northwest of Tehran, Pardisan Park serves as the vital lungs of the city, producing oxygen and purifying air.

The name Pardisan was derived from the Old Persian word "pardis," from which the word "paradise" was also derived. In the time of the Achaemenians (559–330 BC), the first illustrious period of Persian history, the word signified a royal garden, where "all good things the earth provides" might be enjoyed (John-Alder 2016). The name was first

Now the Pardisan park—the lungs of the city—exists half-finished in a city that is struggling with pollution.

TOP LEFT: Khannevadeh Garden, traditional Persian garden found in the desert near the city of Kerman.
BOTTOM LEFT: Golestan, former Royal Palace in the center of Tehran, now open to the general public.

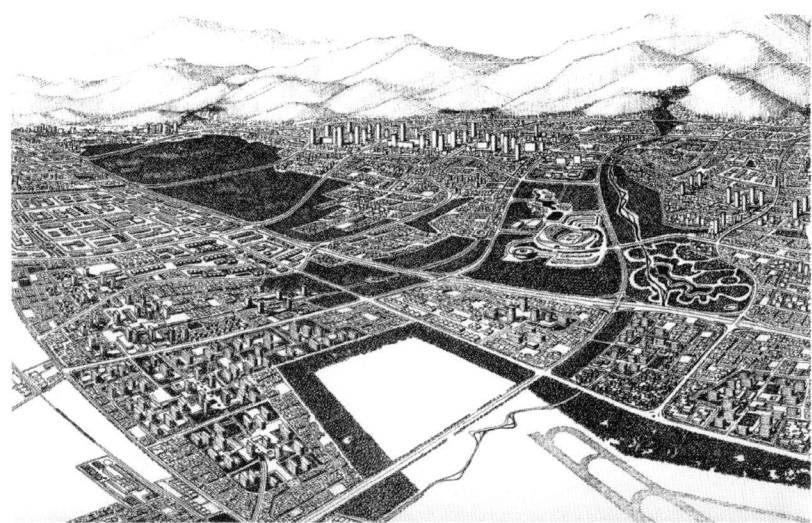

The 1968 model of Tehran expansion towards the west with large green areas and a network of highways. (Mashayekhi 2018)

proposed by Eskandar Firouz[2] following the Stockholm Conference[3] in 1972. The conference conceived of the earth as a mother, nature as sacred, and pollution of the environment as sin. This led Firouz to conceptualize the park as embracing "all known aspects of the universe and the world." It was characterized as ranging "from its beginning; to the evolution of life; to man and his use and abuse of nature; to a future wherein he must find harmony with his world. Venues would display the natural history of the earth, and they would situate the history of human settlement within this context" (McHarg 1975). Ian McHarg,[4] whom Firouz recommended, produced the Pardisan Park master plan in 1975.

The project was halted due to the increasing social unrest in Iran that led to the 1979 Revolution. The Islamic Republic of Iran did not implement the plan but instead followed earlier reforestation recommendations.

The park's official name is now Pardisan Forest Park, commonly referred to as Shahrak-e Gharb Forest Park. A small natural history museum and a recently constructed climate research facility are the only indicators of the original proposal. The area of the park has shrunk by 70 to 80 hectares due to usage of the land by various entities: the Milad Tower complex to the east; a Tehran water treatment plant; various office buildings and museums; and an expressway. A major threat was President Mahmoud Ahmadinejad's plan to build affordable housing in the park, but this was stopped by the Tehran city council.

TOP RIGHT: Nahj-ol-Balagheh, a modern park west of Pardisan park.
BOTTOM RIGHT: Pardisan Park in the northwest of Tehran as seen from Milad Tower, 2015.

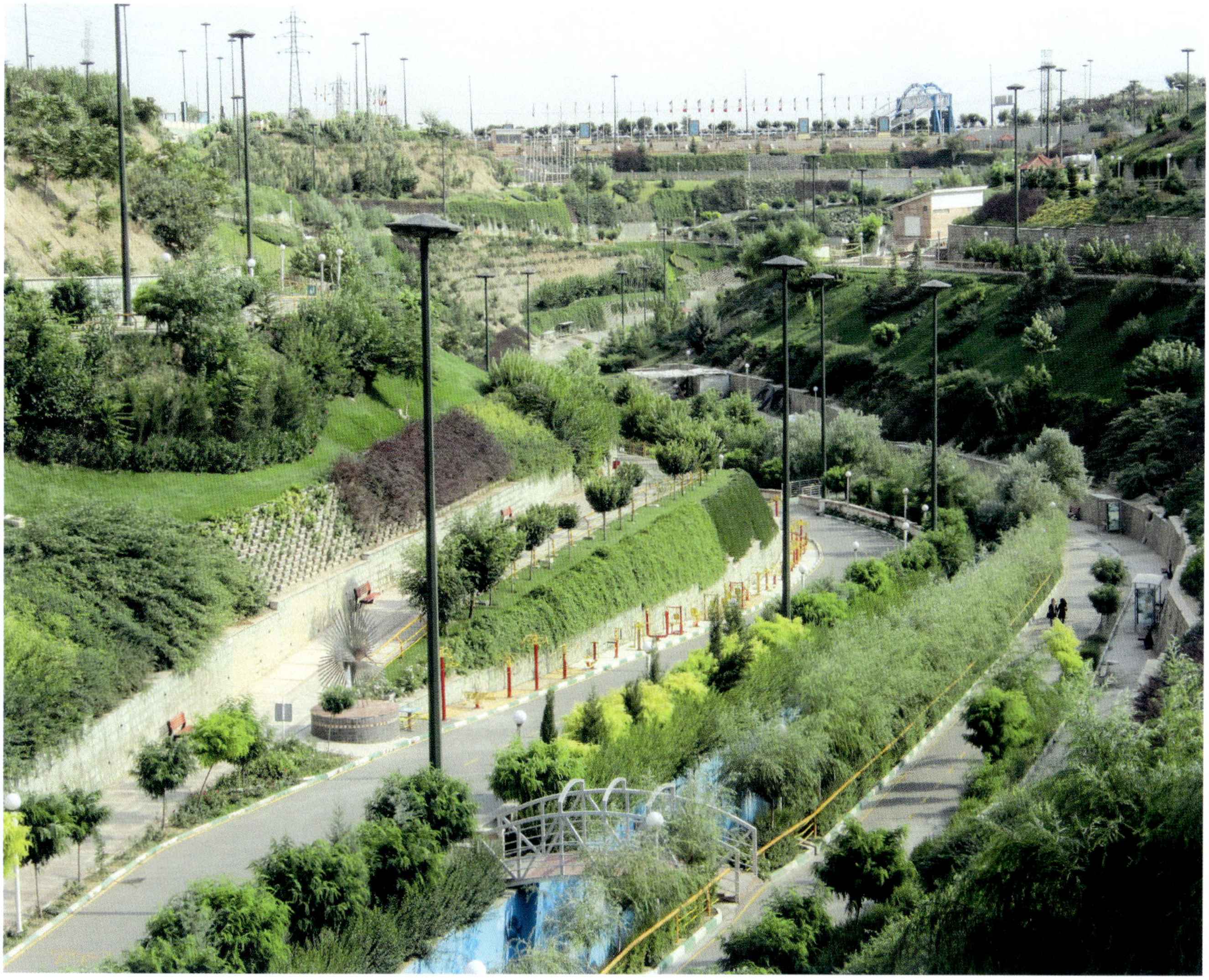

SANAZ SHOBEIRI

PRIVATE SOURCE

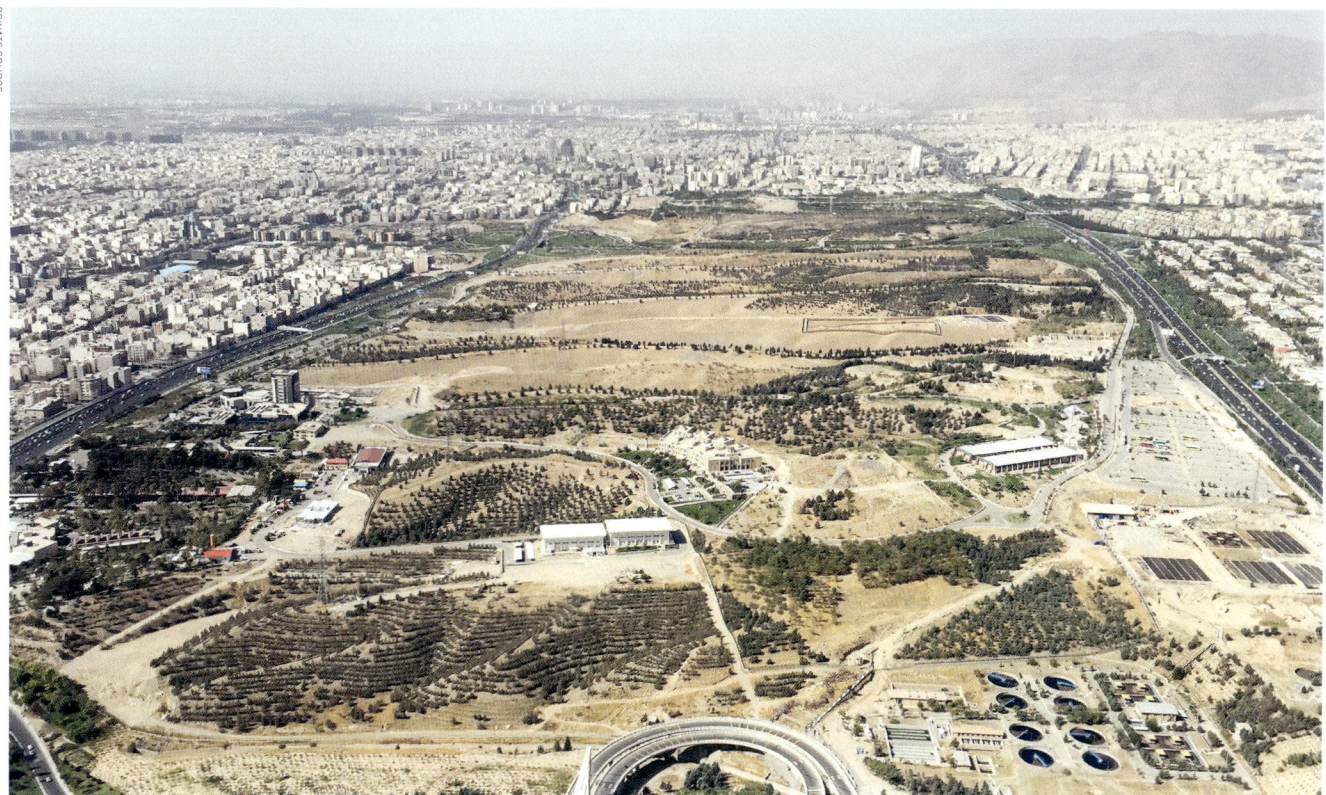

Milad Tower from Pardisan Park on a smoggy day.

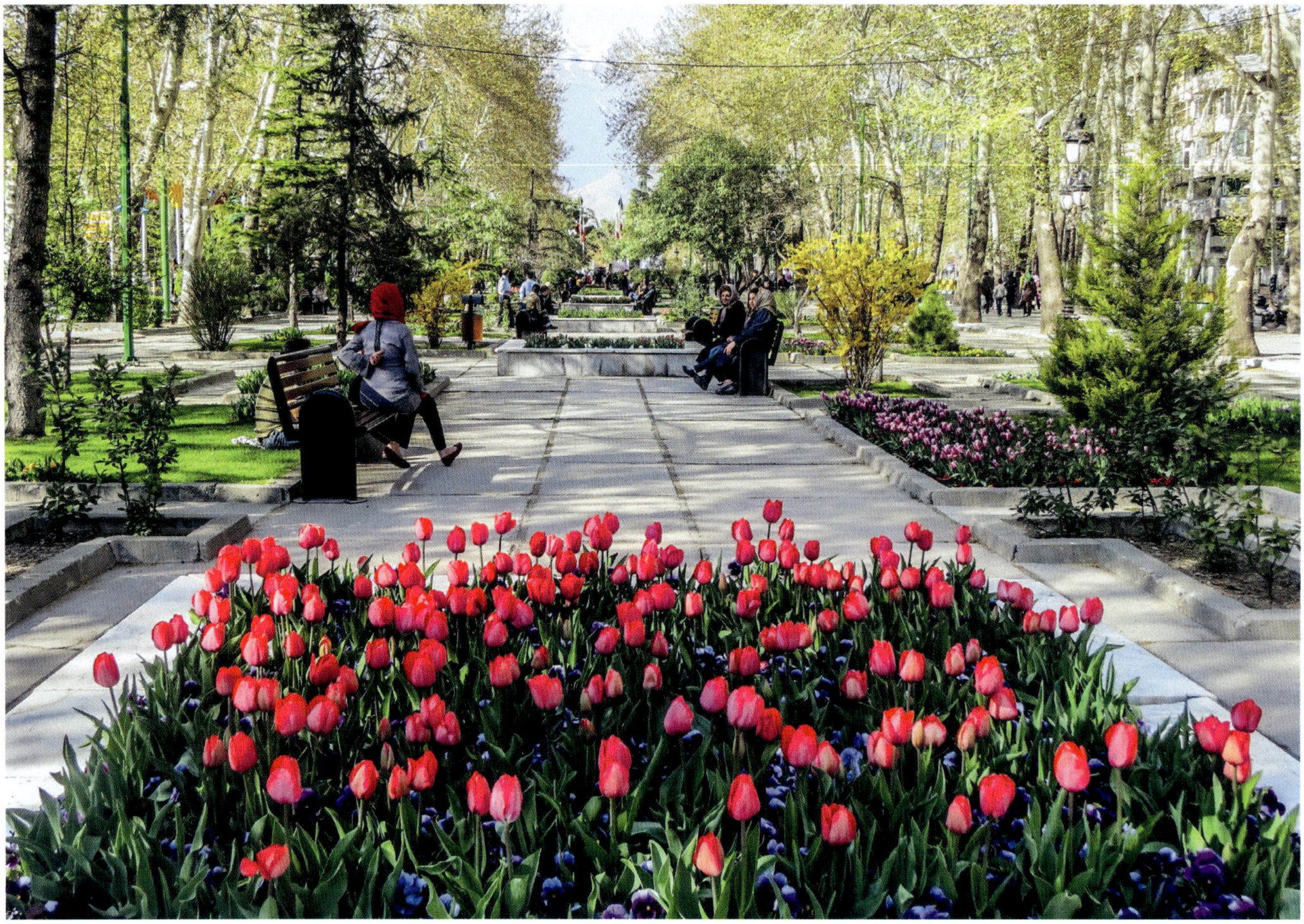

NINARA WIKIMEDIA COMMONS

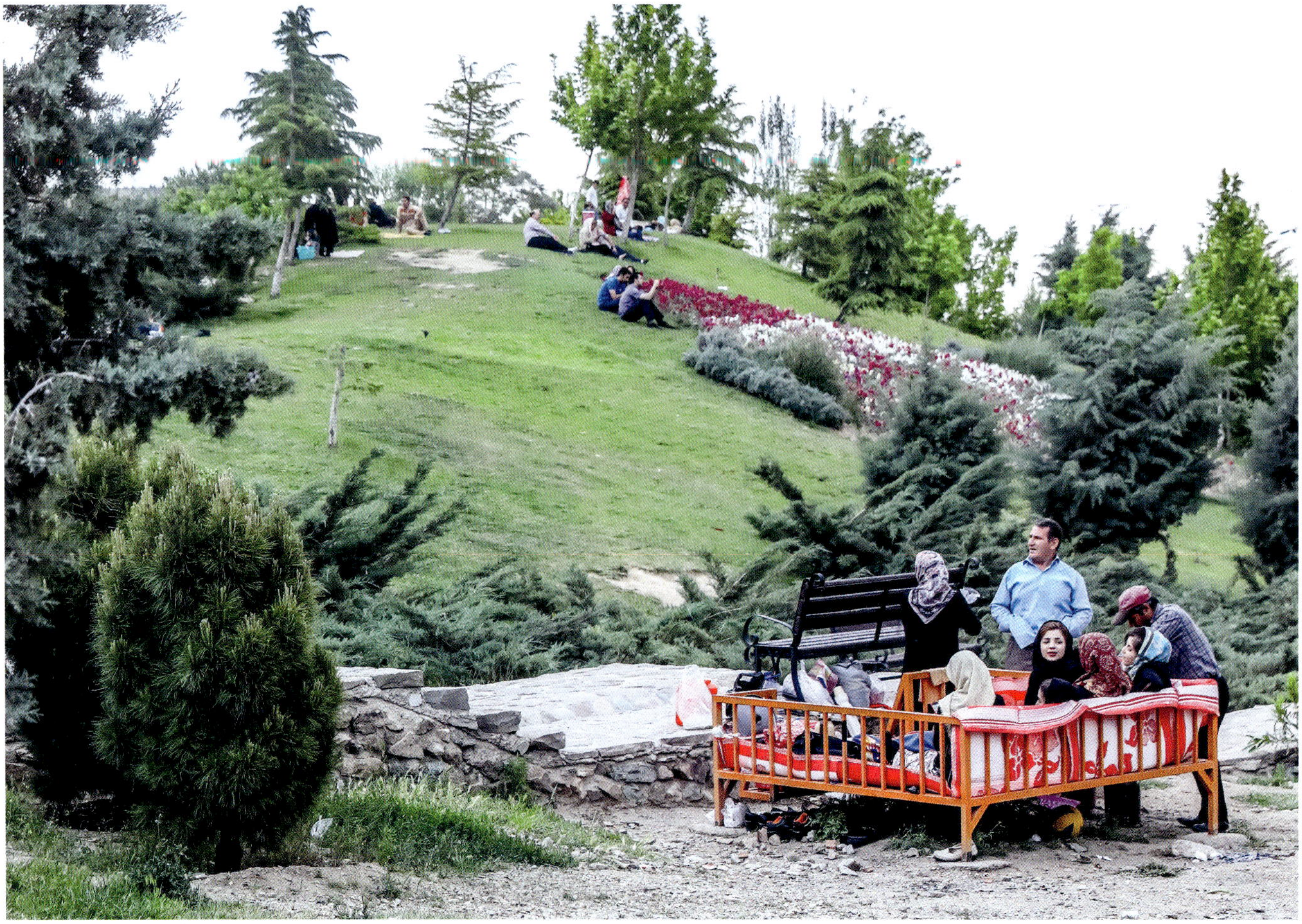

NINARA WIKIMEDIA COMMONS

Now the Pardisan park—the lungs of the city—exists half-finished in a city that is struggling with pollution.

The Pardisan Park is rich in biological diversity due to the broad spectrum of habitats within the park. Over 300 indigenous plant species thrive in the park, many of which are endemic.

Due to its location and accessibility, numerous social activities, events and celebrations are held in the park. These include a summer story-telling festival, Ramadan charity festival, Health Festival during Health week, celebration of Eid-e-Mab'ath[5] and Ramadan, as well as public sports. Because of the vast space of the park, kite flying, picnic, cycling and hiking are popular activities.

The Tehran Municipality is responsible for the restoration and maintenance of Pardisan Park. For the next four years, 3 trillion Rials ($75 million) has been earmarked. The park is in poor condition: "Trees have dried up, fires have occurred, irrigation is failing and there is not sufficient lighting at night," an official said.

In the past the municipality has inflicted harm on the park by taking over the surrounding areas of the park, including the Farahzad valley, and by constructing highways around the park. More than 100 plant species of the Farahzad riverside, including many orchids, were wiped out.

The municipalities in Iran usually consider urban parks as a potential source of economic revenue when exploited. This is a potential threat to the Pardisan Park due to the natural character of the park in comparison with other parks in Tehran.

[1] This may seem paradoxical, since the coup overthrew the democratically elected Prime Minister Mohammad Mosaddegh, but can be explained by the very strong western orientation of the Shah.

[2] The first director of the Department of Environment in Iran in the period of second Pahlavi dynasty. He was known as the father of environmental protection and conservation in Iran.

[3] United Nations Conference on the Human Environment held in Stockholm, Sweden, in June of 1972, commonly referred to as the "Only One Earth" conference.

[4] American urban planner of the University of Pennsylvania.

[5] The day of selection of a prophet by God for guidance of the people. On this day, at the age of forty, Muhammad was chosen in the Hira' cave in Mount Nur (close to Mecca), and the first verses of Qur'an were revealed to him.

TOP LEFT: Mellat Park, Tehran.
BOTTOM LEFT: Picnicking in a park is a favorite pastime.

PHOENIX PARK

DUBLIN

PHOENIX PARK

Dublin
Bay

CASTLEKNOCK

River Road

Rafoath Road

Navan Road

Ashtown Gate

Castleknock Gate

The White Fields

White's Gate

Ordnance Survey Ireland

Farmleigh

Ashtown Castle and Visitor Centre

Chesterfield Avenue

Ratra House

John Paul II Park

CABRA WEST

Nephin Road

Navan Road

Tower Road

U.S. Ambassador's Residence

PHOENIX PARK

Áras an Uachtaráin

Furry Glen

Polo Grounds

Dublin Zoo

Cheshire Home

Playing Pitches

Garda HQ

STONEY-BATTER

Glenaulin

St. Mary's Hospital

Magazine Fort

Wellington Testimonial

Park Gate St. Entrance

CHAPELIZOD

Chapelizod Gate

Chapelizod Bypass

River Liffey

National War Memorial Garden

Dublin Heuston

Guinness Brewery

Le Fanu Road

BALLYFERMOT

The Lawns

Kylemore Road

INCHICORE

KILMAINHAM

BLUEBELL

N

0   500   1000 Meters

0   0,5   1 Mile

© OpenStreetMap contributors

PHOENIX PARK, DUBLIN, IRELAND

SIZE: 7.07 square kilometers.

LEGAL PROTECTION: Phoenix Park Act 1925; Phoenix Park Bye Laws 1926.

OPENED: 1747.

OWNER: The Phoenix Park is owned by the national government.

MANAGEMENT: Office of Public Works.

HOW TO GET THERE: Lots of public transport to the Park including by bus, rail and tram from Dublin city center.

MARGARET GORMLEY
Phoenix Park, Dublin, Ireland

Phoenix Park was established as a royal deer park in 1662 by one of Ireland's most illustrious viceroys, James Butler, Duke of Ormond, on behalf of King Charles II of England. Today Phoenix Park extends to over 700 hectares and represents a unique natural and cultural landscape: both a historic park and an urban park enjoyed by over 10 million visitors annually.

The park provides a setting for a range of activities and amenities as well as acting as a location for a number of important public institutions and residences. As a natural and constructed park, enclosed over 300 years ago by a demesne wall, Phoenix Park is unique in Ireland. Its location (2.3 kilometers from Dublin's center), size, and use can be compared to similar large urban parks in other cities, including Regent's Park in London, the Bois de Boulogne in Paris, and Central Park in New York.

The cultural heritage of Phoenix Park enriches people's lives, providing a deep and inspirational sense of connection to history and landscape. It provides a sense of place, locations for community cohesion and social inclusion, and promotes cross-cultural enjoyment and space for many recreational events. The park is also good medicine in that it provides numerous opportunities for green exercise. It is educational too, with a range of science and learning experiences, and also plays an important role in the tourism economy.

Since the Neolithic and Early Bronze ages, humans have had long associations with the lands that now form Phoenix Park. In 1680 Phoenix Park was reduced to its present size of 700 hectares. In 1747 the Earl of Chesterfield, having considerably improved the park, opened it to the public for the first time. Its present landscape and infrastructure are inherited from designs and managerial decisions taken between 1800 and 1880, mainly by Decimus Burton, who also worked in the royal parks in London, England. The Victorian People's Flower Garden was designed during this period and was noted for its novel horticultural experimentation and floral displays.

In the center of Phoenix Park, dating from 1750, is Áras an Uachtaráin, the residence of the President of Ireland, which also served from 1782 to 1922 as the residence of the British viceroys.

The Duke of Wellington was based in Phoenix Park when he was chief secretary to the Lord Lieutenant of Ireland between 1807 and 1809. Born

Phoenix Park enriches people's lives, providing a deep and inspirational sense of connection to history and landscape.

in Ireland, the Duke later became Commander-in-Chief of the British Army and Prime Minister of England. The Wellington Testimonial in Phoenix Park commemorates his great battle victories, including Waterloo.

Gas lighting was first introduced into the park in 1859 by the Hibernian Gas Company and is a major visual element in Phoenix Park's landscape. It is one of the few remaining public areas in Europe that still relies on gas for public lighting. The decision to preserve this unique system was made to support the conservation of the historic fabric of the park and to retain low levels of light pollution so that it can remain one of the few locations in Dublin where stargazing is possible.

In the early 1860s the Lord Lieutenant of Ireland sponsored a bill in Parliament in London, granting civil servants a cricket ground in the park. The All-Ireland Polo Club was founded in 1873, making it the oldest polo club in Europe. With health and wellbeing to the fore, the park today caters for over 2,300 recreational events including athletics, soccer and the Irish national sports of Gaelic football and hurling (played by men) and camogie (played by women) which are deemed the fastest ball and stick sports in the world.

Phoenix Park is an extremely important site for biodiversity in Dublin. Fifty percent of all mammal species found in Ireland occur within Phoenix Park and over forty percent of all bird species occurring in Ireland have been recorded therein. A herd of over 550 wild fallow deer have roamed the park since the 1660s.

Woodlands and tree-dominated areas cover 31 percent and grasslands cover 56 percent of the area. Twenty-five different types of habitats include six types of woodland, five types of grassland, as well as hedgerows, scrub, ponds, streams, and wet ditches. Among the 351 different plant species to be found there are three that are rare and protected including the Hairy St. John's Wort (*hypericum hirsutum*), Hairy Violet (*viola hirta*) and the grass Meadow Barley (*hordeum brachyantherum*). Almost all the semi-natural grassland in Dublin is found in Phoenix Park.

The park is a complex place comprising many components that serve a variety of functions. It is used by large numbers of people and is also a heavily trafficked route between the center of the city and outlying suburbs. In recent decades, the growth of Dublin has significantly altered the position and use of Phoenix Park and affected its role and potential.

TOP RIGHT: Cricket game at Playing Pitches.
BOTTOM RIGHT: Phoenix Park Tea Rooms.

Phoenix Park with Wellington Testimonial facing Dublin with harbor.

ANDREW MCMILLAN WIKIMEDIA COMMONS

SUPERCHILUM WIKIMEDIA COMMONS

The Park has over twenty-five kilometers of roads, seventeen kilometers of cycle trails, twenty-seven kilometers of surfaced footpaths and eleven kilometers of perimeter wall. The Park caters for an average of nine million car journeys per year, the majority of which are merely passing through. A recent initiative introduced by the Office of Public Works was the closing of a major section of the main road, Chesterfield Avenue, to through-traffic at weekends during the summer months.

Over 260 major and medium scale events take place annually in Phoenix Park, ranging from triathlons to pop concerts to horticultural shows. While these events open the park to a wider audience, they may impact the fabric of the park. Drainage, ground preparation measures, and tree constraints plans[1] are some of the physical measures that have been introduced on site along with penalties for breaches of the rules. An event framework is in place for the park, which limits events on grass from May to September each year and the number of concerts and major events taking place annually, and also specifies their location within the park. The bulk of income derived from fees generated is returned to government. All costs and repairs associated with an event are charged to the organizer of that event.

The sheer scale of Phoenix Park conveys the false impression that it can absorb significant levels of development, whether functional or for amenity, without altering its essential character. This, however, is not the case. Phoenix Park is a finite resource, the integrity of which is dependent on maintaining its historic character and its openness.

Given the high volume of people and vehicles using Phoenix Park daily, its exposure to the spreading of disease within the flora and fauna populations is inevitable. For example, during the foot and mouth outbreak, access to the park was limited and sanitary measures were placed at all the entrances. Plant species selection has been modified to suit the changing weather patterns and strengthen resistance to pests and diseases. The new Commemorative Avenue, opened in 2016, was planted with London Plane trees given our concerns with Oak Processionary Moth, etc.

There are international charters, conventions, etc., which are of relevance to the cultural heritage of Phoenix Park. Although these do not have any legal effect, it is considered good conservation practice to consider the principles contained within them (see references).

The Irish Government's Vision for Phoenix Park is:

To protect and conserve the historic landscape character of The Phoenix

… to retain low levels of light pollution so that it can remain one of the few locations in Dublin where stargazing is possible.

TOP LEFT: Winter in Phoenix park.
BOTTOM LEFT: Fallow deer.

Park and its archaeological, architectural and natural heritage whilst facilitating visitor access, education and interpretation, facilitating the sustainable use of the Park's resources for recreation and other appropriate activities, encouraging research and maintaining its sense of peace and tranquility.

Given the over-riding importance of the historic designed landscape of Phoenix Park, priority is accorded to the conservation of the landscape, even where this restricts or limits the achievement of other objectives relating to the park. For example, requests for formal playgrounds to be introduced throughout the park have been refused as they would materially alter the landscape setting of Phoenix Park. A modern playground catering for multiple age groups was instead built at Phoenix Park Visitor Center close to toilets, car parking and a café.

Phoenix Park Conservation Management Plan 2011 identifies over one hundred specific actions to be undertaken over the next ten years. The first stage of the funding process is the development of the Phoenix Park Tourism Master Plan.

The long-term vision for Phoenix Park combines its protection, conservation, enjoyment and tranquility as an important unique historic landscape for the residents of Dublin and visitors to Ireland. Its historic continuity, together with its vast scale, urban setting and serenity, are the attributes that define Phoenix Park and give it a unique appeal. The essence of managing historic parks and gardens is continuity. That is, management must strive to maintain the valuable inheritance of the past, but must also address challenges and opportunities arising from the inevitability of change.

Management must strive to maintain the valuable inheritance of the past, but must also address challenges and opportunities arising from the inevitability of change.

[1] A Tree Constraints Plan contains the positions and dimensions of all trees on a site, crown spreads, root protection areas, and more.

RIGHT: Gas lamp, still in operation.

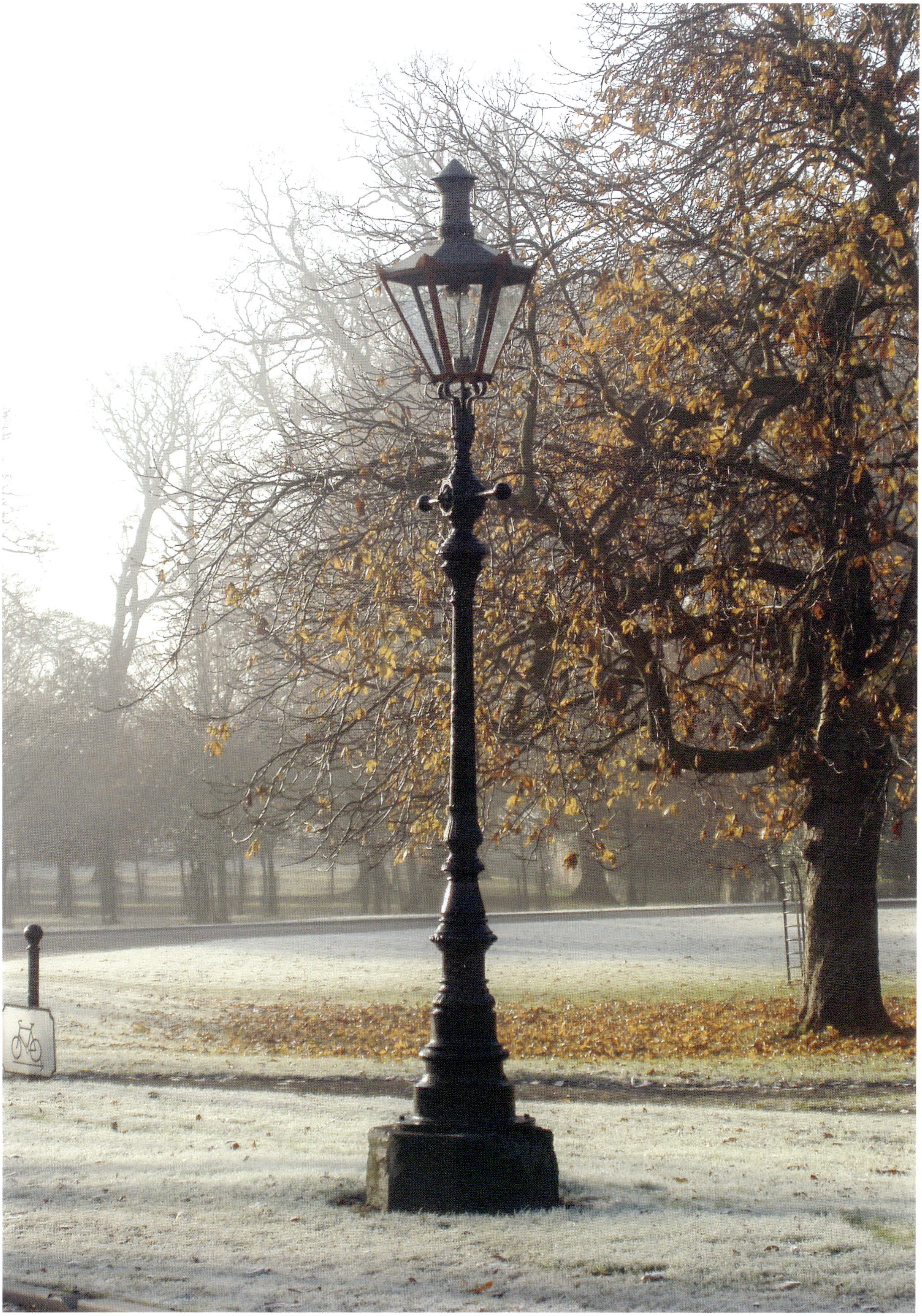

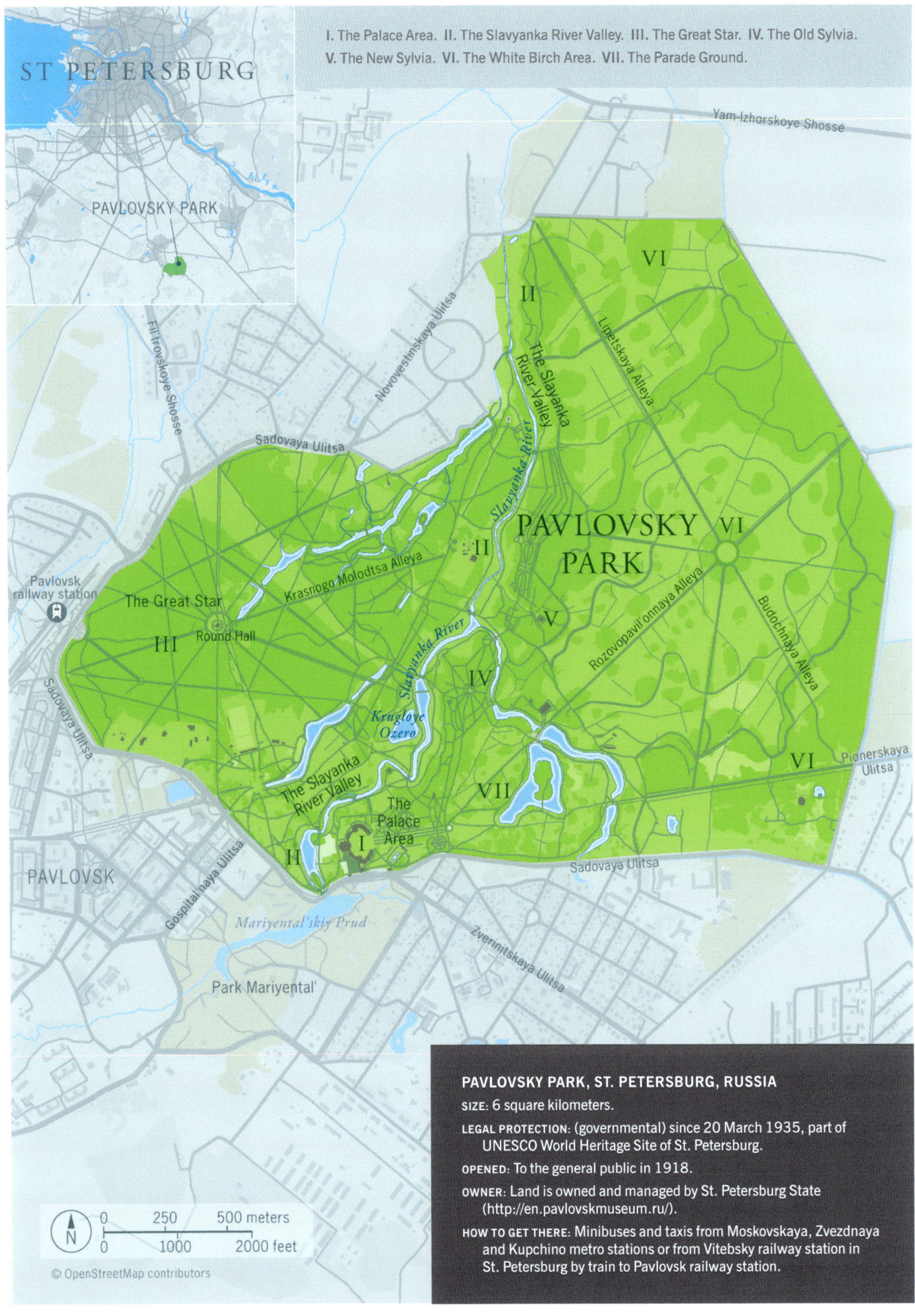

ST PETERSBURG

PAVLOVSKY PARK

I. The Palace Area. II. The Slavyanka River Valley. III. The Great Star. IV. The Old Sylvia.
V. The New Sylvia. VI. The White Birch Area. VII. The Parade Ground.

Yam-Izhorskoye Shosse

VI

II

The Slavyanka River Valley

Novovestninskaya Ulitsa

Filtrovskoye Shosse

Lipetskaya Alleya

Sadovaya Ulitsa

PAVLOVSKY
PARK

VI

Pavlovsk
railway station

The Great Star

Krasnogo Molodtsa Alleya

Slavyanka River

II

V

Rozovopavil'onnaya Alleya

Budochnaya Alleya

III   Round Hall

Slavyanka River

IV

Kругloye
Ozero

Pionerskaya
Ulitsa

VI

Sadovaya Ulitsa

The Slavyanka
River Valley

VII

II   I

The
Palace
Area

PAVLOVSK

Gospital'naya Ulitsa

Sadovaya Ulitsa

Mariyental'skiy Prud

Zverinitskaya Ulitsa

Park Mariyental'

N

| 0 | 250 | 500 meters |
| 0 | 1000 | 2000 feet |

© OpenStreetMap contributors

PAVLOVSKY PARK, ST. PETERSBURG, RUSSIA

SIZE: 6 square kilometers.

LEGAL PROTECTION: (governmental) since 20 March 1935, part of
    UNESCO World Heritage Site of St. Petersburg.

OPENED: To the general public in 1918.

OWNER: Land is owned and managed by St. Petersburg State
    (http://en.pavlovskmuseum.ru/).

HOW TO GET THERE: Minibuses and taxis from Moskovskaya, Zvezdnaya
    and Kupchino metro stations or from Vitebsky railway station in
    St. Petersburg by train to Pavlovsk railway station.

MARIA IGNATIEVA

Pavlovsky Park, St Petersburg, Russia

Pavlovsky Park, which is part of the green "necklace" of St. Petersburg, is one of the finest examples of landscape-style parks in the world and one of the largest European landscape parks (English-style parks).

Pavlovsky Park and palace complex covers 6 square kilometers and consists of seven landscape areas, the main palace (Pavlovsky Palace) and numerous pavilions and bridges.

In 1777, Empress Catherine the Great gifted the land along the Slavyanka River to her only son, Paul, and his wife, Maria Feodorovna, to create a country residence. Catherine the Great invited the Scottish architect, Charles Cameron, to design the country residence in Pavlovsky Park. Cameron worked on the project from 1780 to 1786. He outlined the main areas and their planning structure and designed the compositional centers of the park. He created the Palace Area with the Private Garden, the Three Graces Pavilion, the Lime Avenue, the Slavyanka River Valley and the Great Star, and outlined the main pathways in the White Birch Area. Cameron was the first architect of the Palace and the designer of the Friendship Pavilion, the Apollo Colonnade, and the Aviary Pavilion. His architectural designs were inspired by the beauty of Russian nature and created a new landscape park on the site of indigenous mixed coniferous forest.

When Paul became the tsar in 1796, he appointed the Italian architect Vincenzo Brenna as the main architect for the official imperial residence at Pavlovsk. Brenna rebuilt and significantly extended the palace to reflect the status of a royal residence and gave a more ceremonial character to the surrounding park. He included a formal (geometrical) style with avenue corridors and enclosed green rooms decorated with marble and bronze sculptures. Thus, the park's green areas became a continuation of the palace's halls. Brenna designed the Great Circle and New and Old Sylvia (see map). The planning structure of these areas had a regular character that was skillfully included in Cameron's existing composition and did not destroy his general principles of the park's landscape design (Iliyanskaya, 1993).

After the assassination of Paul in 1801, Maria Fedorovna (Paul's wife) took charge of the estate development until her death in 1828. It was her love, and during her life Pavlosvky Park was "the family poem into

Pavlovsky Park can be considered the gold standard of scientific restoration of historical parks.

which one human life was fitted." During this period, she engaged Pietro Gonzaga, an outstanding Italian theatre and stage designer and decorator, to work on the palace and Pavlovsky Park. He had worked at La Scala in Milan and the theatres of Moscow and St. Petersburg. Gonzaga created the Parade Ground and the White Birch area. Gonzaga's sceneries in the park clearly reflect his background as a theatre designer and perspective painter (Hayden, 2005).

Gonzaga's sceneries in the park clearly reflect his background as a theatre designer and perspective painter.

The main idea of Gonzaga was to create the White Birch Area, almost one-third of the whole park, to glorify northern Russian natural landscapes. There are no ponds, pavilions or sculptures, only Gonzaga's "music for the eyes," a special planting rhythm created by establishing native woodlands, meadows, groups of trees and solo trees. Gonzaga literally transferred the aesthetics of theatre set design to landscape architecture by painting with plants. He had groups of trees planted around the open space to create the appearance of stage wings and an illusion of very deep, long views (Ignatieva, 2005).

Even using a limited number of native species, Gonzaga achieved a great variety of landscapes by using different sizes, configurations, colors and textures of groups, groves and solo trees, and by varying the distances between them. Basically, he "improved" nature using design and art principles.

Meadows in the area were created by clearing the existing forest and sowing meadow plants. The first mixtures came from European countries and probably consisted of various clovers and perennial grasses, such as timothy *(Phleum pratense)* and ryegrass *(Lolium perenne)*. The management of the meadows included hand mowing at least once a season to obtain hay. The mowing discouraged the growth of woody species, which would otherwise grow and replace the meadow grasses.

Another key to the successful development and the decorative quality of the White Birch Area meadows was regular cleaning and maintenance of the drainage system, which also kept the area from reverting to the original damp woodlands of silver birch, downy birch *(Betula pendula, B. pubescens),* Scots pine *(Pinus sylvestris),* and Norway spruce *(Picea abies)*. All planning and plantings were finally completed in 1840. In the 19th and early 20th century, the White Birch Area meadows produced high-quality hay that was used in nearby villages.

After the Russian Revolution in 1917, the government nationalized Pavlovsky Palace and Pavlovsky Park, and since 1918 the park has been

TOP RIGHT: The Pavlovsky Palace.
BOTTOM RIGHT: Area containing formal flowerbeds in the Private Garden next to the Palace.

IGNATIEVA

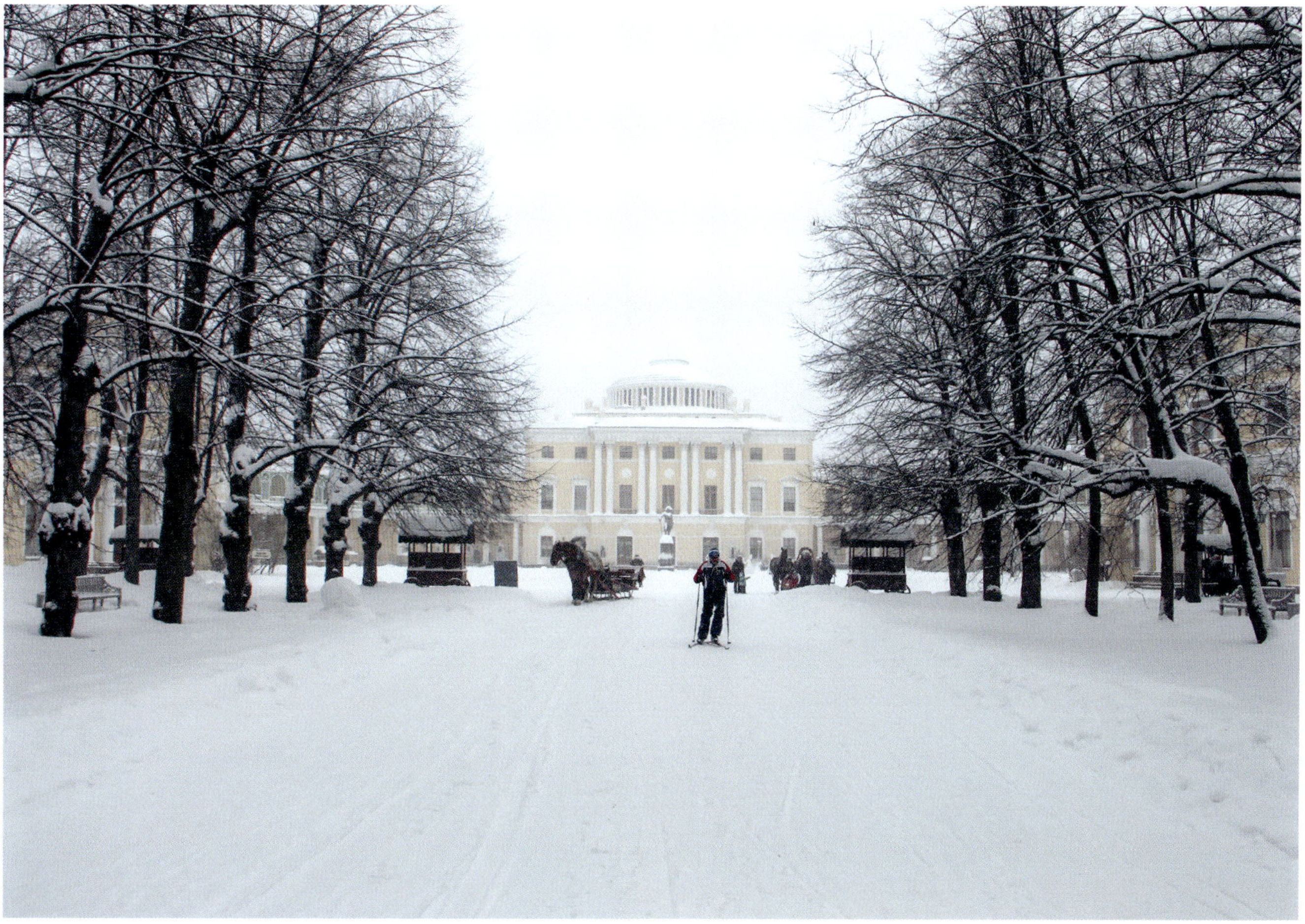

IGNATIEVA

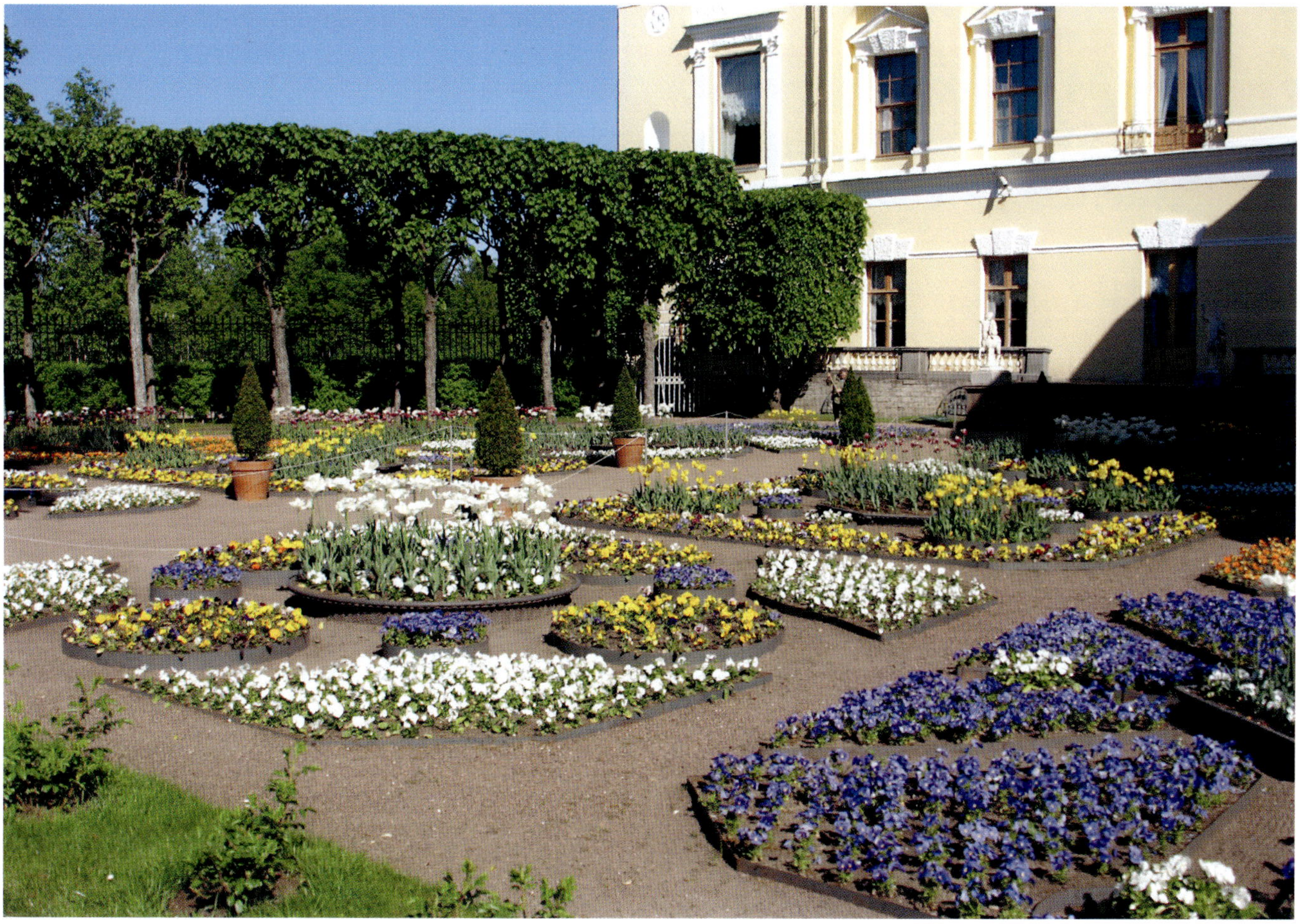

Meadow plant communities cover 120 hectares of the White Birch Area.

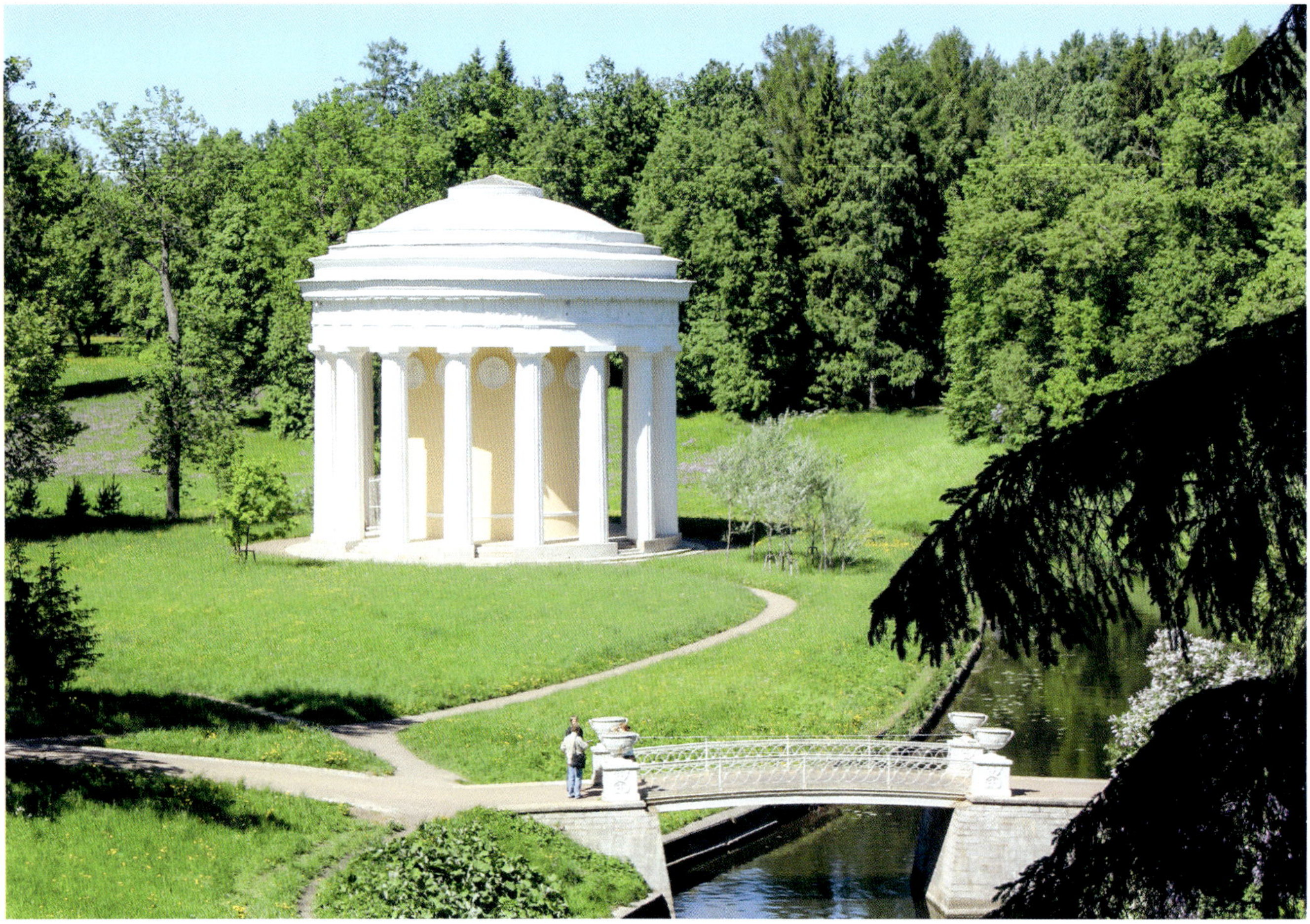

IGNATIEVA

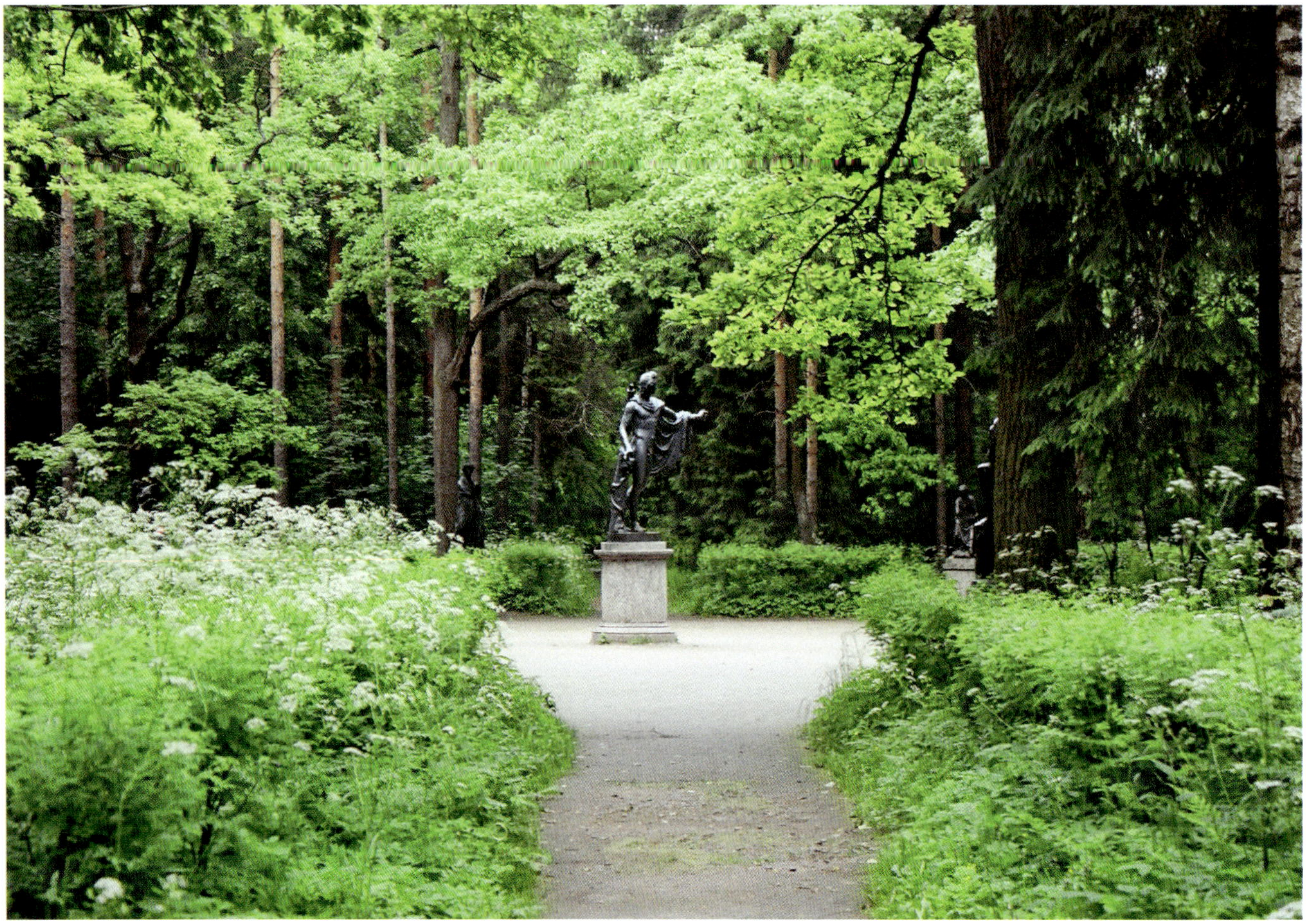

IGNATIEVA

under state protection. The palace was opened for public visits two or three days a week and the park started to be used for public recreation. The boat station, green theatre, skiing lodge and other facilities offered visitors many options.

Both the buildings and the park suffered substantial damage during WWII. The Nazis destroyed the pavilions, blew up the bridges, cut down more than 70,000 trees and, on retreating, burnt the palace. Restoration and reconstruction work at Pavlovsky Park began almost immediately after the end of the war. It was the first of the parks in the St. Petersburg (then known as Leningrad) suburbs to receive this attention, largely because of its extraordinary importance in Russian history and the love of St. Petersburg citizens. The St. Petersburg restoration school of historical parks is now world-renowned for its scientific approach and system of restoration supervision. The project designer is responsible for making regular visits and inspections of park projects, even years after the implementation phase has been completed, to ensure that problems are identified and resolved at an early stage.

In the case of Pavlovsky Park, special acknowledgment should be given to Marina Flit, the main landscape architect of the park. Only recently retired, she worked closely with the garden restorers and supervised the park's maintenance and management for over 40 years. Pavlovsky Park can be considered the gold standard of scientific restoration.

After the War, all landscapes of the White Birch were carefully restored. This included the spatial composition to restore the original relationship between open and closed spaces, and the original kind of trees. This was crucial to maintaining the authenticity of Gonzaga's original design concept.

Today the White Birch Area is composed mostly of Norway spruce, a fair amount of birch and some Scots pine. The flora is rich with over 200 higher vascular plant species. Among the spring flowering herbaceous species *Anemone nemorosa* dominates in many forest and edge plant communities. There are also protected species such as *Lycopodium clavatum*, *Platanthera bifolia*, *Drosera rotundifolia* and *Nuphar lutea*, and rare ones like *Pimpinella major* and *Actaea spicata*. *Pimpinella major* was probably introduced into Pavlovsky Park via seed mixtures. This particular plant dominates in many of the park's meadows.

Restoration and reconstruction work at Pavlovsky Park began almost immediately after the end of the Great Patriotic War because of its extraordinary importance in Russian history and the love of St. Petersburg citizens.

TOP LEFT: The Friendship Pavilion, a key feature in Slavyanka River Valley. This pavilion has been the inspiration for a number of similar garden temples in other Russian parks (Hayden, 2005).
BOTTOM LEFT: The Old Sylvia with the Apollo Belvedere in the center of the composition and twelve radial paths. The Apollo is surrounded by bronze statues of the Nine Muses.

According to a sociological survey in the White Birch Area in which visitors were asked the question "What is important for you in this historic park?", 50 percent responded "the park has a history" and another 13 percent "the old age of the park is important." For 11 percent of the interviewees, the park's protected status had value, while 10 percent loved the park's beautiful landscapes ("admiring the scenery"). People reported that they walk, cycle or ski in the White Birch Area. In wintertime, 40 percent use it for walking and 60 percent for skiing. In summer, 70 percent of visitors walk and 30 percent cycle in the White Birch Area (Emelyanova, 2017).

Because St. Petersburg is growing dramatically larger, Pavlovsky Park is now much closer to the city boundary and more affected by industrial pollution. Another problem is recreational pressure. In 1994, 338,795 people visited the Park. In 2016, the figure was 1.5 million. There is occasional vandalism and violation of the park museum regulations (camping, graffiti) and illegal private building-construction within the park's boundary.

Pavlovsky Park offers year-round recreational activities, including tours of the palace, park pavilions and the park, and horseback riding. Many locals use the forested areas for gathering berries and mushrooms —traditional summer activities for Russians—although this is prohibited.

The park is based exclusively on native forest species such as spruce, pine, birch, willow, oak, maple, and lime. This unity of plant material is supported by architectural homogeneity, including all pavilions and the palace, which are built in the classical style. All the architects working in the park have respected the designs made by previous architects and did not violate their principles. They brought new ideas but kept the goal of creating a united park ensemble. In this sense, Pavlovsky Park is a truly unique landscape park where all the pavilions, bridges, topography, rivers, ponds and carefully located groups of trees and open spaces/ meadows help to create a special landscape rhythm and changing vistas. As the famous Russian poet Vasily Zhukovsky (1783–1852), said about Pavlovsky Park: "with each new step, there is a new picture for my eyes." In 1990, Pavlovsky Park was added to the UNESCO list of World Heritage Sites, as part of "St. Petersburg and related groups of monuments."

TOP RIGHT: Pavlovsky Park has educational facilities and is used for art and music festivals like "Imperial Bouquet" in summer 2018.
BOTTOM RIGHT: A traditional Russian "troika," a carriage drawn by three horses, on its way through the Pavlovsky Park.

E. MALSKAYA

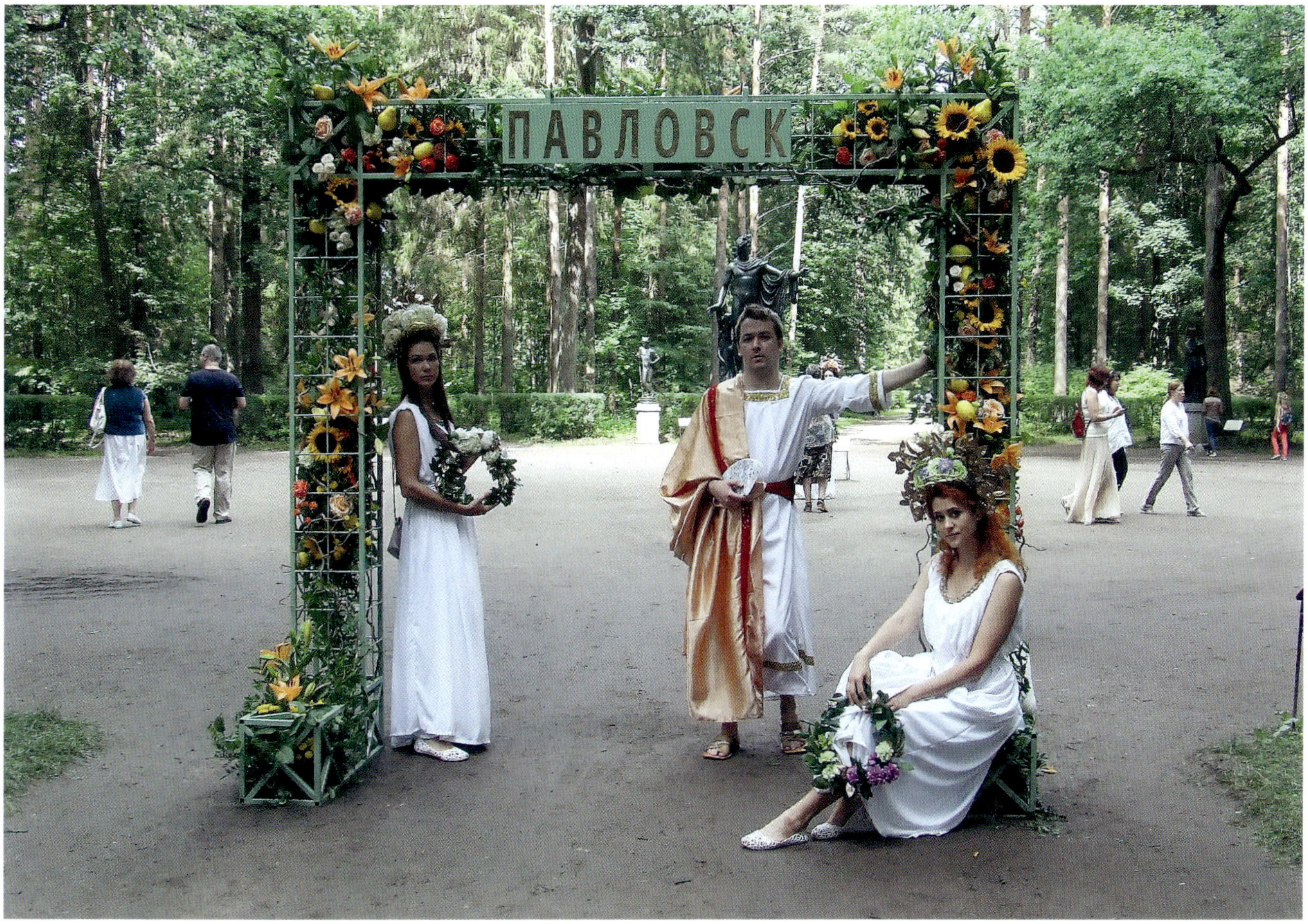

IGNATIEVA

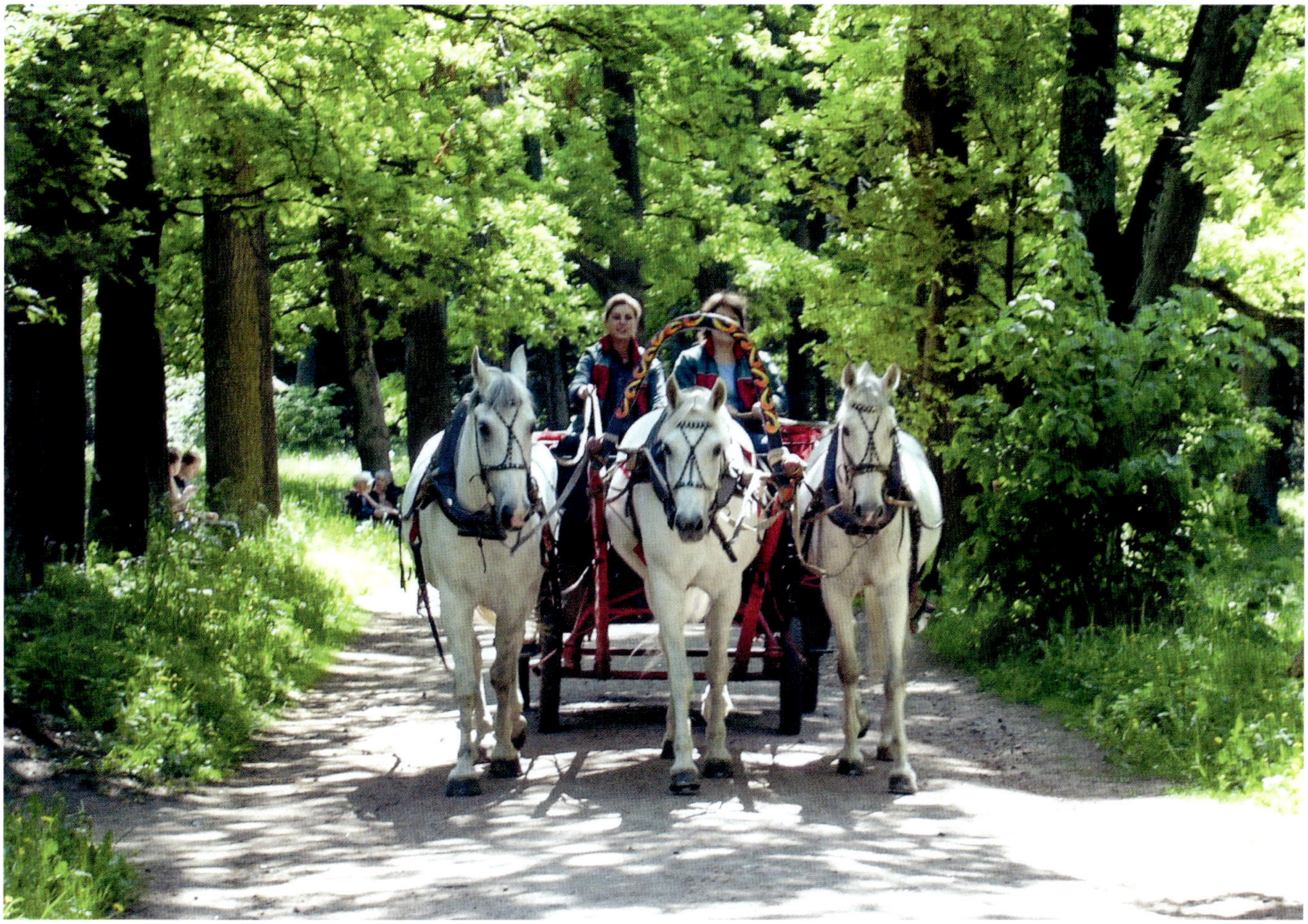

LIVERPOOL

BIRKENHEAD
PARK

West Float

Duke Street

Vittoria
Dock

Corporation Road

NORTH
BIRKENHEAD

Cleveland Street

Laird Street

Birkenhead
Park

Park Road North

Park Road North

Cavendish Road

Upper
Lake

Visitor
Center

Park Drive

Roman Boat
House

The Grand
Entrance

Ashville Road

BIRKENHEAD
PARK

Park Road West

Lower
lake

Rockery

Swiss
bridge

Park Road East

Cricket
Club

Manor Hill

Shrewsbury Road

St. Anselm's
College

Plam Grove

Park Road South

CLAUGHTON

Birkenhead
High School
Academy

0   100   200   300 meters
0   500   1000 feet

N

© OpenStreetMap contributors

BIRKENHEAD PARK, WIRRAL, UNITED KINGDOM

SIZE: 50-hectare site within a 90-hectare Conservation Area.

LEGAL PROTECTION: Grade I on Register of Historic Parks and Gardens for special historic interest. Designated as a Conservation Area under the UK's Planning (Listed Buildings and Conservations Areas) Act 1990 along with adjacent building plots.

OPENED: On Easter Monday 5th April 1847.

OWNER: Wirral Council, manages with the help of Birkenhead Park Management Advisory Committee. This includes local politicians (Councilors), representatives of the Friends of Birkenhead Park, and other stakeholders.

VOLUNTEER ORGANIZATION: Friends of Birkenhead Park.

HOW TO GET THERE: By Merseyrail to Birkenhead Park station.

ELIZABETH DAVEY

Birkenhead Park, Wirral, United Kingdom

Birkenhead Park is located in the town of Birkenhead on the Wirral Peninsular in North West England. Today the park occupies 50 hectares (125 acres). Birkenhead Park was formally opened in 1847 with the specific intention that it was to be open to all sections of society and free to use, that is, no charge for entrance. Significantly, it was created using public funds raised specifically for the project by the area's newly formed municipal body—the Birkenhead Improvement Commissioners. The Commissioners had to apply to the British Parliament in London for permission to undertake the project, a park explicitly for public use and benefit.

Birkenhead Park is designated as Grade I on Historic England's Register of Historic Parks and Gardens for its special historic interest. The entry notes its importance as the first public park established at public expense in the UK as well as being influential in the design of public parks both nationally and internationally. The park also forms the focal aspect of a 90-hectare Conservation Area (established in 1977), which helps protect both the park and the historic housing plots around its perimeter.

The park has a wide variety of uses—both formal and informal. Informal use includes family groups picnicking and playing, commuters using the park as an attractive part of their route walking to and from work or school, and people meeting for a gentle stroll or to take refreshment in the café. The site also has a range of sports facilities including cricket, rugby, football, soccer, green bowling, tennis, and angling. It is also used for an assortment of events—both large and small, for example, marathons, firework displays, sponsored walks and races, festivals (e.g., flower and vegetable shows) and music. The opening of the park's purpose-built visitor center in 2006 has greatly transformed the park's offer by providing a popular café and exhibition space and a base from which the park rangers run events. Today 2 million visits are made in the park annually.

At the dawn of the 19th century, Britain was in the throes of the industrial revolution. This was characterized by rapid and unplanned urbanization. This had been particularly pronounced in the neighboring county of Lancashire where industrialization since the mid-18th century had led to social problems including deficiencies in housing and public health. Manchester, for example, became known as the "shock

A 90-hectare Conservation Area (established in 1977), which helps protect both the park and the historic housing plots around its perimeter.

city of the age." At the same time, there was both a loss of common land and footpaths were closed to the public. This, along with the spread of urban development, meant that town-dwellers were increasingly cut off from easy access to the countryside on the urban fringe.

By 1833 Parliament in London noted that "the great manufacturing towns of the north … had doubled their population in thirty years; and yet there was not one of these … which had a sufficient open space for artisans to take exercise or recreation upon Sundays and holidays." A Select Committee on Public Walks was appointed to … "consider the best means of securing open spaces in the immediate vicinity of populous towns, as public walks calculated to promote the health and comfort of the inhabitants."

The Committee's Report stressed the health benefits of open spaces. Such places offered opportunities for recreation and an alternative to the alehouse; they brought people into contact with nature; and they acted as locations where members of different social classes could mix with each other, thus reducing social tension. In conclusion the Report voiced the "hope that public walks may be gradually established in the neighborhood of every populous town in the kingdom."

Public parks, (rather than 'walks') emerged, where philanthropists or private investment provided sufficient funds. In Derby the Arboretum was the gift of the philanthropist, Joseph Strutt; in Manchester and Salford, Phillips Park, Queen's Park, and Peel Park were paid for by public subscription; and Liverpool's Princes Park was funded by a mixture of philanthropy and private investment.

Birkenhead can be considered to be an industrial-age sensation. It was to grow from almost nothing, a rural hamlet with a population of 105 people in 1811 to a town based on shipbuilding, docks and associated industries with 8223 residents in 1841. However, for Birkenhead there was a desire that it should avoid the unplanned and unsanitary development of those other towns that had emerged and mushroomed during industrialization. It was to be a 'city of the future'.

Rapid growth raised concerns. The Liverpool Courier, on August 10, 1842, expressed that: "Bricks and mortar are there so fast taking the place of green fields and sooty vapors are so thickly mingling with the fragrant breezes that … Birkenhead will soon no longer be in the country." The solution was clear. "It is … good policy to provide a healthy and agreeable place of resort to secure the permanent benefits of pure air and

Such places offered opportunities for recreation and an alternative to the alehouse, they brought people into contact with nature and they acted as locations where members of different social classes could mix with each other, thus reducing social tension.

Select Committee on Public Walks, 1833

TOP RIGHT: Boat house.
BOTTOM RIGHT: An alley, curved.

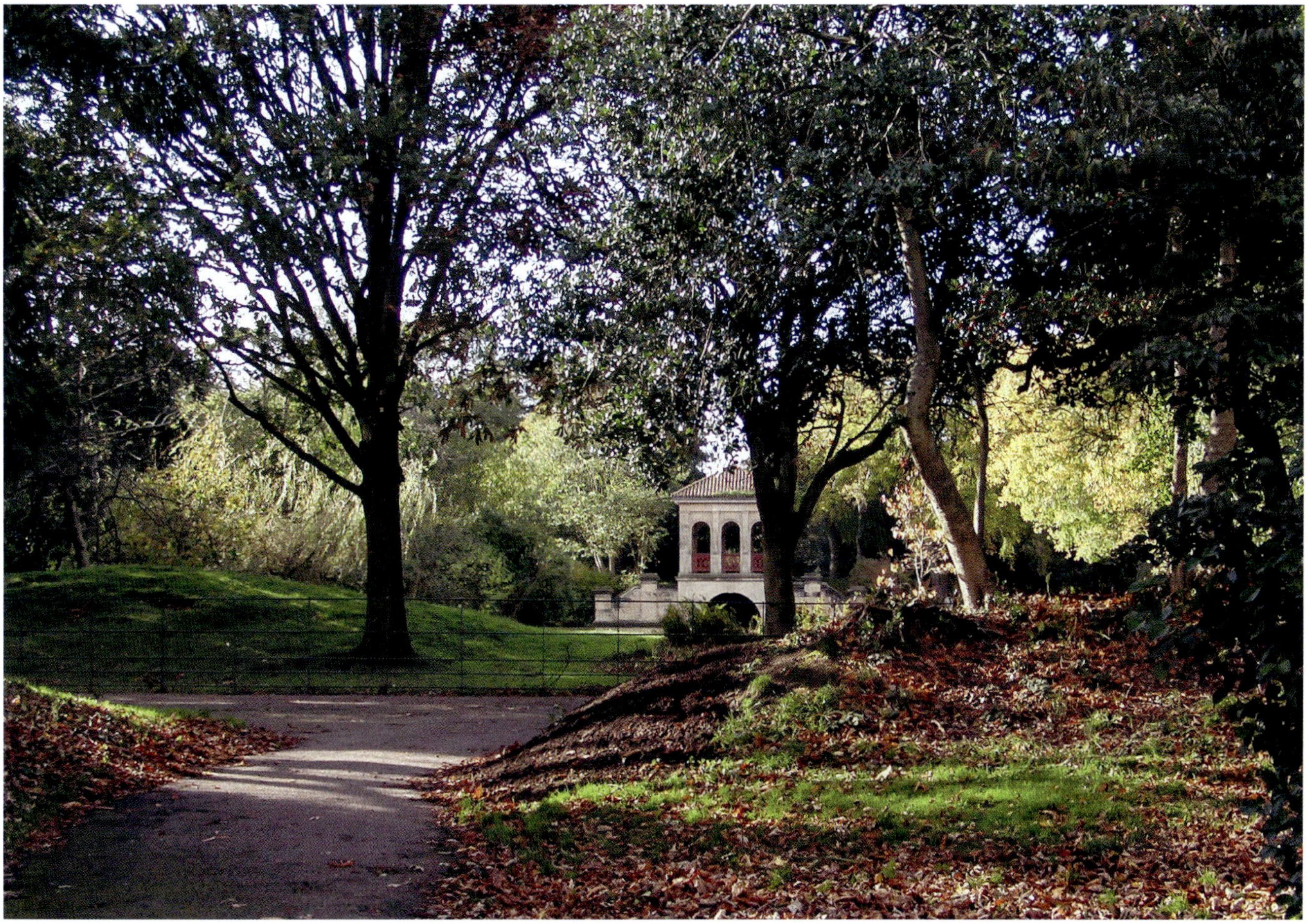

RICHARD MURRAY

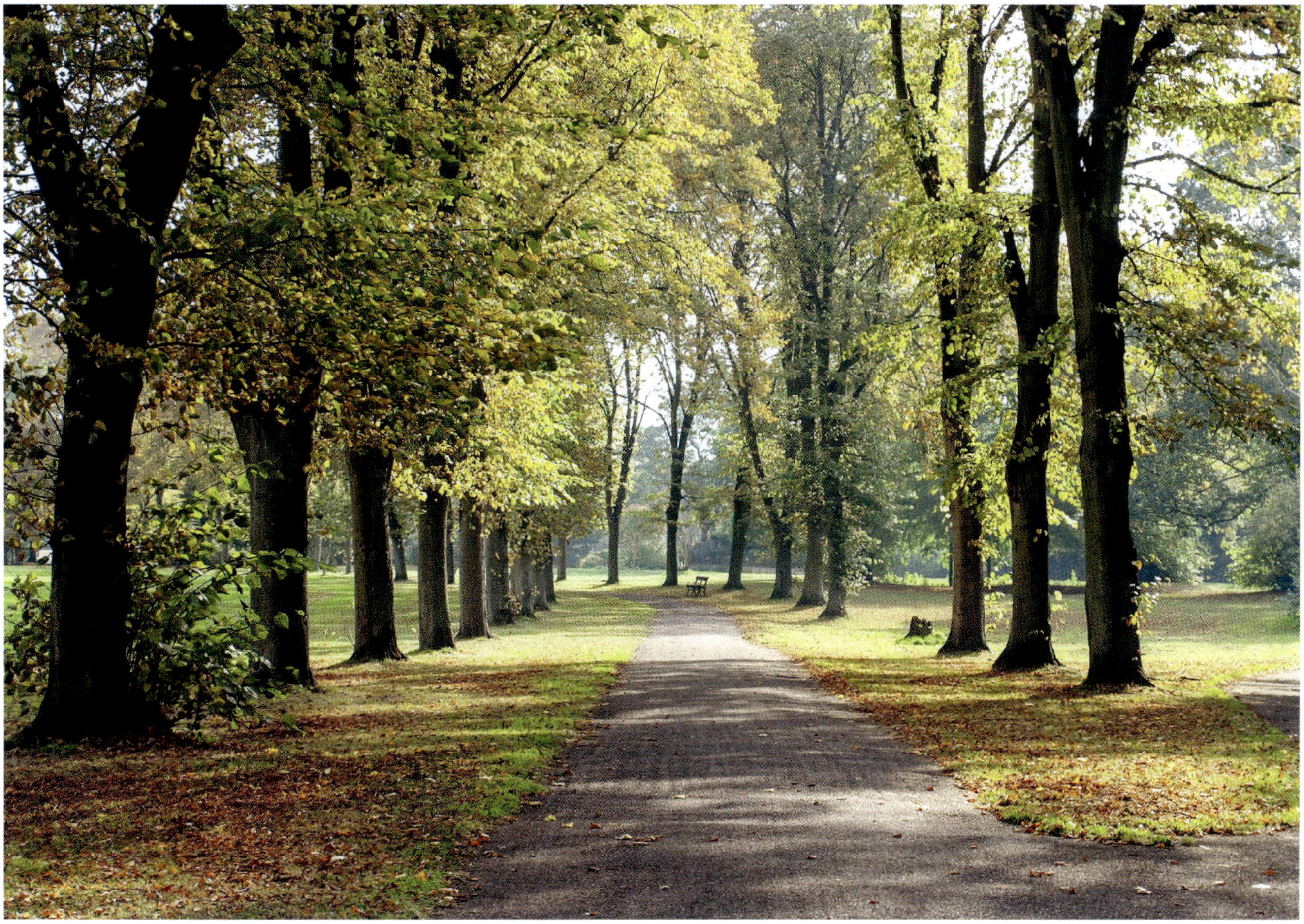

WIRRAL COUNCIL

exercise, at a moderate distance from the town while this can be done at a comparatively trifling cost."

Crucially, the affairs of Birkenhead had been in the hands of Improvement Commissioners since 1833. These were an elected body of men with responsibility for developing the town and with a role very much the forerunner of the modern local authority. It was one of these commissioners, Isaac Holmes, who in early 1841, is recorded as suggesting the creation of a park.

The new park was to be a municipal venture with its cost borne entirely by the Improvement Commissioners, a move unprecedented at the time. The Commissioners sought Parliament's permission and in April 1843 an Act of Parliament (Birkenhead's Second Improvement Act 1843) granted them "full powers … to purchase lands for the formation of a Park for the recreation of the inhabitants." This legislation represented a milestone in the history of the creation, funding, and design of urban public (or people's) parks.

Around 90 hectares of farmland were purchased to the west of the town. The site comprised ground covered by glacial sediments. The upper sections were reasonably well drained and growing a mixture of arable crops but the lower parts were less well drained, indeed marshy, and laid down to pasture and meadow.

Approximately 50 hectares of the site were to be "appropriated for public use in perpetuity." Land round the perimeter was divided into plots to be sold for private residences. The key idea was that this sale would cover the cost of creating the park. However, an economic crisis in the late 1840s left many of these villa plots unsold, with a loss of revenue to the Commissioners.

The Commissioners sought the services of Joseph Paxton (1803–1865), a self-educated man who became a celebrated landscape gardener and designer. He had famously worked for the Duke of Devonshire on designs for the park at the Chatsworth Estate in Derbyshire, England, and was already the author of the plan for Princes Park in Liverpool. Paxton's vision for Birkenhead Park was influenced by his personal experience of a large aristocratic estate. At first, he thought the site unpromising but then realized it would be a challenge that could help demonstrate his ingenuity. Work began in 1843, supervised not by Paxton himself but by Edward Kemp (1817–1891), a gardener trained at the Horticultural Society Gardens in London who became one of the most prolific and influential landscape designers of the Victorian era.

The original site, a 'greenfield' site of low-lying fields and some marshy land was totally transformed. The park, therefore, is a designed landscape and its creation resulted in a greatly modified and distinctive topography—one created very much with the visitor and user in mind.

The landscape which Kemp was tasked with creating included open meadow or 'greensward', interspersed with clumps of trees. Trees and shrubs were also planted on irregular mounds which were artificially created to give the topography more character. Both indigenous and exotic plants were used with most of the existing trees being cleared from the original landscape. Key features in the design are an 'avenue' of rock work, two ornamental lakes (sinuous in outline, with exaggerated bays and promontories plus islands to add variety and to break up the view across the water) and specially-designed structures to provide key focal points for the onlooker. Through the parkland there are curving carriage drives and pedestrian pathways, while the Upper and the Lower Park are separated by a public road.

Six lodges and other structures were designed by John Robertson and Lewis Hornblower in different styles. This included the massive and classical Grand Entrance with a pair of lodges on either side. Structures such as the Roman Boathouse and Swiss Bridge were designed as pictur-esque incidents—focal points to be in turn "hidden and then revealed" within the landscape.

There was a large amount of drainage necessary including inserting large stone agricultural drains and incorporating lakes as part of the water management system. The spoil from creating the lakes was used to form the distinctive mounds which give variety and a sense of enclosure within the park.

Paxton's plan had focused on aesthetics but as Birkenhead expanded and the tide of new housing swept west, the emphasis of the Park shifted. By 1851 the town was recovering from the financial crisis of the 1840s. Its population had tripled over a decade to 24,285 and the provision of recreational space for this growing urban population was increasingly important. There had been provision for leisure pursuits in the park from the start: a cricket club opened in 1846 and individual entrepreneurs soon operated a camera obscura, a refreshment saloon and (for a short time) an aerial suspension bridge. In the years that followed a quoits ground was laid out, provision was made for archery and for fishing in the lakes, and in 1861 the playing of football (soccer) was allowed. In later years a cast-iron bandstand and a Palm House (both now gone), were added in line with the fashion for such features in many urban parks.

The role of Birkenhead Park in shaping both the provision and character of urban parks has been significant, both nationally and internationally. Its influence on Central Park has long been recognized.

All this magnificent pleasure ground is entirely, unreservedly and forever the people's own.

Frederick Law Olmsted, 1850

In 1850 Frederick Law Olmsted, one of Central Park's designers, traveled to England, visiting Birkenhead in May. While in the park he encountered the "head working-gardener," presumably Edward Kemp, and was provided with a detailed description of the park. On his return to the States, Olmsted produced his *Walks and Talks of an American Farmer in England*, published in 1852, which included his account of the park at Birkenhead. However, even before the book's appearance, the importance of Birkenhead had been drawn to American attention, by no lesser an individual than Andrew Jackson Downing (1815–1852), America's leading landscape architect and a dedicated protagonist of public parks. The influence of Birkenhead on the winning design of Central Park in 1858 cannot be underestimated.

Birkenhead Park's impact on the creation and design of municipal parks has been considerable. While parks were still often a gift from local landowners or industrialists (for example the People's Park, Halifax, England, the gift of the Crossley family), municipal authorities were increasingly prepared to follow Birkenhead's example and create parks using their own funds. In 1852 Nottingham Arboretum was laid out on corporation land, and in Ipswich (1853), Leeds (1855), and Blackburn (1857) land was also purchased for the creation of public parks. Much of this might not have taken place had it not been for the pioneering example of Birkenhead.

The park benefited from external funding in 2004–2007 (including the Heritage Lottery Fund grant of £7.4 million). Wirral Council and its key stakeholders are keen to progress further restoration and conservation works, for example restoring structures, addressing occasional flooding, and enhancing biodiversity. Also important is the aim to augment the park's educational offer for both residents and out-of-town visitors (e.g., through developing a Discovery Centre).

TOP LEFT: The Swiss Bridge and the Roman Boat House.
TOP RIGHT: Giant spruce.
MIDDLE LEFT: Grand entrance.
BOTTOM LEFT: Boathouse mosaic.
BOTTOM RIGHT: Rock garden.

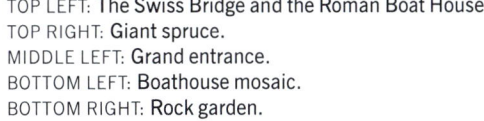

CENTRAL PARK
NEW YORK

BRONX

MANHATTAN

NEW YORK

JERSEY CITY

BROOKLYN

MORNINGSIDE HEIGHTS

HARLEM

Columbus Avenue

Great Hill

Harlem Meer

EAST HARLEM

The Ravine

Conservatory Garden

W97

North Meadow

CENTRAL PARK

E97

MANHATTAN

Broadway

Hudson River

UPPER WEST SIDE

W86

Jacqueline Kennedy Onassis Reservoir

YORKVILLE

MANHATTAN

W81

Great Lawn

E85

Park Avenue

The Metropolitan Museum of Art

The Ramble

The Lake

E79

Central Park West

Bethesda Terrace

Madison Avenue

UPPER EAST SIDE

W66

Sheep Meadow

The Mall

5th Avenue

E66

Central Park Zoo

MIDTOWN WEST

N

0   0,5   1 Km
0   0,5 Mi

© OpenStreetMap contributors

CENTRAL PARK, MANHATTAN, NEW YORK, UNITED STATES OF AMERICA

SIZE: 3,41 square kilometers.

LEGAL PROTECTION: In 1963 as National Historic Landmark.

OPENED: 1858.

OWNER: New York city, managed by Central Park Conservancy (non-profit), which is also a volunteer organization.

HOW TO GET THERE: *Subway stations:* Columbia Circle, Central Park North, Fifth Avenue, 59th Street, Central Park West, 110th Street, and Malcolm X Blvd.

PHYLLIS ELLIN, LANE ADDONIZIO AND
SARA CEDAR MILLER

Central Park, New York, United States

Central Park is America's first and most outstanding example of a major urban park. Created in Manhattan between 1858 and 1873, it has been recognized throughout the world as a masterpiece of landscape architecture and one of the most significant and enduring works of American art. Its creation encompasses significant moments in 19th- and 20th-century American history. In 1963, Central Park was awarded National Historic Landmark status and in 1974 it received designation as New York City's first scenic landmark.

Central Park is 3.4 square kilometers, 4 kilometers long and 0.8 kilometer wide. Its boundaries are Fifth Avenue to Central Park West and 60th Street to 110th Street. It is located in the geographic center of Manhattan Island. The park encompasses one reservoir of 40 hectares and a former reservoir that is now the Great Lawn. It is home to two major cultural organizations: the Metropolitan Museum of Art and the Central Park Zoo, now managed by the Wildlife Conservation Society. It features pre-Park relics from the American Revolution and the War of 1812, an Arsenal from the antebellum period and vestiges of two major pre-Park settlements of Seneca Village (founded in 1825 by free black people) and the Convent and School of the Sisters of Charity of Mt. St. Vincent (established in 1849). By 1859, both the residents of Seneca Village and the Sisters of Charity had moved.

Central Park was created during an era of tremendous social instability and physical urban growth. America's first significant wave of Irish and German immigrants arrived in the 1840s due to a devastating famine and political unrest in Europe. An influx of upstate residents also relocated to New York City, attracted to the new employment opportunities of the industrial revolution. Northern free blacks and manumitted slaves gravitated to New York seeking both employment and community. These recently-arriving groups rapidly altered New York's homogenous and stable population. Threatened and strained by ethnic, racial and religious diversity, economic challenges, poor housing, food shortages and a series of devastating health crises—cholera, yellow fever and a skyrocketing rate of infant mortality—New Yorkers experienced frequent riots in their streets and public squares.

Central Park was the first major park in any world metropolis created solely as "a public place" to be paid for by its citizens through general taxation.

Many progressive politicians and reformers sought ways to ameliorate this frightening gash in New York's social fabric. They envisioned a large public park that would offer city dwellers a healthy and naturalistic environment that would celebrate democracy and offer the growing populace a much-needed public place for social interaction and recreation. It would also give New York City the status of the great cultural centers of Europe that were celebrated for their central parks. As a result, the New York State Legislature in 1853 set aside over 3.1 square kilometers (the land from 106th to 110th was added in 1863) in an unprecedented act for "a public place." Although it took sixteen years to build (1857–1873), it was deemed a success as soon as completed landscapes were opened to the public in 1858.

In the 19th-century major European cities, such as those in London and Paris, were celebrated for their parks and park systems, though they were originally either royal hunting grounds or properties, belonging to the aristocracy, that were given over to the public in the late 18th or early 19th century. Birkenhead Park, a 90-hectare naturalistic landscape in a suburb of Liverpool, was created in 1847 as the first public park in Europe, financed by selling adjacent lands for the creation of private houses and villas. Central Park at 3.4 square kilometers was the first major park in any world metropolis created solely as "a public place" to be paid for by its citizens through general taxation, assessment and municipal bonds. Even though citizens considered the new park as a magnet for uptown growth, no real estate development was involved in its creation. In this way, Central Park became a significant and singular prototype for many American cities.

The design for the park was the result of the 1858 competition won by Connecticut-born journalist and agriculturalist Frederick Law Olmsted (1822–1903) and British architect Calvert Vaux (1824–1895). They named their plan Greensward for their preferred landscapes of sweeping meadows and vast water bodies designed to appear limitless to park visitors. These grand pastoral scenes—Sheep Meadow and the Lake, for example —were carefully juxtaposed with the intimacy of picturesque woodlands of the Ramble and the Ravine, featuring dense plantings, meandering streams, and dramatic rockwork arranged to include naturalistic caves, grottos, and cascades. Together these orchestrated views would be the antidote to the psychologically and physically "cramped and crowded" stresses of urban life. And because the designers recognized a need for

TOP RIGHT: Sheep Meadow (2015).
BOTTOM RIGHT: The Mall.

SARA CEDAR MILLER/CENTRAL PARK CONSERVANCY

SARA CEDAR MILLER/CENTRAL PARK CONSERVANCY

Central Park.

civic socialization, they created a formal Mall, the grand elm-lined promenade and the main architectural feature, Bethesda Terrace, an open-air esplanade featuring elaborate carvings and a central fountain.

Visitors experienced these park scenes through a system of intertwined recreational byways: 45 kilometers of pedestrian paths, 9.6 kilometers of undulating drives to be shared by both equestrians and carriages, and a rural bridle trail exclusively for horseback riding. The park's 9.6-kilometer tree-lined perimeter offered an urban promenade that acted as a buffer between the city and the park. To ensure the safety and psychological peace-of-mind for all park visitors, Calvert Vaux and/or assistant architect Jacob Wrey Mould created a series of ornamental bridges that separated walkways for quiet strolling away from the faster horse and carriage traffic.

The creation of four below-grade roadways are Olmsted and Vaux's most innovative feature.

The design competition required the inclusion of "four or more" transverse roads that crossed the park at intervals, accessible to city traffic both during the day as well as at night when the park was closed. The creation of four below-grade roadways—65th Street, 79th Street, 86th Street and 96th Street—are Olmsted and Vaux's most innovative feature. These external arteries, artfully camouflaged behind dense vegetation, ensured visitors the continuity of a purely rural experience inside the park.

The design for Central Park was born, in part, out of a tradition dating back to the 18th century of equating designed landscapes with landscape painting. The paintings that influenced Olmsted and Vaux were by contemporary artists of the Hudson River School, who celebrated on canvas America's dramatic picturesque scenery of ancient forests, rugged mountains, and majestic rivers, along with its quieter pastoral images of rolling green hills and placid lakes.

The triumph of Central Park's designers was the transformation of their "broken" rocky and swampy "canvas" into one grand composition made up of a sequence of landscapes that contrasted and complemented the infinite with the intimate, the pastoral with the picturesque, and the public with the private.

They called their concept "passages of scenery," with compositions that juxtaposed the seemingly infinite lakes, meadows, and glades with the intimate woodlands. This being a metropolitan park, one major grand formal element—the Mall and its adjacent Bethesda Terrace—added to its complexity.

TOP RIGHT: The Ravine.
BOTTOM RIGHT: The Perimeter Central Park South. The trees and the wall act as a buffer between the city and the park.

SARA CEDAR MILLER/CENTRAL PARK CONSERVANCY

SARA CEDAR MILLER/CENTRAL PARK CONSERVANCY

The Terrace is Calvert Vaux's architectural masterpiece and one of the most significant and elaborate works of American public architecture of its time. With the exception of the United States Capitol, no other work of public architecture was as extravagant or costly and featured a sculptural program that represented such a combination of American cultural, philosophical and political thought.

To foster better relationships between disparate socio-economic groups of park visitors, Olmsted and Vaux designed the best views and the most beautiful landscapes to be enjoyed solely on pedestrian paths. That way the wealthier carriage and horseback riders would naturally mingle with those less economically fortunate, thus reinforcing the democratic impulse behind the park's creation.

For the designers, the purpose of the park was not merely to create visual experience but a deeply emotional and psychological effect that is the result of both great art and being in nature. Olmsted equated the feeling with "music that is of a kind that goes back of thought, and cannot be fully given the form of words."

Central Park became the first major landscape that gave birth to the profession of landscape architecture in America. Olmsted and Vaux, or Olmsted alone through his design firm, went on to create our great American urban parks and park systems: Prospect Park and Eastern and Ocean Parkway, all in Brooklyn, New York (1866); Delaware Park, Buffalo (1869); South Park (later Washington and Jackson Parks and Midway Plaisance), Chicago (1871); Belle Isle, Detroit (1881); Mount Royal, Montreal (1877); Boston's "Emerald Necklace" (1885); Genesee Valley Park, Rochester, New York (1890); Cherokee Park, Louisville (1891); Riverside Park, (1875) Morningside Park (1873 and 1887) in New York and Fort Greene Park (1868) in Brooklyn; Walnut Hill Park in New Britain, Connecticut (1870); South (now Kennedy) Park in Fall River, Massachusetts (1871); Beardsley Park in Bridgeport, Connecticut (1884); Downing Park in Newburgh, New York (1887); and Cadwalader Park, in Trenton, New Jersey (1891).

It was New York's well-publicized achievement that brought the value of all parkland to our national consciousness and indirectly inspired the preservation of our state and national natural treasures. From 1863–1865 while Central Park was still under construction, Olmsted was directing a mine in the Mariposa redwood groves and saw the need to preserve the grandeur of the natural landscape before it was destroyed. Under his influence, Congress passed an act in 1864 providing the preservation of Yosemite Valley and the Mariposa Grove. It was his early plan that led to the formation of the world's first system of national parks.

Central Park was instrumental in the birth of the historic preservation movement in the late 19th century. Starting in the spring of 1892 several influential citizens and state politicians passed a bill transforming the undulating landscape of the Park's west side into a straight 3,2-kilometer horseracing track. Citizens from all walks of life vehemently protested the intrusion into their beloved masterpiece and within six weeks the racetrack law was repealed. As a result, three years later Central Park commissioner and controller, Andrew Haswell Green, created the American Scenic and Historic Preservation Society to rescue endangered parks, landscapes and historic sites, and the preservation movement was born.

In the first half of the 20th century, the population in New York swelled and the definition of a park had changed. A preference for scenic parks and passive recreation became an outmoded philosophy. Public parks around the country were charged with providing more active recreation and specialized recreation facilities than they had in the past. During this period Central Park managers adopted this new definition and gradually expanded the existing menagerie to create the Central Park Zoo. In addition, 30 tennis courts, 21 playgrounds, an outdoor amphitheater, a skating rink, a swimming pool/skating rink, over 25 softball backstops and infields to several park meadows transformed many of the Park's landscapes. These additions represent an important phase of American park thought and Central Park in its entirety contains the comprehensive history of both 19th- and 20th-century park philosophy and design. Furthermore, the past four decades add an even more significant layer to Central Park's history.

Due to New York City's succession of fiscal crises in the 1960s and 1970s, the park's landscapes and structures experienced a serious deterioration, and management of the park was at an all-time low. In an unprecedented call to action in 1980, private citizens came to the park's rescue through a groundbreaking public-private partnership between the City of New York and the not-for-profit Central Park Conservancy. A master plan—sensitively focused on the restoration of the original design while recognizing the reality of the more recent recreational facilities—provided the roadmap for the restoration and maintenance of all 3,4 square kilometers and its component features. In 1998 the City of New York signed the first of three management contracts with the Central Park Conservancy to officially manage, restore and maintain Central Park. Today, through that successful partnership, Central Park has never been more beautiful or better managed in its 160-year history than it is today, and parks throughout the world use the Central Park

Central Park was instrumental in the birth of the historic preservation movement in the late 19th century.

One of the four transverse roads.

SARA CEDAR MILLER/CENTRAL PARK CONSERVANCY

SARA CEDAR MILLER/CENTRAL PARK CONSERVANCY

model for the restoration of their urban parks. The features contributing to the authenticity of Central Park's original design include the circulation system of pedestrian paths, bridle paths and carriage drives, the constructed meadows, woodlands and waterbodies, the locations and routes of water features, and Victorian structures such as buildings, bridges, and rustic architecture. Although some changes have naturally taken place throughout the park's history, the park's original design is well documented through original drawings, written reports, and historic photographs. They provide excellent support to park managers to ensure that critical attributes are maintained or returned as close as possible to their authentic states. As an official landmark of New York City all permanent alterations for Central Park undergo intensive public review and must be approved by the New York City Landmarks Commission and, in many instances, the Public Design Commission.

TOP LEFT: Bethesda Terrace.
BOTTOM LEFT: Cherry Hill.
THIS SIDE: Bow Bridge.

Quality of Life

That nature has a healing effect has long been understood. Fresh air was prescribed for those stricken with tuberculosis. Today, urban life is associated with stress and pollution, while nature is not. Urban green matters to the extent that the lifespan of people living in the same city can differ by 10–15 years depending on the extent of urban green in their neighborhoods. Due to their location, some cities have been endowed with nearby, large green or blue areas, others not. But with clever city planning and bold initiatives, those less geographically fortunate can achieve the same qualities.

PETER SCHANTZ

Can Nature Really Affect our Health?
A SHORT REVIEW OF STUDIES

Memories from early childhood in the center of Stockholm emerge after being invited to contribute this chapter. Some blocks away from where we lived there was a park that was laid out in 1619. Many of the original trees still graced its double avenues and accompanied me in my first encounter with nature. I used the park all year round and in various ways. The skating and tobogganing in wintertime left particularly vivid memory traces, as did my earliest pictorial memory: on a warm summer day, I sit with my beloved mother on a blanket on the lawn, and she serves blueberries with milk. This haven for resting, playing, strolling, running, aesthetic contemplations, and so forth, definitely contributed to an appreciation of my childhood as being a harmonious time. And I sometimes reflect on how that time in my life would have been without the generous park silently waiting for me day after day during those important formative years.

The area alluded to is Humlegården, the largest urban park in downtown Stockholm. It is as well-known, as it is appreciated by many Stockholmers. Hence, it was the perfect setting for an unusual type of advertisement that appeared one morning in a major newspaper. All trees in the park had been sawn off and transported away, but the ground, lamps, walking paths remained, and a statue of the Swedish botanist Carl von Linné was still standing upright. An immediate sense of un-wellbeing overwhelmed me when I saw this image. Luckily the explanation was that it had been photoshopped: "Imagine that the deforestation would occur in Humlegården" was the bold-type message of the advertisement. The aim was to gather support for a tree-planting project around the Victoria lake in Africa.

One might think that this visual experiment demonstrated the importance of the park's trees *per se* for well-being. Alternatively, the effect of the trees was just that they hid stressful settings such as traffic

RIGHT: Humlegården, the oldest park in Stockholm, the capital of Sweden. The image below is from the same viewpoint as the image above, but photoshopped. These images were published in the daily newspaper Svenska Dagbladet on 16 December 2013, with the aim of highlighting the large-scale deforestation of eastern Africa. The text read: "Imagine that the deforestation would occur in Humlegården" and urged the reader to donate money for Vi Agroforestry's (in Swedish: Vi-Skogen) reforestation campaign (www. viagroforestry.org).

VI AGROFORESTRY AND ADVERTISING AGENCY MRM

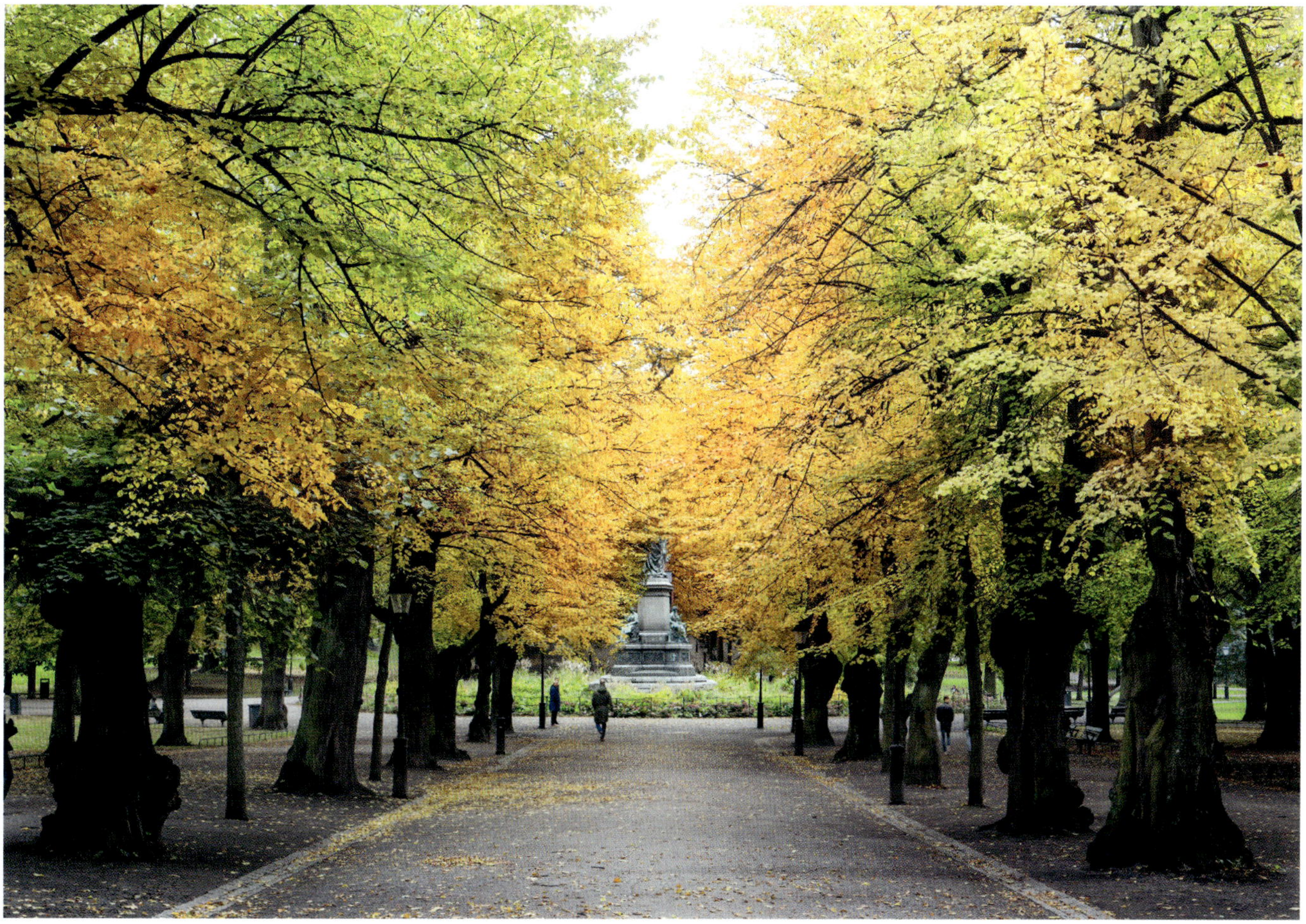

VI AGROFORESTRY AND ADVERTISING AGENCY MRM

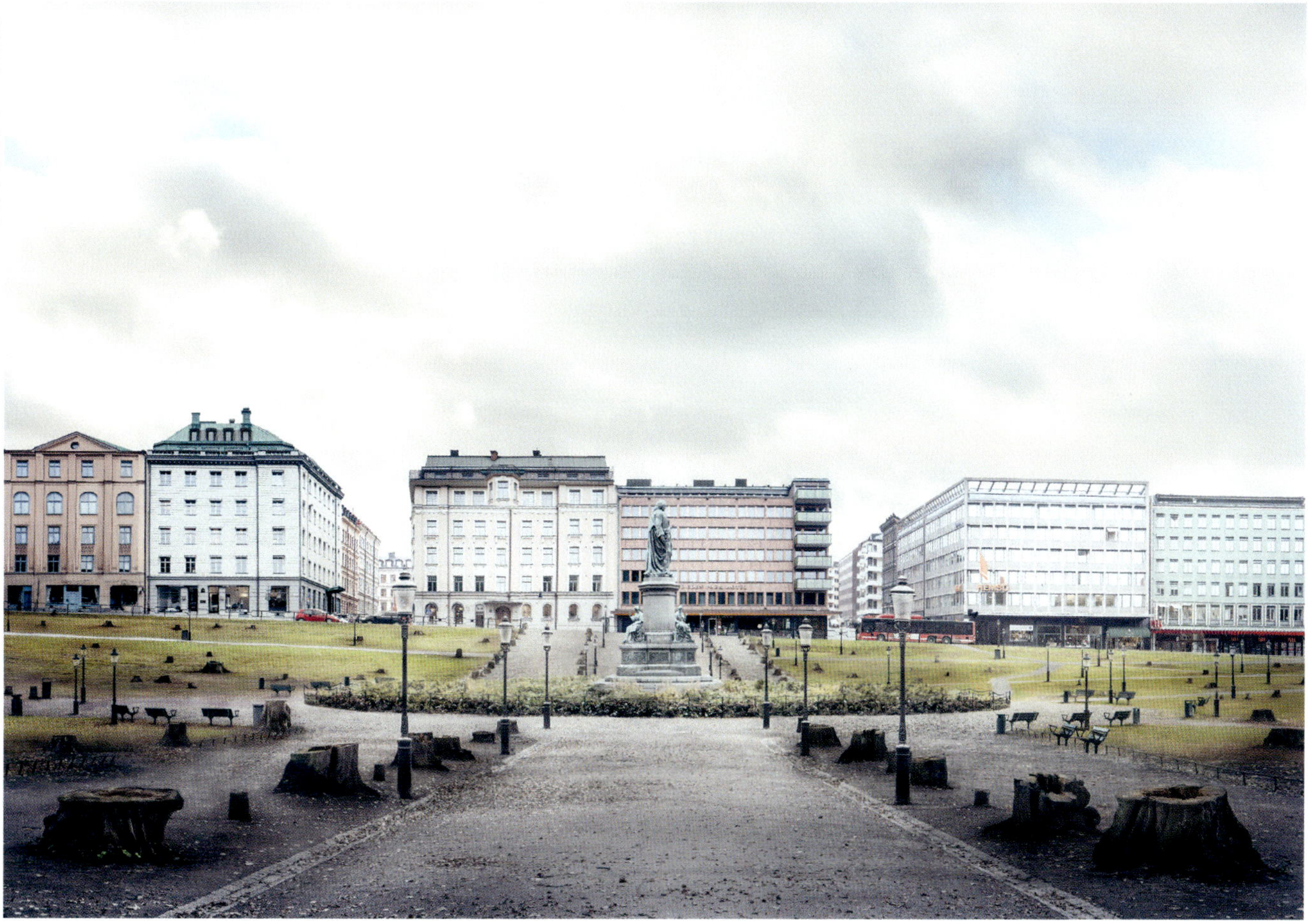

environments and buildings. In my case, the reaction might have been provoked by the juxtaposition of that image upon my childhood memories of a much-appreciated area. It has been through posing such alternative explanatory hypotheses, and testing them, that that knowledge about the effects of greenery on health has evolved.

This chapter will, in chronological order, guide you through some of the more important steps towards advancing our understanding of the relation between nature and the two most central dimensions of health, namely, well-being and lack of diseases (WHO 1946). An intriguing question is by which pathways greenery may act. This will be discussed later in the text. One way could be that it stimulates increased levels of physical activity, an issue that will be illuminated at the end of this short review.

ISOLATING EFFECTS OF GREENERY

The idea that there is a connection between nature and health dates back to antiquity (van den Berg et al. 2019, p. 56). The viewpoint that nature has a calming effect has given rise to embedding mental hospitals within greenery, and such practices may have emerged from empirical experiences of physicians. This idea was probably rather widespread, at least from the time of urbanization. In 1915, the Swedish Tourist Association celebrated its 30th anniversary and, in a speech by Chairman Louis Améen, a rhetorical question was posed: "The fresh winds of nature, the peacefulness of nature, and its beauty, are things that are certainly not in the pharmacopoeia. But is there a better medicine offered for our nervous generation?" (Améen 1915).

It is not an easy task to sort out the effects amongst the diversity of environmental factors and their potential individual actions per se, or in combination with one factor or the other. Still, as indicated above, this is necessary in order to move forward in our understanding of the role of greenery.

At the beginning of the 1980s, concordant images from both North American and European research conveyed the message that most green settings, compared to urban environments without green elements, have the following effects (cf. Ulrich 1984):

- sustain interest and attention
- elicit positive emotions
- reduce fear in stressed individuals
- block or reduce stressful thoughts
- foster restoration from anxiety or stress

Of the above effects, those reducing stress are especially important for health, since a lack of recuperation from stress can, in the long run, affect both physical and mental health negatively. An alternative interpretation of these findings could, however, be that they were caused by physical activity in conjunction with nature encounters. This uncertainty led the American researcher Roger Ulrich to conduct two studies aiming at isolating the effect of nature *per se*.

The first study was based on an analysis of medical records from patients in a Pennsylvania hospital who had undergone standardized gall-bladder surgery (Ulrich 1984). The postoperative treatment is similar in uncomplicated cases, and normally creates a considerable degree of anxiety. Ulrich aimed to study whether the view from the patient's hospital room affected the patient during the postoperative period. One set of rooms overlooked leafy trees; the view from the other rooms was that of a monotonous brick wall without character. To which room the patient was assigned after the surgery had been a matter of chance and the recovery rooms in which each patient had been placed after the operation were identical in other respects.

Each group consisted of 23 patients, who were matched in pairs in relation to relevant background variables. The results were that those who could see the trees had fewer negative notes recorded by nurses in the medical charts in the seven days after the surgery than those facing the brick wall. Furthermore, the amount and strength of pain-reducing medications were lower for those patients. Also, their postoperative hospitalization period was shorter. Thus, this study provides evidence that greenery can positively affect the recuperation after surgery in several ways.

How could the trees have such an effect? Many studies indicate that greenery has the potential to fascinate us more than man-made synthetic settings, and thus greenery may be the vehicle that distracts us and de-emphasizes problems, such as pain and anxiety. A well-designed psychophysical study from Stockholm University is in this respect of great interest (Ceci & Hassmén 1991). In it, healthy runners regulated their own running speeds with a scale for rating perceived exertion. They either ran horizontally outdoors on a trail in a green setting by a lake or indoors on a treadmill in a room without any windows. On each occasion, they regulated their speeds with three levels of perceived exertion. Despite these levels being identical, the participants ran on average 80 percent faster in the green-blue setting and had much higher levels of heart rate and lactic acid in the blood (both are objective measures of physical strain).

The idea that there is a connection between nature and health dates back to antiquity.

The type of outdoor setting at the shore of Lake Brunnsviken in Stockholm used for the study.

The results were interpreted in this way: stimuli reaching conscious levels depend on competition between internal cues from the body and external cues from the experience of the external environment. The maximal capacity of the consciousness to handle information is limited and, as a consequence, the influx of signals from the body may be filtered away or dampened through competition from an influx of external cues, making us experience physical work as considerably easier when running outdoors in nature (Ceci & Hassmén 1991).

In a later study by Ulrich and co-workers, physiological indicators of stress were used to evaluate how natural settings, compared to urban environments, affect stress recovery (Ulrich et al. 1991). A film showing serious injuries, leading to bloodshed, occurring in an occupational context, was viewed by 120 students. Thereafter they either saw films with (1) natural environments, (2) urban environments with people walking in a pedestrian mall or (3) urban environments with a lot of motorized traffic. The film with the occupational injury led to markedly increased stress levels in all students. Among those who viewed afterwards the natural settings, the stress levels lowered rapidly, while on the other hand, they declined more slowly and to a lesser extent in the

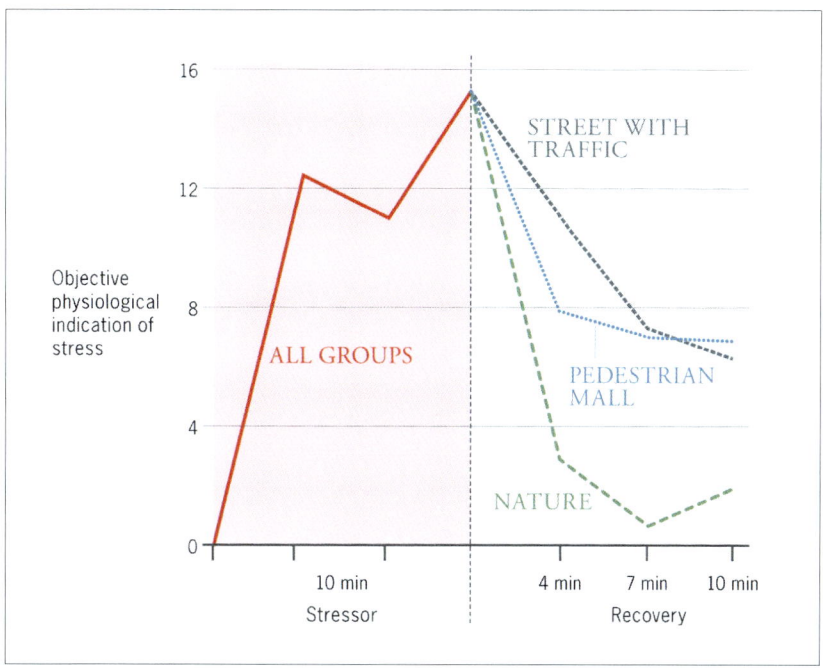

Relationships between recoveries from stress, measured by an objective method, in relation to different environmental factors. The objective method measured the conductivity of the skin by electrical current, which is linked to the secretion of sweat glands. The image is based on Ulrich et al. (1991).

individuals watching the two urban settings. The students' subjective judgments coincided with the objective findings. The researchers interpreted the effect of the natural setting as being due to activation of the parasympathetic nervous system and considered that it was in line with a psycho-evolutionary theory that as humans we are created in consonance with nature, and therefore process that form of visual perception differently than we process visual perceptions of man-made synthetic environments (Ulrich et al. 1991).

GREENERY, MORBIDITY AND PREMATURE MORTALITY

In order to illuminate mechanisms for the development of diseases, study models from epidemiology are often used. They stand for analyses of the distribution (who, when, and where) and determinants of health and disease conditions in defined populations. The first study of that kind, which used objective measures of greenery in the residential neighborhood and self-reported general as well as mental health, was undertaken by de Vries et al. (2003), and noted positive relations.

From 2008 and onward, a number of studies have used objective measures of both greenery and health. This is nowadays possible

RYAN BERNANDINIS SHUTTERSTOCK

OLGA KASHUBIN SHUTTERSTOCK

through geographical information systems that geocode residential addresses and map greenery in the neighborhood based on photos from satellites. The environmental data is then combined with data from socioeconomic and medical databases for statistical analyses of relations. Results from three such studies are presented below.

A pioneering study on 41 million Britons below the age of retirement showed that the differences in health, which are related to socioeconomic status—and reveal themselves as differences in premature mortality—were lower in residential neighborhoods with more greenery. Where greenery was in abundance, the difference in premature mortality for all different causes was 43 percent between those with the greatest socio-economic differences, whereas the difference was 93 percent in areas with little greenery. The corresponding values for premature death due to circulatory diseases were 55 percent and 120 percent, respectively (Mitchell & Popham 2008). Thus, the health differences related to socioeconomic situation were more than halved with more greenery.

A Dutch study made use of a corresponding innovative study strategy, but with morbidity as the outcome (Maas et al. 2009). It was based on about 345,000 persons and used registrations of new diagnoses by primary care physicians during one year. The results were startling. In 15 of 24 disease clusters, a lower frequency of diagnoses was noted with more greenery in the residential settings. Within these 15 disease clusters, the differences were on average 26 percent between areas with the least and those with the most greenery. In absolute numbers, these differences corresponded to 162 new diagnoses per 1,000 inhabitants per year.

Examples of diseases with a lower frequency of diagnoses with more greenery were coronary heart diseases, diabetes, upper respiratory tract infections, asthma, chronic obstructive lung disease, more severe forms of headaches, depression and angst, as well as various forms of musculo-skeletal problems. The authors compared the relations between greenery and health with age and health, and noted that, in general, a one percent decrease in greenery corresponded to the increase by one year in morbidity.

A Danish study analyzed the degree of greenery in different neighbor-hoods of where almost 1 million individuals lived from birth until the age of 10 (Engemann et al. 2019). It showed that a higher incidence of greenery during childhood is associated with a lower risk for psychiatric disorders from adolescence to adulthood. The risk of mental diseases was 55 percent higher for those with the lowest compared to those with the

TOP LEFT: Will this work as well? Skyscraper in Sydney, Australia.
BOTTOM LEFT: Or should it be like this? Greenwich Village, New York, USA.

highest amount of greenery. The relations remained after adjustments for socioeconomic variables, age of parents and if they themselves had a history of mental illness, as well as the degree of urbanity (from rural to capital region) in which the children were raised. The relation was stronger the more years that the children had been living in these environments, pointing to the importance of the totality of the childhood period for these effects.

A one percent decrease in greenery corresponded to the increase by one year in morbidity.

During the eleven years that have passed between these three studies, there have been a number of reports and reviews supporting a relation between greenery and health. However, although disclosure of relations in epidemiological studies is an important step forward along the pathway to gaining knowledge, we still need to search for the causal ways in which nature may act. How can the relations noted be understood? Given the well-controlled study of effects of greenery on stress reduction (e.g. Ulrich et al. 1991), some of the results are not entirely surprising, for example the fact that symptoms of angst and depression were inversely related to residential greenery (Maas et al. 2009). Perhaps nature can act more or less directly in some diseases. In relation to others, the causes can be secondary to the degree of greenery. Potential pathways can in such cases be to: (1) reduce harm, e.g., less air pollution, noise and heat; (2) restore capacities, e.g., physiological stress recovery and attention restoration; and (3) to build capacities, e.g., increase physical activity, facilitate social cohesion, enhance the satisfaction with the residential area and the sense of place (Markevych et al. 2017). Of all these potentially operating pathways, the clearest at this point is the one relating to stress reduction (van den Berg et al. 2019, p. 61). At the same time, a new causal pathway has been suggested recently. It relates to that rumination, which is associated with depression and other mental illnesses, has been shown to be reduced with nature encounters, and that at the same time the neural activity decreases in the corresponding areas in the brain (Bratman et al. 2015).

DOES GREENERY STIMULATE PHYSICAL ACTIVITY?
As indicated above, a fundamental question in this context is whether greenery may act indirectly by increasing levels of physical activity. There is plenty of evidence that green environments are conducive for physical activity (cf. Schantz 2003; de Vries et al. 2011). At the same time, so many variables differ when we compare an urban setting with a green one. Hence, how can it be possible to state anything about the effect of greenery *per se*? Let me illuminate this complexity with a citation from Erik Hohwü Christensen, professor in exercise physiology and hygiene

Cycling through the Royal National City Park in Stockholm to work in the city.

at the Royal Gymnastic Central Institute (GCI, today GIH) in Stockholm, Sweden. In 1945, he wrote the following under the heading "The physiology and hygiene of outdoor life" (Hohwü Christensen 1945; translation by Anders Schaerström):

Sunday after Sunday there is a flow of people from the greater cities by train or bus, by bicycle or by foot, away from the built-up areas out to forests, and open land, or to the coasts. During summer, the roads in the countryside are overflowing with walking and cycling groups of children, youth and elderly, striving with a spirited tempo to the goal of the day: a beautiful camping place, a tourist cabin, a weekend cottage or a room at a hotel. During wintertime, the bicycle is changed to skis, and the ski trails are densely located around the cities, as long as there is any snow left. What is it that drives urban residents out into the open air [...]?[---] The question is not easy to answer, but in the following, an attempt will be made to analyze some factors that may be considered to be decisive for this urge away from the cities, while at the same time some of the benefits will be touched upon, such as that a healthy outdoor life undoubtedly brings with it [---] There is such a close relationship between the bodily and mental functions, that bodily well-being usually also causes mental equilibrium. For this reason, in itself, outdoor life is a mental health factor of paramount importance. But also, in many other respects, it means a psychic stimulus for urbanites. The greatest significance lies perhaps in

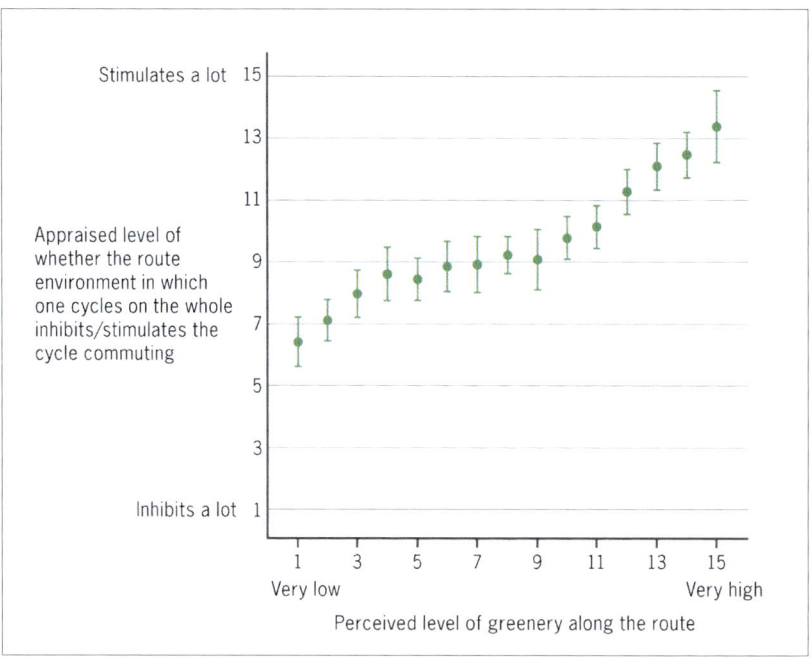

The relationship between commuting cyclists' perceived levels of greenery along their own routes and appraised levels of whether the route's environment as a whole inhibits or stimulates this cycling (n=805, mean and 95% confidence interval). The number 8 stands for a neutral position in both variables studied. All estimates apply to the urban part of Stockholm's inner city. The figure is based on data from Wahlgren & Schantz (2012).

the environmental change and in the relaxation from daily life, and the work, troubles and concerns that this entails.

In suggesting that 'environmental change' might be a driving force for the move out from the cities to the countryside, Erik Hohwü Christensen brought to light the difficulty of assessing the possible effects of the greenery in itself. This is since 'environmental change' in this context includes a number of different opposites that may act: noise/silence, exhaust fumes/fresh air, crowdedness/solitude, everyday life/leisure time life, flows of vehicles/vehicle-free zones, synthetic elements/natural elements, etc.

Given that complexity, a way forward could be to illuminate whether perceived greenery in a normal everyday cycle-commuting context stimulates cycling in itself. For that purpose, a study was conducted in which cyclists were asked to rate their own cycling routes in Stockholm's inner-city during September-October, i.e., when nature is green. They were well acquainted with these environments, having made about 250 commuting trips a year. The cyclists also rated a number of other environmental variables, such as flows of motorized vehicles, exhaust fumes and noise, that they were exposed to along their different cycling routes. The intention was to take into account variations in ratings of other different environmental variables. The figure to the left shows the connection between how much greenery the cyclists perceived along their individual cycling routes and how they considered that these routes, as a whole, inhibited or stimulated their cycling (Wahlgren & Schantz, 2012).

The analyses showed that even when taking into account estimates of a large number of other environmental factors along the routes, including ugliness/beauty, exhaust fumes, noise, flows and velocities of motorized vehicles, the relationship still remained. Similar results were noted when cyclists (n=1056) rated route environments in the Greater Stockholm suburban/rural landscape (Wahlgren & Schantz 2014).

These two studies strengthen the conclusion that greenery seems to stimulate physical activity, and those findings have later been supported in large scale studies in Germany, with a broad representation of socio-cultural backgrounds, and using standardized photo collages showing street scenes with different levels of greening (Nawrath et al. 2019).

CONCLUDING REMARKS

The new knowledge about health and greenery is, as stated above, in many ways enigmatic in terms of causal pathways. The findings are, however, so strong that WHO (2016, p. 41) has recently stated that, apart from the value of direct contact with greenery in our housing environments, there is evidence for "a need for small, local green spaces very close to where people live and spend their day, as well as large green spaces, offering formal provisions such as playing fields, and opportunities to experience contact with nature and relative solitude" (WHO 2016, p. 41). Another recommendation from WHO is that each child should have access to green spaces for play and physical activity (WHO 2016).

Returning to my childhood park memories: having the park close to where I grew up provided me with experiences that have influenced my current values. They were, and are still, reasons enough for advocating for green elements of various kinds in our living environments. Be that as it may, I am pleased in this short review to have presented a landscape of various studies leading up to the fact that today we have a new and strong body of complementary types of evidence supporting the value of nature from the perspectives of well-being and health (van den Bosch & Bird, 2018). It illustrates that a new chapter is being written in the history of health sciences.

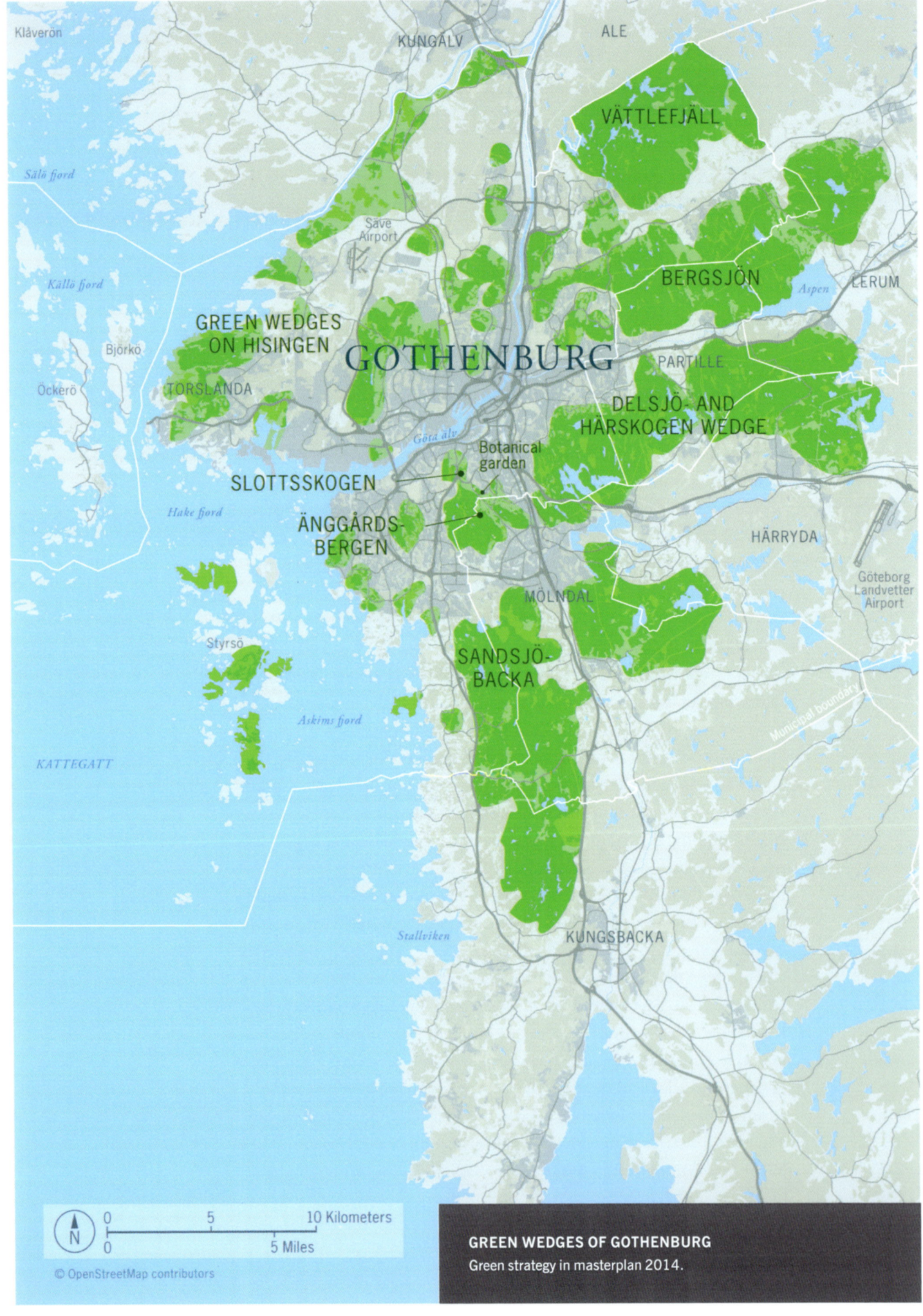

KLÄVERÖN   KUNGÄLV   ALE

Sälö fjord

VÄTTLEFJÄLL

Källö fjord

Säve
Airport   BERGSJÖN   Aspen   LERUM

Björkö   GREEN WEDGES
ON HISINGEN   GOTHENBURG   PARTILLE

Öckerö   TORSLANDA   DELSJÖ- AND
                                                    HÄRSKOGEN WEDGE

                        Göta älv
                                  Botanical
                                  garden
SLOTTSSKOGEN
                                                    HÄRRYDA
Hake fjord
            ÄNGGÅRDS-   Göteborg
            BERGEN   Landvetter
                                      MÖLNDAL   Airport
Styrsö
                        SANDSJÖ-
            Askims fjord   BACKA

KATTEGATT
                                            Municipal boundary

Stallviken   KUNGSBACKA

0   5   10 Kilometers

N
0   5 Miles

© OpenStreetMap contributors

GREEN WEDGES OF GOTHENBURG
Green strategy in masterplan 2014.

PER HALLÉN

A Green Infrastructure Plan for Gothenburg, Sweden

In 2014 the city of Gothenburg approved a green strategy to govern the future of parks and recreation areas (Göteborg 2014). It was given equal status with traffic and housing infrastructure in the planning process. The backbone of the strategy is the creation of green wedges that connect the rural landscape with the city.

The idea of green wedges connecting the most central parts of a city with the surrounding countryside has a long history. In 1829 John Claudius Loudon introduced the concept of green wedges in a new city plan for London. His idea was to bring clean air into the city center.

In 1910 a large conference on town planning was held in London and again green wedges were discussed, by Rudolf Eberstadt, and others. The papers presented at the conference came to international attention but apparently had no known impact on Gothenburg or other cities in Sweden.

The first concrete example of Scandinavian use of Loudon's old idea of green wedges is the introduction of a 'green finger' plan in Copenhagen in 1947. Much later, during the 1990s, Stockholm introduced this concept in its planning and finally the city of Gothenburg in its 2014 green plan. Green wedges are an excellent way to give a large population access to nature and, at the same time, allow for the city to expand outwards. This idea, dating from 1829, can give modern inhabitants of cities a better quality of life.

Green wedges are a recently adopted concept in planning in Gothenburg, but the wedges themselves have been part of the landscape for many years. For the most part, they are not planned but leftovers, land that has been difficult to build on because it is rocky or swampy. The Gothenburg region is an area with fjords, ridges, and grabens. In addition, a very large river, Göta älv, runs right through the city.

Economic crises have on several occasions left land that was purchased for development unexploited. An early example is the speculation bubble in Gothenburg that burst when the Napoleonic wars ended in 1815. During the boom, the old fortifications of Gothenburg were demolished and land was made available for building. However, it never came into use because of the economic crisis. Instead, people started to use the

Green wedges are an excellent way to give a large population access to nature and, at the same time, allow for the city to expand outwards.

PER HALLÉN

Trail markers in the Delsjö-area.

freed land for recreational purposes. Most of the land turned into parks and gardens. The green wedge of Vättlefjäll is another example. During the 1960s large areas were purchased and incorporated into Gothenburg. A new city was planned to house up to 300,000 people. However, the crisis of the 1970s halted the expansion and parts were designated a nature reserve. The green wedges of Hisingen have a similar history. The area was planned for housing and heavy industry. Large areas, especially in the northern and western parts of Hisingen, are still rural.

The large green wedges in Gothenburg have very different forms of protection, if any. The 30-kilometer park (Slottsskogen, Änggårdsbergen, Sandsjöbacka) is regulated in several different ways. Closest to the city center, Slottsskogen, a formal 19th-century city park, was established as a park in 1874, but with no formal protection. Most of the population in Gothenburg, unaware of this, regard it as a highly important green area. South of Slottsskogen is the Botanical Garden, opened in 1923, also without any legal protection. During the post-war economic crisis, some landowners donated "land without any value" to the Botanical Garden. Änggårdsbergen, a forest area south of the gates of the garden was given nature reserve status in 1975. An extension to the west followed in 2016. Änggårdsbergen is cut off from Sandsjöbacka by several

TOP RIGHT: View over the Delsjö- and Härskogen wedge in 2008. A wedge gives many people access to nature.
BOTTOM RIGHT: Delsjö green wedge stretches into the center of Gothenburg. Based on map from City of Gothenburg.

PER HALLÉN

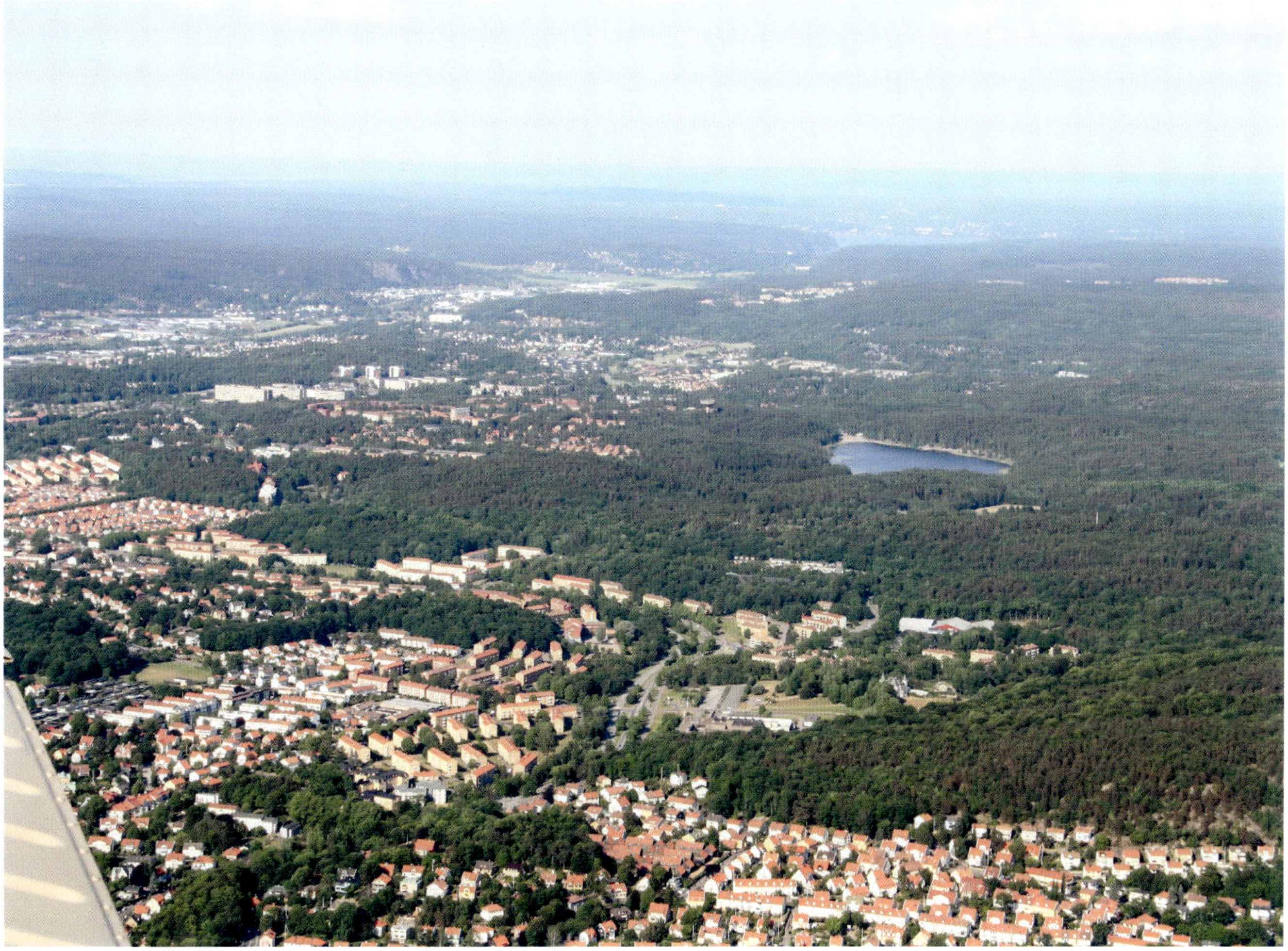

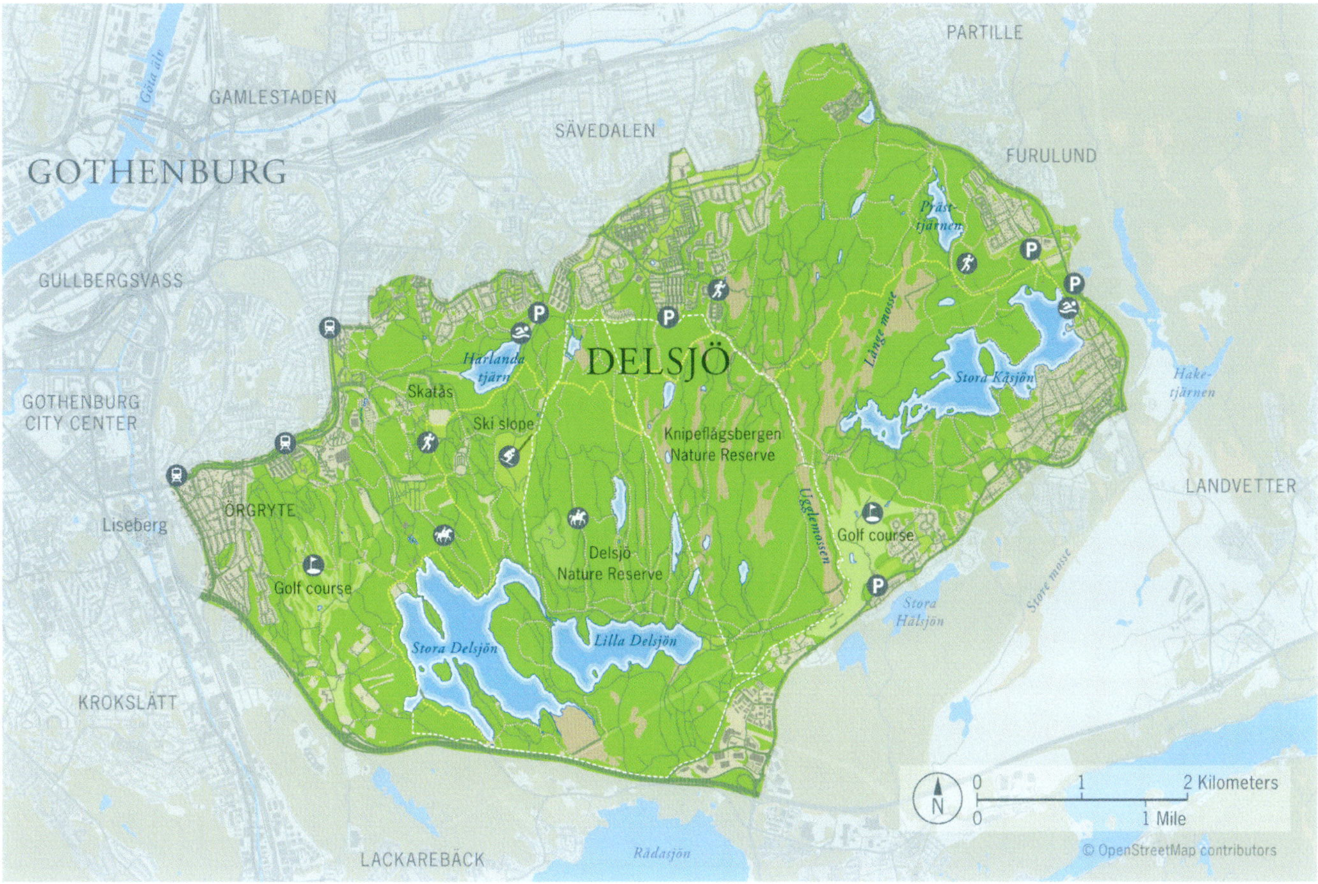

PARTILLE
GAMLESTADEN
Göta älv
SÄVEDALEN
FURULUND
GOTHENBURG
Präst-tjärnen
GULLBERGSVASS
Härlanda tjärn
DELSJÖ
Långe mosse
Stora Kåsjön
Hake-tjärnen
GOTHENBURG CITY CENTER
Skatås
Ski slope
Knipeflågsbergen Nature Reserve
LANDVETTER
Ugglemossen
ÖRGRYTE
Liseberg
Golf course
Delsjö Nature Reserve
Golf course
Stora Hälsjön
Store mosse
Stora Delsjön
Lilla Delsjön
KROKSLÄTT
N
0 1 2 Kilometers
0 1 Mile
LACKAREBÄCK
Rådasjön
© OpenStreetMap contributors

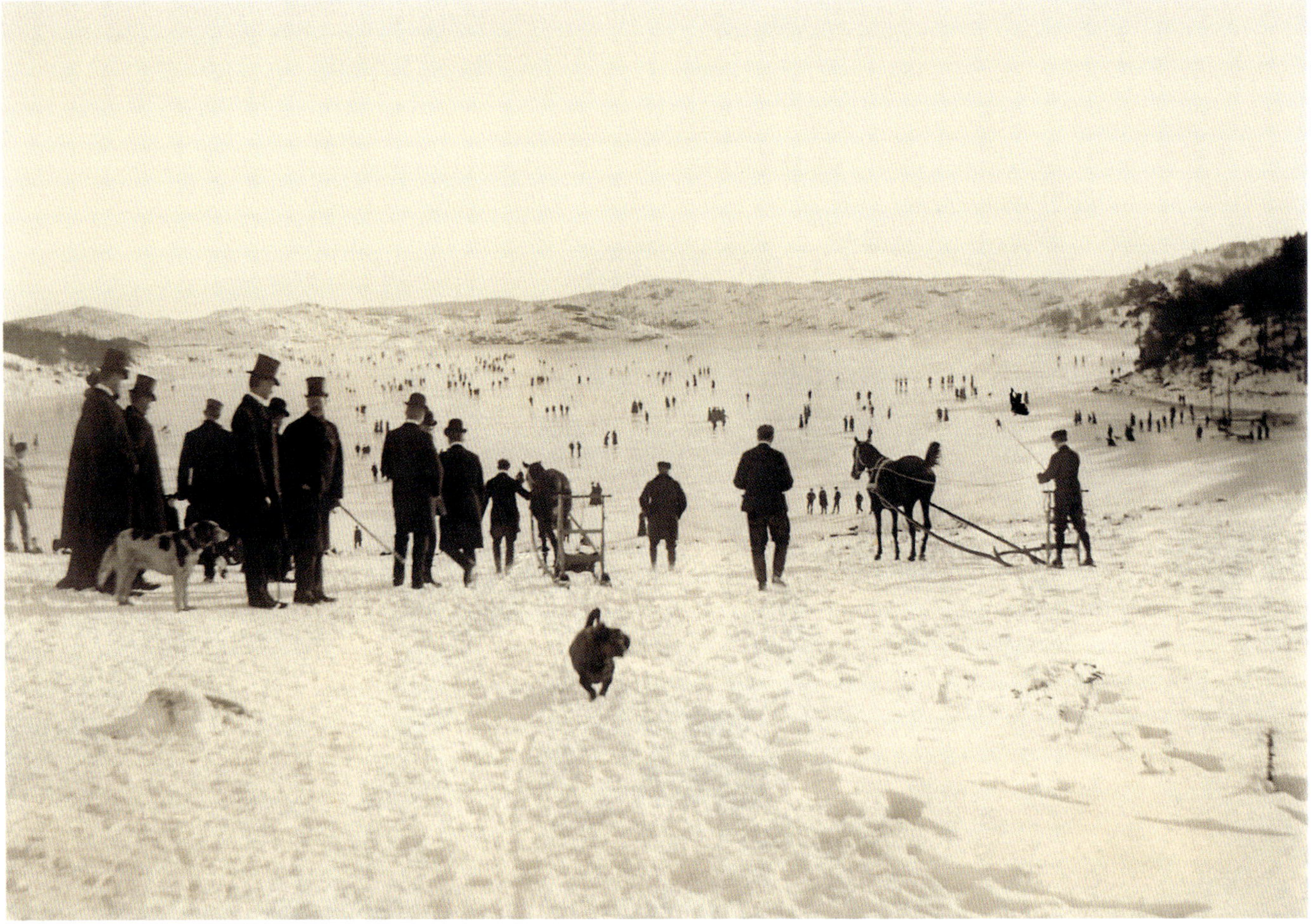

GOTHENBURG CITY MUSEUM

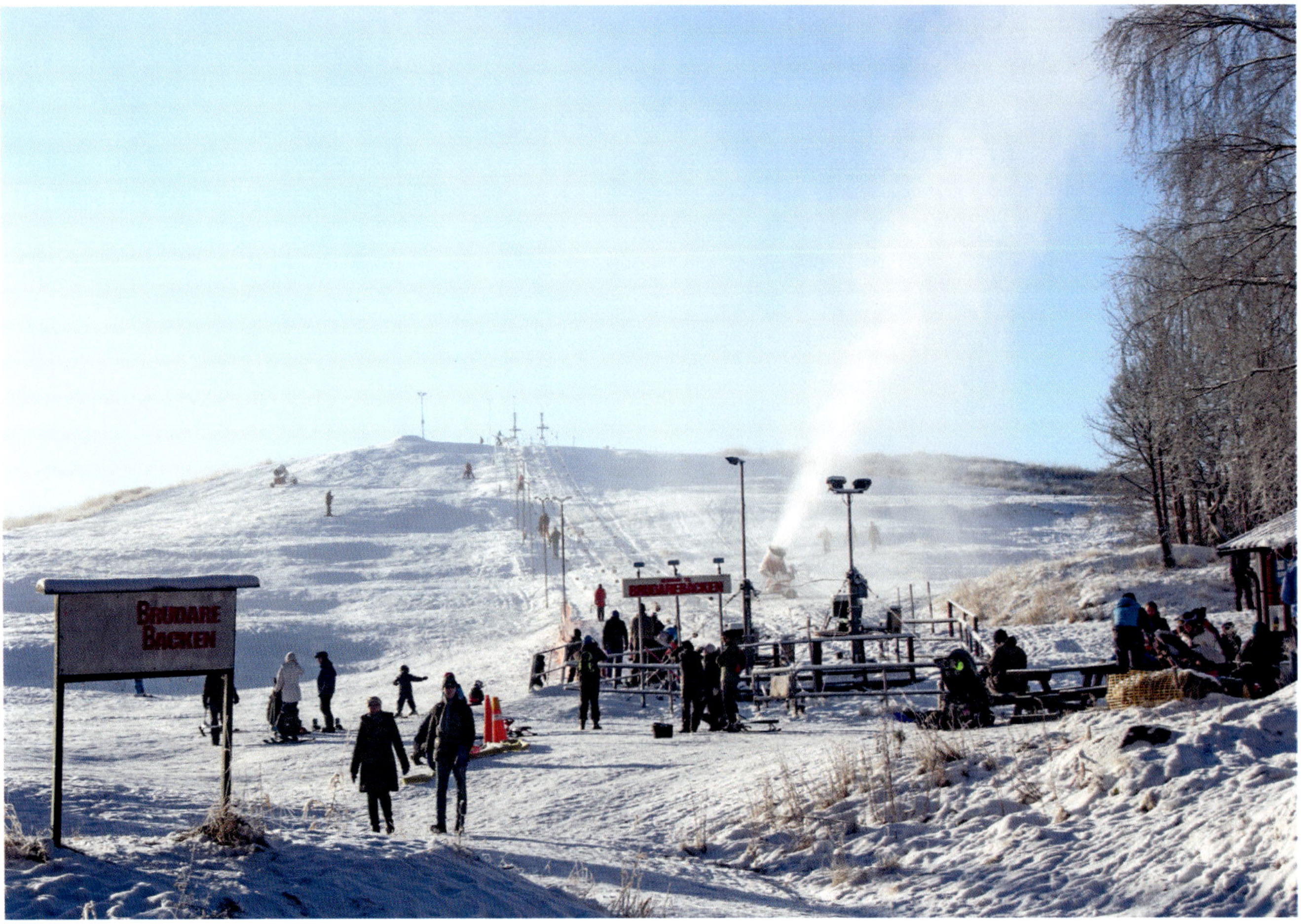

PER HALLÉN

roads and one motorway. Over the motorway an ecoduct has been built to allow wildlife to cross safely. Sandsjöbacka had already been given protection as a nature reserve in 1968 and since then the protected area has been expanded. It is fair to say that without the crisis during the 1920s this park and nature reserve would never have existed. However, the fight for nature protection never ends. The protected area of Änggårdsbergen does not cover the southern part, an area that is now under threat from new housing developments.

Lake Delsjön, 3 kilometers from the city center, attracted city dwellers as early as the 19th century. The parish was incorporated with Gothenburg in 1922 but the city had been buying land in this area since 1868, with the purpose of supplying the city with water.

By studying newspapers from the 19th century I have been able to rewrite some of the assumed history of outdoor activities among the city population. Old assumptions about outdoor activities before the 1900s turned out to be wrong. It was not just the wealthy that enjoyed nature. By the 1860s it was already a well established tradition among all ranks of the city to walk to Delsjön. Swimming and fishing attracted visitors to the lake. After the opening of the new waterworks swimming was banned, but violations were frequent. The attractiveness of Delsjön remained despite the regulations. Riding clubs arranged races and during winter ice-skating and ice-sailing were popular. Late 19th century cross-country skiing started to have a growing number of practitioners.

Skatås, a former military facility and refugee camp, used during and after WWII, was in 1947 transformed into a facility with dressing rooms, dining room, and a hostel. It was an immediate success. In 1962 Skatås had the highest number of visitors to any outdoor facility in Sweden.

The Delsjö-area remains the most popular area for outdoor recreation in Gothenburg. A survey in 2006 concluded that 90 percent of Gothenburg's population had visited Skatås and the area of Delsjö, more than any other green area in Gothenburg.

Delsjö is only partly protected as a nature reserve. The western parts, close to built-up areas, were declared an 'outdoor recreation reserve' in 1970, to be protected from future development. Since the late 1980s the whole Delsjö-area has been deemed an area of 'national importance to outdoor recreation'. Such considerations, however, do not offer the same level of protection as nature reserve status, they are mere guidelines for urban planning. In 2007 the county administrative board suggested

If green wedges are well understood and included in the planning of the city—on par with other forms of infrastructure—they will make a valuable contribution to life in the city.

TOP LEFT: Winter sports on Lake Delsjö in 1902.
BOTTOM LEFT: Winter sports on the old city dump of Brudaremossen, the Delsjö-area 2016.

an expansion of the nature reserve to include also the area of the 'outdoor recreation reserve', but the city of Gothenburg refused to protect the area.

In 2016 a city development program, BoStad2021, was proclaimed with the aim of building 7,000 new apartments by 2021 to mark the city's 400th anniversary. The guidelines of BoStad2021 state that the buildings should be erected on, for example, parking lots and similar hard surfaces. Green space should be added and developed. However, research by Chalmers University of Technology found that most construction projects in BoStad2021 will remove green space rather than hard surfaces.

One of the projects consisting of 260 apartments and a school was completely located on green space at one of the entrances to the Delsjö-area. Indeed, the school and some houses were planned inside the nature reserve which had been suggested in 2007. The 'green strategy' from 2014 was not invoked. City administrations, the county administrative board, and environmental courts did not object to the suggested project, and in Spring 2020 parts of the forest, suggested as a nature reserve in 2007, were cut down.

History has shown a strong connection between general economic development and the fate of green areas in the city. Economic downturns during the 1920s, 1930s, and 1970s were important in preserving land that today is regarded, though not always protected, as green wedges. Strong economic growth can, on the other hand, be a threat. If green wedges are well understood and included in the planning of the city—on a par with other forms of infrastructure—they will make a valuable contribution to life in the city.

By the 1860s it was already a well-established tradition among all ranks of the city to walk to Delsjön.

TOP RIGHT: Camping in the woods circa 1900, the "gentlefolks" of the country house Stora Torp.
BOTTOM RIGHT: Midsummer celebration in Delsjö allotment gardens in 2015.

GOTHENBURG CITY MUSEUM

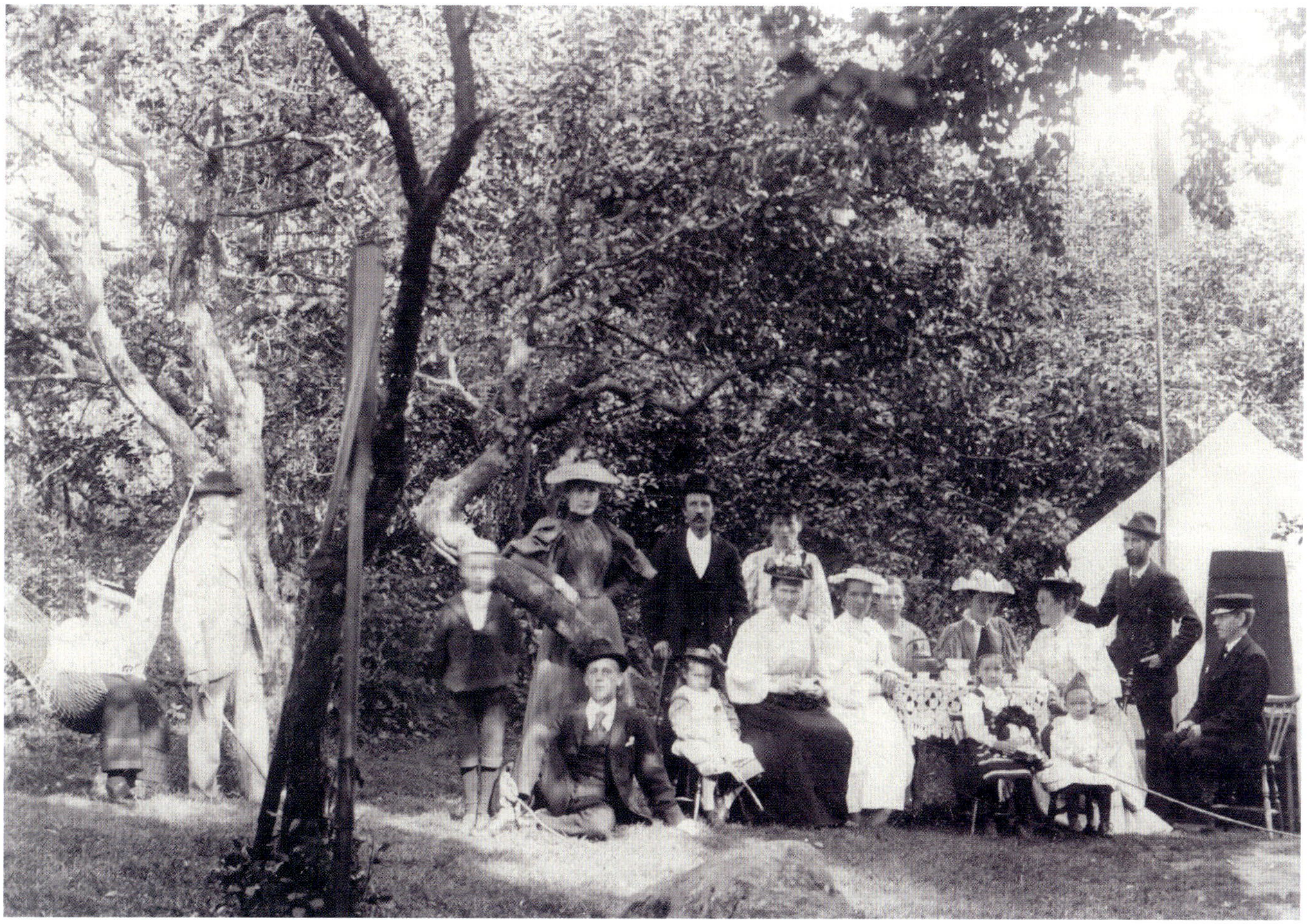

PER HALLÉN

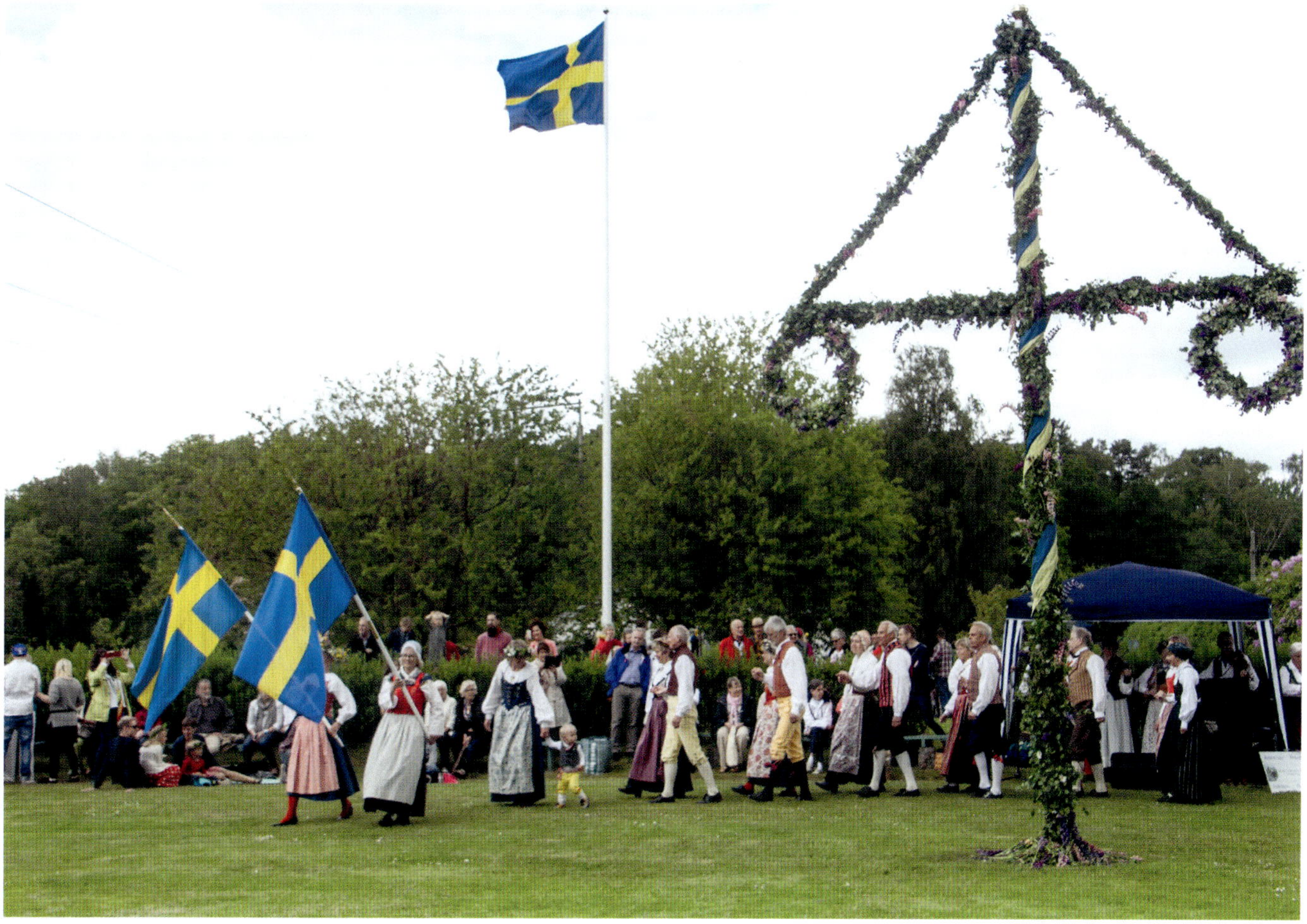

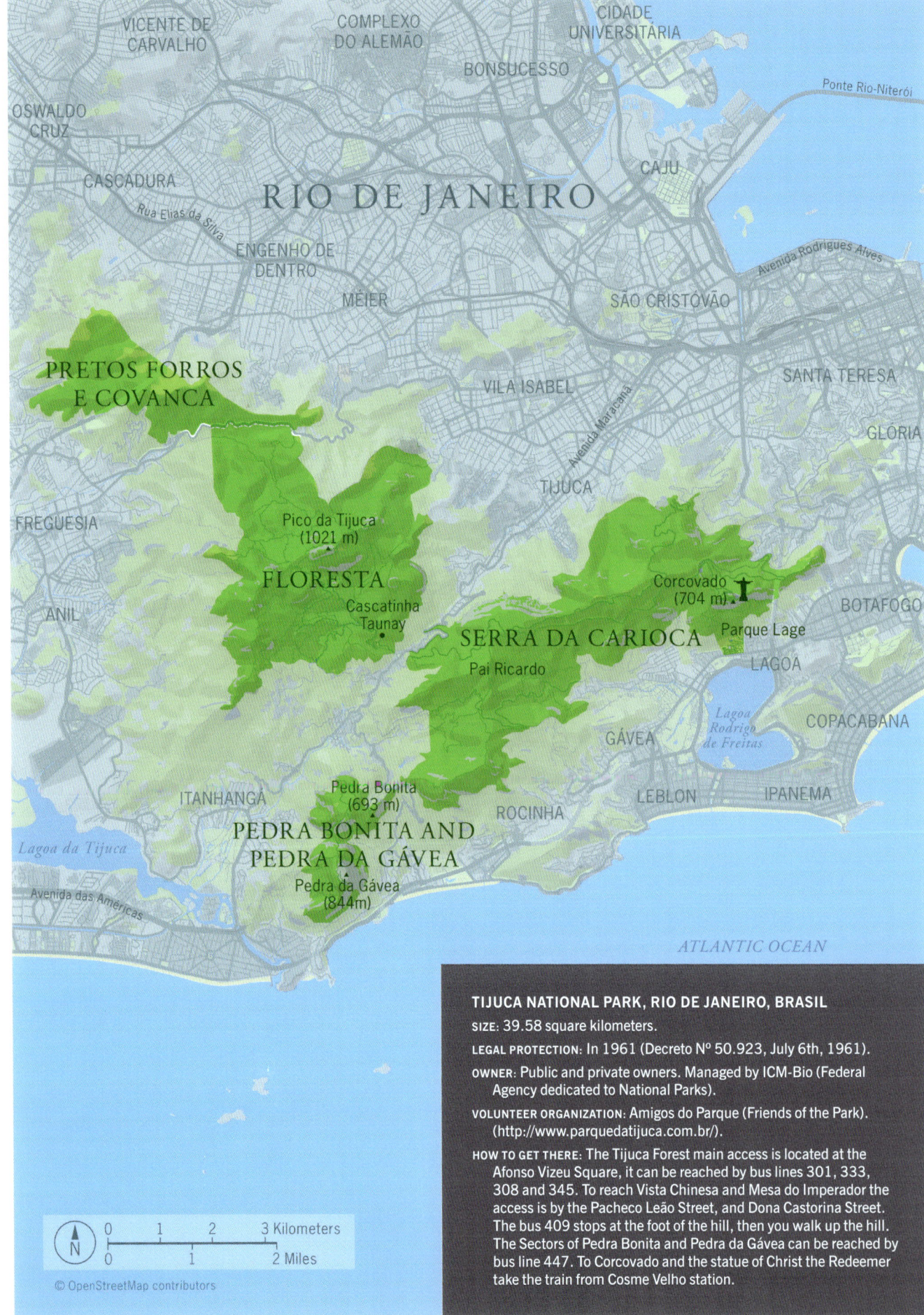

VICENTE DE CARVALHO

COMPLEXO DO ALEMÃO

CIDADE UNIVERSITÁRIA

BONSUCESSO

Ponte Rio-Niterói

OSWALDO CRUZ

CAJU

CASCADURA

RIO DE JANEIRO

Rua Elias da Silva

ENGENHO DE DENTRO

Avenida Rodrigues Alves

MEIER

SÃO CRISTÓVÃO

SANTA TERESA

PRETOS FORROS E COVANCA

VILA ISABEL

GLÓRIA

Avenida Maracanã

FREGUESIA

TIJUCA

Pico da Tijuca (1021 m)

FLORESTA

Corcovado (704 m)

BOTAFOGO

Cascatinha Taunay

SERRA DA CARIOCA

Parque Lage

ANIL

Pai Ricardo

LAGOA

Lagoa Rodrigo de Freitas

COPACABANA

GÁVEA

ITANHANGÁ

Pedra Bonita (693 m)

LEBLON

IPANEMA

Lagoa da Tijuca

PEDRA BONITA AND PEDRA DA GÁVEA

ROCINHA

Avenida das Américas

Pedra da Gávea (844m)

ATLANTIC OCEAN

0 1 2 3 Kilometers
0 1 2 Miles

N

© OpenStreetMap contributors

TIJUCA NATIONAL PARK, RIO DE JANEIRO, BRASIL

SIZE: 39.58 square kilometers.

LEGAL PROTECTION: In 1961 (Decreto Nº 50.923, July 6th, 1961).

OWNER: Public and private owners. Managed by ICM-Bio (Federal Agency dedicated to National Parks).

VOLUNTEER ORGANIZATION: Amigos do Parque (Friends of the Park). (http://www.parquedatijuca.com.br/).

HOW TO GET THERE: The Tijuca Forest main access is located at the Afonso Vizeu Square, it can be reached by bus lines 301, 333, 308 and 345. To reach Vista Chinesa and Mesa do Imperador the access is by the Pacheco Leão Street, and Dona Castorina Street. The bus 409 stops at the foot of the hill, then you walk up the hill. The Sectors of Pedra Bonita and Pedra da Gávea can be reached by bus line 447. To Corcovado and the statue of Christ the Redeemer take the train from Cosme Velho station.

CECILIA HERZOG

Tijuca National Park, Rio de Janeiro, Brazil

Rio de Janeiro is known as the "Marvelous City" because of its astonishing landscapes. The city is located in a biodiversity hotspot: the Atlantic Forest. One of its most precious assets, which enabled the city to be recognized as a UNESCO World Heritage site, is the Tijuca National Park. The contrast between the coast and the massifs is underscored in the appellation that UNESCO gave the city: *Rio de Janeiro: Carioca Landscapes between the Mountain and the Sea*. The park consists of 39.5 square kilometers in four different sectors: (a) Carioca; (b) Floresta; (c) Pedra Bonita and Pedra da Gávea,; (d) Pretos Forros e Covanca. The main peaks of the mountain range are Corcovado (704 m), Pedra Bonita (693 m), Papagaio (987 m), and Pedra da Gávea (844 m).

It is an impressive historical and biological resource inside a dense megacity.

HISTORY OF EXPLOITATION

The hills covered by Rain Forest have fascinated visitors since the 16th century when the French explorer Nicolas de Villegaignon occupied the area so as to install the colonial capital of France Antarctique. At that moment the exploitation of the forest began: valuable trees were cut down and shipped to Europe. The Portuguese countered and expelled the French in 1567. They then, in turn, exploited the natural resources of the forest, at first with emphasis on the tree that gave the name to the country—*Pau Brasil* (Brazil Wood—*Paubrasilia echinata*), used for dyeing.

The pillaging of the forests went on in cycles, starting in the lower areas until it reached almost the entire massifs. There are still a few original remnants with trees that are estimated to be hundreds, even thousands, of years old. There was a giant several thousand-year-old Jequitibá (*Cariniana legalis*), more than 40 meters high, that was broken by strong winds in 2013. As a result, space was opened to allow light into areas in which trees that were dormant now are growing fast. This fallen giant generated an impressive boom of micro-biota that recycles an extraordinary amount of organic matter that is transformed in nutrients. Thus, through this process the biodiversity is increasing at a rapid pace. The Pai Ricardo forest, covering 2 square kilometers, is considered the ecological jewel of the Tijuca National Park, with trees that are taller than 20 meters high, and have more than 3 meters in trunk diameter.

The ecosystem services that this remnant provides are many, such as, protecting water sources, moderating climate, helping clean the air. It is an impressive historical and biological resource inside a dense megacity. The area is one of the most visited spots by researchers, birdwatchers, and nature lovers.

Until the 19th century the source of water was the rivers and creeks located in what today is the Tijuca National Park. In the mid-18th century the rivers Paineiras and Carioca[1] were diverted to an aqueduct to serve the city.

The magnificent trees were cut down, some shipped to Europe, others were used locally to make furniture and boats. Charcoal was essential for the city, and its source also was the forest. Traces of the coal-pits are to be found in many sites of the forests. Furthermore, coffee-growing depleted the soils. However, the Pai Ricardo area remained thanks to the large rocks and steep slopes.

In the 19th century, when the city was growing and water was getting scarce, Emperor D. Pedro II appointed Major Manuel Gomes Archer to start the afforestation of the hills to regenerate the water flows. From 1861 to 1874 some 80,000 trees were transported by mules from a distant preserved forest and planted in the slopes. In addition, Baron Gastão d'Escragnolle had another 20,000 planted from 1874 to 1888, including fruit trees and the invasive Jake Tree (*Artocarpus heterophyllus*), which nowadays dominates large extensions of the park and requires adequate management to allow native trees to grow. Also, the Baron hired French landscape designer Auguste Glaziou to give the Tijuca Forest (Carioca) a European look.

During this period the Tijuca Forest and the Alto da Boa Vista neighborhood attracted the Europeans and the nobles because of the lower prices of land and, being about 300 meters above sea level, its better climate. The legacy of the 19th century is a rich cultural heritage expressed through pathways, sightseeing points, fountains, ponds, works of art, and bridges.

Eventually, the forest regenerated itself well, especially on the southern slopes because of the moisture that comes from the ocean. But in the 20th century it began to suffer from illegal occupation. Slums now cover some slopes, even though they are protected by law. There are encroachments in many areas vulnerable to erosion and landslides. Furthermore, some upper-class residences are located inside the park. Wildfires also

The Tijuca National Park is a democratic space where people from different parts of the city gather and socialize, no matter their social status.

TOP RIGHT: View over Rio de Janeiro from Tijuca National Park.
BOTTOM RIGHT: Giant (possibly 3,000 years old) Jequitibá.

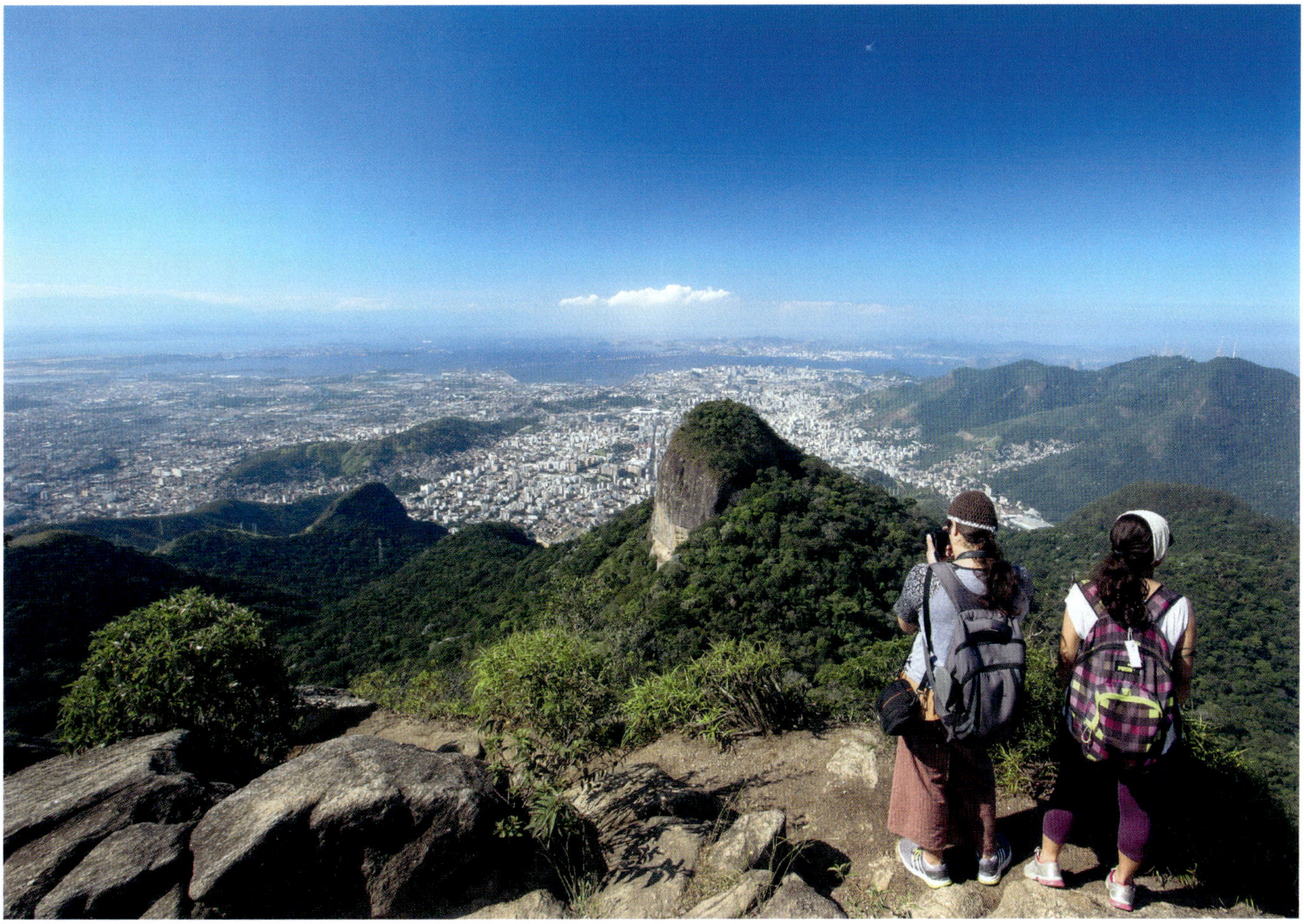

GUSTAVO PEDRO

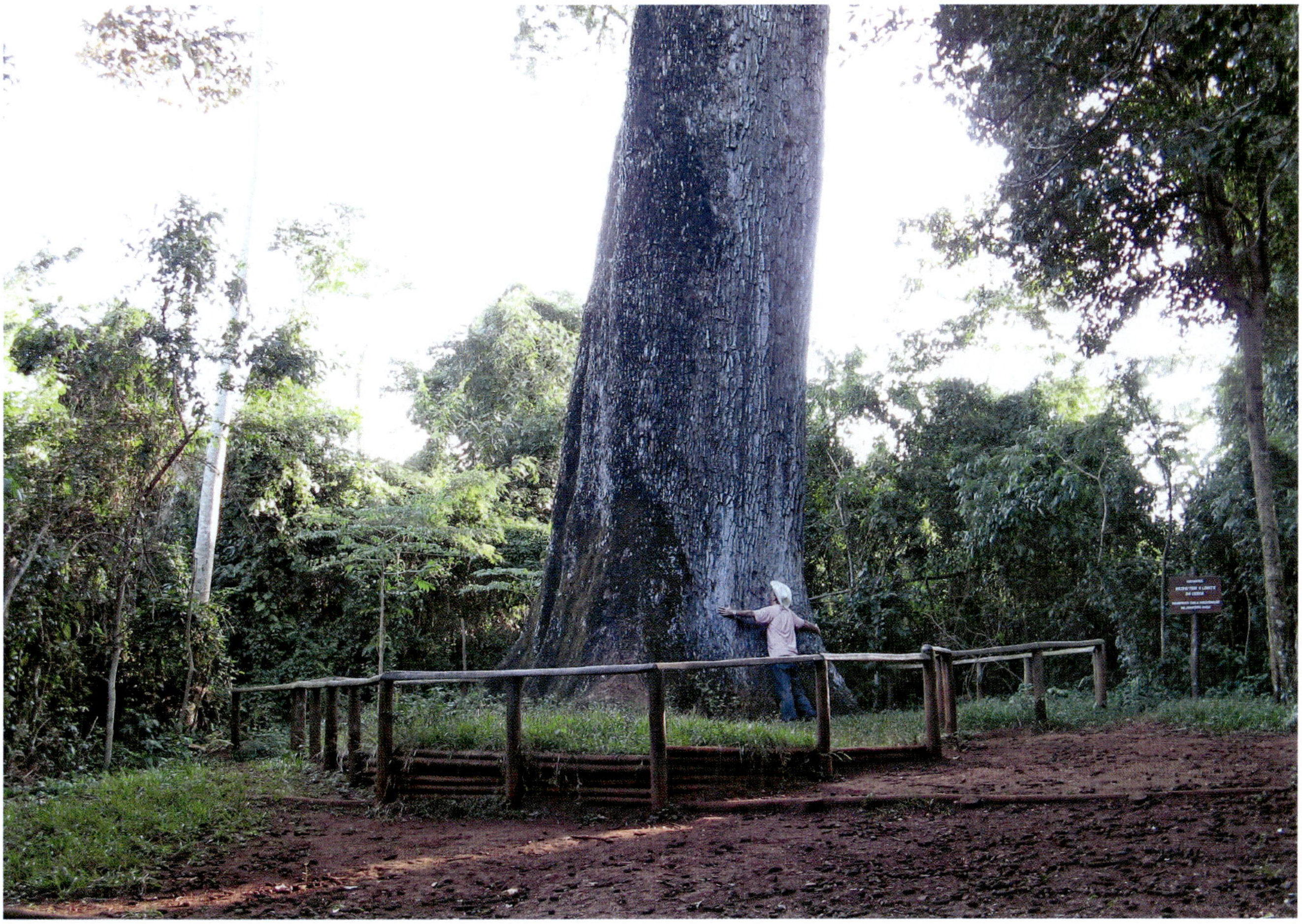

MAURO HALPERN WIKIMEDIA COMMONS

Christ the Redeemer, Corcovado mountain in Tijuca National Park.

TOP LEFT: Ariel tucan (*Ramphastos vitellinus ariel*). 231 bird species are found in the park.

TOP RIGHT: Chorinea amazon.

BOTTOM LEFT: Green cobra.

MIDDLE RIGHT: Birdwatchers in Tijuca Park.

BOTTOM RIGHT: Ring-tailed coati connecting with the city in Rio.

cause devastating impact, mainly on the northern side of the hills, where the sunshine is stronger and where African grasses (mainly *Panicum maximum*) dominate the landscape.

Since 1984 a city program called Mutirão Reflorestamento (Afforestation Collective Action) has been planting and managing trees on the slopes to prevent landslides and enhance the ecosystem services. After more than thirty years of reforestation the program is an inspiring example of community engagement and ecological restoration. The planting was done by residents of the local communities who were trained by the city's forest engineers. Public social workers were in charge of engaging the residents in the process. The forest is now more than 160 years old, in large parts it is at an advanced secondary-successional stage of recovery.

Tijuca was proclaimed a National Park by law in 1961 to celebrate the afforestation program that was started in 1861. This status gave national and international visibility to the precious area that has been protected thenceforth. In 2004, a federal law incorporated the Parque Lage, a historical and cultural site that borders the original park and reaches the very center of the city.

The Conservation Unit still has challenges in its efforts to combat illegal and predatory activities, such as housing, hunting, and fires. Such a huge extension of land also presents limitations to implementing the management plan, due to lack of proper funding to enhance, monitor, and manage the biodiversity.

The flora comprises 1,619 species of trees, bushes, bromeliads, and herbaceous vegetation. Among them, 433 are listed as threatened species. The fauna is also remarkable, composed of insects, spiders and other arthropods, frogs, fishes, snakes, lizards, birds, and mammals: in all, 328 species, 16 considered threatened by extinction.

The forest has a history of religious rituals, mostly performed by African people who arrived to be slaves in Brazil, and their descendants. Trees were sacred, and many were conserved because of believers' care.

The forest is a haven in the densely urbanized area, where more than 6 million people live inside the urban perimeter, and more than 11 million in the Metropolitan Region. The city of Rio de Janeiro is located in the Tropic of Capricorn, where the temperatures frequently go beyond 40°C, and tropical storms are frequent. For instance, 123.6 millimeters of rain hit the city in one hour on 14 November 2018.

The geomorphology of the city is one of the most beautiful in the world but, at the same time, this makes the city vulnerable to destructive floods and landslides. The forests are essential to preventing disasters and lowering the risks that menace people, material goods, and infra-

After more than thirty years of reforestation the program is an inspiring example of community engagement and ecological restoration.

structure. The park also moderates the urban climate: it is estimated that without the forest the temperature in the city would be some degrees higher.

The park offers many cultural services, such as recreation, exercise, contemplation, connecting with nature, learning about history and culture.

The most visited tourist site in Rio de Janeiro is the Corcovado Peak, where the Christ the Redeemer monument was opened in October 1931. In 2007, the 38-meter high statue was deemed one of the Seven Wonders of the Modern World. You can access the hilltop by a train line that dates from 1884. Every year more than 2 million visitors ride up the hill to appreciate the spectacular views. During weekends the road is closed to vehicles and people can safely walk, run or bike to exercise, contemplate the forest and the views, take a natural shower, and enjoy nature.

There are several trails that attract hikers. The 180-kilometer long TransCarioca Trail (Trilha TransCarioca) crosses the hills, going from Copacabana to Grumari (another protected area by the sea). It is maintained largely by the determination and mobilization of the civic society.

The Tijuca National Park is a democratic space where people from different parts of the city gather and socialize, no matter their social status. The park is fundamental to guaranteeing the quality of life and well-being of the residents of the city.

[1] The most common version of the meaning of "Carioca" (in the native language of the Tupi-Guarani Indians who lived in the region), is "house of the white men," but in another more recent version it is the name of the home of the indigenous tribe that lived in the area. Thus, the river was so important that it was named after the city's original inhabitants.

TOP RIGHT: Vista Chinesa, commemorating the Chinese workers who came to plant tea for the emperor in the 19th century.
BOTTOM RIGHT: Cascatinha (small waterfall) Taunay.

GUSTAVO PEDRO

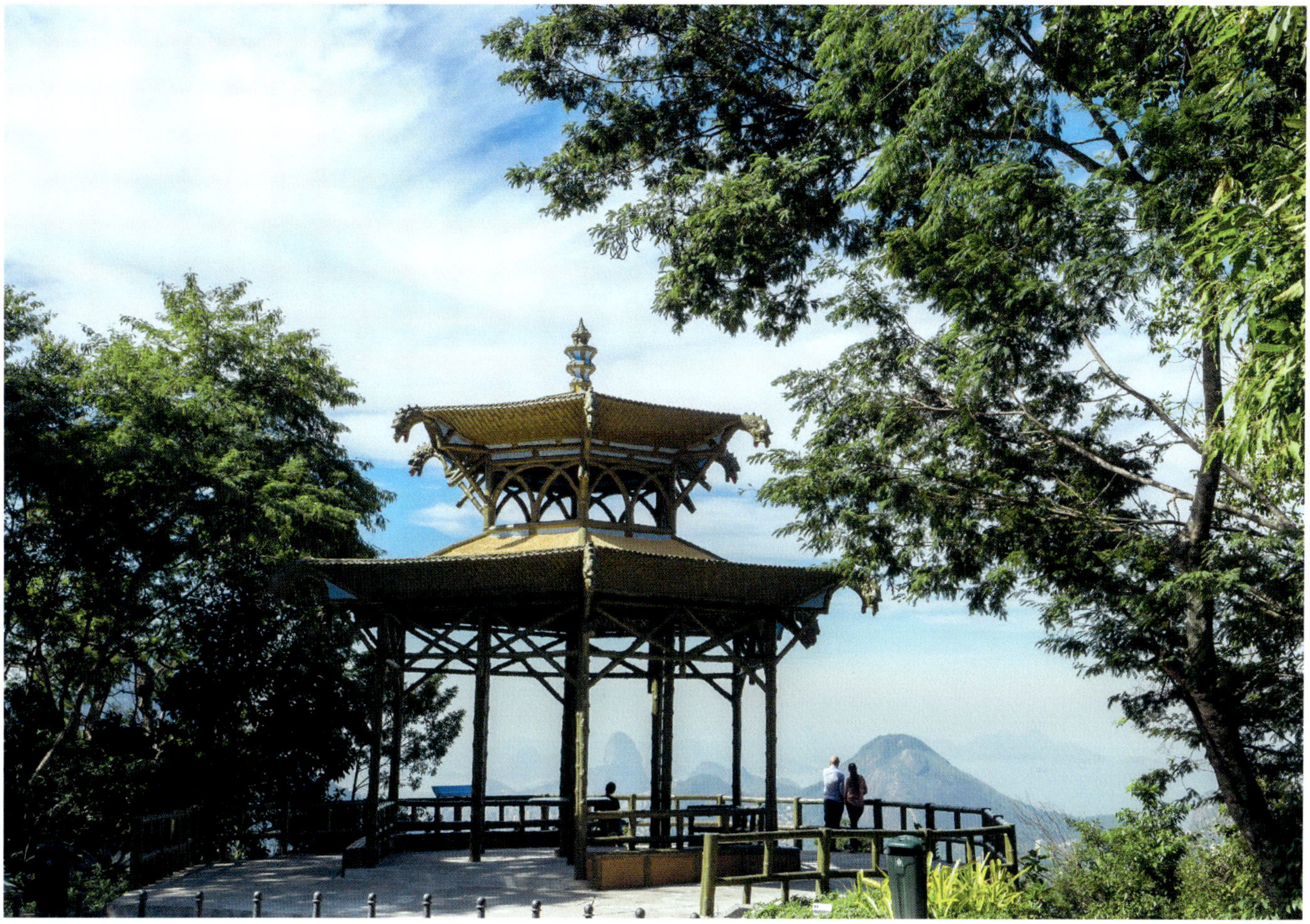

CECILIA HERZOG

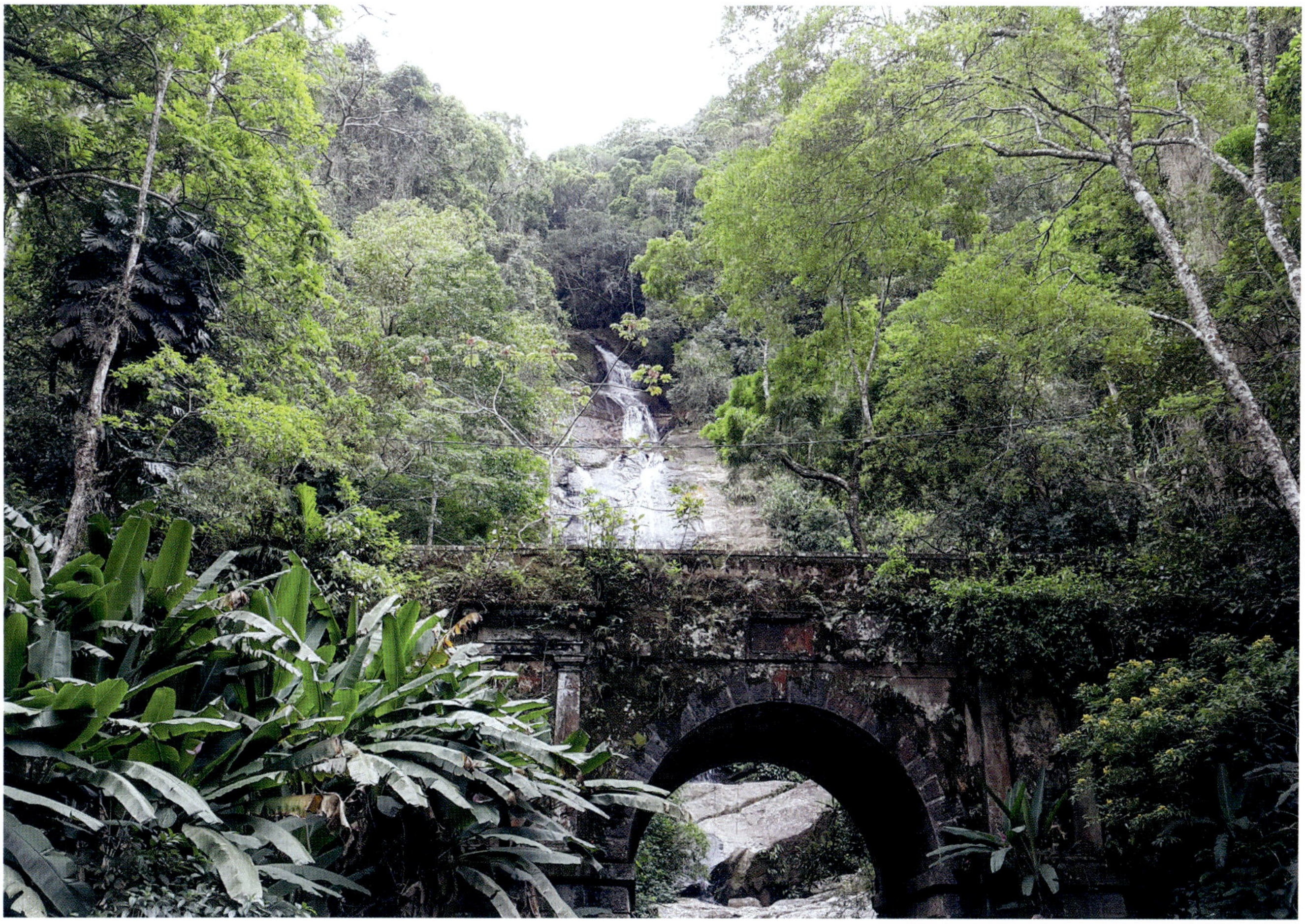

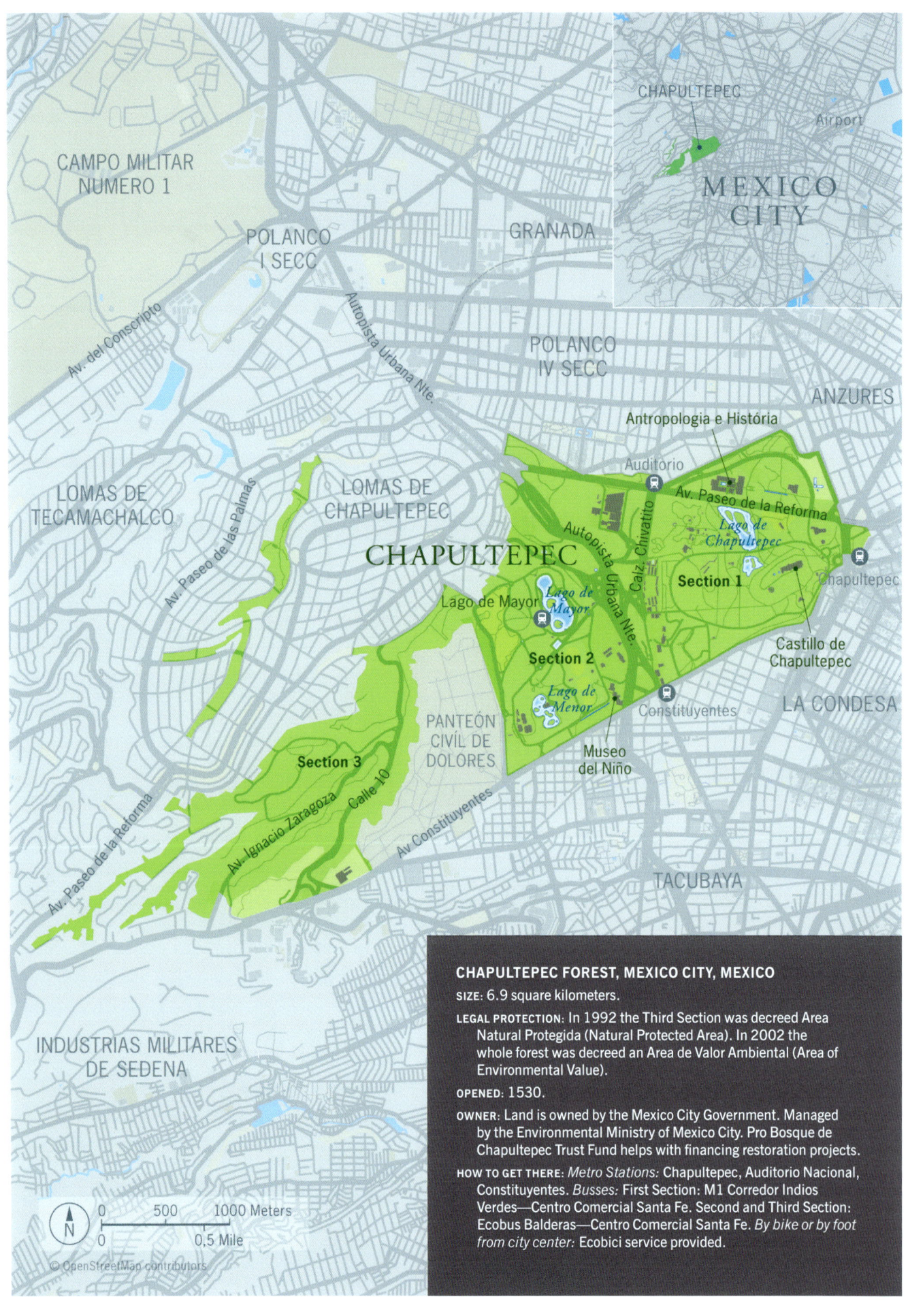

CAMPO MILITAR
NUMERO 1

POLANCO
I SECC

GRANADA

Av. del Conscripto

Autopista Urbana Nte.

CHAPULTEPEC

Airport

MEXICO
CITY

POLANCO
IV SECC

ANZURES

LOMAS DE
TECAMACHALCO

LOMAS DE
CHAPULTEPEC

Antropologia e História

Auditorio

Av. Paseo de la Reforma

Av. Paseo de las Palmas

CHAPULTEPEC

Autopista Urbana Nte.

Calz. Chivatito

Lago de
Chapultepec

Section 1

Chapultepec

Lago de Mayor

Lago de
Mayor

Section 2

Castillo de
Chapultepec

LA CONDESA

PANTEÓN
CIVIL DE
DOLORES

Lago de
Menor

Constituyentes

Av. Paseo de la Reforma

Section 3

Av. Ignacio Zaragoza

Calle 10

Av Constituyentes

Museo
del Niño

TACUBAYA

INDUSTRIAS MILITARES
DE SEDENA

N

0   500   1000 Meters

0   0,5 Mile

© OpenStreetMap contributors

CHAPULTEPEC FOREST, MEXICO CITY, MEXICO

SIZE: 6.9 square kilometers.

LEGAL PROTECTION: In 1992 the Third Section was decreed Area
Natural Protegida (Natural Protected Area). In 2002 the
whole forest was decreed an Area de Valor Ambiental (Area of
Environmental Value).

OPENED: 1530.

OWNER: Land is owned by the Mexico City Government. Managed
by the Environmental Ministry of Mexico City. Pro Bosque de
Chapultepec Trust Fund helps with financing restoration projects.

HOW TO GET THERE: *Metro Stations:* Chapultepec, Auditorio Nacional,
Constituyentes. *Busses:* First Section: M1 Corredor Indios
Verdes—Centro Comercial Santa Fe. Second and Third Section:
Ecobus Balderas—Centro Comercial Santa Fe. *By bike or by foot
from city center:* Ecobici service provided.

LILIA HAUA MIGUEL

Chapultepec Forest, Mexico City, Mexico

Chapultepec is the oldest urban forest in the Americas, and is among those claiming the greatest historical legacies. It is comprised of three sections spread over a total area of 686 hectares: the first section spans 274 hectares, the second, 168 (considered a recreational, playful and cultural area), and the third, 244 (declared in 1974 as a recreational area for the expanding city). All three sections were declared as an Area of Environmental Value, Urban Forest category, in 2003. The will to expand the forest by another 100 hectares to create a fourth section has now been officially announced by the Mexican Federal Government. Recent statistics put the number of visitors at some 19 million annually. Chapultepec alone constitutes 52 percent of the green areas in the metropolitan zone. Its importance cannot be overstated in serving as rainwater catchment and filtration, generating oxygen, processing contaminants, and protecting plant and animal biodiversity.

22 million people live in Mexico City, the largest city in North America. It is located in the Valley of Mexico (Valle de México), a large valley in the high plateaus in the center of Mexico, at an altitude of 2,240 meters. It is surrounded by thin strips of highlands, separating it from other adjacent metropolitan areas which togther comprise the Mexico City megalopolis. Mexico City lacks green areas, as do most South American cities. Hence the importance of Chapultepec Forest, which many call "the city's great lung." At the same time, with almost seven square kilometers of forest it is an important part of the hydrological cycle in the Valley of Mexico.

Foremost, though, the story of Chapultepec is bound to the history of Mexico. It is seen as the "heart of the country," playing a central role in its identity. Representing some thirty centuries of history it is appreciated as a place of immense historical and psychological value. It is the site of the foundations of the Aztec capital Tenochtitlán.

The First Section of Chapultepec Forest is a green area filled with long-lived trees in the central zone of Mexico City. In imperial Aztec times, the *tlatoanis*, as the emperors were known, used it as a place of repose, closed off to the general public. Strategically located, with abundant water and favorable conditions for observing the natural phenomena of the sky, it became a sacred place for taking decisions. Emperor Moctezuma llhuicamina (1466–1520) ordered the creation

The story of Chapultepec is bound to the history of Mexico. It is seen as the "heart of the country," playing a central role in its identity.

153

of an extensive botanical garden, the construction of a zoo, and an aquarium with 10 pools for fishes brought from different areas controlled by the Aztecs.

Chapultepec (Grasshopper Hill) has been a strategic place since pre-Hispanic times and a setting for many battles. It has witnessed fundamental episodes of Mexican history: the arrival of the Mexicas around 1250; the construction of a hydraulic system to bring water to the island city of Mexico—Tenochtitlan—in 1466; and the arrival of the Spaniards, when it was designated to form part of the lands of the Marquis del Valle, the title bestowed upon conquistador Hernán Cortés. With Emperor Charles V and the Spaniards, a new era had begun. On 30 June 1530, in an almost unprecedented decree, Charles V established the Bosque de Chapultepec as property of Mexico City, designating the area as a place of recreation for the inhabitants of the city in perpetuity.

At the start of independent Mexico (1821), the forest suffered from total neglect. In 1833 a Military School was established on its summit in Chapultepec Castle (built as a summer home for viceroy Bernardo de Gálvez between 1778 and 1788). It was shortly after the site of the cruelest battles waged in the United States Intervention of 1847. The Castle and the park were badly damaged. Part of Alcázar castle was later restored and served as the royal palace, then the official residence of presidents of Mexico. This lasted until 1939, when it was turned into the National Museum of History.

During the empire of Maximilian (1864–1866), the works that were done in the Bosque de Chapultepec covered mostly the castle and its surroundings. Thus, in the Chapultepec Castle, Italian style ornamental gardens and some surrounding roads were built to allow the castle grounds to continue serving as a recreational area for the upper class, especially for Sunday walks.

The period from 1884 to 1911 represented a prolific time for the Forest, as it grew to the extent that we now know as the First Section of Chapultepec Forest, establishing amenities such as Café Restaurant de Chapultepec (now the Museum of Modern Art). Other museums followed during the 20th century.

Chapultepec Forest suffered badly from the economic crises of the 1980s. Unemployed and homeless people invaded the park. Litter and fires and general neglect of maintenance made the park unsafe and

TOP RIGHT: Aerial view of Chapultepec Forest, Mexico City.
BOTTOM RIGHT: On days off people swarm into the park. Composers' Passage and Xochipilli Monumental Fountain.

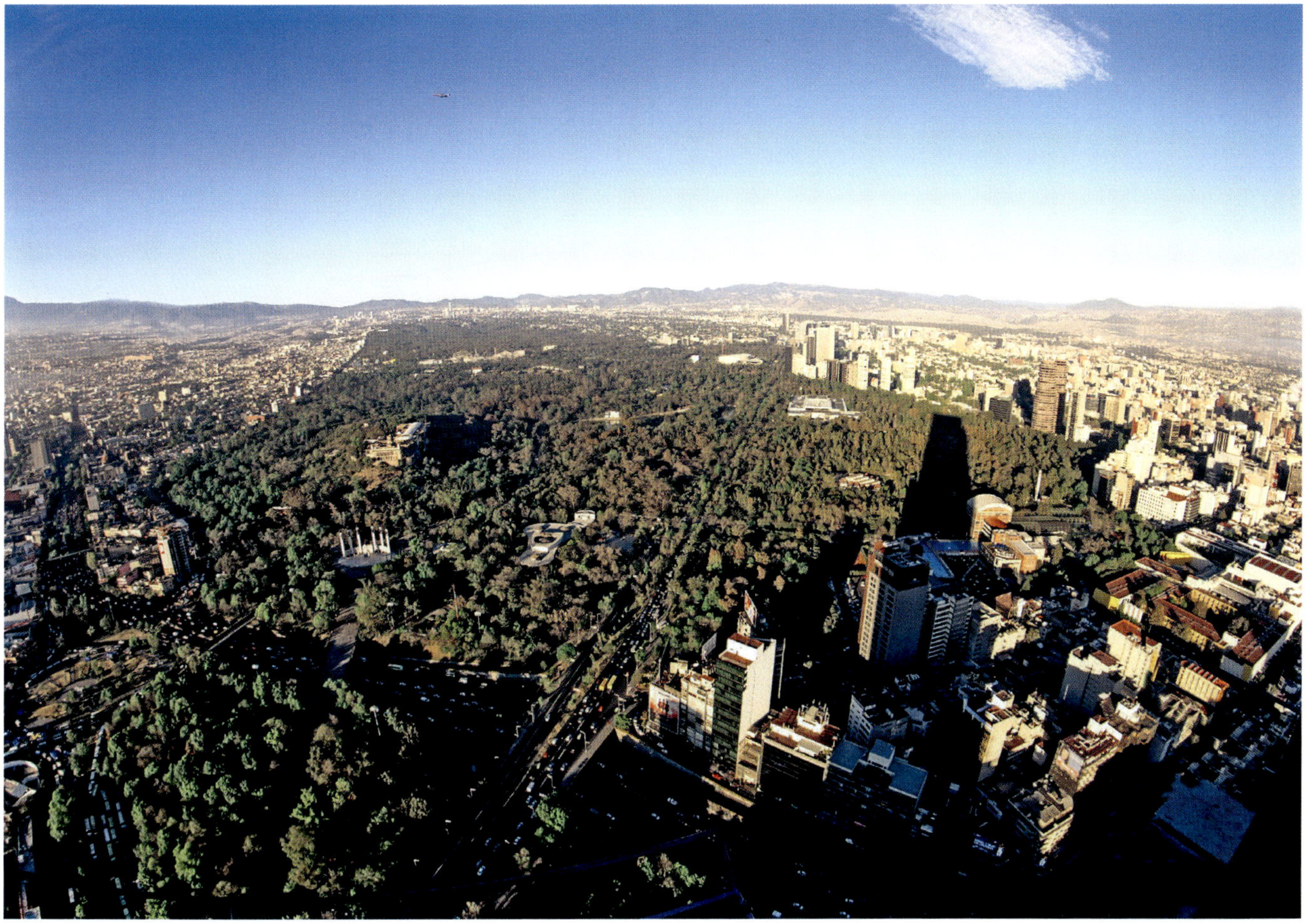

CARLOS HAHN

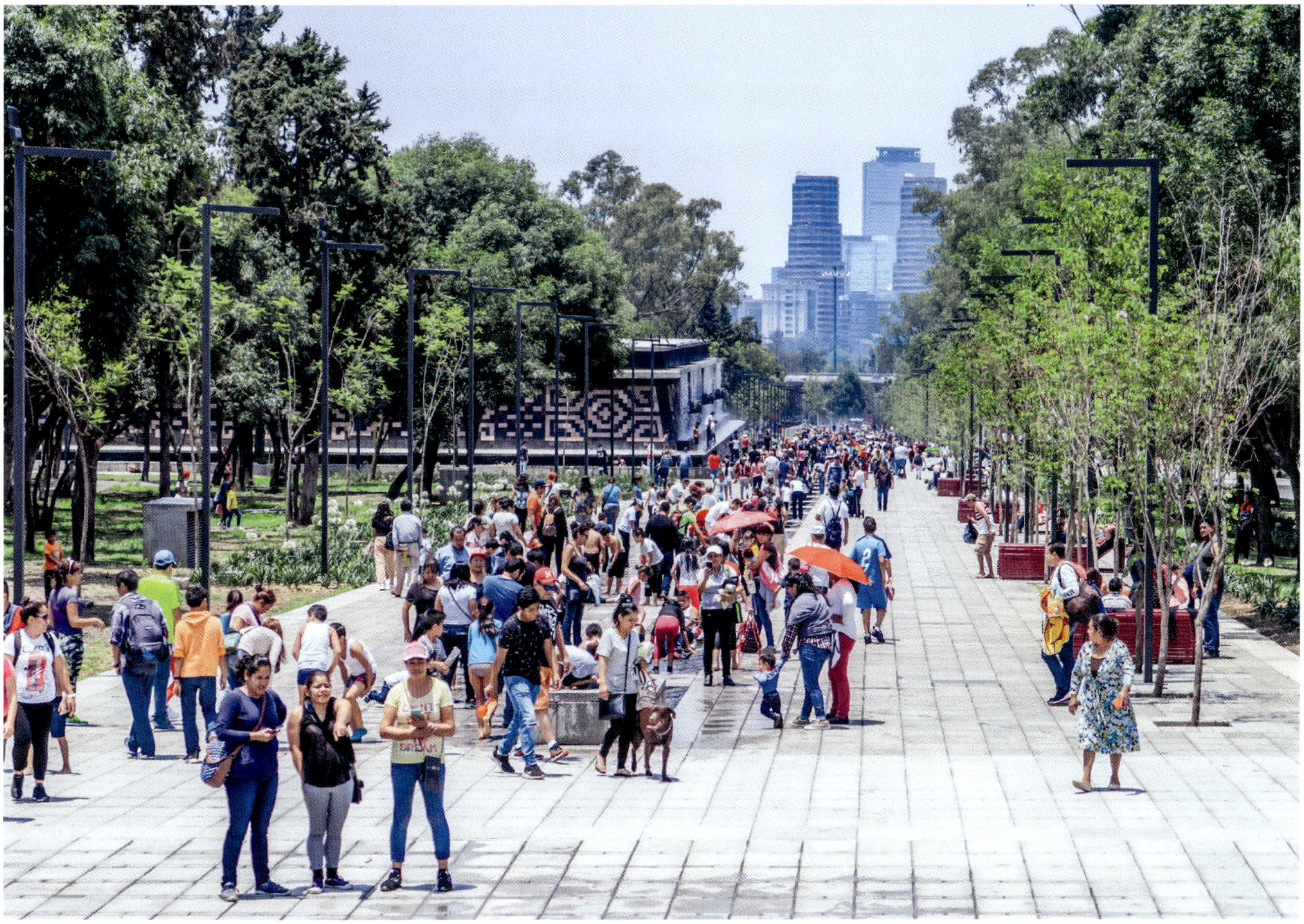

ENRIQUE ABE

untidy. These problems lasted until the late 1990s when a group of citizens, concerned about the deterioration affecting Chapultepec, decided to draft a Park Rehabilitation Master Plan with the support of the City Government. A group of Mexican entrepreneurs dedicated to the rehabilitation and recovery of the forest reactivated the Pro Bosque Trust Fund in 2004 to help finance the restoration plan. For more than fifteen years, the projects that have been undertaken have been carried out in coordination with the Consejo Rector Ciudadano (Citizen Governing Council, a consultancy body for discussion and decision-making)[1] and with the committed support of the Fund. For each peso raised by the Fideicomiso Pro Bosque de Chapultepec, the Mexico City Government is committed to matching the amount invested.

In this tripartite system, preservation and rehabilitation work has been carried out in the Forest through Master Plans. The restoration process began in 2004 with the Master Plan to Regenerate the First Section and continued in 2013 with the respective Plan for the Second Section—both completed according to aims and schedule—to give rise today to the Master Plan for the Third Section, which is in its initial stages.

Volunteers play a very important role in caring for the park. Over the years many thousands of volunteers have helped carry out various projects.

Chapultepec Forest has gone from being a sacred place to being the foremost setting for culture and the arts in Mexico. It houses the nation's greatest concentration of museums, not only in numbers, but also in the magnificence and breadth of their collections. These include the National Museum of Anthropology (1964), the Museum of Modern Art (1964), and the Tamayo Museum (1981).

In the central part of the First Section rises Grasshopper Hill, which gave name to the Chapultepec Forest, a natural stronghold in the ancient landscape of the Valley of Mexico. It has survived in the heart of the megalopolis, together with several natural vestiges of archaeological or historical significance. At its peak stands Chapultepec Castle.

As for the second section, one of the motives in establishing it was the desire to curb urban sprawl. The government ceded, expropriated, and regulated the lands that form it. This beautiful section of the forest was inaugurated in 1964. It offers to its visitors peace, recreation and sports, sculptured fountains, and the wondrous Papalote Children's Museum (1993). Furthermore, in this section are four tanks storing 200 million

A group of Mexican entrepreneurs dedicated to the rehabilitation and recovery of the forest created the Pro Bosque Trust Fund in 2004 to help finance the restoration plan.

LEFT: Altar to the homeland, monument to Niños Héroes (the Six Heroes) from the defense of Chapultepec Castle (in the background) in 1847.

FRANCISCO GÓMEZ SOSA

liters of water that supply a third of the city. The walls of an unassuming building are covered with the mural *Water, Origin of Life*, by Diego Rivera, who also created the Tlaloc Fountain opposite, depicting the pre-Hispanic god of water in a colossal relief made of stone mosaic.

The third section of Chapultepec Forest was inaugurated in 1974 to provide recreational areas to the nearby neighborhoods, as well as to expand the urban green zone in order to ameliorate environmental concerns. This section of Chapultepec is of greatest importance in recharging the Valley of Mexico aquifers. Seventy percent is composed of forests and canyons that provide for the natural runoff. The vegetation includes an important variety of flora and fauna, in addition to contributing to soil conservation, water filtration and circulation, and storage for groundwater sources. Today, work is underway to draft a master plan

ABOVE: The third section was inaugurated in 1974 to provide neighboring areas with recreational opportunities and to deter urban sprawl.
TOP RIGHT: Central courtyard at the National Museum of Anthropology, designed by architect Pedro Ramírez Vázquez, houses archaeological pieces from pre-Hispanic cultures and modern ethnographic groups from throughout the country.
BOTTOM RIGHT: "La Milla," one of the most important pathways in the First Section.

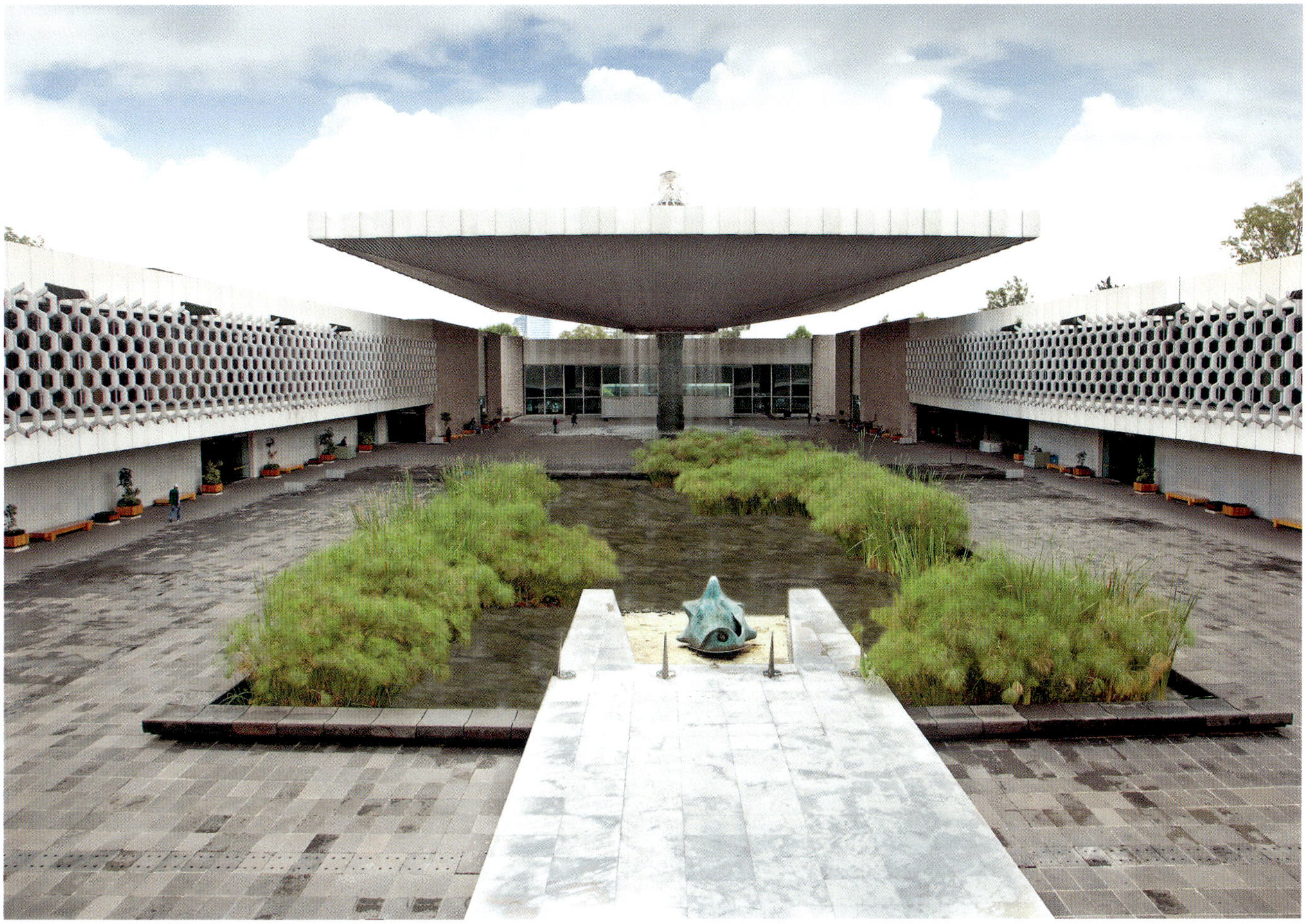

NATIONAL MUSEUM OF ANTHROPOLOGY

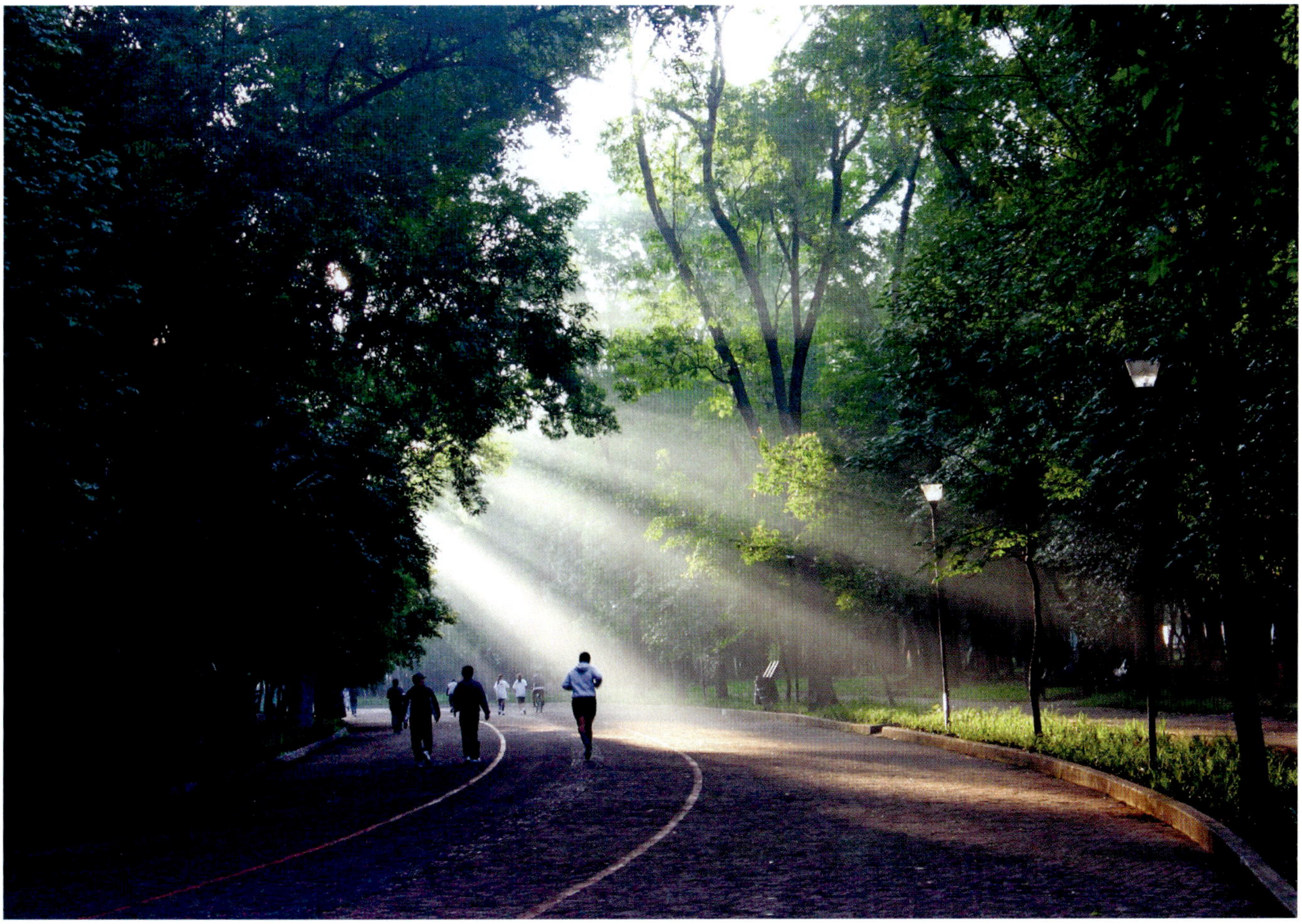

FRANCISCO GÓMEZ SOSA

This (the third) section of Chapultepec is of greatest importance in recharging the Valley of Mexico aquifers.

for the third section, part of an effort to return the region to nature, gradually replacing the forests of eucalyptus trees—not endemic to the zone—with oak stands, similar to those already found in the mountain ranges. The benefits should be reflected in a myriad of ways: lower maintenance costs and decreased water usage; mitigation of erosion; recovery of flora and fauna; and the overall health of the forest and soil.

The three sections of Chapultepec Forest constitute a much-valued biological corridor. It is an ideal refuge for birds, of which sixty-four species have been identified—some of them migratory—including mockingbirds, owls, cardinals, woodpeckers, hummingbirds, crows, various types of hawks, as well as eighty species of butterflies. Twenty mammal species have been detected—squirrels, ringtails, rabbits, bats, Central Mexican broad-clawed shrews, opossums, rats, and mice—and thirty-seven vertebrate species, among them three types of reptiles, serpents, and lizards, Similarly, it is a key area for the preservation of flora. Outstanding are the ancestral Montezuma bald cypresses, although also visible are ash, white cedar, some types of oak, Montezuma pine, piñon pine (among other pine species), sacred fir, acacia, as well as colorful jacaranda and bougainvillea and fruit trees such as lime, lemon, orange, peach, medlar, and plum.

In this great city … light penetrates, it illuminates our trees, flowers, lakes, plants, and is reborn in each of us and it invites us to enjoy its splendor. From our pre-Hispanic past to the present, Chapultepec Forest is and will always remind us who we are and where we came from, and above all, to celebrate our own humanity.[2]

[1] The objective of the Consejo Rector Ciudadano, created in 2002, is to foster citizen involvement in government decision-making: (1) to watch over the appropriate use of the forest budget and provide follow-up on its management; (2) to protect the extension of wooded areas, maintain their integrity, and make an effort, to the extent possible, to expand the forest's green border; (3) to holistically watch over the forest to preserve and strengthen environmental, social, cultural, and historical values, to preserve the integrity of the landscape; and (4) to ensure the cleanliness and good maintenance of the forest's infrastructure and work for the recovery, remodeling, and concessions in existence.

[2] Sharon Fastlicht Kurian de Azcárraga, in *Bosque de Chapultepec*.

TOP RIGHT: Dolores Waterworks and Tlaloc Fountain by Diego Rivera.
BOTTOM RIGHT: Mexico City Skyline.

ENRIQUE ABE

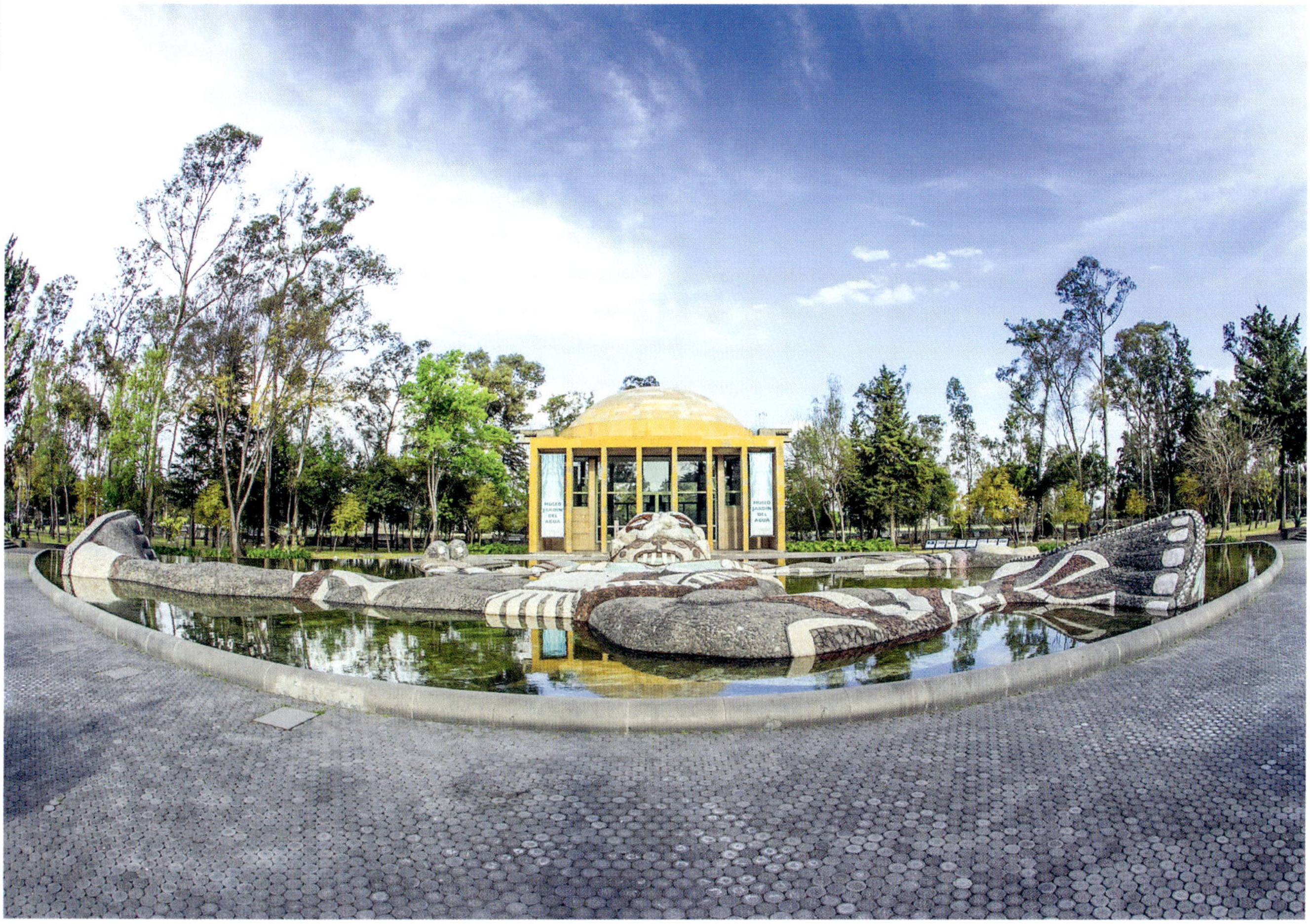

ENRIQUE ABE

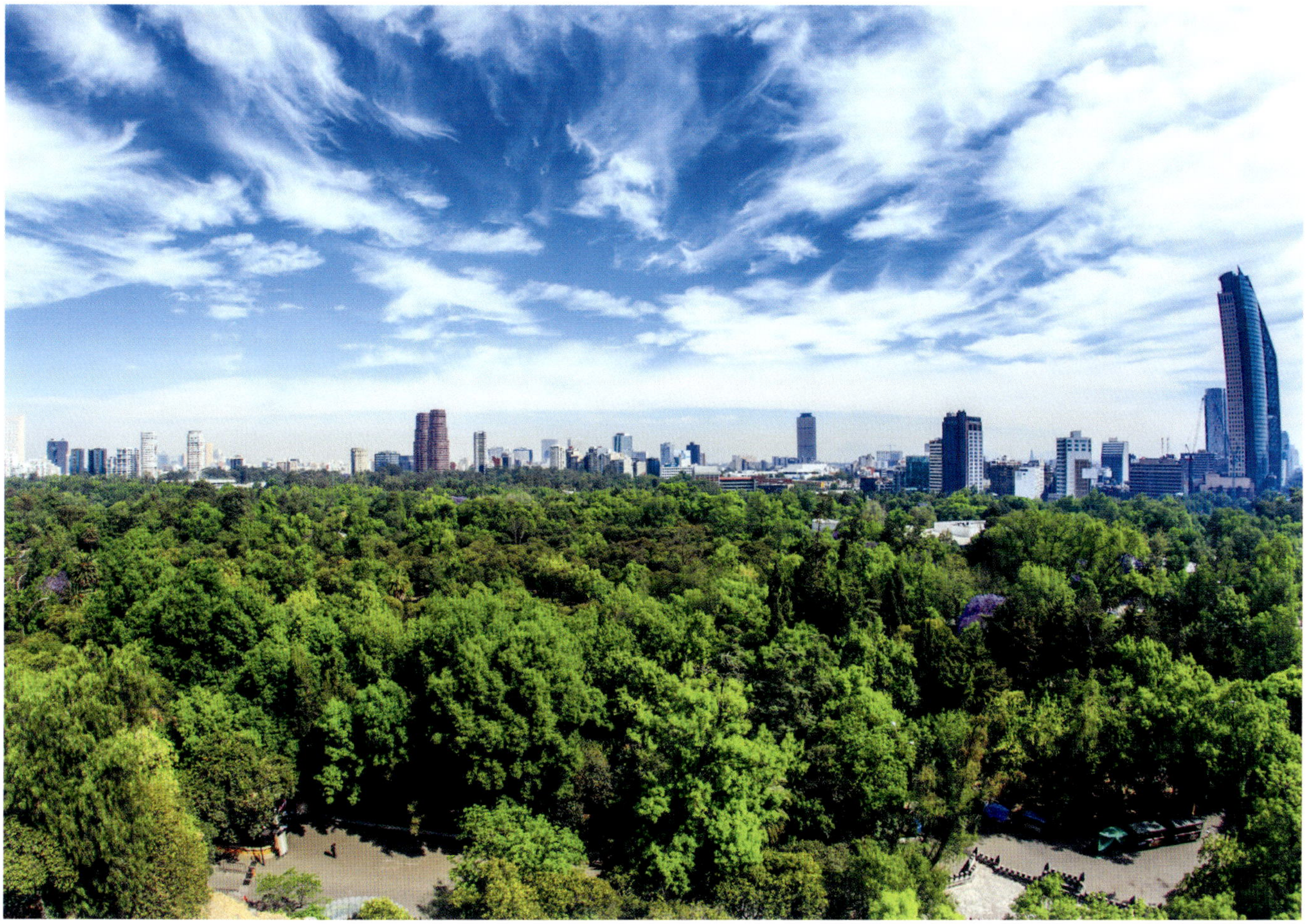

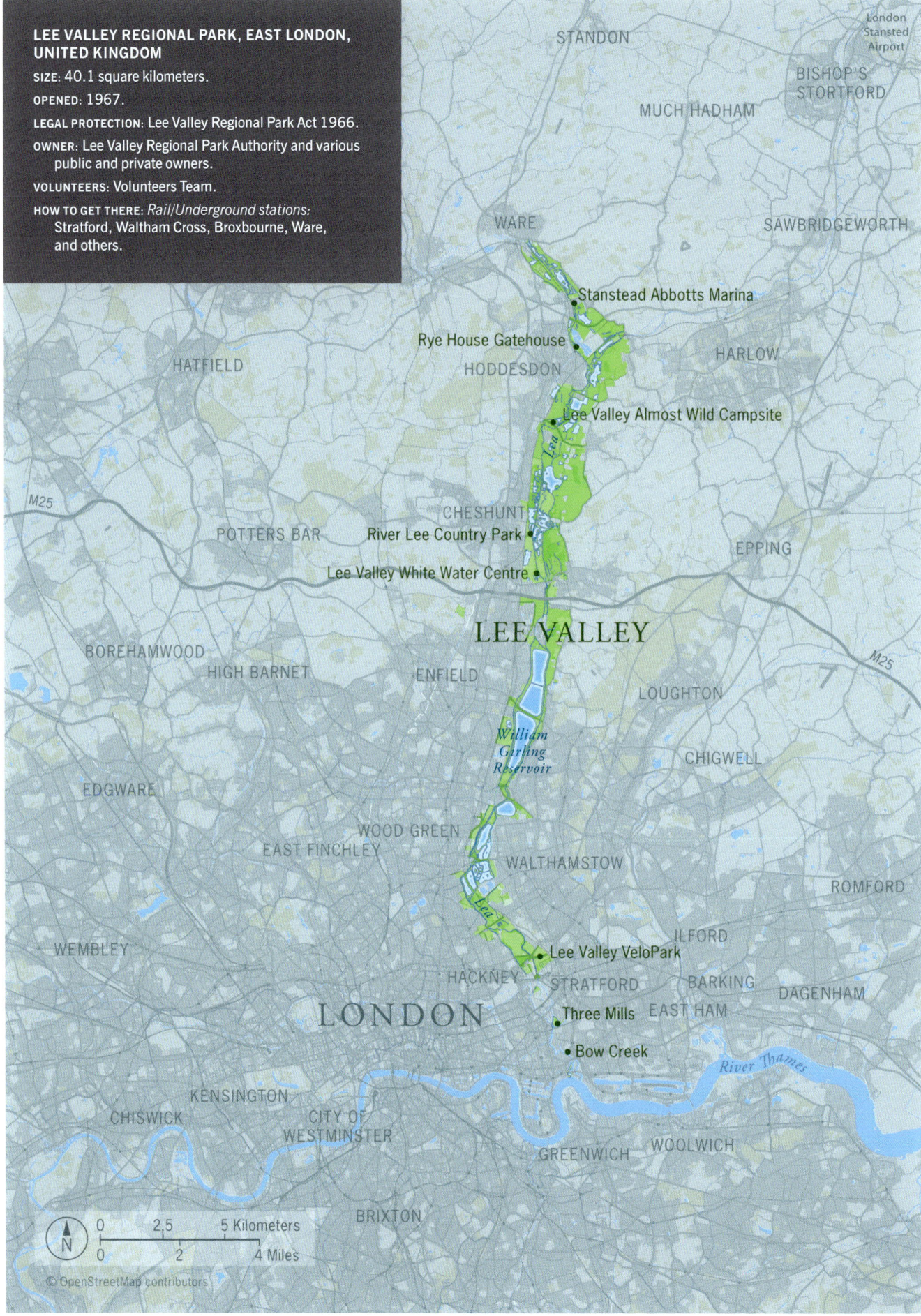

LEE VALLEY REGIONAL PARK, EAST LONDON, UNITED KINGDOM

SIZE: 40.1 square kilometers.

OPENED: 1967.

LEGAL PROTECTION: Lee Valley Regional Park Act 1966.

OWNER: Lee Valley Regional Park Authority and various public and private owners.

VOLUNTEERS: Volunteers Team.

HOW TO GET THERE: *Rail/Underground stations:* Stratford, Waltham Cross, Broxbourne, Ware, and others.

STANDON

London Stansted Airport

BISHOP'S STORTFORD

MUCH HADHAM

WARE

SAWBRIDGEWORTH

Stanstead Abbotts Marina

Rye House Gatehouse

HODDESDON

HARLOW

Lee Valley Almost Wild Campsite

HATFIELD

Lea

CHESHUNT

River Lee Country Park

Lee Valley White Water Centre

POTTERS BAR

EPPING

M25

LEE VALLEY

BOREHAMWOOD

ENFIELD

LOUGHTON

HIGH BARNET

M25

William Girling Reservoir

CHIGWELL

EDGWARE

WOOD GREEN

EAST FINCHLEY

WALTHAMSTOW

ROMFORD

Lea

ILFORD

Lee Valley VeloPark

WEMBLEY

HACKNEY

STRATFORD

BARKING

DAGENHAM

Three Mills

EAST HAM

LONDON

Bow Creek

River Thames

KENSINGTON

CITY OF WESTMINSTER

CHISWICK

GREENWICH

WOOLWICH

BRIXTON

N

0   2,5   5 Kilometers

0   2   4 Miles

© OpenStreetMap contributors

STEPHEN WILKINSON

Lee Valley Regional Park, London, United Kingdom

The Lee¹ Valley Regional Park presents a rich tapestry of open spaces, nature reserves and major sporting venues originally designed to serve the leisure and recreation needs of Londoners. It runs for 26 miles broadly aligned with the natural course of the River Lea from its southern tip, the East India Dock Basin on the River Thames, through East London and Essex to Ware in Hertfordshire.

The idea for a large park along the River Lea, a main tributary of the River Thames in London, was originally conceived in the 1944 Greater London Plan. The Plan was one of many documents written during the Second World War when decision-makers and politicians turned their attention to post war reconstruction. The Plan identified the Lea valley as an "… opportunity for a great piece of constructive, preservative and regenerative planning."

At that time the valley was marked by industrial decline as London's economy underwent structural change. Further north in Hertfordshire and Essex, sand and gravel quarrying continued but the greenhouse vegetable industry was in decline.

During the early 1960s the idea for the Park was re-invigorated and in 1963 a group of local politicians commissioned a study by the Civic Trust, an amenity society, into the creation of a Regional Park along the Lea valley with recommendations for a dedicated public authority to manage and develop the whole area.

The Trust's report was accompanied by bold and exciting new ideas, reflecting the spirit of the 1960s, which challenged convention. Ideas included major sports stadia and large new venues.

The result of this impetus led to considerable support from local politicians who sponsored in Parliament a Private Act, the 1966 Lee Valley Regional Park Act, creating the Lee Valley Regional Park Authority which came into existence on 1 January 1967. Although the Park's statutory boundary overlies the administrative boundaries of the riparian boroughs, the Act requires that those areas are the subject of the Authority's dedicated remit for sport, leisure, nature conservation, and recreation. A tension has always existed between the Authority with its dedicated statutory remit and the wider remit of the riparian boroughs.

The Act stipulates that funding would come from contributions in

The idea for a large park along the River Lea was originally conceived in the 1944 Greater London Plan.

the form of a 'levy' from each resident of Greater London and of the two counties of Essex and Hertfordshire.

However, over time the amount of the contribution derived from the levy to support the Authority's management and development of the Regional Park has been reduced and today amounts to just 30 percent of the total budget required to manage the Parklands and venues with the other income derived from fees and charges. The Park is administered by a Board comprised of local politicians who represent the constituent bodies of its funding base.

Today the Authority employs a team of staff covering property, planning, legal, venues, and parkland management services. The venues include 3 'legacy' venues from the London Olympics and include the Lee Valley VeloPark, the Lee Valley White Water Centre, Broxbourne, and the Lee Valley Hockey and Tennis Centre. The other venues include the Lee Valley Riding Centre, Lee Valley Ice Centre, 2 marinas, 4 campsites, and an Athletics Centre and a golf course in Edmonton. These venues attract 2.5 million visits each year out of a total of approximately 7 million visits to the whole Regional Park; and 4.5 million visits to the parklands each year.

In addition, the Authority owns and manages a commercial dairy and arable farm that allows children to learn about food production.

Although the Park extends to 4,010 hectares, it includes 250 hectares devoted to housing and employment. As the Authority owns just 2,300 hectares, its role is to work with the different landowners through a series of formal and informal partnerships, which include power, rail, water companies, and private landowners, and to implement the broad vision for the whole Regional Park. This can be quite a challenge.

The Regional Park's defining feature is its landscape which, unusually for such a large park, is largely man-made and interspersed with hard infrastructure, including railways, roads, canals, and pylons. These features divide and reduce the parklands into distinct areas. This provides a varied backcloth and physical context for the kaleidoscope of activities which occur. The use of many sites is multi-layered.

Broadly these landscapes can be distinguished by their geomorphology and industrial history. They include areas as diverse as, its rural valley floor, wetlands, industrial landscapes, reservoirs, historic gardens, and farmlands.

TOP RIGHT: Bow Creek, looking towards Canary Wharf.
BOTTOM RIGHT: Hall Marsh Scrape.

LEE VALLEY REGIONAL PARK AUTHORITY

LEE VALLEY REGIONAL PARK AUTHORITY

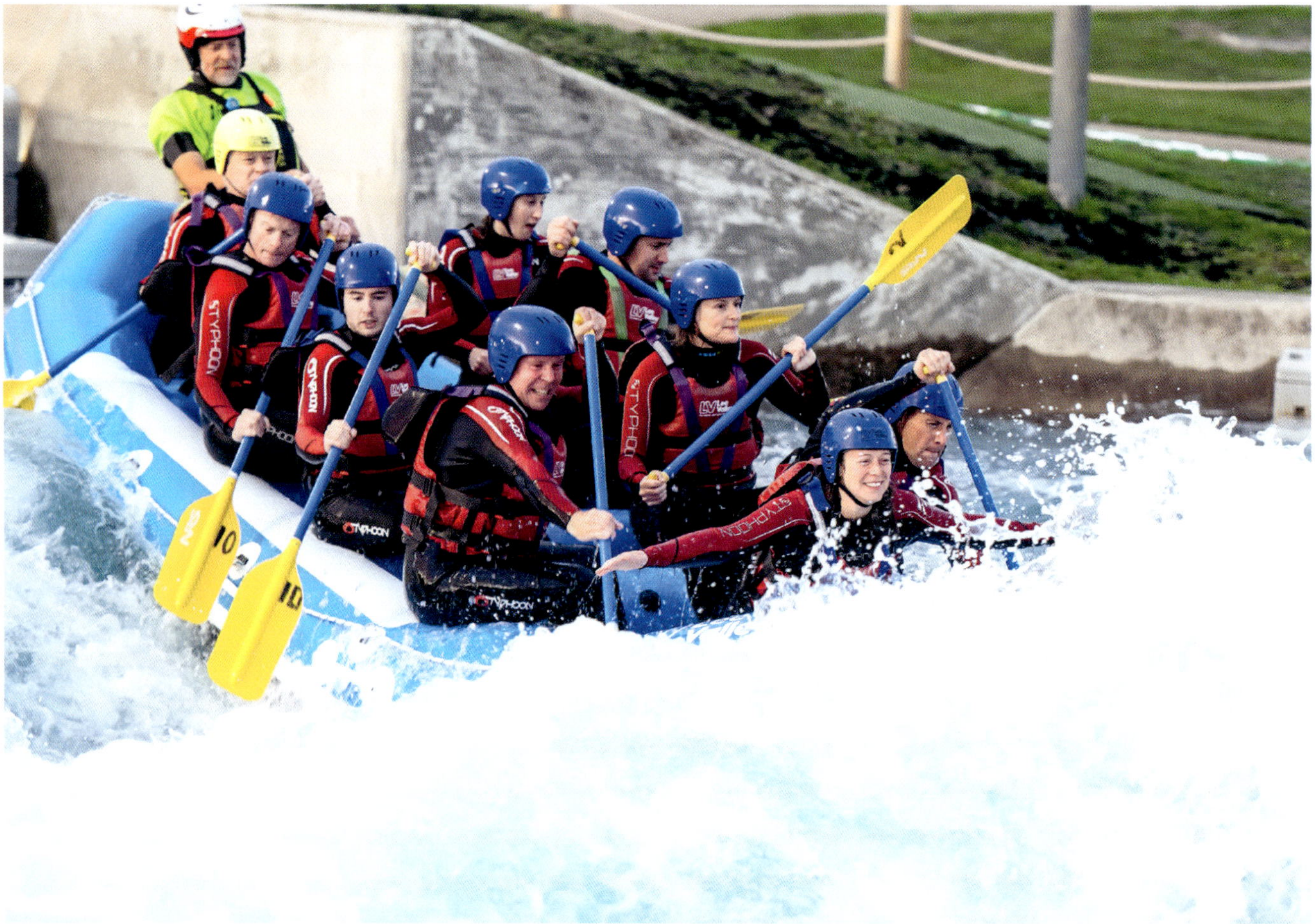

LEE VALLEY REGIONAL PARK AUTHORITY

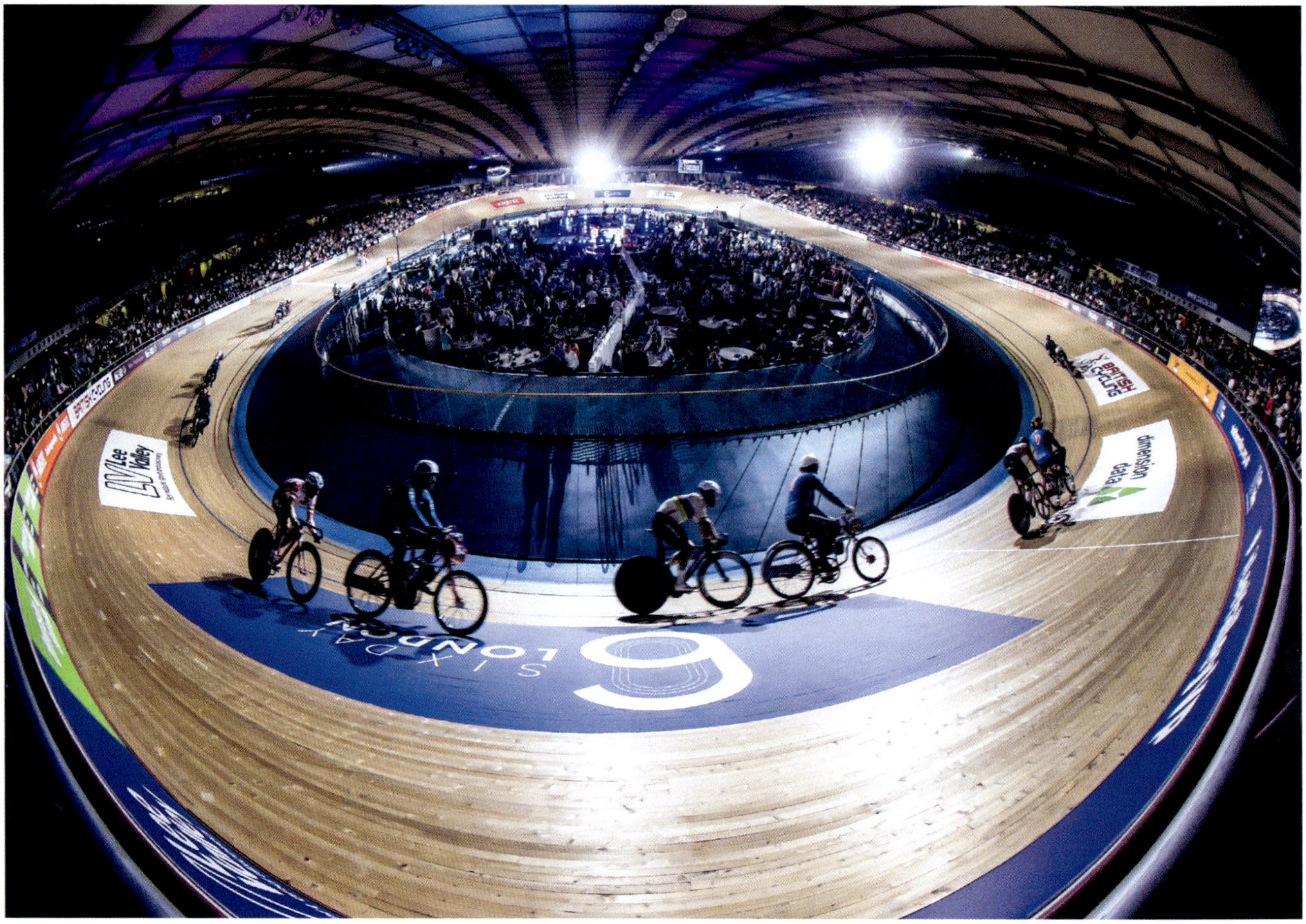

SIX DAY SERIES

The valley of the River Lea forms part of the London Basin. The northern section around Ware and St Margaret's is underlain by chalk which shapes rolling open farmland. Further south, London clay predominates, resulting in a series of shallow ridges, and in East London, a large flood plain.

The hydrology of the valley is complex. The Lea is a tributary of the Thames and has an extensive dendritic pattern which has been 'complicated' in the south with a series of flood defenses.

The Regional Park is rich in biodiversity. It has sites of international, national and regional importance. The Park is made up of a patchwork of habitats resulting from centuries of changes due to the development of agriculture and industry. Whilst areas of habitat with some degree of naturalness can still be found in the valley it is evident that the vast majority has been altered over time by the actions of man. The key habitats are rivers and streams, open water, grassland and fen, and post-industrial. The landscape includes remnants of historic wetlands and marsh at Cornmill Meadows, Waltham Abbey, and Walthamstow Marshes that historically characterized the valley floor but which during the last two centuries have been drained and built on.

The diversity of habitat is protected through statutory designation including eight Sites of Special Scientific Interest. Four of these form the Lee Valley Special Protection Area and Ramsar site. They provide habitat for over 300 species of birds, 32 species of mammal and 900 species of flowering plants. The valley is recognized as an important site for the wintering birds, such as the Gadwall, Shoveler, and Bittern species, and provides a good habitat for wetland mammals such as the water vole (Britain's fastest declining mammal).

Balanced against the need to protect the natural environment is the Authority's role in creating opportunities for people to gain access to nature. With over 4.5 million visits each year to the parklands the Park includes several very popular sites, including nature reserves at Rye Meads and Amwell, Stanstead Innings, and Silvermeade. A further site at Walthamstow Wetlands in London opened in 2017 and attracts over 250,000 visits each year.

The Park is a rich heritage resource. This in part reflects the importance of the valley for early settlement. There are a number of important old houses including the Grade 1 listed Rye House Gatehouse, Hoddesdon, one of the first brick-constructed houses in the country and Myddelton

As the Authority owns just 2,300 hectares, its role is to work with the different landowners through a series of formal and informal partnerships, which include power, rail, water companies, and private landowners, and to implement the broad vision for the whole Regional Park. This can be quite a challenge.

TOP LEFT: Rafting at the Lee Valley White Water Centre.
BOTTOM LEFT: The velodrome, centerpiece of Lee Valley VeloPark

House, Enfield, a late Georgian mansion with historically important gardens, which today is occupied as the Authority's HQ.

One hundred and fifty years ago the valley was home to a considerable number of mills which used the river for power. Wright's flour mill at Ponders End is still working and there is a fine example of an early Georgian mill complex at Three Mills, Bow, in East London.

The relative isolation of the valley in the past led to several sites being used for the armaments and explosives industry. Two sites in Waltham Abbey were used for the manufacture of explosives before being decommissioned in the 1990s. These include the site of the Royal Gunpowder Mills that, from the sixteenth century, was used for the manufacture of explosives. The site is now a visitor-attraction containing over 20 historic buildings and structures. The site also includes a rich diversity of plants and habitat for water voles and invertebrates. To the south was an 'experimental' explosives site developed during the nineteenth century. Since decommissioning this site has been managed by the Park Authority, landscaped, and created into the large Gunpowder Park.

Throughout the valley there are a range of listed structures and buildings, reflecting the valley's legacy of water storage and flood control. These include old lock systems and water pumping stations.

The Regional Park is a multi-layered venue for a range of activities. Some of these can be formal, others are informal, with visitors allowed to wander, either on foot or by bike, along its extensive lattice of paths.

The venues balance a mix of sessions for first-timers, schools, people with disabilities, community groups and hard-to-reach groups with corporate events, birthday parties, clubs and leagues, and extensive venue hire, plus major events. This follows the Authority's community-focused and commercially-driven ethos. This twin approach places community use at the heart of the venues alongside a focus on commercial opportunities to guarantee their financial future.

The three London 2012 venues have hosted four World Cups, two World Championships and ten other major international sports events since London 2012. In 2022, Lee Valley VeloPark will make history, becoming the only venue in the world to host an Olympics, World Championships, and Commonwealth Games in the same sport. The Lee Valley White Water Centre will host the World Championships in 2023. Other events are organized by a variety of organizations including local clubs and charities. The Park is a great venue for sports such as rugby,

TOP RIGHT: Three Mills Bow.
BOTTOM RIGHT: River Lee Country Park.

JEAN CUOMO / SHUTTERSTOCK

LEE VALLEY REGIONAL PARK AUTHORITY

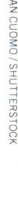

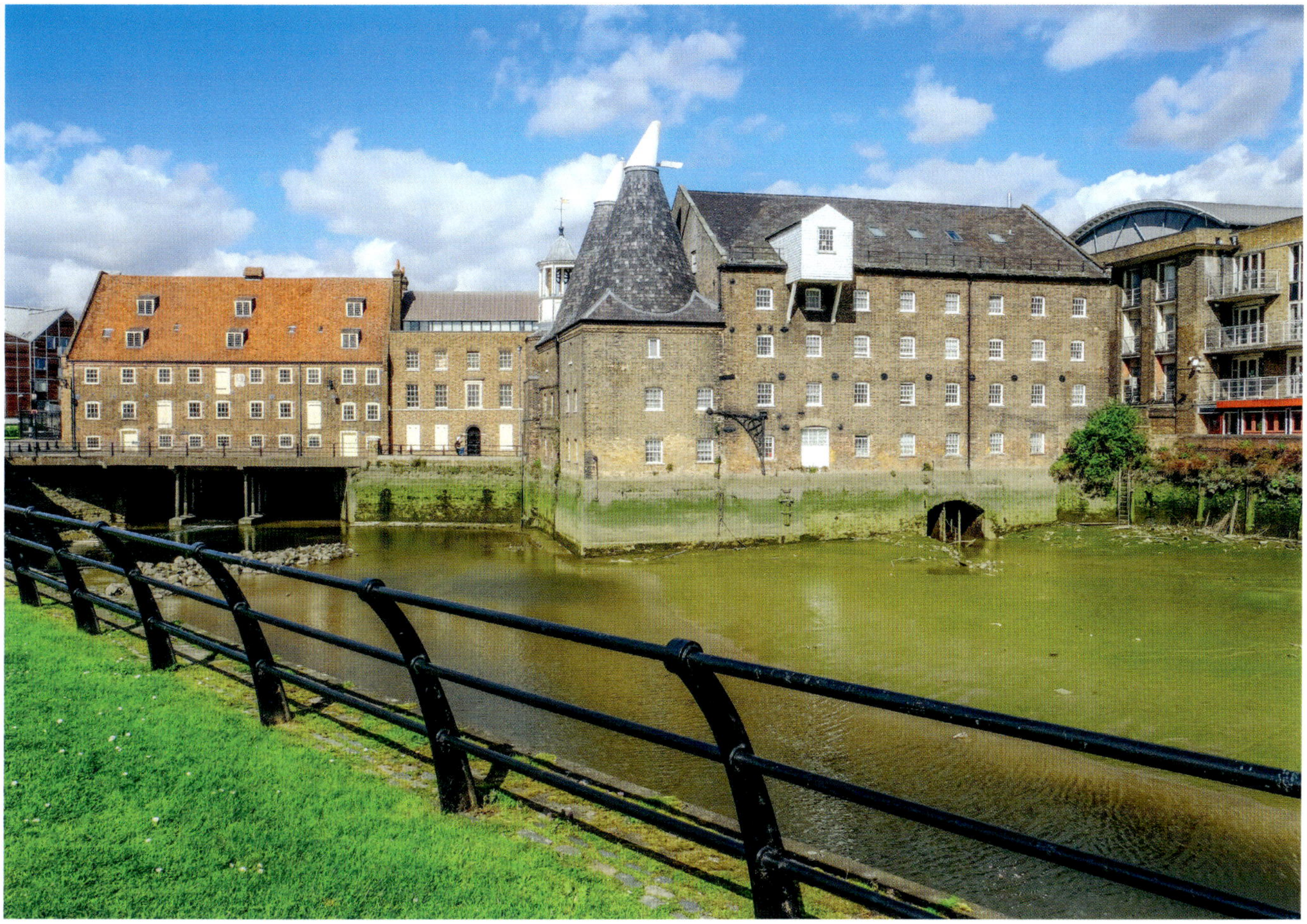

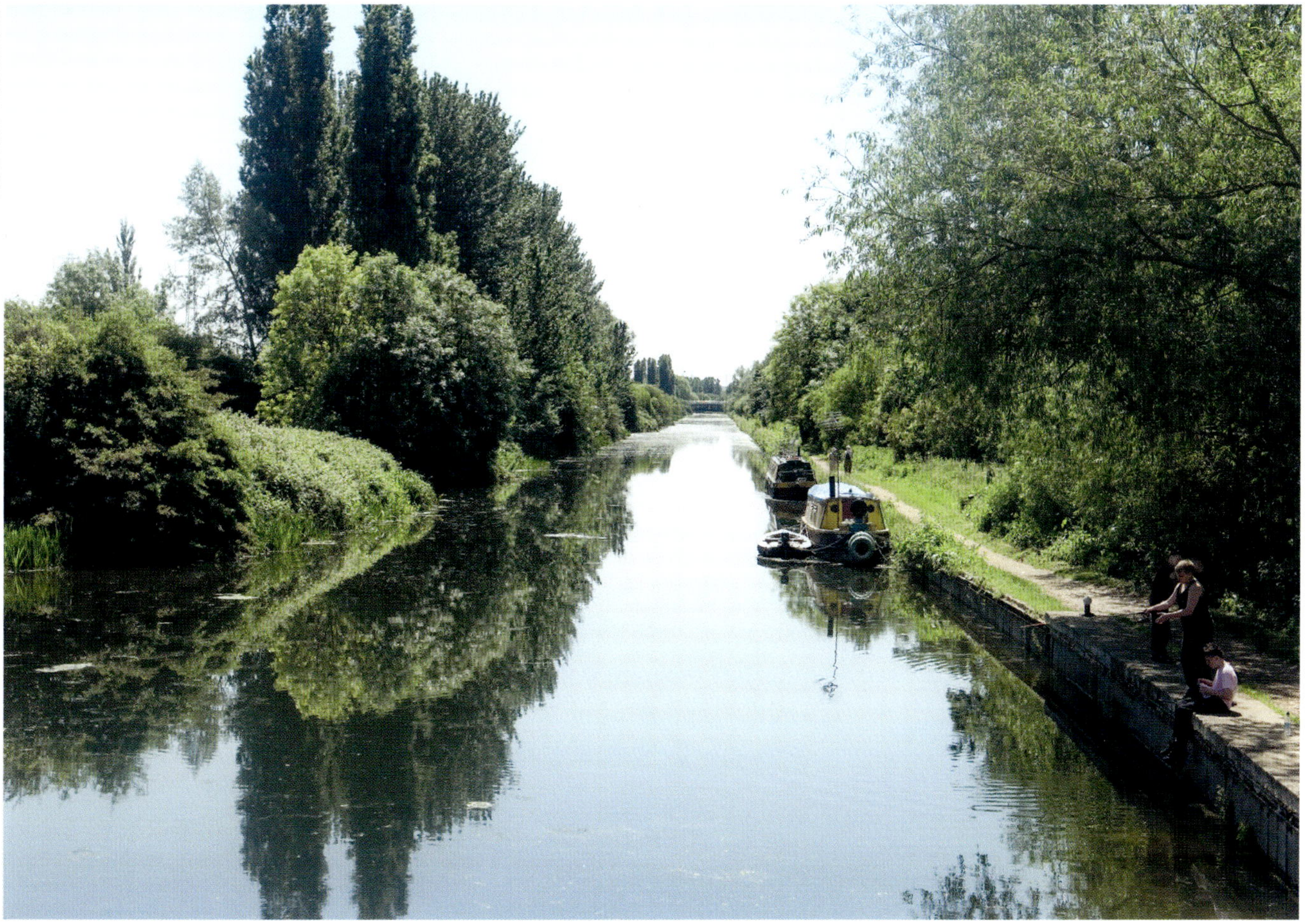

LEE VALLEY REGIONAL PARK AUTHORITY

A lattice of bike trails.

rowing, and other waterside activities. Every weekend there are guided walks, fishing competitions, outdoor theatre, football, rugby and cricket matches, family orienteering, competitive rowing, and boat cruises.

The Park is an enormous venue bringing communities together through an extensive volunteering program which includes 450 volunteers. Collectively they contribute through a broad range of roles either as stewards at major sporting events or by restoring sites by clearing out rubbish and weeds, including invasive species such as Japanese Knotweed.

Furthermore, the location of the Regional Park, which is near major business clusters in London, provides great opportunities for corporate team building events where team meetings can be energized in combination with white water rafting, cycling, or orienteering.

Being in such close proximity to a large population, the Regional Park is the ideal venue for many schools to pursue outdoor learning. The curriculum includes courses that cover geography, physical education, history, and art. Outreach programs reaching over 20,000 people support people with disabilities, and other programs are targeted at women and girls and for people from ethnically diverse backgrounds who unfortunately can face difficulties in accessing opportunities for fitness and well-being.

There are over 100 kilometers of walking and cycling routes through the Regional Park. These include a number of strategic routes including the Lea Valley Walk (extending from Bow Locks to the river's source in Leagrave near Luton), the London Outer Orbital Path, Capital Ring, and the New River Path. The main north-south route follows the towpath along the Lee Navigation. Linked to each of these is a lattice of short routes, which link the parklands and venues. However, the extent of roads and railways is a challenge for the Park Authority in improving the legibility and permeability of its open spaces throughout the Regional Park.

The waterways of the Park are a defining feature with three marinas, three rowing clubs and two boating clubs along the Navigation.

There is a broad range of visitor accommodation which includes a youth hostel at Cheshunt with several caravan and camping sites at Dobbs Weir Hoddesdon, Picketts Lock Enfield, and Sewardstone. In 2017 the Authority opened an "almost" wild camping site that allows people to experience the delights of the natural environment living in clean but basic conditions at a site on the banks of the Lee Navigation.

Recent announcements for a major surfing and outdoor activity center proposed for the Lee Valley Leisure Complex, Picketts Lock, and the Authority's plans for the redevelopment of the existing Ice Centre as a 'twin pad', demonstrate the important contribution the Regional Park will continue to make to meet the recreation and leisure needs of London and surrounding communities. Of increasing importance, though, will be its role in addressing the resilience of these areas through flood mitigation, reducing air pollution, addressing climate change, and the protection of ecology and the natural environment.

... ideal venue for many schools to pursue outdoor learning.

... great opportunities for corporate team building.

[1] The Regional Park Authority uses the spelling "Lee" but the river is "Lea." It is understood that the distinction results from "Lee" used to refer to Lee Navigation and "Lea" to the natural course of the river.

MOKOLA HILL

Obasa St

Ajibade St

Oyo Road

Dejo Oyelese Street

R. Ogunpa

IBADAN

AGODI PARK

Airport

Francis Okediji

Forest path trail

Golf course

Hotel Chalets

Forest area with sculpture displays

Canopy walk

AGODI PARK

Zip line & climbing activities area

Animal enclosures

Water park extention

Phase 1 Water park

Health Spa

Events center

Onigegora St

IBADAN

Sport facilities

R. Ogunpa

Picnic area

Botanical Gardens

Premier Hotel Drive

P

Boating piers

Lake

Island

P

Restaurant

Entrance to Phase 1 from Parliament Road

AGODI

Entrance to Retail Node from Premier Street

Retail node

Parliament Road

Craig Street

Cultural Centre Road

R. Ogunpa

Queen Elizabeth II Road

0   100   200   300 meters
0   500   1000 feet

N

© OpenStreetMap contributors

AGODI PARK AND GARDENS, IBADAN, NIGERIA

SIZE: 0.6 square kilometers.

LEGAL PROTECTION: zoning.

OWNER: Oyo State Government. Managed by a private company.

OPENED: 1967, reopened in 2014 after redesign by Earthworks Landscape Architects.

HOW TO GET THERE: Bus to Secretariat bus stop, or you can take a Micra taxi, go with a Okada motorbike, or walk from State secretariat, Premier Hotel, or University College Hospital (200–600 m).

ADAM VAN NIEUWENHUIZEN AND TESSA TOERIEN

Agodi Park and Gardens, Ibadan, Nigeria

In Africa, with its great national parks, people often focus on wildlife preservation, while urban parks are rare, and often not well managed. It is therefore of great interest to present some examples of a renewed interest in urban parks in Africa. Colonial governments sometimes created and cared for urban parks. But since then, many of the parks created at the time have deteriorated or disappeared altogether as they reflected European values rather than African. With the tremendous growth of African cities during the last half-century, it seems that there is a new-born interest in redeveloping urban parks. An overview is given in the chapter Urban parks in Africa.

This, then, is the story of one of these urban parks, still in its making. Within the sprawling city of Ibadan, Nigeria, lies Agodi Gardens. Ibadan is the third-largest city in Nigeria, and the largest in Oyo State, with over 3 million inhabitants. The name Ibadan means "at the edge of the savannah" in the Yoruba language. Agodi Gardens is located on the border between two biomes[1]: savannah and tropical forest, and this formed one of the key spatial concepts for the redevelopment of the park.

The park is centrally located, providing the bustling city with otherwise-lacking green open space. It also serves as an important environmental reserve for indigenous forest and savannah species of fauna and flora. The surrounding land is hilly, 191 meters above sea level. The park occupies approximately 58 hectares of land and is an ex-situ conservation site (a non-natural site created especially for preservation). The Dandaru River surrounded by floodplains and wetlands runs through the park and the park provides an important ecosystem service of flood attenuation. The expansive space, in contrast with the noisy city life outside, showcases nature and freshness, and provides a place of tranquility, which is highly valued among the people of Ibadan.

Agodi Gardens was created after Nigeria gained independence in 1967. It was formerly called Agodi Zoological and Botanical Gardens. The park contained a fully-fledged zoo during this period. It was at that time a major center for recreation in Ibadan. The number of visitors to the park grew steadily for many years, including large numbers of tourists.

Over the years, however, Agodi Gardens fell into disrepair. The path-ways became muddy, bridges eroded, and the lake was completely over-

In Africa, with its great national parks, people often focus on wildlife preservation, while urban parks are rare, and often not well managed. It is therefore of great interest to present some examples of a renewed interest in urban parks in Africa.

The objective was to unlock economic development opportunities within its borders to help maintain the park.

grown with invasive aquatic plants. Rivers and streams were choked with bamboo and debris. A devastating flood in 1980 destroyed much of the infrastructure, and sadly swept away many of the zoo animals, most of them drowned. Thereafter, there was a lack of programmatic activity and no repair, which meant that the park was seldom used; the animal cages became smaller and more dilapidated. Less than 200 people per year visited the park in 2012 when Gbemi Adio-Moses, managing director of a big construction company, who grew up in this area, discovered its potential and set about transforming it into an accessible park for the people of Ibadan. In 2013 the state administration contracted a landscape architecture firm to do the job.

The park was designed to highlight the transition between the main biomes in the region. Like all ecotones[2], Agodi Gardens is rich in biodiversity despite its location in the midst of a large city. The forest, wetland, and savannah areas are bursting with a biodiversity that had been long left alone. The design thus aimed to disturb as little of this as possible and to extend the natural habitats while sensitively re-introducing people to the park and removing invasive alien plants.

The objective was to unlock economic development opportunities within its borders to help maintain the park, including restaurants and water park. Later phases will include waterfront areas, a hotel, other commercial developments, and rebuilding of the zoo. The natural aspects and processes of the site were to be conserved.

All aspects of the development were based on sensitive and responsive landscape design principles: being guided by natural systems and using wood and lightweight construction materials rather than the locally-favored steel and concrete. In addition, the forest area contained a diversity of very old trees, which needed to be conserved at all costs, in part due to migratory fruit bats that settle in these trees. The bats migrate between West and South East Africa, the largest mammal migration within Africa. Dominant plant species are *Tectona grandis, Khaya senegalensis, Azadirachta indica, Gmelina arborea* and *Eucalyptus torelliana*.

Phase 1 focused on creating healthy outdoor-activity spaces within the savannah and forest environments. This focus ranges from playground areas for younger children to large swimming pools and water slides. Their placement within the park creates an experience of the natural environment in a fun and exciting way, while at the same time creating aesthetic place-making through materials and form.

TOP RIGHT: The church and mission in Ibadan in 1877. Drawing by the missionary Anna Hinderer.
BOTTOM RIGHT: Ibadan today, badly needing green space.

ANNA HINDERER WIKIMEDIA COMMONS

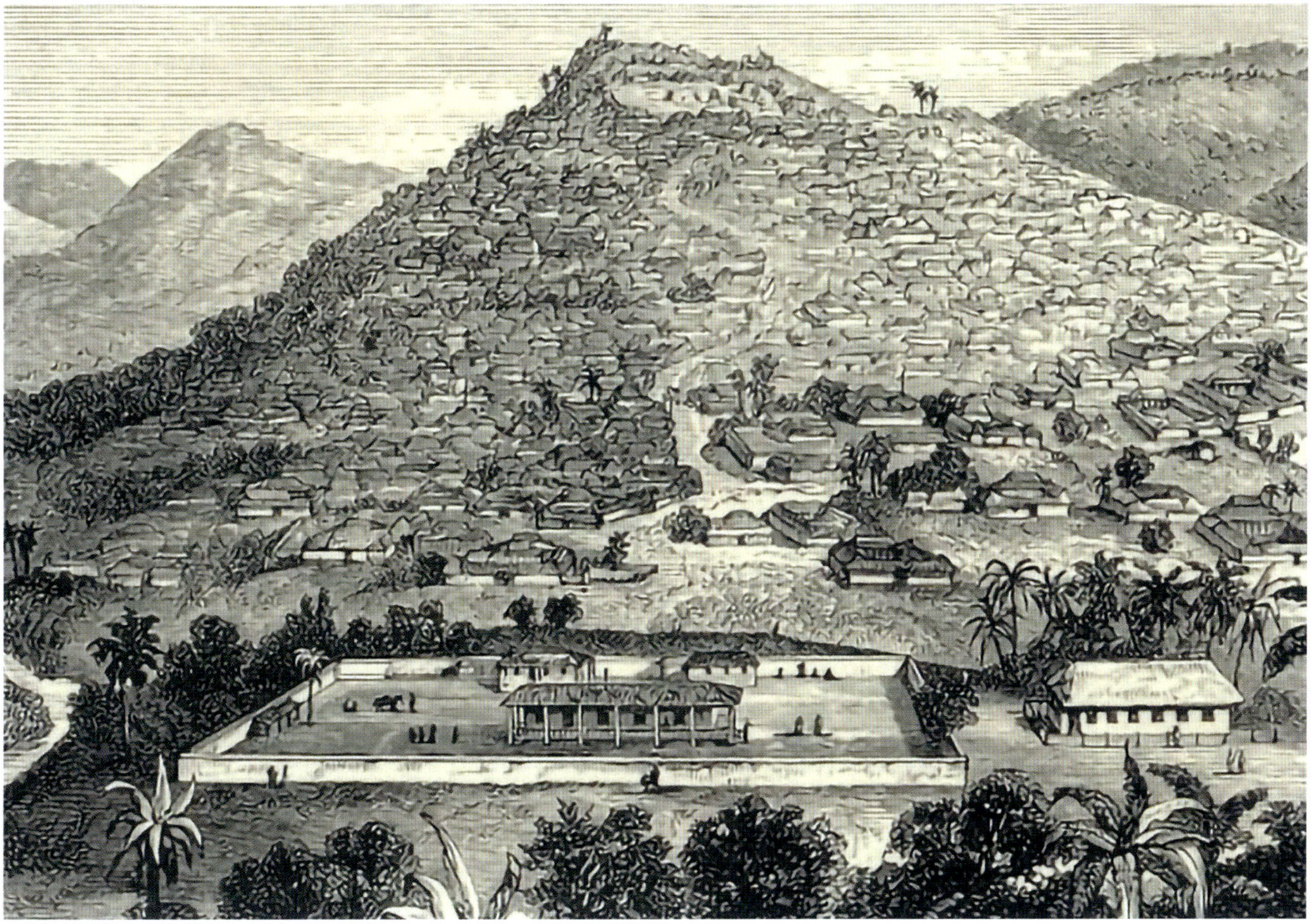

TOLUADE SHUTTERSTOCK

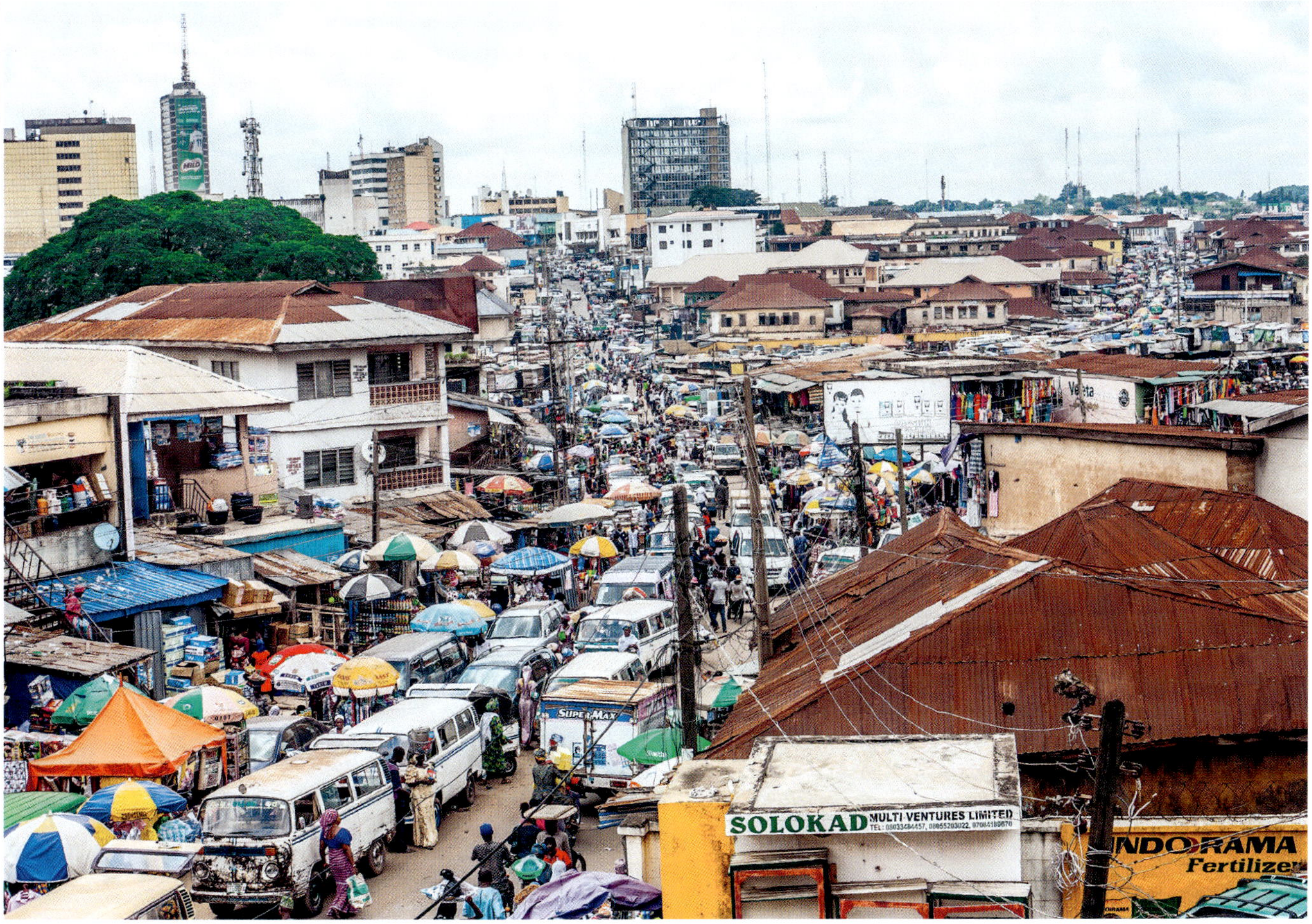

EARTHWORKS LANDSCAPE ARCHITECTS

EARTHWORKS LANDSCAPE ARCHITECTS

The first priority involved repairing the ecological functioning of the lake, which had been severely compromised by the invasive aquatic plant growth, including extensive bamboo. This vegetation choked the lake, leading to the destruction of the aquatic habitat and loss of oxygen. The invasive aquatic vegetation was removed by hand and by mechanical means, to restore the body of open water.

Bamboo created a visual and physical barrier to the usability of the site, whilst water hyacinth clogged the water channels and increased the impact of flooding. The cleared bamboo was then used as construction material. In time the lake ecology will recover and its recreational potential will be realized. The lake was integrated into the various activity areas, and the scenic views were maximized in order to maintain the natural park setting as an escape from the surrounding city.

The park is in a floodplain, so flash flooding and its associated ecosystem services were important considerations in the design. Original water channels were cleared and additional water channels were added to assist in drainage of the wetland. New infrastructure was placed higher up along the forest edge.

The bamboo marsh was developed into large grassed savannah areas, providing for active and passive recreation and event spaces, whilst also allowing for floods to occur without causing damage to infrastructure. Natural materials such as timber and stone were used to blend the structural developments sensitively into the park. The park is hopefully inspiring local designers and builders to be more responsive to nature.

The gateway and access to the park were rebuilt, providing better visibility to the existence of the park from the road edge, and improving the sense of a threshold to the park area.

All construction and activity areas were positioned within previously disturbed areas, or along existing roads. These became the new major pathways within the park, to avoid doing more damage to the fragile ecosystem.

Project developments were planned in phases; phase 1 is to date completed. This phase included clearing out alien vegetation, providing bridges and pathways, construction of the restaurant, construction of a water-play area with water slides, and provision of areas for people to picnic. Future plans include rebuilding the zoo to improve conditions for the animals, in accordance with their natural habitats, and adding a hall, chalets, a golf course, and a canopy tree walk. The vision for the zoo

Meandering streams and open lawn areas contrast with dense indigenous forests.

TOP LEFT: Clearing bamboo.
BOTTOM LEFT: Natural local materials are used like wood and stone.

EARTHWORKS LANDSCAPE ARCHITECT

EARTHWORKS LANDSCAPE ARCHITECT

Fruit bat.

is to move away from the old, outdated model, where animals sit in cages to be viewed by people. The contemporary zoo design accommodates animals in areas resembling their natural habitats.

The redevelopment of Agodi Gardens started in May 2013, and the first phase was completed in August 2014. The park was inaugurated in December 2014 by the Oyo State Governor. This development birthed a new life for the people of Ibadan and today, there are more than 4500 visitors to the park per day on weekends. The park is fenced in and there is an entrance fee. This is necessary to prevent encroachments.

The natural environment of the park has been rehabilitated, while at the same time an important economic and cultural asset for the city was created. In creating a space where the importance and value of public open space, in a natural environment, could be experienced, the park created the potential for a new appreciation of the value of public space.

[1] Climate regions.

[2] Transition area between two biomes.

TOP LEFT: Flash flooding, considered in the plan, happened during construction. The park sustained almost no damage as the floods were contained within the floodplain and savannah areas.
BOTTOM LEFT: The Lake area was cleared of hyacinth to allow fishing stocks to recover, which is now a resource used by local fishermen.

Ecosystem Services

Is nature friend or foe? The Anthropocene has dawned upon us. After centuries of trying to master nature, man has affected the biosphere to the point of losing control. Living in increasingly densely packed cities, but with a larger ecological footprint than ever. Paradoxically, those cities are now, like Noah's Ark, harboring threatened species. Having chased nature away, perhaps, by inviting nature back in, we can learn how to live more harmoniously with nature.

ANNA SOFIE PERSSON

Why Large Parks Matter for Biodiversity and Ecosystem Services

Large urban parks and green spaces with natural or semi-natural vegetation constitute important and often unique habitats for biodiversity. In addition, they provide city dwellers with opportunities to experience and interact with nature. Smaller fragments of urban green spaces, whether private or public, cannot provide the same benefits and can therefore not replace large parks. As cities grow, by densification through infill development and by sprawling peri-urban growth, we must take care to preserve and manage large parks for the future benefit of biodiversity and human experiences of nature.

Some of the species still existing in urban areas may be subject to a so-called "extinction debt."

CITIES CAN BE BIODIVERSITY HOTSPOTS

Several recent research studies show that urban areas, and the green spaces within and surrounding them, can be home to many threatened and declining species. For example, in Australia, around 30 percent of red-listed species occur in the most urbanized regions of the continent (Ives *et al.* 2016). In fact, the urban areas of Australia contain more threatened species per unit-area than the other parts of the continent do.

Urban areas seem to be particularly beneficial for important pollinating insects, such as wild bees, exemplified by a study from the UK that found that urban areas and nature reserves contained similar numbers of species (Baldock *et al.* 2015). This may seem quite contradictory; urbanization is after all one of the main global driving forces of biodiversity loss, together with the intensification of farming (Foley *et al* 2005, CBD 2012). However, some parts of urban areas, for example residential backyards, allotment gardens, and green spaces with native vegetation, can provide substantially better habitats than most of the surrounding agricultural and forestry areas, where large scale losses of natural habitat have occurred during the past century due to intensive mono-culture crop and timber production. Urban areas can also provide novel habitats for species. For example, ex-industrial brown-fields can provide resources for species with particular habitat requirements (Aronson *et al.* 2017), such as rare plants and insects that benefit from nutrient poor, dry and sun-exposed soils.

Historically, many cities were founded in regions with high biodiversity. Being rich in natural resources, these locations were beneficial both

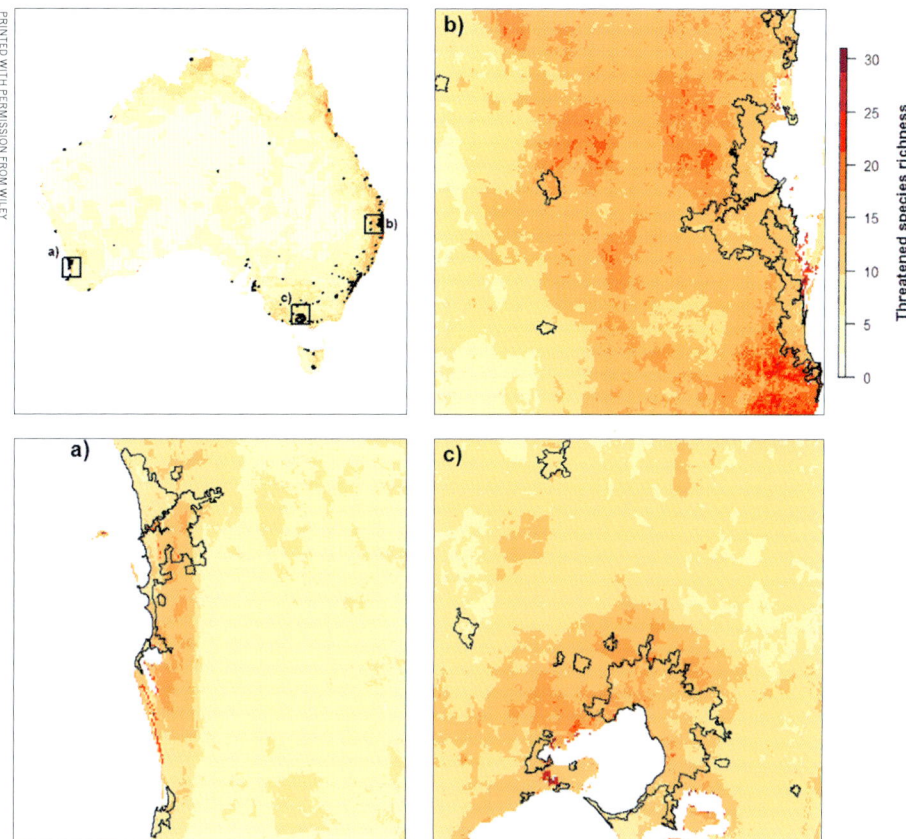

Figure 1. Richness of threatened species across Australia, with darker colors/shades representing greater richness. Urban areas are outlined in black. Cities shown in greater detail in boxes are (a) Perth, (b) Brisbane, and (c) Melbourne. Threatened species are over-represented in urban areas. From Ives *et al.* (2016).

for biodiversity and for human settlements. Today, a large part of the global human population lives in biodiversity hotspots and population growth is particularly high in these regions (CBD 2012). This means that the urban regions of the world can harbor high biodiversity, for example in remnant natural habitats, and can simultaneously exert a threat to biodiversity through further destruction of habitats, pollution and other disturbances resulting from human activities.

Some of the species still existing in urban areas may be subject to a so-called 'extinction debt'. This means that they no longer reproduce at the rates necessary for the populations to survive in the long term, and that their habitats are so isolated that no new individuals can reach and re-colonize them once the existing population is gone. For long-lived species, such as trees, it may take a long time before the loss of new recruitment becomes evident. By that time, there is likely only a scant

Nature invited into the middle of the city.
Wells Cathedral, United Kingdom.

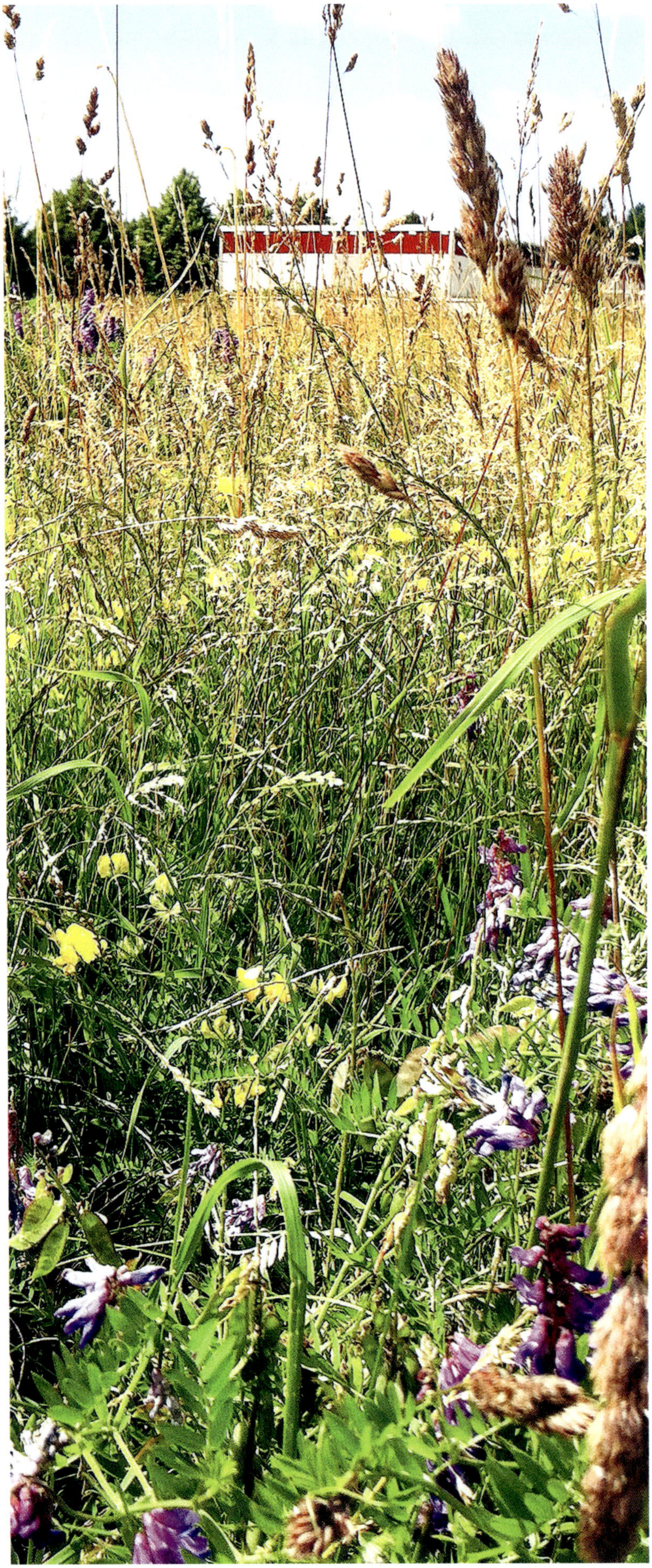

ANNA SOFIE PERSSON (ALL PICTURES ON THIS PAGE)

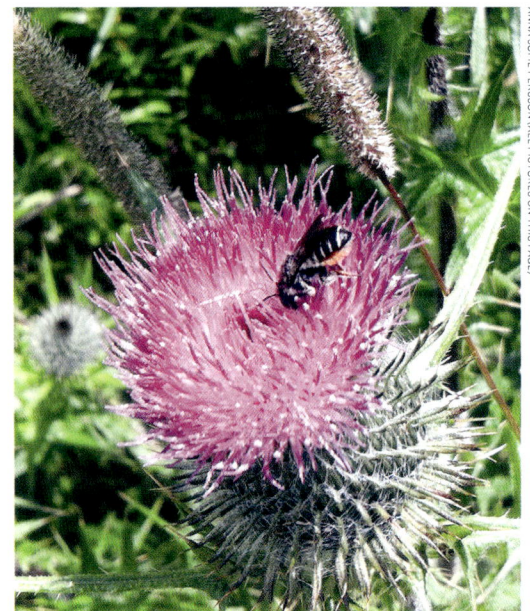

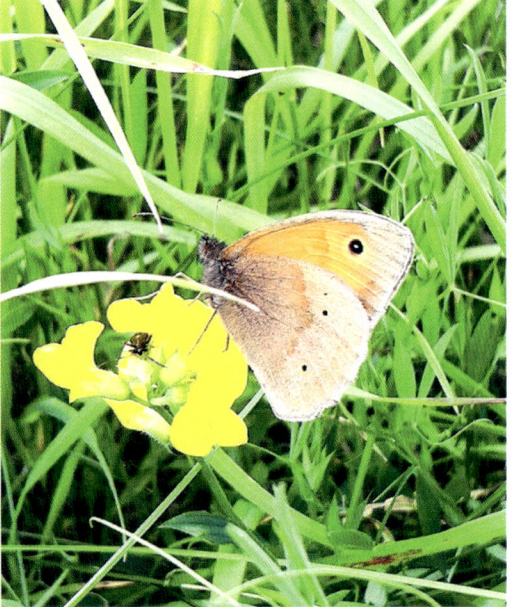

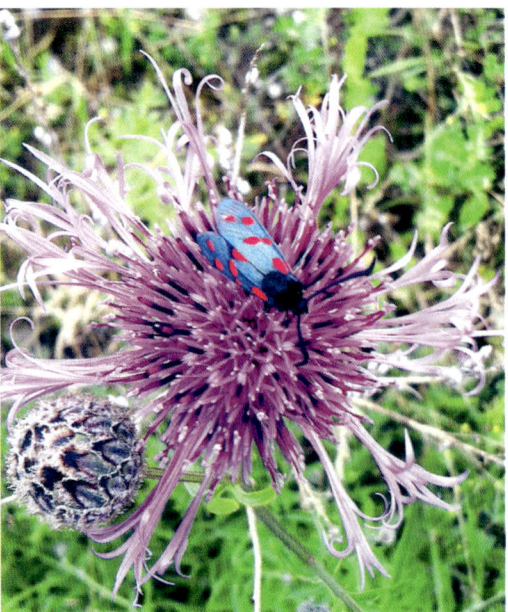

possibility to turn the tide and increase the amount and quality of habitats.

Species that have their home in cities and towns face threats by urbanization through two different pathways (Lin and Fuller 2013). Firstly, as cities grow at the outskirts, relatively pristine areas are converted into urban fabric. This fundamentally reduces the quality of these areas for the vast majority of species and introduces new forms of disturbances, for example from humans, free-roaming pets, traffic, and light pollution. In some instances, peri-urban regions surrounding cities have been reserved for future urban development, but remained unexploited, making them especially important for biodiversity (Hedblom and Söderström 2008). Secondly, urbanization today often occurs through densification, for example through infill development on large properties, or using abandoned ex-industrial land and urban green spaces. This leads to loss of green spaces *per se*, particularly of valuable so-called ruderal or brownfield sites, but it also risks leading to a reduced value of the remaining green spaces for biodiversity. This can happen when native vegetation is lost, when more intensive green space maintenance is introduced, and because the few remaining green spaces are more frequently used by city dwellers, resulting in more wear and tear (Haaland and van den Bosch 2015). Some popular parks are today partly covered in Astro-turf and other types of plastic ground surfaces, to avoid their being turned into dirt or a pool of mud from use by numerous visitors.

LAND SPARING OR SHARING?

Urban green spaces come in many different shapes and sizes. The size of an urban park or green space will affect its value for biodiversity, but also for human recreation and other ecosystem services. In addition, the distribution of green spaces over the city-scape, that is, the layout of a city's 'green infrastructure', will influence both its value for biodiversity and the accessibility for the urban human population. One way to conceptualize this is through 'land sparing—land sharing'—a sparing scenario allows for large areas of urban nature and parks to be kept intact, and housing and other development is of high density, while a sharing scenario is when green space is intermixed with low-density housing and development.

Studies that have looked into urban biodiversity report very different responses of species, or groups of species, to these two scenarios of urban

LEFT: Urban grasslands in Malmö, Sweden (left) contain wild bees such as leafcutter bees (*Megachile* sp., upper right); butterflies such as the Meadow brown (*Maniola jurtina*, middle right); and burnet moths such as the six-spot burnet (*Zygaena filipendulae*, lower right).

development. The diversity of birds benefits from a sparing development, because even species that are sensitive to human disturbance and that need larger territories can survive in large cities if large parks and nature areas are retained in and around the city (Sushinsky *et al.* 2013, Geschke *et al.* 2018). Such species are called urban avoiders, and their existence in an urban park signifies low levels of urban impact. Such species can, for example, be woodpeckers or passerines dependent on dead wood or patches of old-growth forest, or in Australia, species such as the laughing kookaburra or the superb fairy-wren. As parks are chipped away and only smaller pieces are retained, urban avoiders disappear while already common species move in. Such species are sometimes called urban adapters because they can tolerate disturbance from human activities, while species that clearly thrive in human-made habitats are known as urban exploiters. These groups can be exemplified by species such as the house sparrow and the ubiquitous rock dove. It is important to note that other species groups, for example insects, may respond in different ways than birds do to green space distribution (Soga *et al.* 2014). Therefore, neither sparing nor sharing is the optimal solution to benefit all parts of biodiversity, and a backbone of large parks and nature areas needs to be complemented by smaller green spaces in and among residential and commercial development.

A backbone of large parks and nature areas needs to be complemented by smaller green spaces in and among residential and commercial development.

URBAN GREEN SPACE MANAGEMENT FOR BIODIVERSITY

Urban parks and green spaces not only come in various shapes and sizes, they also vary greatly in content, meaning the type of vegetation and intensity of maintenance of that vegetation. Several studies have shown that retaining native vegetation and lowering the frequency of management—such as the cutting of herbaceous vegetation—benefit biodiversity. This is, for example, true for butterflies, where parks dominated by exotic plants and short cut lawns will not provide host plants for the larvae to develop (Chong *et al.* 2014, Aguilera *et al.* 2018). Even if adult butterflies can feed on nectar from a multitude of both native and ornamental plants, the larvae are often much choosier, because their metabolism may be adapted to feeding on the leaves or stems of just a few plant species or genera, and rarely exotic ones. The lack of preferred host plants for insect larvae can, in turn, reduce populations of insect feeding birds in urban areas (Narango et al. 2018). Insectivorous birds are heavily dependent on large numbers of insects to feed their chicks during nesting and cannot replace this food source with something else without negative effects on chick survival and, ultimately, population persistence.

Today, lawns cover vast areas of public and residential urban green spaces. In Sweden and the USA, lawn comprises about 23 percent of city and town surface (Ignatieva and Hedblom 2018).

Not only does this type of land use provide very scant value to biodiversity and to the experience of nature for park visitors, it is also very costly to manage, in both monetary terms and in energy for machinery, fertilizers, etc. (Ignatieva and Hedblom 2018). Re-thinking urban vegetation, such as the domination of lawns, may be a way to both benefit biodiversity and human experience of nature, and save money for city authorities and property managers. In temperate regions, lawns can be replaced with meadow-like vegetation of mainly native species, in combination with paths of cut grass to allow people to move freely. Another approach would be to retain and upgrade existing, spontaneously established vegetation on vacant lots and brownfields when former industrial sites are developed. Such sites often contain high biodiversity of plants and insects, but this is lost when they are converted and designed into generic parks and backyards. Just as for other 'modern' inventions, it will require a combination of wise design, solving the practicalities of management, and dissemination of information about the benefits, to increase acceptance among citizens and avoid conflicts around the expectations of park and green space appearance (Ignatieva *et al.* 2017).

PARKS AND NATURE ARE CRUCIAL FOR OUR WELLBEING

In addition to harboring a diversity of wild species, urban green spaces provide a number of other benefits (ecosystem services) to urban human populations. The most obvious may be to create good environments for physical and mental recreation for urban populations. Evidence for the therapeutic properties of nature are overwhelming (WHO 2016). As a response, guidelines for adequate availability of nature close to dwellings and workplaces are being introduced in city planning, for example in Sweden where a maximum distance of 300 meters to a park is advocated (Boverket 2007). Another crucial benefit from urban parks and green spaces is to provide opportunities to experience nature and learn about species and ecosystems (Sushinsky *et al.* 2017). As an example, research shows that young people do not have the same ability as older generations to detect species loss, reduced water quality, and climate change, because the nature they experience and use as their "base-line" is already impoverished (Soga and Gaston 2018). Today around half the global population lives in cities, and as the world becomes increasingly urbanized, humans become more and more alienated from nature and

Young people do not have the same ability as older generations to detect species loss, reduced water quality and climate change, because the nature they experience and use as their "base-line" is already impoverished.

risk losing the immediate understanding and appreciation of nature. This can lead us into a downward spiral, characterized by a reduced willingness to visit parks and nature areas, which is detrimental for our health, and to lower support for nature conservation, further reducing the availability of nature and green spaces.

While the main health risks in cities historically have been poor sanitation, air pollution, and low water quality, today they are obesity and stress. Researchers report that currently only 13 percent of urban dwellers in 245 cities worldwide live in neighborhoods that have more than 20 percent tree canopy cover, which is the threshold found necessary to protect against depression, stress, and anxiety (McDonald *et al.* 2018). Mental wellbeing for the global population therefore partly depends on retaining and managing urban nature and, importantly, making it accessible to all people, not only those who have the money and time to get to it (Shanahan *et al.* 2015), or can afford to live in expensive garden city parts of the urban fabric.

CLIMATE ADAPTATION AND MITIGATION

Urban areas are heavily transformed compared to nature and to most rural surroundings and often constitute quite extreme habitats. This means that the effects of climate change are in many cases exacerbated in urban areas. The urban heat island effect, caused by the extensive cover of impervious surfaces that trap heat in combination with low vegetation cover that reduces the potential for cooling, leads to higher temperatures in urban areas compared to rural surroundings. In combination with increased mean temperatures and more frequent extreme weather events like heat waves, this has already made some cities inhospitable and will in the near future affect regions where we have, until now, neither experienced nor contemplated this. The vast space allocated to sealed ground surfaces, where water cannot infiltrate, has made many cities prone to flooding. These risks will increase further with more frequent extreme rainfall events caused by climate change. There are technical solutions to these problems, such as improving air conditioning and construction of piped sewer systems for a quicker removal of excess water. However, these systems are not only costly, they also require energy for construction and maintenance and if, or when, they fail, the consequences may be catastrophic (The Royal Society 2014).

An alternative method for climate change adaptation is to harness the

RIGHT: Trees in cities are crucial for a good micro-climate. These large trees in New Farm Park in Brisbane, Australia, welcome visitors to escape the strong sun.

ecosystem services produced by living systems, either on their own, or in combination with 'grey' technical systems (EEA 2015). Each such 'nature-based solution' can contribute to meeting the multiple challenges posed by the urban environment and climate change, while grey solutions often only meet a specific challenge. Urban parks and other green spaces can be considered nature-based solutions. For example, urban vegetation such as grassy areas, shrubs, and urban forests contribute to both temperature control and storm water management through the shade from trees, evapotranspiration of plants, and water infiltration of the vegetated and porous ground surface (Gómez-Baggethun and Barton 2013).

A global research study has shown that investing in restoration and maintenance of large green spaces in and around cities always pays off economically, even when calculations are based on the limited number of ecosystem services that allow monetary valuation (Elmqvist *et al.* 2015).

In addition, urban green spaces contribute to climate change mitigation, as they keep carbon locked in both above- and below-ground living and dead matter, something that grey solutions cannot do (The Royal Society 2014). Using green spaces to adapt to climate change also means providing more space for biodiversity and recreation. However, careful planning, design and management are crucial to avoid trade-offs leading to suboptimal functioning of the desired ecosystem services. For example, potential conflicts of interest can appear between creating habitats for biodiversity, areas for recreation, and storm-water management within the same park. This can be solved by good planning and design but will require that planners and other officials are aware of the potential problems. Unfortunately, it seems that many city authorities have not yet seized the opportunity to combine biodiversity conservation with actions for climate adaptation and mitigation (Butt *et al.* 2018). This is unfortunate, since it is in many cases the very same green spaces that must deliver all of these services, and because cities do not have enough space to provide separate solutions for biodiversity and climate-related actions. There is a great risk that the seemingly more immediate and costly threats from climate change will trump the risks posed by species loss, leading to suboptimal solutions for urban biodiversity conservation. A good example of a successful combination of climate adaptation and biodiversity conservation is the Kyrkparken park in Järfälla, Sweden. A group of landscape architects and ecologists designed a pond and wetland system with the aims of: creating a park

Unfortunately, it seems that many city authorities have not yet seized the opportunity to combine biodiversity conservation with actions for climate adaptation and mitigation.

LEFT: A pond and wetland system designed to both take care of storm-water and benefit biodiversity in Järfälla, Sweden.

for recreation; managing increased levels of storm-water from a newly built neighbourhood; and simultaneously providing habitats for biodiversity. (See the chapter on Qunli Storm Water Park for more examples of smart planning.)

ECOSYSTEM SERVICES AND PARK SIZE

The 'sparing-sharing' concept introduced above is not merely a tool for understanding how urban planning effects biodiversity. Research has shown that ecosystem services respond differently to the distribution of the urban 'green infrastructure'. For example, a city with larger green spaces will provide better opportunities for local food production (which could be important in megacities in the developing world), water infiltration, and summer temperature regulation. In contrast, multiple small green spaces will benefit air purification and the immediate experience of seeing something "green" from your window (Stott *et al.* 2015). In other words, having few but large green spaces provides benefits that can be transported to people (for example when air or water moves) or that can be accessed when people travel to visit green spaces. Many scattered, small green spaces instead provide benefits that are both produced and used where people live and work, such as shade or the view of a tree. Sparing and sharing thus largely provide complementary benefits. It is therefore of great importance to both keep the existing large parks and green spaces and plan for new ones in order to create a web of greenery throughout cities. Protecting existing green spaces is crucial, because once lost, it is often lost forever. Smaller pockets of vegetation are easier to recreate in already developed neighborhoods, for example in the form of pocket parks, vegetation in backyards, street trees, and technical solutions such as green roofs and other vegetation on joists. In an era of urbanization, a strong backbone of large parks is, nevertheless, needed to provide the benefits that cannot be provided by smaller green spaces.

CONCLUSIONS

Large parks should not be chipped away as cities grow inwards through densification. The values they provide, both for humans and as habitats for many other species, cannot be replaced by smaller pockets of green space. In light of this, as cities expand into surrounding landscapes, it is therefore crucial to plan for new large parks and to retain nature areas in the urban periphery. If not, then in the future human populations will be largely without access to the many benefits that nature provides, including the opportunity to experience a diversity of species in the

NATALIA GOLOVINA SHUTTERSTOCK

Laughing, blue-winged kookaburra (*Dacelo novaeguineae*).

wild. It is high time to re-think the ideal park appearance, from short-cut lawns and evenly aged, manicured trees into something else—something wilder—in which ecosystems are allowed to exist, and where we can experience more of nature in our everyday lives.

BRISBANE

KARAWATHA
FOREST RESERVE

Nemies Road

KURABY

Compton Road

STRETTON

KURABY
BUSHLAND
RESERVE

Fauna passage

Compton Road

CALAMVALE

Acacia Road

Karawatha Forest
Discovery Centre

WOODRIDGE

KARAWATHA
FOREST RESERVE

Gowan Road

Gateway Motorway

Beaudesert Road

Trinder Park
station

LOGAN
CENTRAL

Illaweena Street

Logan
Gardens

Karawatha
Forest South

Wembley Road

Logan Motorway

DREWVALE

BERRINBA

0   0,5   1 Kilometer
0   0,5 Mile

N

© OpenStreetMap contributors

KARAWATHA FOREST, BRISBANE, AUSTRALIA

SIZE: 1.2 square kilometres.

LEGAL PROTECTION: Conservation reserve in 1975 and National Estate
in 1997.

OWNED AND MANAGED: By Brisbane City Council.

VOLUNTEER ORGANIZATION: Karawatha Forest Protection Society.

HOW TO GET THERE: Commuter train to Trinder Park, walk down
Elisabeth street where you enter the reserve. It is 1.5 kilometers
to the Discovery Centre. By car and bike you can enter via Acacia
Road. Kuraby officially has no access, although pedestrian access
is now available across the overpass.

MEL MCGREGOR

Karawatha Forest, Brisbane, Australia

In an increasingly busy, modern, and urbanised world, the importance of preserving natural landscapes within cities is becoming far more widely recognized. The benefits of green space are becoming more accepted, as are the broadening presence of natural features within the built environment: trees lining busy streets, carefully tended gardens, and tree-clad, open spaces are often cherished by cities lucky enough to possess them. In Australia, as well as globally, it is becoming increasingly clear that people are not the only beings to benefit from natural landscapes within urban environments; diverse wildlife can be supported, preserved, and protected with the introduction or retention of green space. Parks, golf courses, and recreation areas have all been shown to enhance wildlife diversity within busy cities, but the highest value green space is most often provided with the retention of remnant (naturally occurring) habitat such as forest reserves. The expression of the concept of a National Park, State Park, or forest reserve (recreational park) varies greatly on a global scale, as is evident within the pages of this book. However, in Australia, these terms quite often stimulate images of re-tained natural landscapes, often with modifications which allow people to take full advantage of the natural beauty of the area. These modifica-tions, typically in the form of cycle paths, hiking trails, and picnic areas, provide a valuable capacity to escape to the 'bush' without leaving the city—a genuinely important experience.

Globally, Australia is considered a land of vast natural landscapes and abundant wildlife (whether terrifying or cuddly). We consider ourselves the 'lucky country', possessing plenty of diverse landscapes, juxtaposed with bustling, world-class metropolises, often within close proximity to one another. Increasingly, the value of retaining forest in particular, but also naturally occurring grassland, marshlands, and other habitats, has been recognized by governments, and is becoming increasingly valuable to ever-expanding urban areas, particularly for wildlife. Often, parks may have restricted access or purpose retained areas to allow the persis-tence of wildlife, but what is less common is the existence of modified, implemented structures which not only support wildlife persistence, but actively encourage and assist wildlife presence within a landscape which may otherwise prevent it. Karawatha Forest Reserve and the adjacent Kuraby Bushland Reserve, uniquely exemplify the dedication, care, and ingenuity of Australians, in this case both citizens and scientists, to support a diverse natural landscape within expanding cityscapes.

Karawatha Forest and the Kuraby Bushland exist as forest reserves either side of Compton Road, a major urban arterial road (70 kilometer per hour speed limit, approximately 10,000 cars per day) on the border between Brisbane, the State of Queensland's largest city (2.2 million residents), and Logan, the neighbouring regional city to the south (home to just over 300,000 residents). Combined, the forest reserves comprise approximately 1200 hectares of remnant woodland forest and recreational parkland, within close proximity to surrounding suburbs. Historically, Karawatha Forest has existed as a Brisbane City Council conservation reserve since 1975, gaining Federal heritage status in 1997. Today, two organisations are primarily responsible for Karawatha and Kuraby: Brisbane City Council and the Karawatha Forest Protection Society, a community-based organisation which provides advice based on intimate local knowledge, particularly concerning conservation matters. Karawatha is a typical example of an urban conservation reserve in Australia; recreational use is restricted to small groups, with no access for vehicles, while Kuraby is strictly no access for people, with conservation being the primary concern. Recreational activities such as cycling, hiking, and picnics are encouraged and celebrated within Karawatha, with the reserve hosting its own "family fun day" as an annual occasion. Karawatha is now also home to a Council-sponsored Discovery Centre, aimed at encouraging visitors to learn about and interact with the forest, and its role in conservation within and around Brisbane.

Karawatha and Kuraby are considered regionally significant ecosystems within the Greater Brisbane Area, as they are home to over 200 fauna and over 300 flora species. The reserves also provide important ecological linkages to other regional forests, forming the northern-most point of the Flinders-Karawatha Corridor, the largest continual bushland corridor in south-east Queensland. The corridor is considered critical to the surrounding landscape, as it provides an extensive passageway between many reserves and green spaces throughout the local region. The reserves support the highest frog diversity in Brisbane, as well as populations of wallabies, kangaroos, koalas, gliders, possums, bats, and

TOP RIGHT: Overpass connecting Karawatha forest and Kuraby bushland reserves.
MIDDLE LEFT: Overpass with glider poles.
BOTTOM LEFT: Lace monitor using underpass as seen on CCTV.
BOTTOM RIGHT: Karawatha forest, eucalyptus bushland.

MEL MCGREGOR

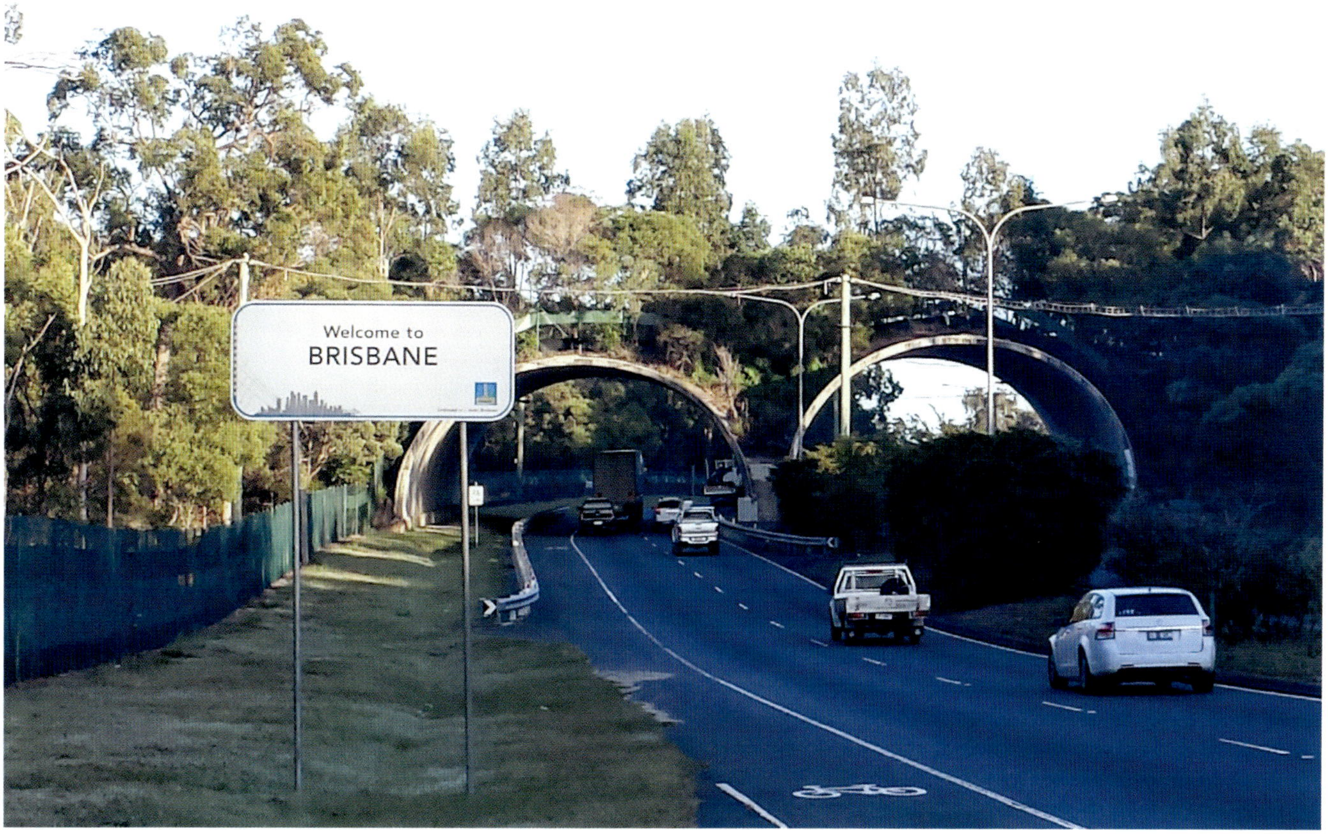

DARRYL JONES

RAYMONDE DE LATHOUDER

MEL MCGREGOR

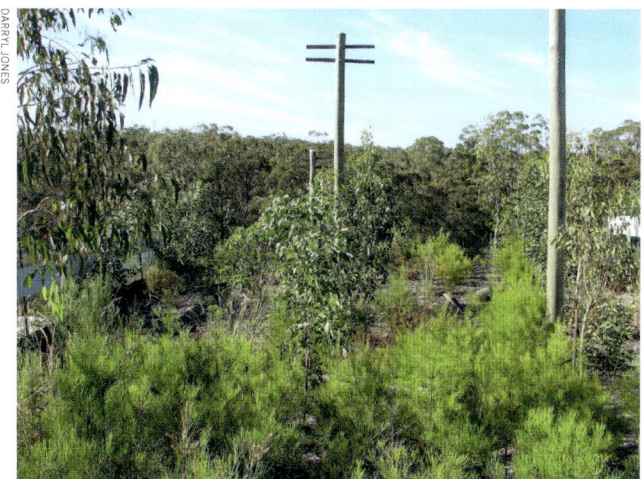

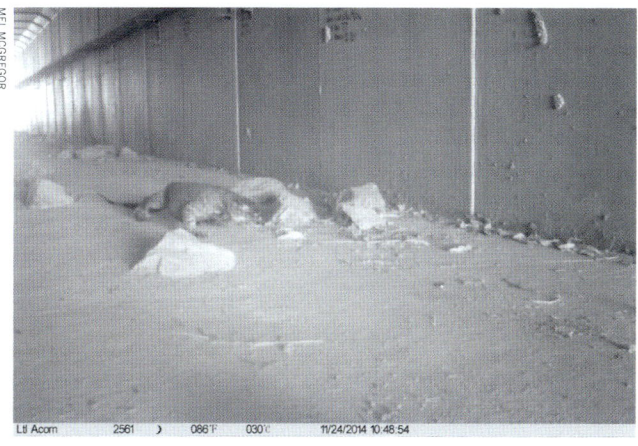

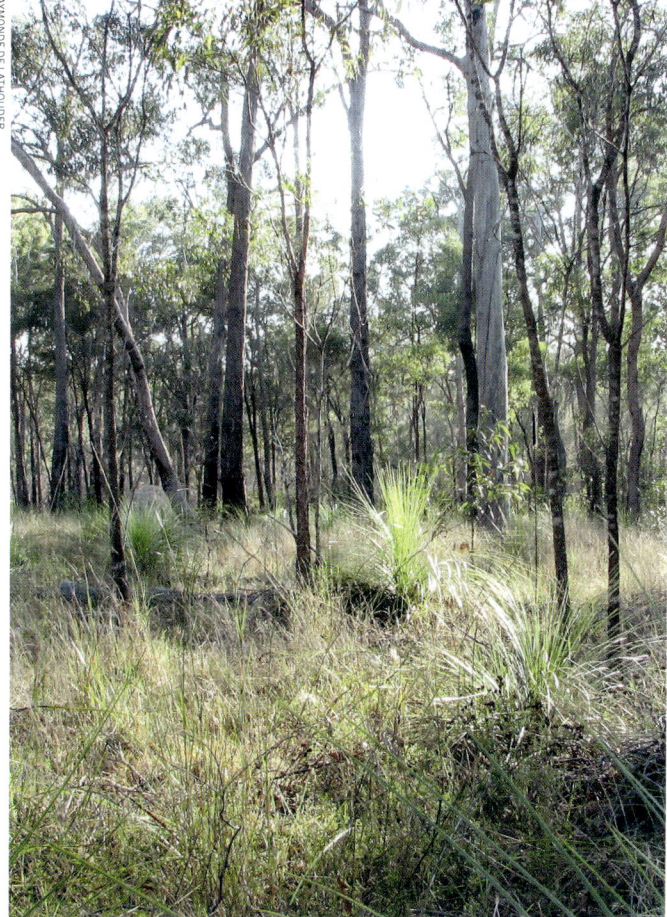

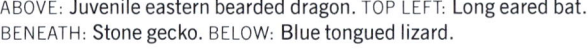

STEVE WILSON

MEL MCGREGOR

ABOVE: Juvenile eastern bearded dragon. TOP LEFT: Long eared bat.
BENEATH: Stone gecko. BELOW: Blue tongued lizard.

MEL MCGREGOR

MEL MCGREGOR

an astounding array of birds and reptiles. The urban location of the reserves, however, is not without consequence: invasive predators, fire, and damage from trespassers with vehicles are among the reserves' biggest threats.

While Karawatha and Kuraby are immediately adjacent, they were separated by busy Compton Road. The safeguarding of wildlife and habitat retained within the reserves became a matter of major concern when the single lane, dual carriageway road was to be duplicated in 2005, further isolating the already segregated forest reserves. As the duplication plans progressed, local naturalists and scientists fought arduously to include multiple wildlife crossing structures, now collectively known as the Compton Road Fauna Array. The ambitious plan outlined the construction of four different types of wildlife crossing structures, establishing what was at the time the most complex array of wildlife crossing structures in one place anywhere in the world. The array included two underpasses, rope bridges, glider poles and, most spectacularly, a vegetated overpass, which - theoretically - would allow a diverse range of wildlife to safely cross Compton Road. In addition to these passages, wildlife fencing was erected to prevent wildlife from accessing the roadway.[1] The array aimed to enhance wildlife movement, increasing the likelihood of population persistence between the reserves, while the fauna fencing would reduce the high number of wildlife deaths resulting from collisions with cars. It was an ambitious project, and the first of its kind in Australia, but would it work?

Since completion of the Compton Road Fauna Array in 2005, multiple researchers have investigated how well the passages are performing, and which species are benefiting. Many of Australia's mammals are recognised globally, with the kangaroo and koala leading the unequivocal status of fauna emblems. Naturally, mammals were the first research targets, primarily kangaroos, koalas, and wallabies, with arboreal (climbing and gliding) mammals instigated shortly after. A number of research publications were produced over the next decade, detailing the success of the glider poles in particular, but also the overpass, underpasses, and rope bridges for providing a safe passage for mammals across Compton Road.

The incredible diversity of wildlife contained within Karawatha produced equally diverse research; not too long after the initial success was realised, researchers began targeting more cryptic species, including native rats, marsupial mice, snakes, and the remarkable array of lizards existing within Karawatha. Birds would also become a hot topic on the Compton Road Fauna Array, and microbats would eventually follow.

The most complex array of wildlife crossing structures in one place anywhere in the world.

The Compton Road overpass became a key aspect of the success of the array, with possums and bandicoots shown to frequent the structure, while echidnas often utilised the underpasses. Large reptiles were also seen using both the overpass and underpass, most prominently the unmistakable lace monitors, which were consistent visitors to the array. As research progressed, it was realised that the key to the success of the overpass was likely the native vegetation, which had been carefully planted across the overpass and had seemingly become not just habitat, but an extension of the natural forest landscape. Lizards, snakes, and frogs had come to colonise and live atop the overpass, while birds and bats had been recorded utilising the structure as foraging and shelter habitat in their hundreds. The crossings had worked better than expected. The overpass had provided new natural habitat and permanently connected Karawatha Forest with Kuraby Bushland (Bond and Jones 2008, Taylor and Goldingay 2012, McGregor *et al.* 2015, Pell and Jones 2015, McGregor et al. 2017).

The current loss of natural ecosystems within Australia, and the species which rely on them, is quite alarming. Regarded by many as a source of national shame, Australia has one of the highest rates of mammal extinction worldwide. However, this is not a prominently known feature of what is often regarded as one of the most 'wild' countries on Earth. Urbanization and population growth within south east Queensland is higher than ever before, and that area has often had the most rapid growth in Australia. Habitat and diversity loss, which inevitably occurs with increased urbanization, means parks and reserves such as Karawatha and Kuraby are priceless and entirely irreplaceable natural assets for the surrounding region. Due to their locality within the urban landscape, continual intensive impacts from human activity can, and do, seriously impact the longevity of such reserves; however, the closeness we enjoy with these parks also provides us with a number of unique opportunities.

The most obvious, and arguably most utilized, aspect of parks such as Karawatha is the capacity for outdoor recreation for the local community. It must not be understated, or in fact underappreciated, that being able to visit a natural area such as a park or reserve is an important benefit of living in an Australian city. It is essential to note, however, that alongside the capacity for recreational enjoyment, unique parks such as Karawatha, which include novel approaches to applied conservation, present an outstanding opportunity for educating the surrounding community. This is not restricted to providing community-accessible information on the incredible success of wildlife crossing such as the

Compton Road Fauna Array, but also simply engaging the community within the natural environment and encouraging a broad respect, understanding, and emotional connection to local parks and their wildlife. With this, the opportunity to encourage increasing conservation measures with the introduction of parks and reserves, not just for use by the public, but also to ensure the survival of urban and peri-urban wildlife, becomes far more achievable.

City parks in particular, but parks and reserves in general, are key avenues by which we can encourage the community to respect, understand, and support biodiversity, as well as enjoy outdoor recreation. With increasing urbanization and the ever-expanding sprawl of city borders, it is becoming increasingly important for us to appreciate, and indeed advertise, the importance of parks within the urban landscape. Karawatha Forest stands as a fantastic example of a conservation reserve which is enjoyed and cherished by the local community, while the addition of Kuraby Bushland and the Compton Road Fauna Array are globally unique contributions of conservation within an urban context. I suspect that the residents of Brisbane and Logan already understand the unique contribution Karawatha and Kuraby both make to the surrounding region, and I would hope that, in this increasingly busy, modern, and urbanised world, Karawatha Forest Reserve and Kuraby Bushland Reserve are and will continue to be cherished by the cities lucky enough to possess them.

[1] The cost was approximately $750,000 for the overpass and $1.4 million for all passages (Australian dollars 2005).

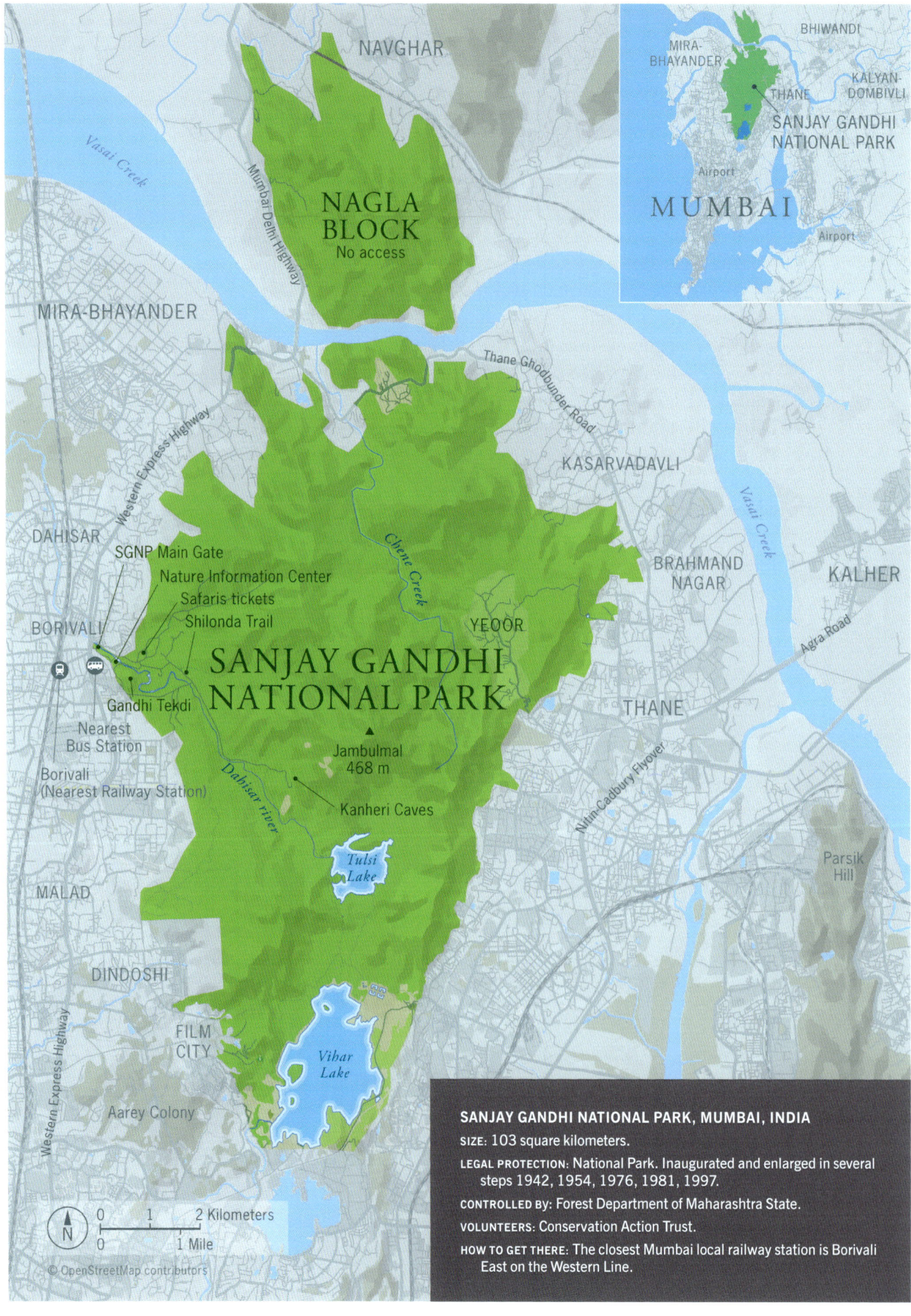

BHIWANDI

MIRA-BHAYANDER

THANE

KALYAN-DOMBIVLI

SANJAY GANDHI
NATIONAL PARK

MUMBAI

Airport

Airport

NAVGHAR

Vasai Creek

Mumbai Delhi Highway

NAGLA
BLOCK
No access

MIRA-BHAYANDER

Thane Ghodbunder Road

Western Express Highway

KASARVADAVLI

Chene Creek

Vasai Creek

DAHISAR

SGNP Main Gate
Nature Information Center
Safaris tickets
Shilonda Trail

BRAHMAND
NAGAR

KALHER

Agra Road

BORIVALI

YEOOR

SANJAY GANDHI
NATIONAL PARK

Gandhi Tekdi

Nearest
Bus Station

Jambulmal
468 m

THANE

Borivali
(Nearest Railway Station)

Dahisar river

Kanheri Caves

Nitin-Cadbury Flyover

MALAD

*Tulsi
Lake*

Parsik
Hill

DINDOSHI

FILM
CITY

*Vihar
Lake*

Western Express Highway

Aarey Colony

N

0   1   2 Kilometers
0   1 Mile

© OpenStreetMap contributors

SANJAY GANDHI NATIONAL PARK, MUMBAI, INDIA

SIZE: 103 square kilometers.

LEGAL PROTECTION: National Park. Inaugurated and enlarged in several
steps 1942, 1954, 1976, 1981, 1997.

CONTROLLED BY: Forest Department of Maharashtra State.

VOLUNTEERS: Conservation Action Trust.

HOW TO GET THERE: The closest Mumbai local railway station is Borivali
East on the Western Line.

DEBI GOENKA AND SUNJOY MONGA
Sanjay Gandhi National Park, Mumbai, India

The Sanjay Gandhi National Park lies in the midst of the metropolises of Mumbai and Thane which have over 17 million inhabitants. It lies at the ecological crossroads of two key bio-geographical zones—the Malabar coast and the Western Ghats, a mountain range that runs along the west coast of India.

This entire valley, with its mangroves and forests, was first recognized as a green oasis in the late sixties when it became a National Park. At that time, the sparsely populated suburb of Borivili lay on Mumbai's distant horizons, and Thane was a different town altogether. Today, at the turn of the 21ˢᵗ century, the park is hemmed in by bursting Mumbai suburbs to the south, east and west, and by the expanding city of Thane in the northeast. Within its boundaries, it comprises of 103 square kilometers of forest, earning its nick-name of City Forest.

Topographically, the park appears as a hill range. Indeed, that's what it seems like when viewed from afar in real space. The terrain is largely undulating, the main hills running in a north-south axis, with little side-shoots in either direction. Much of the terrain averages 100 meters and reaches its highest point at a site named Highest Point, 486 meters above sea level.

The setting is punctuated by lowland areas, the most extensive being the bowl-shaped Tulsi valley in the southern half of the park. This stretch, extending through almost a quarter of the park, includes the Tulsi and Vihar lakes. The rest of the hilly sprawl is dotted with numer-ous narrow and serpentine ravines that are the conduits for seasonal rivulets, most of which either drain into the two lakes, or merge with the Dahisar river that drains out into the western suburbs of Mumbai and eventually merges with the Arabian Sea.

A string of hill-ranges extends northward to the Tungareshwar Wildlife Sanctuary. That is the main connection with wild nature beyond the city.

The Park has always been a favourite spot of Mumbai residents for their family picnics. It is surrounded by millions of land-starved citizens and two rapidly expanding cities. Yet, its preservation has often been re-legated to the toil of small groups and sometimes even lone individuals.

The great conservation stalwart, Humayun Abdulali (1914–2001), spent and eventually gave his life to defend the park. Between 1940 and

The Park has always been a favourite spot of Mumbai residents for their family picnics. It is surrounded by millions of land-starved citizens and two rapidly expanding cities. Yet, its preservation has often been relegated to the toil of small groups and sometimes even lone individuals.

the early 1950s, while he was the Honorary Secretary of the Bombay Natural History Society, Abdulali brought attention to the wilderness north of the city. "Let it be declared a National Park to be preserved in its entirety," he advocated.

The Bombay National Park was enacted in 1950 and thereafter an area of approximately 20 square kilometers was declared the Krishnagiri National Park. This area was subsequently increased to over 68 square kilometers with the inclusion of adjoining forested tracts. It was called Borivali National Park. Years later, in 1981, the site was re-christened the Sanjay Gandhi National Park, in memory of the Prime Minister Indira Gandhi's younger son. In 1997, 87 square kilometers were finally secured for the National Park.

Without Humayun Abdulali's relentless attention, the park might never have been preserved. He single-handedly convinced the Bombay High Court to prevent the construction of a highway which threatened to slice right through the park. When he was trying to stop some illicit tree cutting in the park, he was attacked and grievously injured. Unfortunately, the injury to his head hindered his memory functions severely and he could no longer protect the park like he had before.

The fight for the park went on for many years. The boundaries of the park were indiscriminately nibbled away for years—first by quarry operators and then by slum lords, aided and abetted by local politicians and the land mafia. Every attempt to remove these encroachers by the handful of activists and the Forest Department was invariably foiled by higher authorities. It was as if the words 'National Park' meant nothing. A cancerous growth of slums encroached into the park. Forest patches were shorn of all their vegetation. Every water hole and stream was contaminated with human waste. Trees and bamboos were ruthlessly chopped to extract housing material by the encroachers. Over time, the slum lords became so bold as to enter into 'agreements of sale' on stamp paper and thousands of families were tricked into thinking that they had legally bought their patches of land from the actual owner! Forest fires became a daily affair. Bootlegging within the forest became rampant.

Before the landmark decision by the Bombay High Court on 7 May 1997, there was not one place along the entire periphery of the National

TOP RIGHT: One winter morning we embarked on this invigorating trek, now an annual or even a bi-annual "pilgrimage" for some of us. We hiked up to the summit for a breath of the freshest, purest air possible in the entire region, sipping piping hot coffee and eating sandwiches that never tasted better. Arriving at the rocky edge of the promontory to the south of this summit plateau, we felt that nowhere was sheer exhaustion so refreshing!
BOTTOM RIGHT: The Shilonda stream.

SUNJOY MONGA

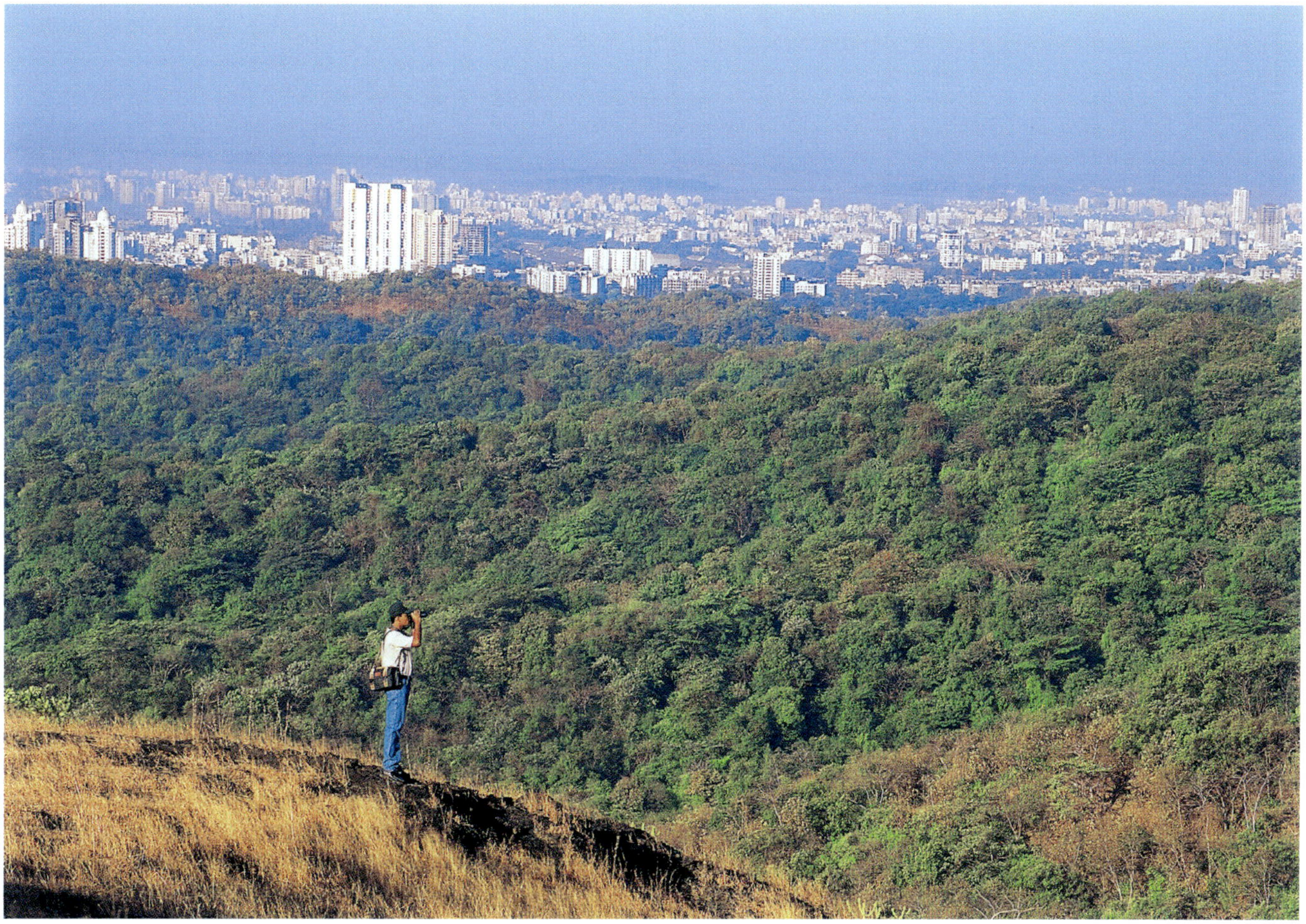

SUNJOY MONGA

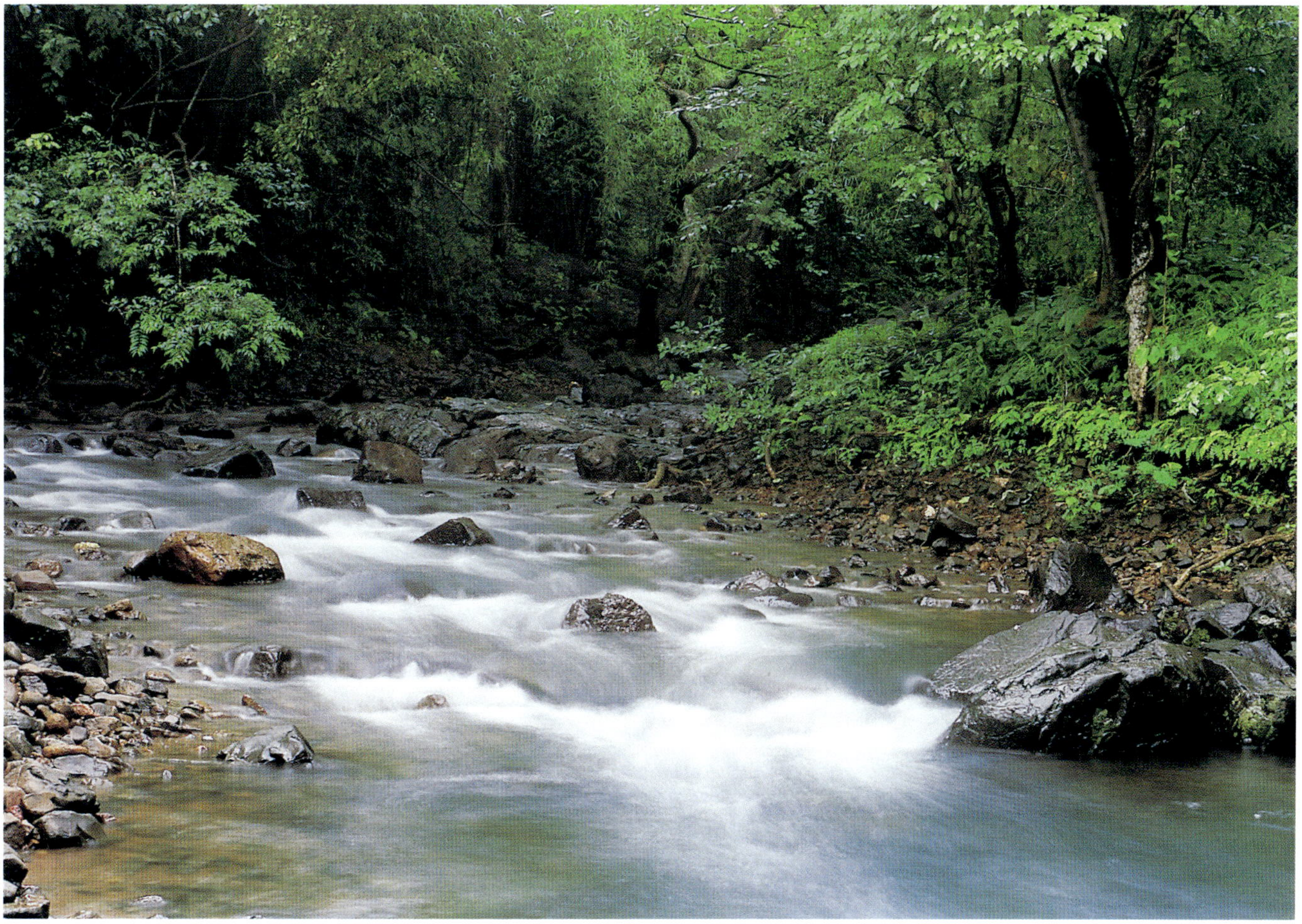

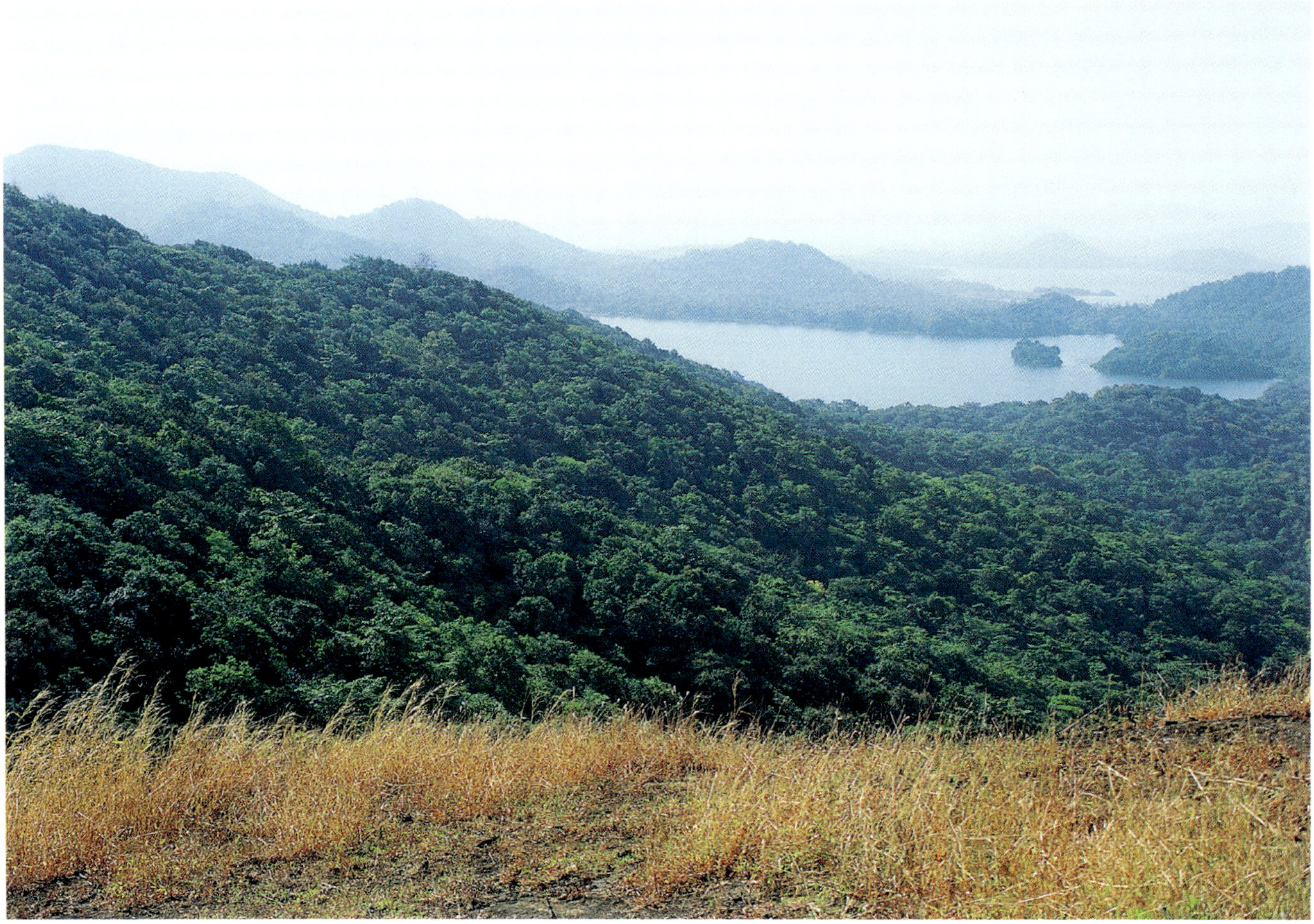
SUNJOY MONGA

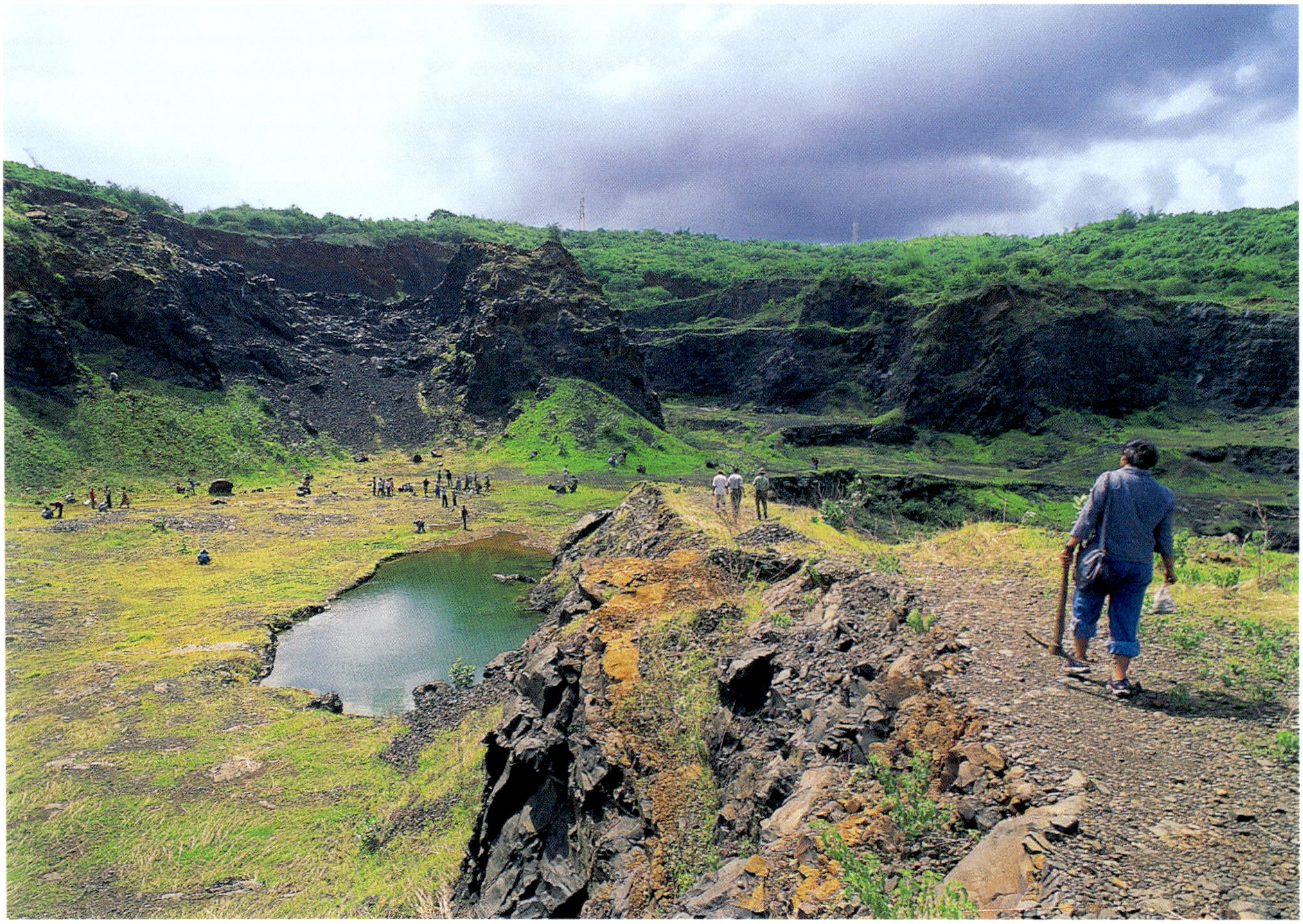
SUNJOY MONGA

Park that was free of human disturbance. The larger mammals bore the brunt before other animals and began to perish in high numbers. The last hyenas to be seen were a pair shot dead just adjacent to the southern entrance to the National Park in the mid/late-1980s. Much of the wild ungulate population had been wiped out.

The wild animals released into the Park did not survive human poaching for very long. It was at that point that the Bombay Environmental Action Group (BEAG) filed a Writ Petition in the Bombay High Court highlighting the magnitude of the problem. Eventually, the Bombay High Court ordered the setting up of a committee to investigate and solve the issues pertaining to the park. However, its recommendation was that part of the land of the National Park should be used for constructing multi-storeyed buildings under the Slum Rehabilitation Scheme—an action quite contrary to the need to protect the park!

Finally, in May 1997, the Bombay High Court issued the following directives for the protection of the forest:

- all encroachments should be removed within 18 months (eligible encroachers were to be rehabilitated outside the Borivali National Park)
- all commercial activities should be stopped
- all quarrying should be stopped

It led, among other things, to the removal of about 50,000 illegal structures that had been erected within the National Park after 1 January, 1995.

Still, the battle over the park was far from over. BEAG continued to push for the protection of the Park, especially the relocation of a bacon factory and the approximate 30,000 remaining dwellings.

On 16 July 1999, in response to further legal proceedings by the BEAG, the Bombay High Court passed another landmark order. It specifically directed the State Government to relocate the bacon factory within the next two years and ordered a prompt rehabilitation of all encroachers. Furthermore, BEAG was to be represented on one of the committees set up to monitor implementation.

Following the orders of 1999, the park saw 50 new posts being created within the Forest Department with the intent of beefing up the park's protection. Two contingents of the State Reserve Police were permanently deployed to help the Forest Department and the park also received six new patrol vehicles. Legal proceedings followed but again in 2003 the

TOP LEFT: Tulsi and Vihar lakes.
BOTTOM LEFT: The Shaik Quarry, a site saved after timely conservation action.

Bombay High Court painstakingly heard all the cases and passed a final Order for the Protection of the park on 15 September 2003. Fortunately, in its ruling the High Court maintained its previous stand and required the Government to remove all the encroachments in a timely manner and rehabilitate those who were eligible for rehabilitation as per the governmental criteria. A direction was also issued to construct a boundary wall to protect the National Park from encroachments. This was indeed a great legal victory.

The High Court required the Government to remove all the encroachments in a timely manner and rehabilitate those who were eligible for rehabilitation.

A little beyond the entrance to the park is the Nature Information Centre. Carefully thought-out walking trails now exist throughout the Park, and every week, in every season, sees various groups visit the forests. A couple of these trails can be navigated by bicycle too.

The Shilonda trail is an eastwards winding one, beginning as scrub land and then thickening with vegetation, ending in a tall, evergreen forest. It takes one through tribal lands that used to be occupied before but have now grown back into forests. Here and there, this trail is dotted with rapid forest streams, at times knee-deep. Wading through these streams is like chicken soup for the city-weary soul!

Some trails lead to watering holes. The Malad Trail Line runs along the western edge of the park. Teak, red silk, cotton, palash, laburnum, and bamboo lend a dry-deciduous character to nature. The trail runs close to the suburb of Malad and is known to bear the most paw prints of leopards.

If you're a history buff, the Upper or Kanheri Trail might interest you. It takes one through the forest to the Kanheri caves. In undertaking this trail, you'll see many birds and a waterfall. The Kanheri caves, built by the Buddhists between 1st century BC and 9th century AD, were part of a larger system of scattered cave complexes established in the western region of India, along the then prosperous trade route between the Deccan plateau and the coastal ports north of the present-day Greater Mumbai region. The caves were mostly small, modest chambers or monk cells. There is a network of narrow channels that collect water into underground tanks.

Over the centuries, a rich ecosystem evolved from this basalt rock and water landscape. Today, there are armadas of nightjars, short-eared owls, peregrine falcons, oriental honey-buzzards, and blue rock-thrushes, geckos, black-naped hares, and a variety of herpetofauna that forage a living amongst the rock, cactus, and grass.

Up until 1940, the people inhabiting the forests claimed to have seen

TOP RIGHT: A stray Leopard seeking refuge in a villa garden next to the park.
BOTTOM RIGHT: Tigers are no more in the park, but panthers, or leopards, are.

SUNJOY MONGA

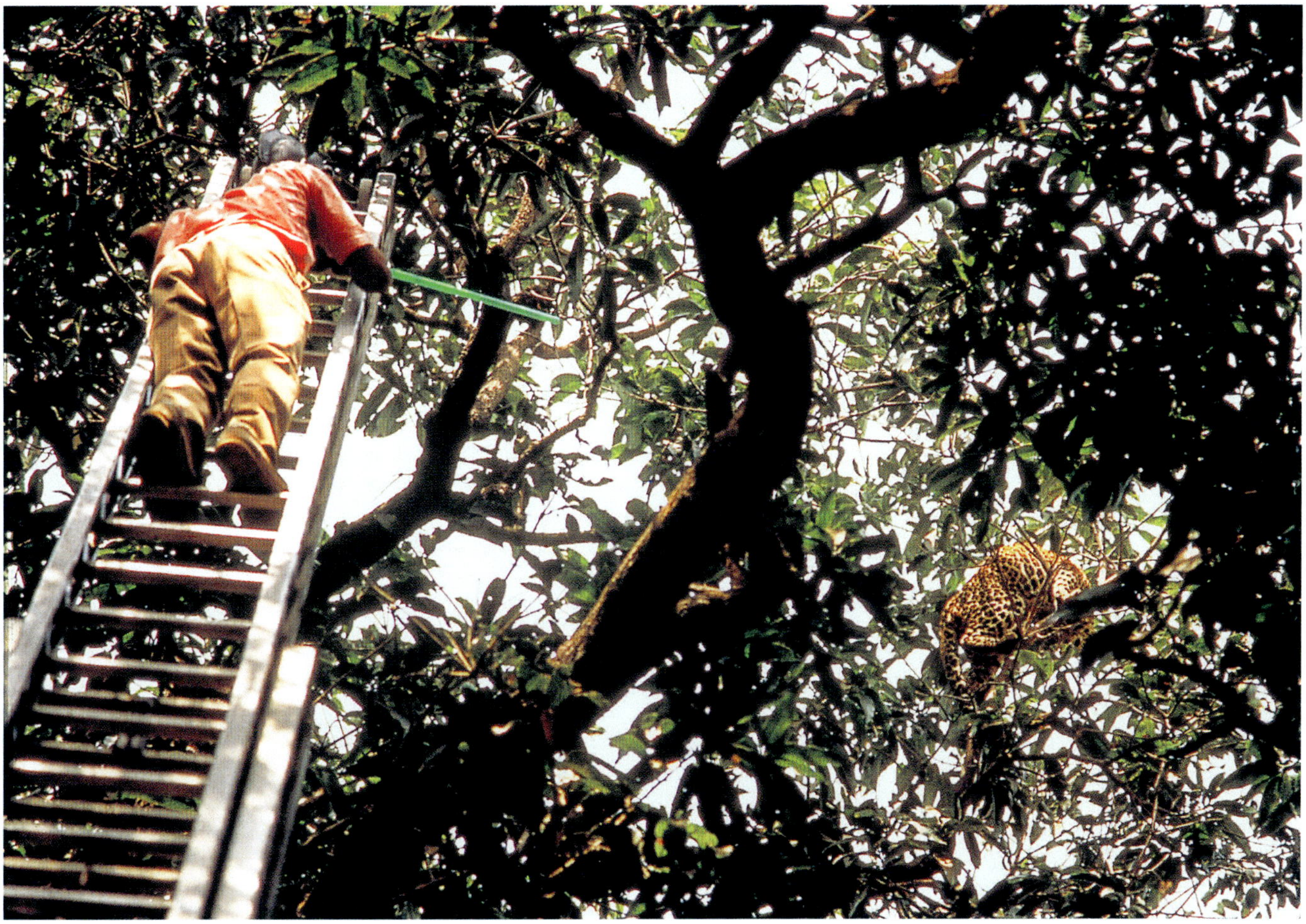

SUNJOY MONGA

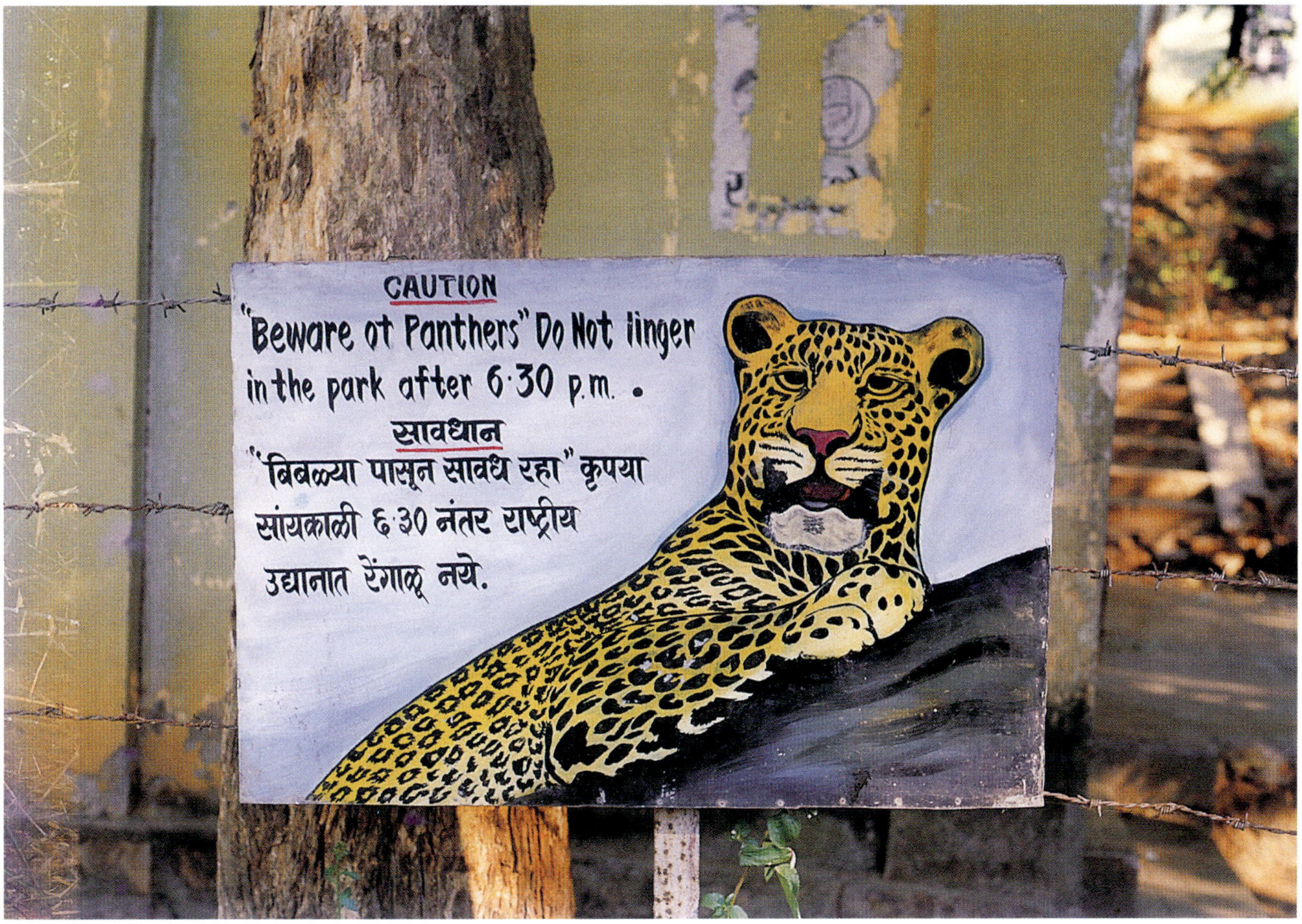

CAUTION
"Beware ot Panthers" Do Not linger
in the park after 6·30 p.m. .
सावधान
"बिबळ्या पासून सावध रहा" कृपया
सांयकाळी ६·३० नंतर राष्ट्रीय
उद्यानात रेंगाळू नये.

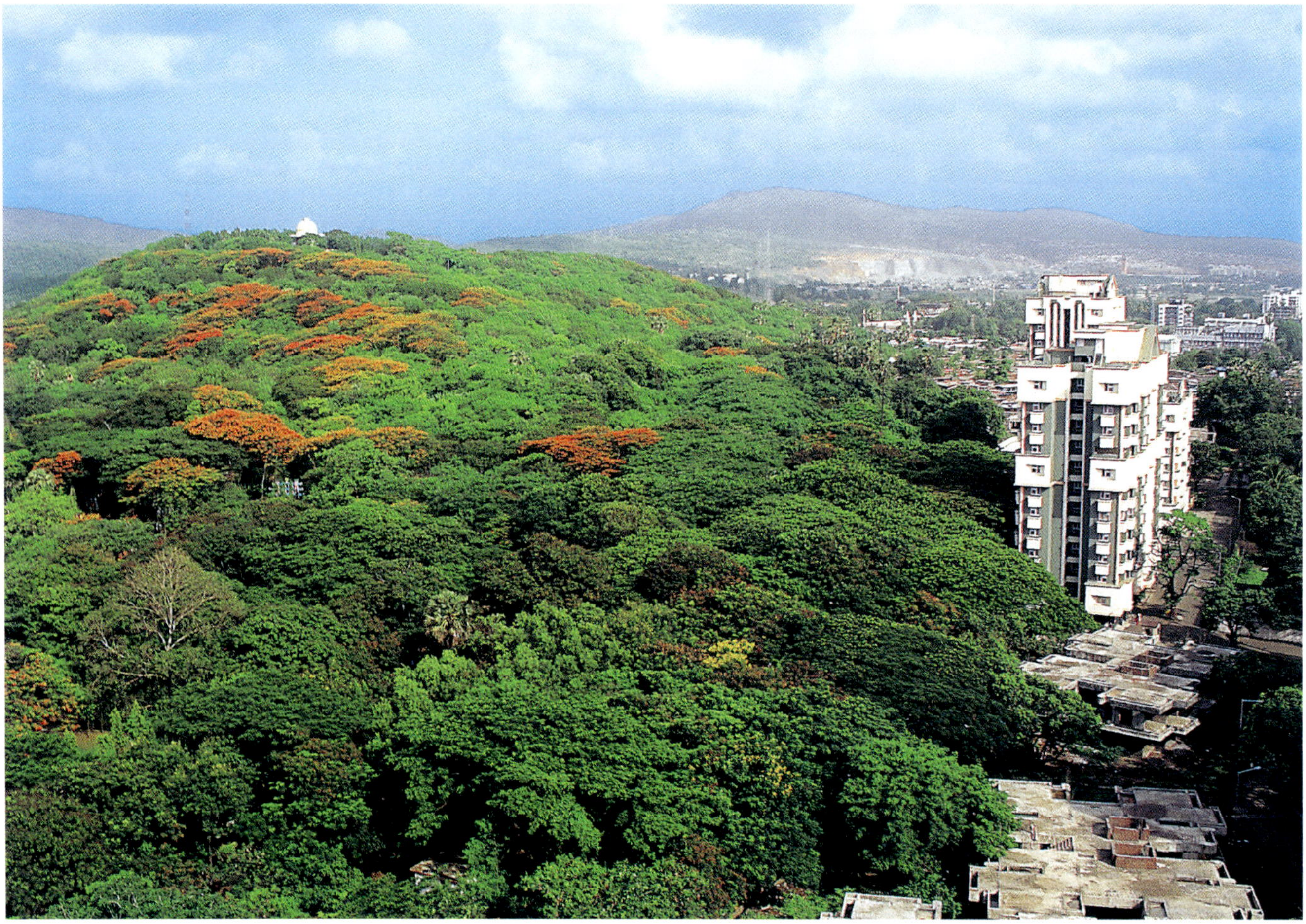

SUNJOY MONGA

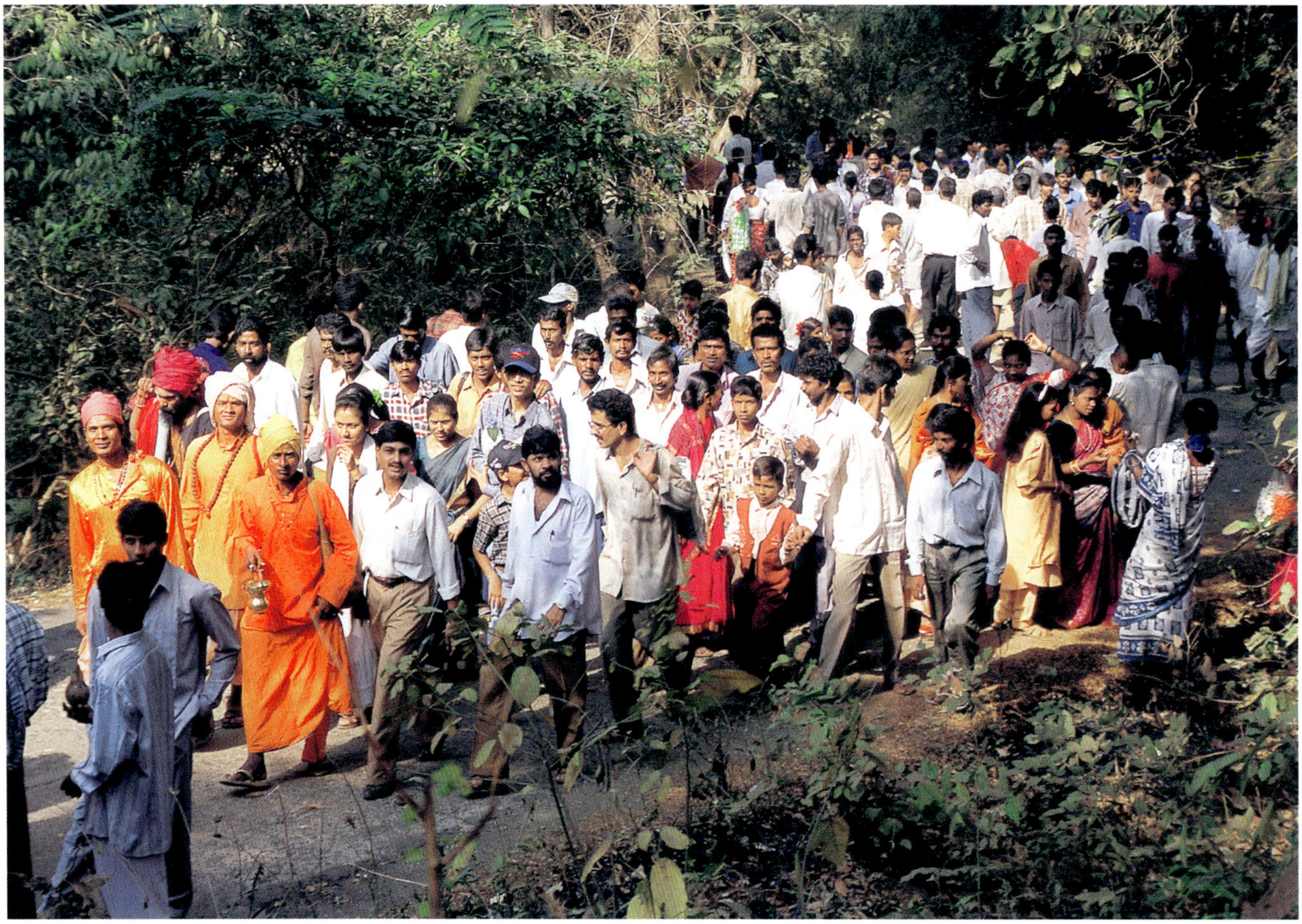

SUNJOY MONGA

tigers in the wilderness too. In the 1800s the Thane District Gazetteer reported that the area is "infested with tigers who came freely down to the plains." Today tigers are found in separate enclosed landscapes.

Amongst the park's deer, the Indian muntjac is the smallest—a secretive tawny brown creature you are more likely to hear than see. Its call is a sharp and far-reaching bark-like cry, hence its name "barking deer."

Add to that the mysterious creaking sound when the wind blows and the soaring bamboo stalks bend and sway, and as junglefowl, spurfowl, racket-tailed drongo, black-naped monarch, tickell's flycatcher, and jungle owlet scream, whistle, and hoot!

Nature's working is supreme, and there are lessons at every turn, along every water course and up and down the hill slopes. There is the interplay between light, altitude, and moisture so distinctly influencing not just the dispersal of species but the entire ecological design and framework of the National Park. Therefore, the next chapters of the park's existence, which are still being written, will be concerned with the issues of how to preserve the park and its inhabitants (flora and fauna), while still safeguarding the people living around it. But, there are several positive signs that nature is at work again, and there is abundant information still to be gleaned about its dynamic environs.

Nature's working is supreme, and there are lessons at every turn, along every water course and up and down the hill slopes.

TOP LEFT: View over Sanjay Gandhi National Park.
BOTTOM LEFT: Hordes of devotees throng to the forests to celebrate Mahashivratri every year.

UXBRIDGE

WHITCHURCH-STOUFFVILLE

ROUGE NATIONAL
URBAN PARK

TORONTO

Lake Ontario

19th Avenue
Day Use Area

Reesor Road Day Use Area

ROUGE NATIONAL
URBAN PARK

48

Train

7

407

7

Bob Hunter
Memorial Park

Little Rouge

MARKHAM

7

407

AJAX

Train

PICKERING

401

Toronto
Zoo

Glen Rouge Campground

Rouge

Zoo Road Welcome Area

Train

Rouge Beach

Lake Ontario

401

2A

SCARBOROUGH

TORONTO

N

0   2,5   5 Kilometers

0   2,5 Miles

© OpenStreetMap contributors

**ROUGE NATIONAL URBAN PARK, TORONTO,
ONTARIO, CANADA**

SIZE: 79.1 square kilometers.

LEGAL PROTECTION: National urban park.

OPENED: Officially established in May 2015, preceded by a smaller
park established in 1995.

OWNER: Land is owned by the federal government and managed by
Parks Canada, Canada's national parks service.

HOW TO GET THERE: Train (Rouge Hill GO station), and a combination
of subway and buses, bike.

KAITLYN CHOW
Rouge National Urban Park, Toronto, Canada

As the second largest country in the world by area, with a population of just over 36 million people, Canada is well known for its vast open terrain. Canada's national parks system includes many examples of the country's most famous landscapes, from towering mountain peaks to rugged coastlines and sweeping arctic tundra. Parks Canada, the country's national parks service, is responsible for managing these special protected areas. But despite an abundance of land, much of Canada's population is concentrated in relatively small, highly urbanized areas, where accessing these wild places can be challenging.

Rouge National Urban Park is Canada's first national park in an urban area. Created in 2015, the park covers 79.1 square kilometers and is located in the densely populated Greater Toronto Area, in close proximity to over six million Canadians. The park's boundaries span the cities of Toronto, Markham and Pickering, and the township of Uxbridge. It stretches from the Oak Ridges Moraine in the north—an important geological landform and the source of the region's drinking water—to the shores of Lake Ontario in the south. The park's landscape includes forests, meadows, rivers, wetlands, and some of the most fertile farmland in the country.

The Rouge, as it is known to long-time visitors, is an oasis of nature within the city. When walking under the shady canopy of a mature forest, or paddling down the tranquil Rouge River, it is easy to forget that you are in one of Canada's most densely populated areas. The park's close proximity to Canada's largest city and to 20 percent of the country's population makes it uniquely accessible, giving more people the opportunity to connect with nature and learn about protected areas.

Urban Canadians, new immigrants, and other groups who may not traditionally be exposed to outdoor recreation opportunities have a chance to explore the wilderness in their very own backyards. Rouge National Urban Park is accessible by bike and public transit, including a free shuttle from downtown Toronto, making it possible for people from all over the Greater Toronto Area to visit. A new initiative announced in April 2018 will see the creation of a 16 kilometer linear park along an underutilized hydro corridor, linking downtown Toronto with Rouge National Urban Park and further increasing accessibility.

Rouge National Urban Park is Canada's first national park in an urban area.

For many people, the Rouge acts as a 'gateway park', giving them the skills, confidence, and desire to venture into more remote areas. The park is home to the only campground in the City of Toronto and park programs and events include Learn-to-Camp workshops, free weekly guided nature walks, and celebrations of the park's vibrant farming community. Hiking, running, cycling, fishing, canoeing, and bird-watching are all popular activities.

The creation of Rouge National Urban Park was made possible thanks to the collective efforts and cooperation of community leaders, conservation authorities, all levels of government, First Nations and farmers over several decades. Efforts to protect the Rouge Valley were first prompted by concerns over increasing urban development. Toronto is Canada's largest city and a growing population has created intense development pressure along its outer fringes. Farm fields on the outskirts of the city have steadily disappeared, replaced with new housing developments. This pressure has long been a source of concern for conservationists seeking to protect these lands.

In 1975, a group of local activists created an organization called Save the Rouge Valley System in an attempt to protect the area from development. In 1988, the group proposed a plan to preserve a large tract of land in the valley. They focused on promoting the significant environmental and recreational benefits that a wilderness preserve in the city would provide. Their plan soon gained traction with the public and in 1990 the provincial government committed to creating a park in the Rouge Valley. The federal government also committed $10 million to this initiative. In 1995, Rouge Park was officially opened as a partnership park co-managed by several different partners and levels of government. The park initially spanned 23 square kilometers. Over the following years, the park was expanded twice to a total size of 40 square kilometers. One of these expansions, in 2006, created Bob Hunter Memorial Park within the Rouge, commemorating Robert Hunter, a leading Canadian environmentalist and one of the founders of Greenpeace.

A transformative change for Rouge Park came about in 2011, when it was announced that the park would be brought under federal protection and management. Formally established in 2015, the new Rouge National Urban Park would eventually add 39 square kilometers of lands to the park and nearly double its size to 79.1 square kilometers, making it one

TOP RIGHT: Hiking in Carolinian forests close to the megalopolis.
BOTTOM RIGHT: Restored meadow in Bob Hunter Memorial Park.

SCOTT MUNN / ©PARKS CANADA / ROUGE NATIONAL URBAN PARK

HEIKE REUSE / ©PARKS CANADA / ROUGE NATIONAL URBAN PARK

Toronto viewed from Rouge National Urban Park.

of the largest urban parks in the world. The establishment of the new national urban park also strengthened legislative protections for the area, ensuring that the park will remain protected in perpetuity. For many in the local community, Rouge National Urban Park represents a huge victory in the constant push and pull between development and conservation.

Interestingly, much of the land that now makes up Rouge National Urban Park was originally intended for a different purpose. In the 1970s, the provincial and federal governments acquired large areas of land in the eastern part of the Greater Toronto Area in preparation for the construction of a new airport. For various reasons, to date the airport plans have never come to fruition. Large portions of the acquired lands were later committed to Parks Canada for the creation of Rouge National Urban Park.

Although the park was only established in 2015, the history of the area stretches back over 10,000 years. Indigenous Peoples have inhabited the Rouge Valley for millennia, using these lands for hunting, fishing, farming, and an important travel and trade route. The park is located within the ancestral and traditional territories of the Anishnaabe, Haudenosaunee, Huron Wendat, and the Mississaugas of the Credit. Parks Canada works closely with a First Nations Advisory Circle consisting of 10 First Nations with an expressed interest and cultural connection to the parklands. Some of Canada's oldest known Indigenous sites are found in the Rouge. Archaeological fieldwork undertaken by First Nations and Parks Canada can help tell the stories of these early First Nations communities. One particularly prominent site is Bead Hill (Gandatsekiagon), a 17th century Seneca village located on the banks of the Rouge River. It was designated as a National Historic Site in 1991.

The first European settlers arrived in the Rouge Valley in 1799. These early pioneers worked hard to clear the land and establish small farms. Some of the original families who settled in the area in the early 1800s still live and farm in the park and many historic barns and houses remain standing today. The park preserves large tracts of Class 1 farmland—the richest, rarest and most fertile soil in the country—as well as the last remaining working farms in the City of Toronto. Farms in the park provide a source of locally grown food for the Greater Toronto Area. Rouge National Urban Park is unique among national parks in Canada in protecting agriculture alongside nature and cultural heritage.

When walking under the shady canopy of a mature forest, or paddling down the tranquil Rouge River, it is easy to forget that you are in one of Canada's most densely populated areas.

TOP LEFT: Farming is part of Rouge National Urban Park.
BOTTOM LEFT: Eastern coyote.

Farming has played an important role in the history of the Rouge Valley and will continue to be an important part of Rouge National Urban Park into the future.

In terms of biodiversity, the Rouge holds its own against its larger siblings in the Canadian national parks system. The park features mature forests, meadows teeming with wildflowers, and the largest remaining coastal wetlands in Toronto. More than 1700 species of plants and animals have been documented in the park, including over 1000 plants, 200 birds, 50 fish, 40 mammals, and 20 reptiles and amphibians. This tally includes 27 species-at-risk, classified as either endangered, threatened, or of special concern, as well as many locally rare species. One example of a species-at-risk found in the park is the Blanding's turtle. Parks Canada has been working in collaboration with partner conservation organizations to re-introduce Blanding's turtles into the park through a 'head-start' program in which baby turtles are hatched and raised at the Toronto Zoo and then released into local wetlands. Some of the larger mammals that can be found in the park include white-tailed deer, coyotes, red foxes, beavers, and river otters.

The relatively high biodiversity observed in Rouge National Urban Park is a result of its location and its many variations in topography, micro-climates, soil type, and land use. Heavily shaped by glaciation during the last ice age, the park features many hills, ridges, and deep river valleys that provide a multitude of different habitats for plants and wildlife. The park also lies at the northern edge of the Carolinian Life Zone, a type of forest ecosystem characterized by a predominance of deciduous trees. Carolinian forests are rare in Canada, existing only in a small area of Southern Ontario. However, these forests are home to more species of plants and animals than any other ecosystem in Canada, including many rare and endangered species. The edge of the Carolinian Life Zone represents the northern limits of many of these species, which are found nowhere else in Canada.

The size of the park is important for animals that require large areas of continuous habitat. For example, bird species such as the ovenbird or scarlet tanager need large tracts of undisturbed forest to thrive. Within the broader conservation landscape, the Rouge provides an important link between other green spaces, allowing wildlife to move freely along protected corridors. The park is also part of the Greenbelt, a 7,200 square kilometer area of protected land in Southern Ontario

The relatively high biodiversity observed in Rouge National Urban Park is a result of its location and its many variations in topography, micro-climates, soil type, and land use.

RIGHT: First Nations Advisory Circle representative for Scugog Island First Nation, Dave Mowat, holds a baby Blanding's Turtle before its release.

Exposed glacial sediments at Glen Eagles Vista.

designed to shelter farmland, forests, and watersheds from urban development.

Although Rouge National Urban Park is now well protected under federal legislation, the area has a long history of mixed uses and much of the land has been altered or disturbed in the past. Roads, railway tracks, and power lines crisscross the park, posing unique challenges to park management. Past disturbances have allowed invasive species to obtain a foothold in some areas. One particularly vigorous invader is known as dog-strangling vine, a perennial native to Eurasia. It forms dense stands that crowd out native plants and young trees, reducing biodiversity. Dog-strangling vine is a member of the milkweed family and its resemblance to native milkweeds makes it a threat to the iconic monarch butterfly. The butterflies will lay their eggs on the plant, but they cannot use it to complete their life cycle.

Balancing recreational usage and ecological protection is another concern. Limiting high-impact activities, such as mountain biking, and encouraging visitors to use the area responsibly is one way to reduce these issues. Educating visitors about invasive species, wildlife, and the various ways they can help preserve the park will be an important factor in ensuring that these unique natural landscapes remain intact for future visitors to enjoy.

In the face of these challenges, significant efforts are now underway to restore native ecosystems, including meadows, wetlands, and forests. These efforts are the result of collaboration between Parks Canada, conservation groups, farmers, Indigenous partners, municipalities, schools, and community members. From 2015 to 2018, 52 ecological restoration and farmland enhancement projects were initiated and completed throughout the park, resulting in the restoration of more than 86 hectares of wetlands, forests, and riparian habitat. Over 100,000 native trees, perennials, shrubs, and aquatic plugs were planted as part of these efforts. The Rouge also provides an interesting model for how agriculture and nature can thrive alongside each other. Farmers often bring forward ideas for restoration projects on their lands, including the creation of wetlands in low-lying areas of their properties.

Restoring ecological integrity is a noble goal in itself, but it is also directly beneficial to the millions of local residents who rely on the important ecosystem services that these natural systems provide. These services include carbon sequestration, water and air purification, flood mitigation, and climate regulation. Restored habitats provide excellent educational and outreach opportunities and often feature prominently in the park's free weekly guided hikes.

As Canada's most accessible national park, the Rouge is a hotspot of activity. The park's urban setting poses challenges to conservation, but also great opportunities to connect more people with the natural settings and outdoor recreation opportunities so often missing from urban life. With careful management, the unique natural, cultural, and agricultural landscapes of Rouge National Urban Park will be preserved for the enjoyment of visitors for generations to come.

LIVINGSTON

HAMMOND

PONCHATOULA

COVINGTON

MANDEVILLE

PICAYUNE

PEARL RIVER

SLIDELL

BILOXI

GULFPORT

Saint Louis Bay

Mississippi Sound

Lake Maurepas

Lake Pontchartrain

CONVENT

LAPLACE

NEW ORLEANS

Mississippi River

METAIRE

ST BERNARD BASIN

ALGIERS POINT

MARRERO

THIBODAUX

Lake Salvador

HOUMA

Little Lake

Mississippi River

PORT SULPHUR

GULF OF MEXICO

CHAUVIN

DULAC

West Bay

East Bay

0   20   40 Kilometers

0   20 Miles

N

© OpenStreetMap contributors

NEW ORLEANS, UNITED STATES

Situated at the estuary of Mississippi, surrounded by the river, Lake Pontchartrain, and the Gulf of Mexico.

SIZE: 19 square kilometers urban parks (48 square meters per inhabitant).

ERIC TAMULONIS

The Parks of New Orleans, United States: Living with Water

When it rains five days, and the skies turn dark as night,
There's trouble taking place in the lowland that night.

"Backwater Blues" by Bessie Smith and Jimmy Johnston

New Orleans is an American city in the deep southern state of Louisiana. Known for its history and architectural distinction, it is celebrated and beloved for the rich cultural diversity of its people, cuisines, and musical traditions. Public parks are an important part of the city's cultural context and they enrich its physical, social, and ecological fabric. Parks play a key role in *The Greater New Orleans Urban Water Plan*, (Waggonner & Ball 2013) which calls for the most sweeping change in the city's recent development. Spurred by the devastating impact of Hurricane Katrina, it is the city's response to the effects of climate change and sea level rise. The city's park heritage, as well as its geography, history, social conditions, and park system informed the Urban Water Plan.

The Mississippi River system is the largest in the U.S., making its deep-water port at the confluence with the Gulf of Mexico an obvious crossroad of commerce. In the late seventeenth century, French trappers and settlers occupied a natural levee formed by the river's sharp bend. France, which had colonized much of the Americas at that time, founded New Orleans there in 1718.

New Orleans is a water city close to the Gulf and its massive storms. 1,500 mm of rain falls on the city each year, and 140,000 cubic meters of Mississippi River water flow by it every second in high flow. From the beginning people knew how to "live with water": they built on the high ground; they built on poles in lakes; and they dug drainage canals that crisscrossed the city.

As the city grew, people with low incomes, largely minorities of mostly African American heritage, settled in the lowlands, which in the delta region is only a few feet below high ground. African Americans, who comprise 60 percent of the population, are most affected by flooding and have suffered first from slavery and then the abusive practice of 'Jim Crow' segregation. This has left a legacy that continues to shape present day concerns about social equity in the city. New Orleans' poverty rate remains very high: 28 percent, compared to 15 percent nationally.

AVERAGE FLOW

The Mississippi River's flood control system involves 41 percent of the land area of the U.S. The width of the rivers is proportional to the flow of water. Map created by M. Heberger, Pacific Institute, 2013. Used with permission.

As in many southern cities, New Orleans' segregated parks were among the proving grounds of the African American community's campaign that led to passing the 1964 Federal Civil Rights Act.

Through the 19th century, and especially after the 1913 invention of a highly efficient screw pump, the city was able to occupy the flood-prone lowlands that were increasingly being filled by people. Drainage channels were filled in and dirt roads paved. These actions reduced the capacity of floodwaters to be absorbed. New Orleans was no longer living *with* water; it was living *despite* water.

The patchwork flood control system developed up to the early 20th century relied fully on levees and pumping. The system broke down in the disastrous 1927 Good Friday Flood, the most destructive in the history of the United States (Barry 1997). In response, during the following decades the world's largest levee system was built throughout the Mississippi River's watershed, resulting in another 2,230 kilometers of levees, dams and reservoirs, and altering the river's drainage basin, which encompasses 41 percent of the nation's land area. After Hurricane Betsy in 1965, a system of levees and walls was constructed specifically to protect the city from storm surge from Lake Pontchartrain.

Yet the integrity of the post 1927/1965 flood engineering solutions was only as strong as its weakest point. Tragically, on 29 August 2005, Hurricane Katrina swept into New Orleans. It breached Lake Pontchartrain's protective levees, flooded 80 percent of the city, caused the loss of 1,500 lives, and inflicted $160 billion in economic damage.

Katrina and other catastrophic storms have triggered a major re-thinking of the public response to climate change and the type of protective public infrastructure needed. In 2013, the synergy of local, national, and international (primarily Dutch) expertise produced the *The Greater New Orleans Urban Water Plan*. The plan presents a novel approach to stormwater management that integrates innovative planning and hydraulic and ecological thinking with the traditional 'hard' engineering protection of levees, pipes, and pumps that remains central to the solution.

The traditional approach to managing stormwater in New Orleans was to pump every drop of rainfall out of the levee enclosure. In practice, this had lowered the water table under the city, causing an imbalance of soil and water in which the land sank, roads cracked, and pipes broke. This pumping action made nearly 50 percent of the city drop below sea level. The new approach, "Living with Water," outlined in the Urban Water Plan is based on the premise that water should be part of the urban ecosystem of natural and public spaces, including the essential stabilization of soil.

The plan is designed to slow and temporarily store thousands of cubic meters of stormwater, and then to divert it with small-scale retrofitted storm drainage infrastructure, waterways, and canals. Water will be stored and infiltrated in new strategically located parks and wetlands, and at a smaller scale, on commercial and residential properties. The water can then percolate into the ground, keeping the water table intact and the soil stabilized, and also be repurposed for irrigation and other uses. This system will reduce the load on existing drainage systems and reduce flood risks. The presence of water in these urban wetland ecosystems also will foster biodiversity not found in purely engineered systems of pipes and pumps. These new parks promise a striking new urban aesthetic that blends formal and natural imagery.

Both in the city's considerable existing system of parks and those proposed in the plan, parks will function as stormwater sponges important to the plan's flood control strategy. Though addressing flooding is the plan's focus, it also addresses social justice and equity, and will significantly enhance the quality of life in the city. An unintended consequence is the potential gentrification of areas where people with low incomes live, making life more difficult for them by raising rents. The

The water can then percolate into the ground, keeping the water table intact and the soil stabilized, and also be repurposed for irrigation and other uses.

city is well aware of this and anticipates steps to counteract the effect.

The city's existing park system is robust, if underfunded. It ranks 20th in The Trust for Public Land's assessment of the top 100 US cities by population, a ranking based on several criteria such as access, facilities, budget, and land area. Public parks occupy 110 square kilometers, almost a quarter of New Orleans' land area, although the immense Bayou Sauvage National Wildlife Refuge makes up roughly 83 percent of that total. This leaves 19 square kilometers in urban parks, providing approximately 48 square meters of parks and public space per resident (not counting the Refuge). Approximately 80 percent of residents live within a 10-minute walk to parks, which are fairly equally distributed for different income brackets and races.

This abundant and accessible parkland includes several signature legacy parks: City Park (almost 5 square kilometers), Audubon Park, and Lakeshore Park (both at 1.6 square kilometers). Beyond these parks are Jean Lafitte National Park and two state historic parks. The city's public open space also includes over 425 kilometers of publicly-accessible green areas called "neutral grounds" along streets, 54 active pre-Katrina community gardens (of which 25+ have currently been re-established), and bike and pedestrian paths on levee sections along the river and lakefront.

Although New Orleans is already comparatively well-endowed with parks, the Urban Water Plan calls for more. New parks are located at key points in the drainage system and designed specifically to aid in flood mitigation as well as for habitat creation and visual and recreational amenity.

Homes and businesses, streets and highways, waterways, parks, and wetlands are integrated into a living water network suited to the demands of the 21st century. The resulting urban environment will be greener and cooler, with more trees and water-loving plants, and with acres of parklands that safely store stormwater and allow it to soak into the ground. (Waggonner & Ball 2013, p 30).

Proposed projects illustrate the breadth of the plan's vision. The Lafitte Blueway project in which the Carondelet Canal is proposed to be reopened, recalls the historic waterway that once connected Lake Pontchartrain, through Bayou St. John, with the French Quarter.

TOP RIGHT: Scenes such as this inundated Live Oak tree in Audubon Park illustrate the compromise of accepting the presence of water as a necessity of flood mitigation.
BOTTOM RIGHT: Parks contribute to equity and social life. Many, like Congo Square seen here, have been central to the city's African American story, both its jazz roots, and the historical—and present day—locus of civil rights advocacy efforts in the city.

JF HARMAN SHUTTERSTOCK

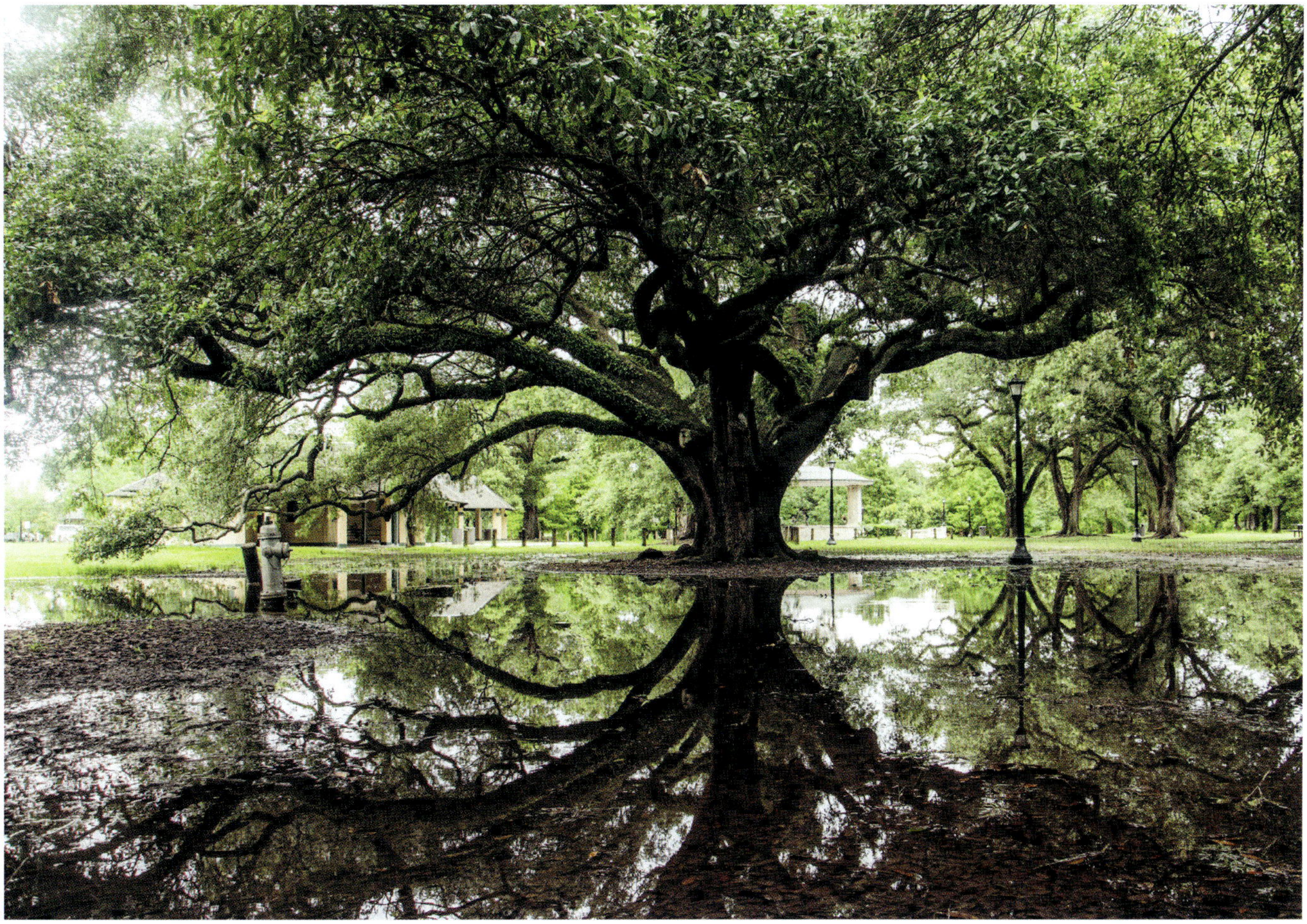

BART EVERSON CREATIVE COMMONS

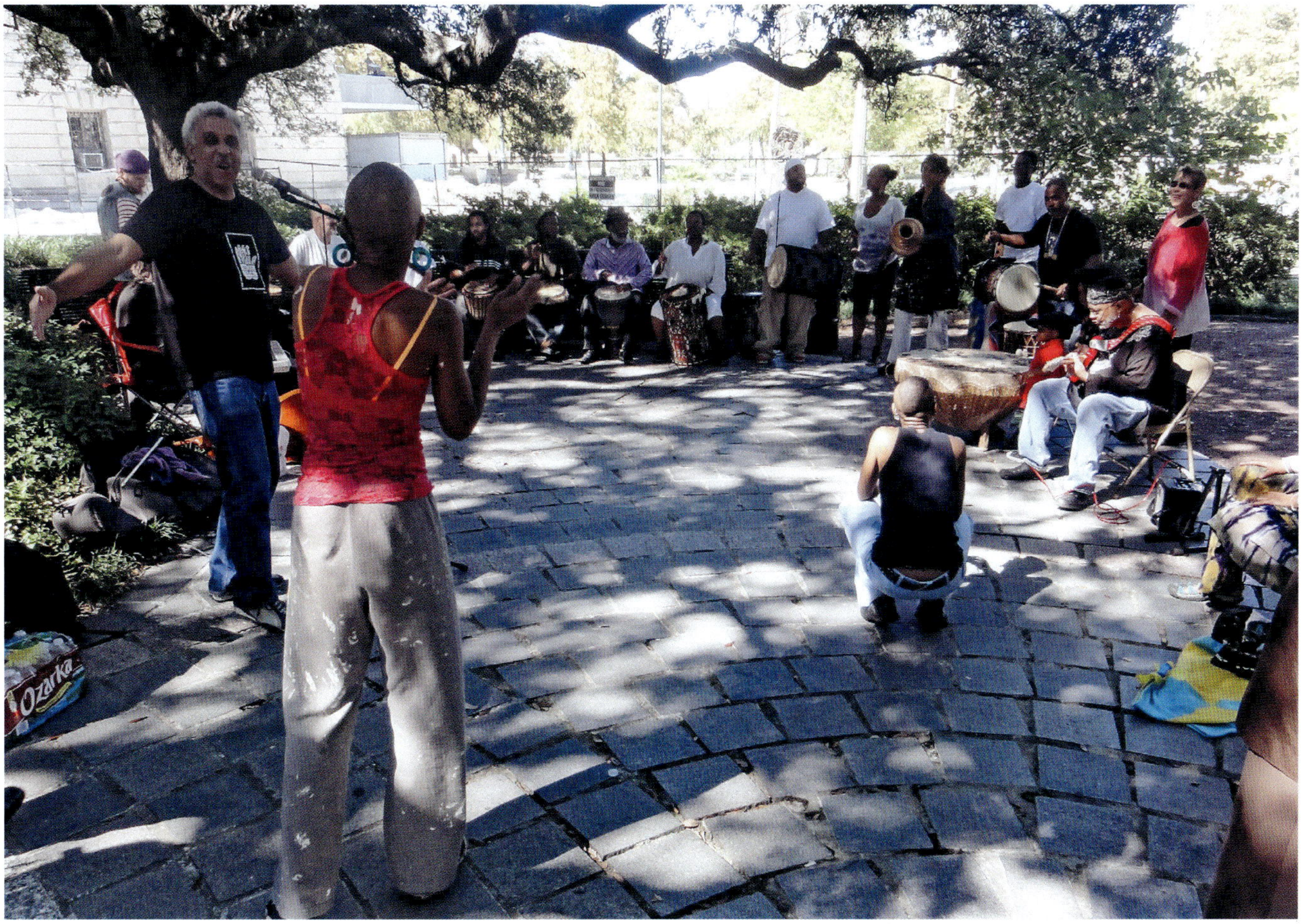

The city and its river: New Orleans is indivisible from its setting along the Mississippi River.

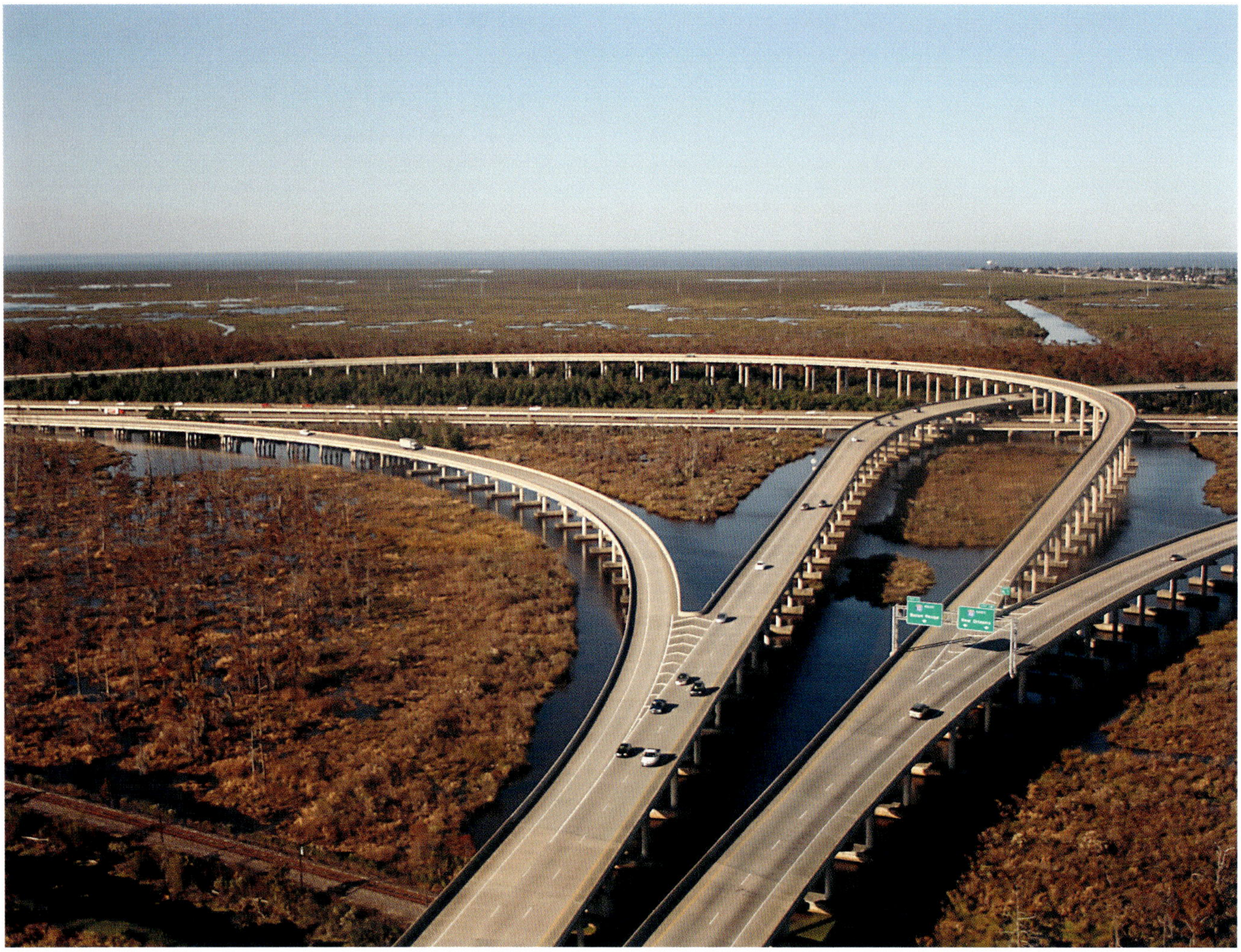

STEPHEN FINN SHUTTERSTOCK

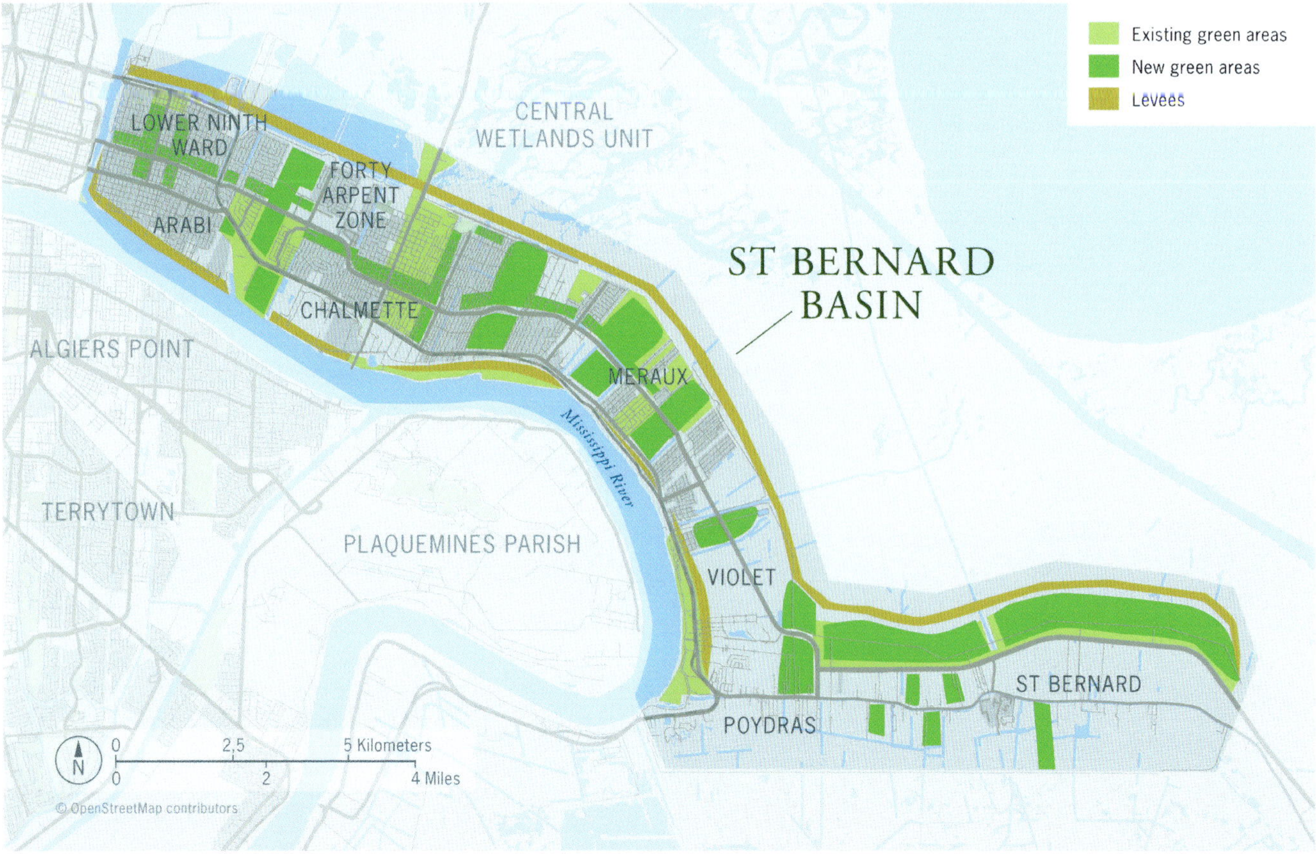

Existing green areas
New green areas
Levees
CENTRAL
WETLANDS UNIT
LOWER NINTH
WARD
FORTY
ARPENT
ZONE
ARABI
CHALMETTE
ST BERNARD
BASIN
ALGIERS POINT
MERAUX
Mississippi River
TERRYTOWN
PLAQUEMINES PARISH
VIOLET
ST BERNARD
POYDRAS
N
0
2,5
5 Kilometers
0
2
4 Miles
© OpenStreetMap contributors

WAGGONNER & BALL 2013, P.37, ADAPTED BY MARTIN THELANDER

Several other canals will be reopened or repurposed. These canals will bring new recreational or ecological amenities to the neighborhoods. New or retrofitted parklands play an important role. In the Lakeview District, streets will be reconstructed as "floating streets" that store stormwater in the roadway subbase and in curbside bioswales, while conveying excess water towards large storage areas.

The scale and complexity of the Urban Water Plan dictate a long-term implementation timeline. So, to address immediately available 'soft' and small-scale flood control opportunities, New Orleans also prepared a citywide green infrastructure plan in 2014, adding bioswales, infiltration basins, rain gardens, and community gardens. The city's major parks play a role in this plan as well. City Park, with 55 hectares of lagoons, is key to flood management in the city center. Lowering the lagoons' water levels in advance of storms creates room for greater flood water storage; design is underway to enlarge their capacity even more. The basic site maintenance of stormwater management facilities in the new parks can be defrayed by separate funding dedicated to public safety.

The Urban Water Plan is a bold green infrastructure plan that will considerably increase green and blue space in New Orleans. It will make the city more flood resilient, improve living conditions throughout the city, and promote cultural and biological diversity. New Orleans' population has dropped from 627,000 fifty years ago to a present-day 389,000. The 'new' New Orleans will be positioned to retain current residents, bring back some of those who left, and attract new residents. In addition to New Orleans' efforts, an integrated multi-jurisdictional regional program is underway to cope with the effects of climate change and sea-level rise. Like the UWP, this program is dependent on the fragile political will and federal funding to implement it. Nevertheless, there remains great hope to flourish safely, even "when it rains five days, and the skies turn dark as night."

Though addressing flooding is the plan's focus, it also addresses social justice and equity, and will significantly enhance the quality of life in the city.

TOP LEFT: Interstate Highway 10 near New Orleans shows the infrastructural challenge of access through the Mississippi delta's watery environment.
BOTTOM LEFT: One example of the Urban Water Plan's approach at a neighborhood scale is the green infrastructure plan for St. Bernard Basin, a third of central New Orleans. Light green is existing green. Dark green is proposed green. Other sections of the central city follow the same pattern.

Social Cohesion

The big take-off in designed green space in cities was during the 19th century (Peter Clark). Sanitary concerns mixed with social, when many of the most treasured large urban parks were created. Present day city growth makes these concerns just as vital. Parks serve as melting pots not only for biodiversity but also for social and cultural diversity. Concepts like "people's parks" and "the great outdoor living room" capture this idea. *The disease* [Covid-19] *requires social distancing; the racism requires social action. Parks can accommodate both* (Michael Van Valkenburgh*).

Landscape architect in New York, Birthe & Per Arwidssons stiftelse, June 2020.

PATRICIA M. O'DONNELL

Inclusive Large Parks for Everyone: Fostering Social Cohesion and Environmental Justice

Historic and contemporary parks of all sizes and types serve nature and people in cities, towns, and villages around the world. As protected areas, set aside as parts of human settlements, that provide habitat, oxygen, cooling, water absorption, and places for people to engage with nature and each other, urban parks are vital to the health and well-being of many life forms, including humanity. Public parks are the commons of urban living, places shared by people, plants, animals—they are assets of our human settlements. Large parks in cities and towns are places of heritage, set aside through advocacy, philanthropy, and/or civic leadership for the good of all urban dwellers. They are local and regional assets. Today we are beginning to understand the broad benefits and responsibilities that large parks embody in making progress toward equitable, resilient, and sustainable cities. As vessels of culture and nature, they uplift daily urban life. This chapter explores the importance of parks as places of inclusion and justice, accessible to all. Parks provide a common ground and foster social cohesion as equitable meeting places in cities.

In order to fully understand the significance of parks and especially that of large urban parks, we must recognize that landscape intertwines and entangles biological and cultural diversity, natural and human habitats, tangible and intangible heritage. These aspects of landscape overlap each other in ways which are important to understand. Biodiversity has a counterpart in cultural diversity, connoting the variety of people within a society. For ecosystems: diversity of plants, animals, insects, etc. supports a wider spectrum of life forms and heightens productivity of the biosphere as a whole. For cultures: the differences in traditions, practices and, identity offer richness and breadth that enrich life and encourage innovations and cultural developments. Cultural diversity enriches as global migration shapes daily life with an array of traditions, beliefs, and world views. Parks provide a welcome platform for expressing this cultural diversity, tolerance, and mutual respect. The pressing challenges of the 21st century can be better addressed by embracing diversity through opening large parks as common grounds to meet, exchange, express differences and similarities. Parks provide a ground for everyone's use without censor or selection. Large parks, as democratic gathering space open to all, are an incredible asset to our cities, towns, and villages as shared places of biotic and cultural diversity.

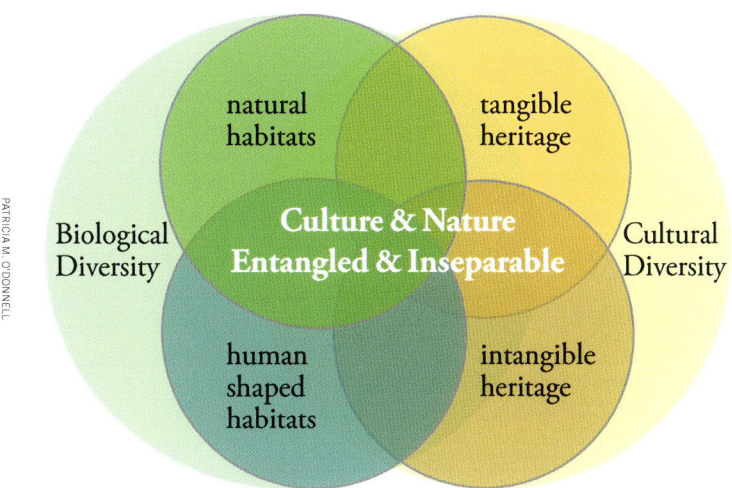

Bio and Cultural Diversity overlap in many ways.

In 1962 the clarion call sounded by Rachael Carson with her book *The Silent Spring* began to raise awareness about environmental degradation caused by pesticides. Initially the environmental movement looked toward decreasing agricultural chemical usage on crops and spurred a growing drumbeat to conserve wildness in places far away. Taken up by local people, that movement soon turned to concern for the environmental quality of their city and neighbourhood parks. For example, in the USA city park protection and heritage preservation were widespread vectors for action beginning with curation and preservation planning applied to Central Park in New York in the early 1970s, at that time in a sorry state. In the decades following, local governments, philanthropists, and citizen advocates have embraced their urban park heritage with partnerships for park revitalization, continuing to this day. Large parks are the iconic shared urban oasis. This important work is often aligned with research and understanding of historic continuity, respecting the inheritance of this generation and the need to pass on this legacy to future generations. Central Park remains a place to gain greater social cohesion in a highly diverse city. The hope for all large parks is that a large green shared-place of beauty helps people meet, increases tolerance, and aids in setting conflicts aside.

Rio de Janeiro, Brazil. The massifs rise out of the Tijuca forest. This large urban park provides accessible public spaces and defines the visual character of the city.

In 1987 the Brundtland Commission defined sustainability as balancing benefits to society, environment, and economy, which are often termed the three pillars. Large Parks provide a strong platform to achieve sustainability that favours societal benefits while it also includes environmental and economic ones. The year 2015 witnessed the ratification of United Nations Sustainability Development Goals (UN SDGs). Parks play an invaluable role in implementing these goals.

In recent years there have been a series of global doctrines and actions that recognize the values of large parks as cultural landscapes, the combined works of humanity and nature. In 1992 the World Heritage guidelines were expanded to include listings of cultural landscapes. Several of the listed landscapes are urban, including large parks, such as: Royal Botanical Gardens of Kew, United Kingdom; Singapore Botanic Gardens; Rio de Janeiro's Tijuca Park and botanical garden, Brazil; and West Lake Cultural Landscape of Hangzhou.

Rio de Janeiro: Carioca Landscapes between the Mountain and the Sea, World Heritage nomination was inscribed in 2012, to include the Tijuca National Park and Rio de Janeiro Botanical Garden, Copacabana Beach, Flamengo Park, and the adjacent sea. For Rio de Janeiro these are the iconic public landscapes and seascapes that encompass the identity of this international city. This nomination noted that the inscribed property

encompasses all the key natural, structural elements that have constrained and inspired the development of the city. These stretch from the highest points of the mountains of the Tijuca National Park with its restored Atlantic forest, down to the sea, and include the Botanical Gardens established in 1808, Corcovado mountain, with its statue of Christ, and the chain of dramatic steep green hills, Sugar Loaf, Pico, Leme and Glória, around Guanabara Bay, as well as the extensive designed land-scapes on reclaimed land along Copacabana Bay which, together with Flamengo and other parks, have contributed to the outdoor living culture of the city.

The Carioca Mountain Range of Tijuca National Park and the Botanical Garden covers 1,897 hectares. For a city that prizes outdoor living, the vast expanse of the mountainous, second-growth woodlands of the Tijuca Forest park, dotted with iconic steep massifs, is an unparalleled asset for the well-being and equity afforded Rio's citizens and visitors.

With an ever-advancing percentage of urban dwellers across the globe, large urban parks are instrumental in reaching several of the 17 SDGs, for example:

• Goal 3:   Ensure healthy lives and promote well-being for all ages.
• Goal 10: Reduce inequality within countries.
• Goal 11: Make cities and human settlements inclusive, safe, resilient, and sustainable.
• Goal 13: Take urgent action to combat climate change and its impacts.
• Goal 15: Protect, restore, and promote sustainable use of terrestrial ecosystems.
• Goal 16: Promote peaceful and inclusive societies for sustainable development, provide access to justice for all, and build effective, accountable, and inclusive institutions at all levels.

The 17 global sustainable development goals (SDGs).

The strategy focuses on: creating partnerships to enhance the city's biodiversity; increasing tree canopy; replacing city streetlights to reduce light pollution and enhance the night sky; as well as increasing ecological literacy.

An important target under Goal 11 is target 11.7: "By 2030, provide universal access to safe, inclusive, and accessible, green and public spaces, in particular for women and children, older persons, and persons with disabilities." Likewise, target 13.1, to "strengthen resilience and adaptive capacity to climate-related hazards and natural disasters in all countries," can be applied to parks that foster enhanced resilience for communities. Target 15.9 is: "By 2020 integrate ecosystem and biodiversity values into national and local planning, development processes, poverty reduction strategies, and accounts," which can be applied to urban parks as human shaped ecosystems that provide important habitats.

Much of this has to do with green infrastructure and large urban parks. At the 2016 United Nations conference, Habitat III, a rich dialogue took place about public landscapes, parks, and conserved lands for city dwellers, and the New Urban Agenda was approved. This lengthy document incorporates a broad global discussion about the future we want in our cities. Paragraph 37 is highly relevant as it states (bold text added by author):

> *We commit ourselves to promoting safe, inclusive, accessible, green and quality public spaces, including streets, sidewalks and cycling lanes, squares, waterfront areas, gardens and parks, that are multifunctional areas for social interaction and inclusion, human health and well-being, economic exchange and cultural expression and dialogue among a wide diversity of people and cultures, and that are designed and managed to ensure human development and build peaceful, inclusive and participatory societies, as well as to promote living together, connectivity and social inclusion.*

A report on the state of heritage entitled *Culture Urban Future* was launched at Habitat III. It includes a chapter entitled "Enabling Access to public spaces to advance economic, environmental and social benefits" (O'Donnell 2016). The take-away messages for public spaces include:

- Urban public spaces form a valued cultural heritage for everyone.
- Public spaces are vehicles for community tolerance and cohesion as they welcome everyone.
- Urban public parks and protected areas nearby add to a city's climate change resilience.
- Traditions and practices of a diverse urban populace can be expressed in public spaces.
- Public space additions and upgrades in low-income areas of cities foster social justice and equity.

In 2011, supported by worldwide discussions, the UNESCO Recommendations on the Historic Urban Landscape (HUL) were approved in the form of a soft law that the signatories of the convention in 193 countries are requested to embrace and apply. This tool seeks to apply a landscape approach to sustaining urban identity by addressing preservation of historic cities while daily life, continuity, and change proceed. In terms of public landscapes, HUL sees cities as built environments with parks as an important component. Like other public resources, parks should be identified, conserved, and managed by recognizing and integrating their multiple values. HUL "is rooted in a balanced and sustainable relationship between the urban and natural environment, between the needs of present and future generations and the legacy from the past."

It is of specific interest in the context of large urban parks that the urban landscape is seen not only as built-up areas, buildings, and roads but as a "relationship between the urban and natural environment." UNESCO has delivered recommendations on the preservation of the historic urban landscape. A first step is "to undertake comprehensive surveys and mapping of the city's natural, cultural, and human resources." Planning in cities around the world can benefit from applying this recommendation, which integrates cultural heritage and development, especially in those countries experiencing rapid city growth and densification. Large park advocacy for retaining the precious ground we inherit must push back against efforts to nibble away park acreage for institutional and residential uses—these landscapes of inclusion are irreplaceable.

An interesting example embracing all these ambitions is Pittsburgh, United States. The city has long been known for its steel production, and today some of that rich industrial history is being repurposed into new forms of urban nature. The city has constructed riverfront trails and green space, including the new South Shore Riverfront Park, and is re-energizing former industrial spaces with all manner of innovative uses, among them gardening, farming, and art installations. From reimagining its industrial past to becoming a member of the Biophilic Cities Network in 2016, Pittsburgh is connecting residents with nature in diverse ways. The city's main biophilic endeavours have focussed on improving both water and air quality while also increasing residents' engagement with the natural world. The strategy focuses on: creating partnerships to enhance the city's biodiversity; increasing tree canopy; replacing city streetlights to reduce light pollution and enhance the night sky; as well as increasing ecological literacy. As a member of the Biophilic Cities Network, Pittsburgh plans to measure success through several measures: tree canopy coverage change over time; inventory of

Percentage of tree canopy is one of the indicators of the Pittsburgh biophilia city program. More tree planting in all parts of cities strengthens environmental justice. Jacaranda lined avenue in Houghton, South Africa.

biodiversity; percent of city budget devoted to nature conservation, restoration, education; and other relevant indicators.

The biophilia hypothesis suggests that humans possess an innate tendency to seek connections with nature and other forms of life. Edward O. Wilson introduced and popularized the hypothesis in his book, *Biophilia* (Wilson 1984). In biophilic cities, large public parks fulfil an especially important role as a commons that welcomes everyone and at the same time supports well-integrated habitat and environmental quality. Grand old trees play a very special role in the minds of people. They convey the idea of continuity, of present generations standing on the shoulders of predecessors, and the promise of a future life. It is very easy to fancy Thomas Jefferson addressing a gathering under the old oak in Williamsburg, Virginia. Therefore, it causes great upheaval when old trees are cut down. Citizen outrage followed the removal of a 600-year-old oak in the midst of a city street in Stockholm in 2011, without public notice or cause given. People connect to and care deeply about large, old trees. Large parks are one of the few places in cities where trees can age gracefully and be cared for adequately.

The interplay between natural and human-shaped habitats is essential to understanding and appreciation. In caring for a large tract, not only the natural prerequisites but also the human-shaped changes, additions, alterations, and improvements must be noted to account for and respect the cultural heritage of the landscape. For several decades the citizens of Seattle, Washington, USA have valued Gas Works Park, designed by Richard Haag, an important landscape architect. Haag sought to shape a park on Lake Union employing the former gasification plant that fuelled lighting, operating from 1906 to 1965. Encountering much resistance, he persevered, working with the emerging field of bio-remediation, to resolve toxicity issues. Haag was able to retain some plant towers and tanks as ruins while others were stabilized and painted bright colours to enliven this historical layer in the public landscape. Connected to a network of city bicycle trails for easy access, this highly visible park symbolizes local identity, serving as the site for annual July firework displays and summer solstice celebrations, a democratic gathering of all citizens and visitors; and as a place for political gatherings, peace vigils, and protests. Likewise, citizens of the Ruhr Gebiet are able to understand and reflect on their predecessor's constant focus on coal mining as traces of the local coal industry are integrated into the Emscher

LEFT: This giant, 600-year-old oak in Stockholm was cut down in 2011, raising public outcry. People care about trees.

The wilderness in St. James's Park in London as depicted by Richard Wilson 1770–1775.

(Duisburg-Nord) Landschaftspark in Germany. Today these factory remains serve as historical texts for visitors; as places to climb and view; and as bird nesting sites.

Tangible heritage is more easily noticed—parks, mounds, houses, burial places, canals, etc.—than the intangible heritage of meanings and values expressed through traditions and practices. This immaterial heritage of practices and traditions, such as art, literature, music, cuisine, ceremony, and worship, plays an important role in shaping public identity, fostering mutual respect, and developing community cohesion. Landscapes, tracts, and specific places have been depicted in water-colour, oil or by pen, extolled by poems, novels or music. The Royal National City Park in Stockholm has been the object for literary and artistic searches for the soul of the landscape over the centuries. Likewise, contemplation about the complex interdependence of humanity and nature takes place in an urban park.

The central axis of New Delhi, the capital of India, is an iconic open space serving as a public park for gathering, protest, and daily recreational use. New Delhi was laid out in the early 20th century with about one-

TOP RIGHT: A valued industrial history transformed. Seattle's Gas Works Park.
MIDDLE RIGHT: Chapultepec Park in Mexico City is a great meeting place.
BOTTOM RIGHT: A place for people and free speech, India Gate, New Delhi.

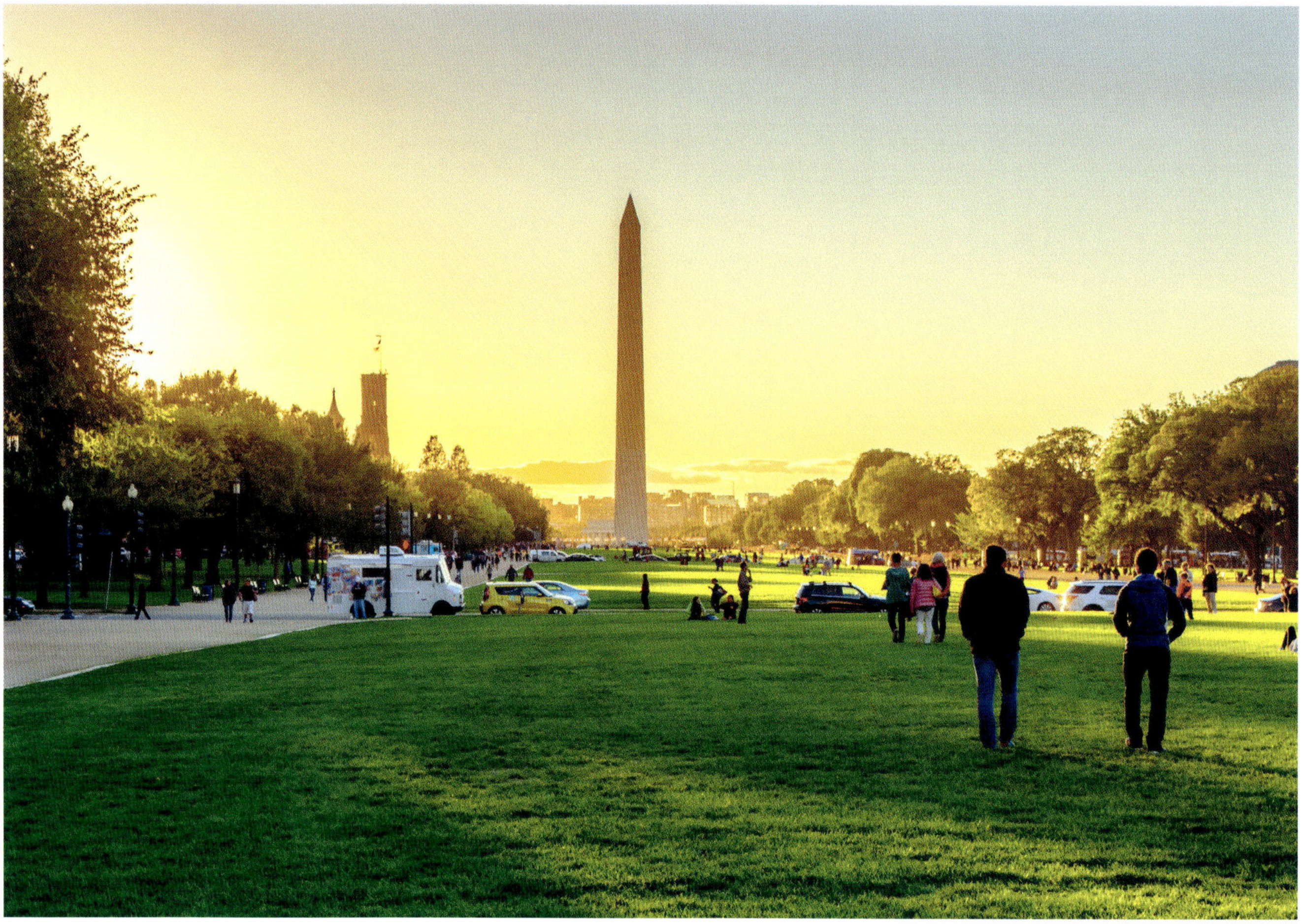
OOMKA SHUTTERSTOCK

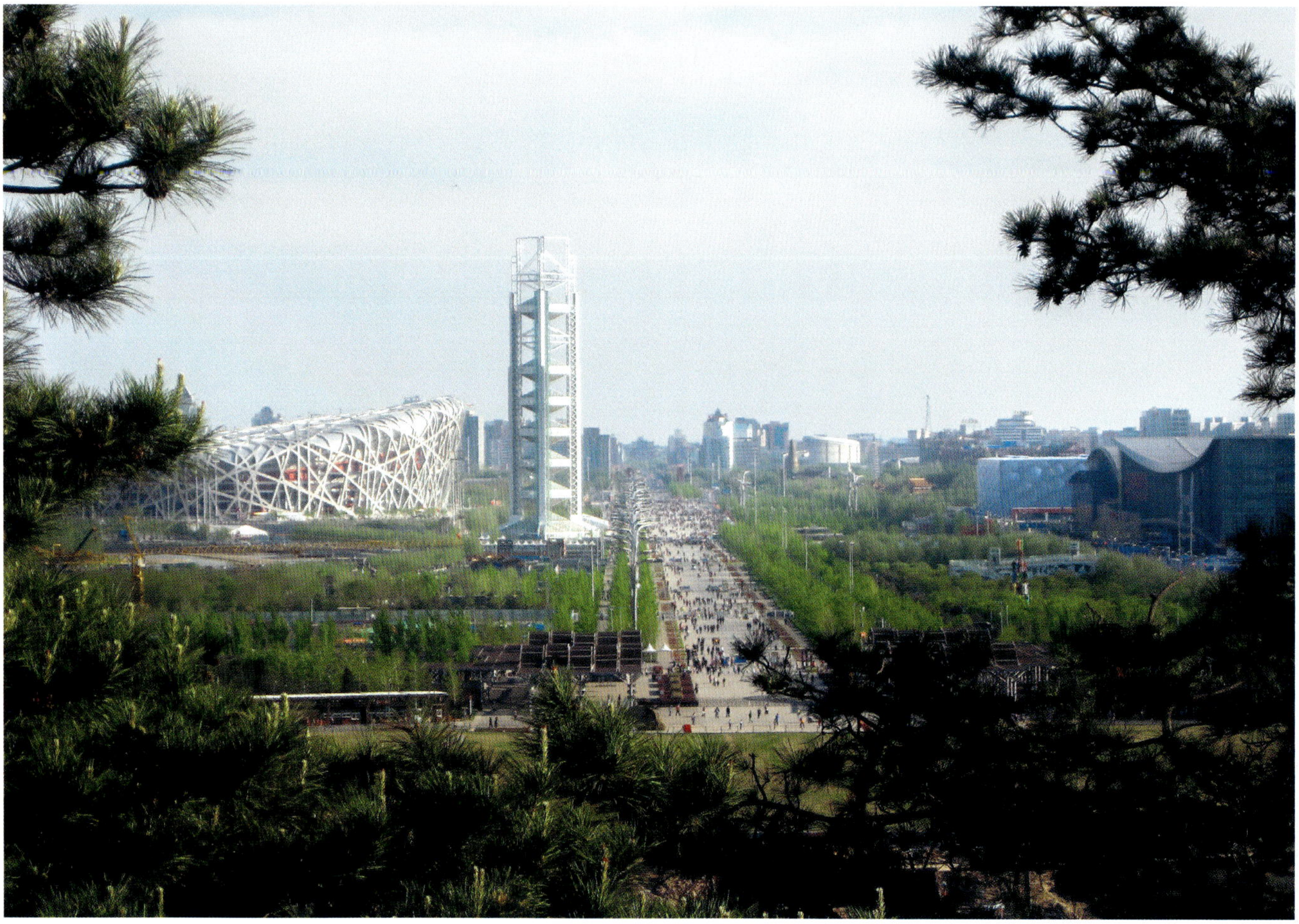
MAJ0191 SHUTTERSTOCK

third of the area as a landscape designed by a team led by Sir Edwin Lutyens. He drew on ideas of the Garden City movement of the late 19th century that continue to be a reference today, to permeate the area with green space. The open space of the central mall with long views of the India Gate was the setting for Gandhi's protests and the starting place for marches. These gathering places play a tremendous role in not only holding the society together but also in reminding citizens of their common history, hence helping to make democracy work.

The National Mall in Washington DC is a public park, essentially the front lawn of the United States of America. Originally it was planned as a grand avenue, but was later designed as a park by Andrew Jackson Downing in the early 1850s and Frederick Law Olmsted Jr. in the 1930s. From 2009 to 2015 the National Park Service, working with an interdisciplinary design team and contractors, completed an impressive green infrastructure upgrade to the 1-mile (1.6 kilometer) ribbon of green lawn that extends from 3rd Street to the obelisk of the Washington Monument. The National Mall and Memorial Parks area covers more than 4 square kilometers in the heart of the capital. With 25 million visitors and over 3,000 free speech and public events each year, the lawn of the Mall was battered, brown, and as dense as concrete. The lawn improvement project mixed sand and compost into the soils to improve fertility and percolation; constructed curbs with stormwater inlets piped to cisterns to harvest rainwater; added industrial strength deep irrigation that uses cleansed rainwater; and regrew the lawn for healthy cover.

The National Mall in Washington was created as part of an experiment to shape a democratic society. It has had and still has a tremendously important role as a shared outdoor public space of that society, where all citizens and visitors enjoy equal access. In China, in a different cultural context, Beijing's Olympic Forest Park was developed as a public space in 2008. It is positioned on the central axis of the city north of the Forbidden City and is bisected by the fifth ring road. The two parts are connected via a green bridge over the motorway. This large contemporary park employs 'Feng Shui' traditions to merge traditional Chinese landscape concepts and elements shaping an extensive modern oasis of 6.8 square kilometers. A swift tower was constructed as a habitat for a locally endangered bird. Multiple sports facilities and buildings constructed with energy-efficient technologies are well integrated within the park landscape.

The National Mall in Washington was created as part of an experiment to shape a democratic society.

TOP LEFT: US National Mall in Washington DC.
BOTTOM LEFT: A view from the Olympic Park of Beijing towards Bird's Nest.

To address the degraded and unsafe Jackson Park—a 2.2 square kilometer park on the shores of Lake Michigan in Chicago, United States, and the site of the 1893 World's Columbian Exposition—a talented interdisciplinary team interwove biodiversity and cultural authenticity in a US Federal Great Lakes Fisheries and Ecosystem Restoration (GLFER) project. Enabled through a private-public-civic partnership, the ecological health, historic integrity, and performance of this public landscape were upgraded. Project objectives required simultaneous rehabilitation of the historic park in accord with the Fredric Law Olmsted design; and restoration of terrestrial and aquatic habitats for pollinators, amphibians, small mammals, and more. The team of landscape architects, planners, ecologists, engineers, and construction specialists developed construction documents that rehabilitate the landscape's historic Olmsted character, improve habitat for a range of diverse species, welcome a broad range of daily park uses by people, and address management needs for a sustainable twenty-first-century urban park. Climate change amelioration aspects will enhance resilience. As landscape functions and scenic qualities improve, upgraded perceptions and uses will foster positive effects within adjacent urban neighbourhoods. The results are already receiving wide acclaim, not to mention more birds, insects, and increased usage. The construction and 3-year maintenance efforts of the project were completed in 2019.

Joining these recent documents, a wealth of studies at global and local levels aid in communicating the values of urban parks to the swelling

Jackson Park reconstructed after the 1893 Columbian Exposition, and in 2015 revitalized with the integrated access, habitat, and preservation project, Chicago, United States.

urban populace. For example, the World Health Organization (WHO) addresses the relevance of urban green spaces for public health, in a recent report titled "Urban green spaces: a brief for action." (WHO 2017) The WHO report positions urban green space as part of the green infrastructure of a city that should be widely accessible to all urban dwellers, noting in a headline that: "Contact with nature is an essential component of healthy cities."

The presence and stewardship of city parks are outward signs of good governance. Local governments, agencies, and elected officials hold responsibility for protecting and maintaining large public parks and reservations. In the most liveable cities public landscape is abundant. A growing trend is public-private partnerships for parks in the USA.

Recognizing that good governance supports parks underscores the importance of citizen advocacy and support for visionary leaders who value, protect, and support large parks.

In conclusion, large parks are social platforms, family-friendly and intergenerational. As urban assets that are free for all, varied cultural expressions can be welcomed, political voices can be raised, and citizen participation in management can be fostered. Large parks are cornerstones of social cohesion and inclusion that provide a foundation for a peaceful and pluralistic society to flourish. As the 21st century progresses and our Earth becomes an increasingly urban planet, large parks are irreplaceable resources of nature and culture for all people and the well-being of our planet.

Monnickendam

Zaandam

AMSTERDAM

AMSTERDAMSE
BOS

Diemen

Schiphol
Airport

Amstelveen

Weesp

Aalsmeer

A4

Tuinpark
Ons Buiten

Nieuwe Meer

Amstelveenseweg

Vrije
Universiteit

Boerderij
Meerzicht

Bosbaan (Rowing course)

Bosbaanweg

Amstelveenseweg

BUITEN-
VELDERT

Bostheater

*Grote
Vijver*

Grote
Speelweide

A9

Schipholweg

AMSTERDAMSE
BOS

Sportpark

De Braak

ELSRIJK

*Kleine
Vijver*

Schiphol
Airport

A9 Burgemeester van Sonweg

AMSTELVEEN

Rembrandtweg

STADSHART

A9

Bosrandweg

Schinkel
Bos

*Amstelveense
Poel*

KEIZER KARELPARK

Handweg

Noordpolderweg

Sportlaan

AALSMEER

Aalsmeerderweg

N

0   0,5   1 Kilometer

0   0,5 Mile

© OpenStreetMap contributors

AMSTERDAMSE BOS, AMSTERDAM, THE NETHERLANDS

SIZE: 10 square kilometers.

OPENED: 1934–1970.

OWNER: Municipality of Amsterdam. Management; Directorate Sport
and Forest, Department Amsterdam Forest.

VOLUNTEER ORGANISATION: Vrienden van het Amsterdamse Bos.

HOW TO GET THERE: *Metro station:* Amstelveenseweg (line 50 and 51).
Bus stops: Amsterdamse Bos (lines 242, 347, 348, 357 and 358)
Camping Amsterdamse Bos (line 199) and Nieuwe Meerlaan (line
186, weekdays only).

JAN HEEREN
Amsterdamse Bos, The Netherlands

The Amsterdam Forest, or Amsterdamse Bos, is completely man made and is a "child of its time." With 6 million visits yearly, a history of 85 years and a surface of 1,000 hectares, this large urban park has iconic value for the city of Amsterdam and possibly for The Netherlands as a whole. It is part of the social history of the city. Even its old historic name Boschplan is still used among older inhabitants of Amsterdam. The park is owned and run by the city.

Eighty-five years ago, the area of what is now the Amsterdamse Bos was agricultural land. As is the case in all of northwest Holland, it was a marshy area when people settled there 900 years ago. The soil was drained with a system of dykes and ditches that caused land to sink. Later, peat (for fuel) was exploited, causing further subsidence. 'Polders' (land gained from marshes or lakes with a controlled water level) were established with the building of larger dykes. The land was used for agriculture, mainly cattle as the soil was considered of a low quality.

Elements of this earlier landscape are still recognisable: old dykes along the eastern edge and other dykes, sometimes for only a shorter stretch, running roughly north-south through the park. An old polder (Polder Meerzicht) is still present in its original location with meadows for grazing cattle.

Water is abundant in the park. Nieuwe Meer, which has grown from a stream to a lake through erosion, is now quite prominent. In the south we find Amstelveense Poel.

One old farm, Boerderij Meerzicht, has survived and now serves as a pancake restaurant. Only one historical tree (a horse chestnut) from the pre-Amsterdam Forest period can be found.

The establishment of the Amsterdamse Bos is connected to the rapid growth of cities in Europe and America. Living conditions in the cities were poor and diseases and health issues were important challenges. At the turn of the 19th century this led to a debate in The Netherlands about the "parkenquestie" (the question of parks) given that in 1870 the Beekbergerwoud—the last natural forest in The Netherlands—was erased. A central theme in the discussion was that green spaces and parks were needed for the general well-being of the population. Thus, from the start, the impetus to establish the Amsterdamse Bos was driven by cultural and social motives.

This large urban park has iconic value for the city of Amsterdam and possibly for The Netherlands as a whole. It is part of the social history of the city.

ARCHIVE AMSTERDAMSE BOS

Milking 1934.

Jac. P Thijsse, a famous naturalist at the time, played an important role and took part in the Boschplan-commission, set up in 1928 by Amsterdam City to develop a large park.

The realization of the Amsterdamse Bos became part of a large national system of public works. The City of Amsterdam, having a high number of unemployed in the economic depression of the 1930s, directly took this opportunity and decided to have a large rowing course dug, 2.2 kilometers in length, and to raise the land and have the first trees planted. Subsequently, in April 1934, the realization of the park started. "Work for 5,000 unemployed for 5 years" was the propaganda slogan.

The first landscape design was more or less a classical English park with meandering paths and lanes. Its broad design is owed to Cornelis van Eesteren, director of the city planning department, and Jakoba Mulder, one of the first female architects from Delft Technical University. Mulder was the one who actually detailed the design and later succeeded Van Eesteren.

Van Eesteren and Mulder went abroad to pick up ideas, came back and wrote *Report of the Study trip for the benefit of the Forestplan to England, Belgium and Germany in October 1935.* What they had seen was sometimes directly copied and realized in the Amsterdamse Bos! Famous examples are the "rotten row" copied from Hyde park (in the Amsterdamse Bos this is "de galopeerbaan" or horse gallop bridle path)

TOP RIGHT: Polder Meerzicht.
BOTTOM RIGHT: Nieuwe Meer with Schiphol flight traffic control tower in the background.

MARIJKE MOOIJ

MISIGN, SHUTTERSTOCK

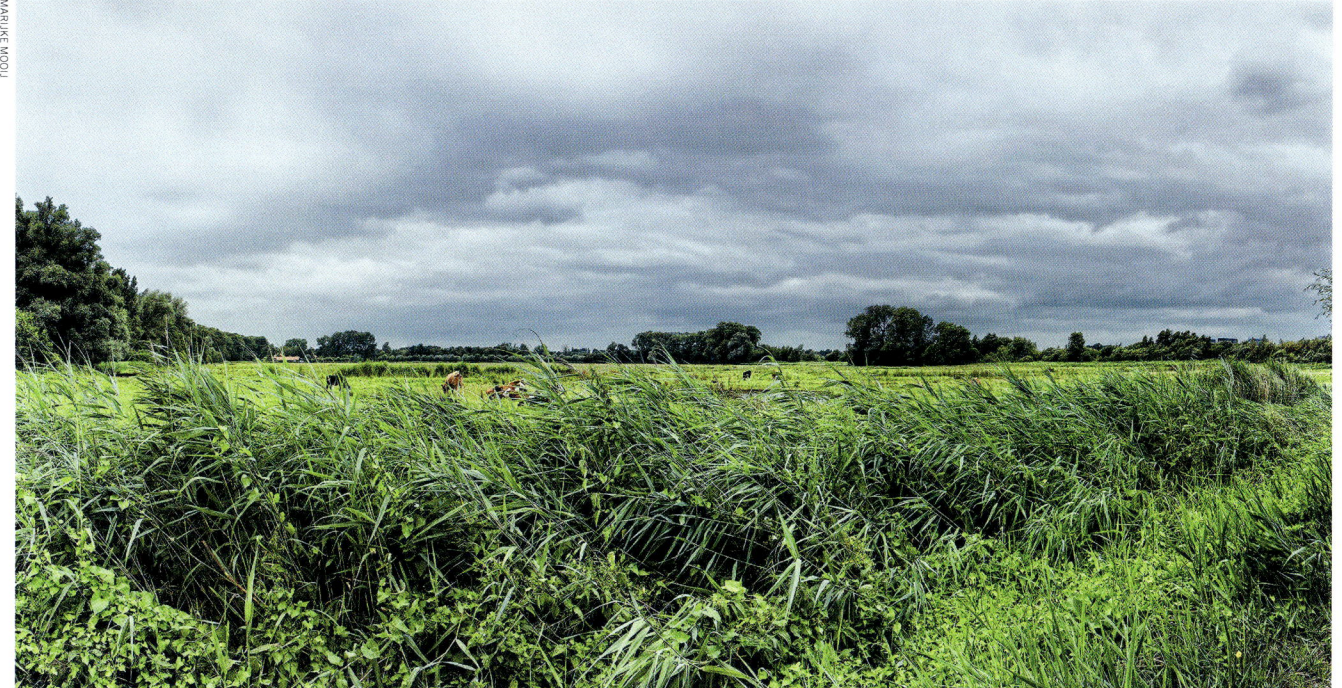

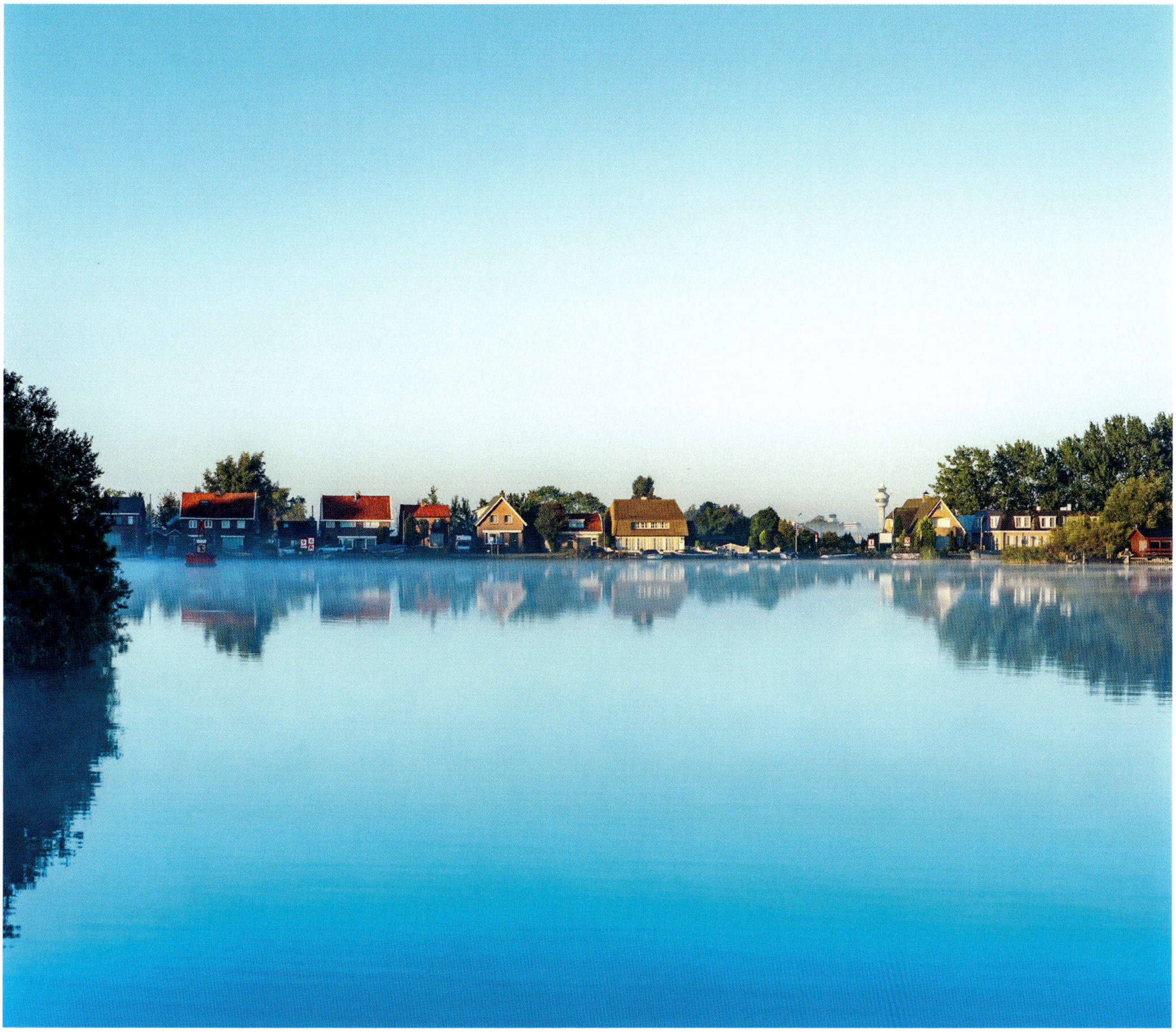

Amsterdamse Bos, Autumn.

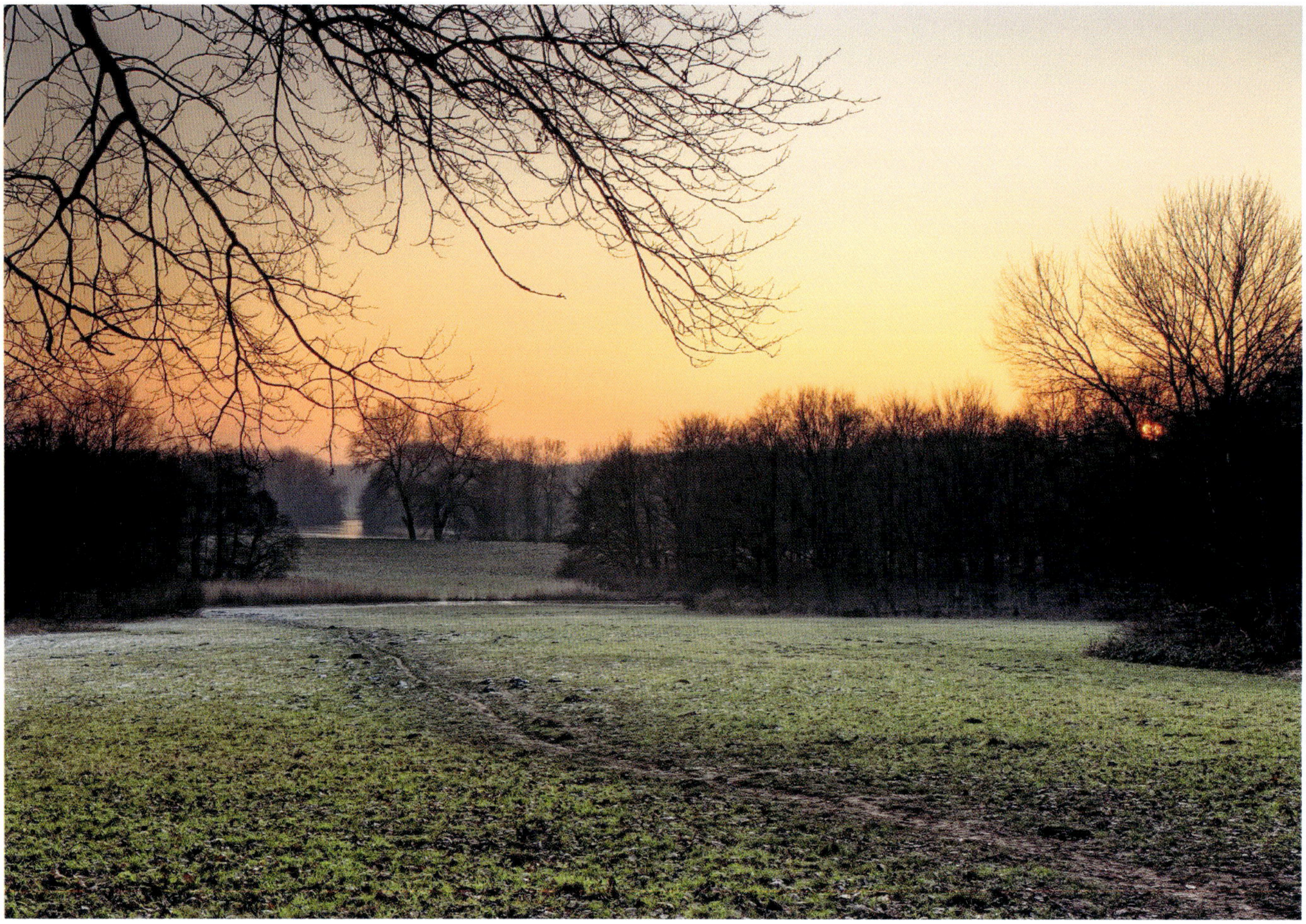

MARIJKE MOOIJ

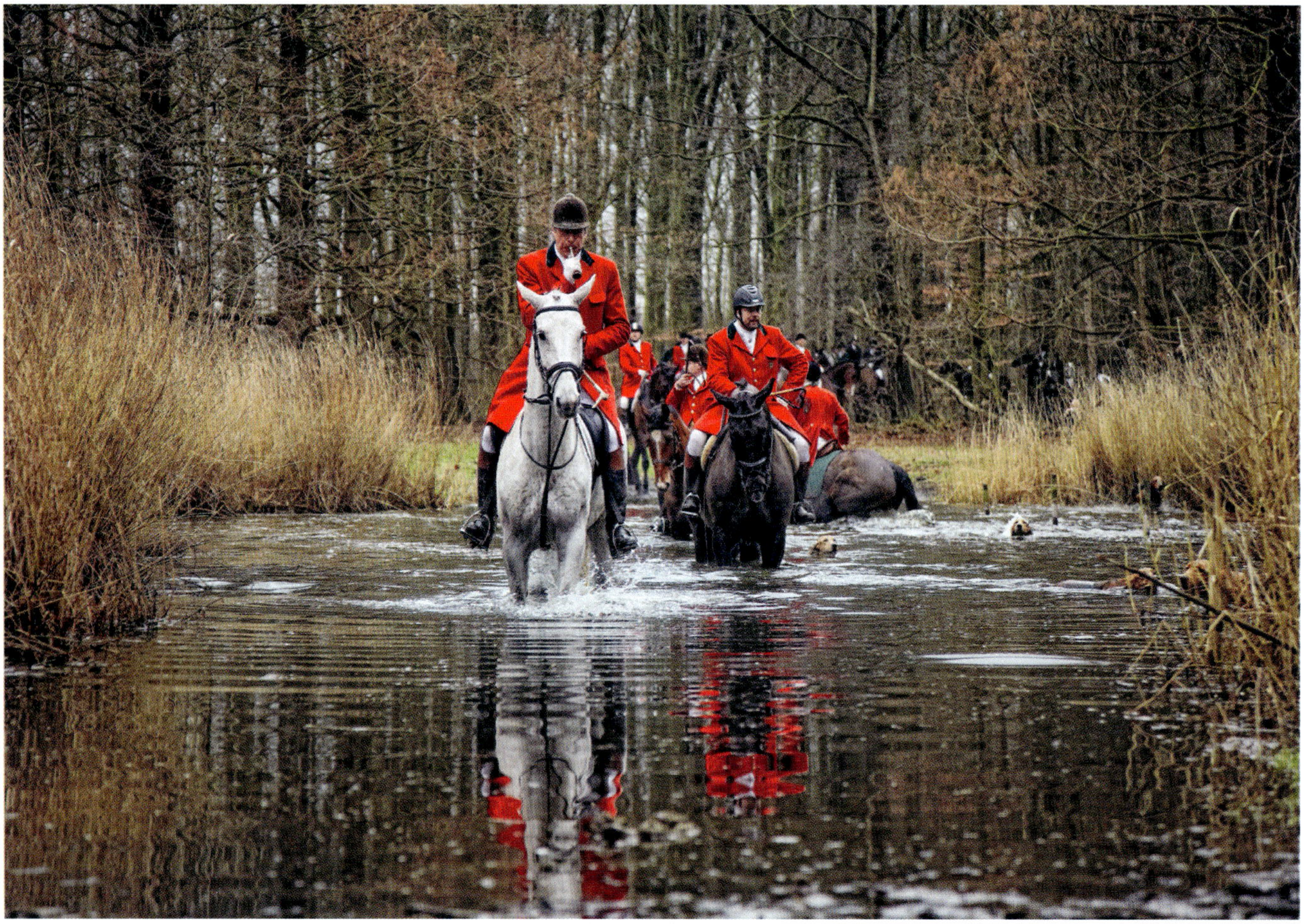

MARIJKE MOOIJ

ARCHIVE AMSTERDAMSE BOS

Relief work was organised on a massive scale to create Amsterdamse Bos.

as well as the view over the rowing course that was inspired by Hampstead Heath gardens.

A comprehensive plan was developed to meet all social and recreational demands. The following needs/desires were mentioned in 1934: fields for 'organized and unorganized' sports, meadows for sunbathing (even nude), a campsite, a youth hostel, an arboretum, children's playgrounds, a tree-nursery, lakes for canoes, an open-air theatre, flower gardens, restaurants, and a labyrinth. All these functions were carefully planned, although some were never realized. Others have disappeared in the course of time.

The emphasis on social functions of the park was directly inspired by the German example of the "Volkspark," especially the sports facilities. Mulder preferred 'unorganized' sports to have priority and created large, open grass meadows for play and recreation. Currently a professional hockey stadium has reduced the public space of the park to some extent.

From the start attention was paid to art. An open-air theatre with 1,500 seats has performances throughout the summer. Some 15 years ago two major works of art, sculptures by Uhlrich Rückriem, were erected in the forest. Temporary sculpture exhibitions have been organized in the forest, mainly with art made out of wood from the forest.

The park's landscape design followed the German example, namely,

TOP LEFT: Amsterdamse Bos, Heuvel.
BOTTOM LEFT: Hunting with scent in the Bos.

a large "activity axis" as an organizational backbone. It runs from the northeast corner to the "grote speelweide" (large playing meadow) in the heart of the northern half of the park. Vistas and perspectives were carefully planned. There are, for example, the large vistas of the children's big play-pond overlooking the great meadow and the large pond to the other shore, or smaller vistas over ditches and open waters. An important organizing element is an artificial hill, 16 meters high. Debris from the city was used as well as soil dredged in creating ponds and lakes. This hill is more or less the pivot of the wheel organizing the forest—its three spokes are the main vistas, one roughly to the north, one west, and one south. Elevation differences were another important aspect in the design of the park. In addition to the hill, elevation features were realized by shaping the slopes of the dykes around the forest quite gently. A significant aspect was the "alternation of dark and light." This was realized with small, carefully planned areas of woods, alternating with meadows and sweeping views. Sometimes a clear tunnel effect was achieved, out of the cool dark forest into the warm light of the meadows.

The forest was planted with trees specifically from the northern hemisphere. Large amounts of oak, beech, ash, and maple can be found, as well as some linden, alder, birch, and elm trees. All in all, some 200,000 trees can be found. At the start quite a few poplars were planted to protect the young trees against the wind. Dutch-elm tree disease and now ash dieback disease, the latter considered quite serious, have been detected in the Amsterdamse Bos. However, due to the mixed variety of trees, the diseases have fortunately not been disastrous.

The Arboreta as originally planned have disappeared, but sometimes a rare tree can be found, such as a giant Sequoia. The words "the Balkan" and "America" are still used for the areas that were originally intended for the Arboreta.

Roughly one third of the park is water, one third woodland, and one third meadow, a scheme that was also used later in large recreational areas in other parts of The Netherlands. Water is an integral part of the design and forms a sort of water network. It is necessary to manage water as the general water level in the park is −4.5 meters below the average. A system of sluices and pumps ensures a fixed water level, which ensures that the trees develop healthily and the roots do not rot.

In 1970 the forest was finished. In 2005 the Amsterdamse Bos was enlarged by adding the Schinkel Bos (35 hectares) in the southwest. This was in compensation for turning the "Vietnam Meadow" into tennis courts. The layout of the addition reflects changed ideas on landscape and park design. The idea is a design that rhymes with the traditional North Holland landscape of water and reedbeds, lined with willow, alder, and ash. No dramatic landscaping, no sports fields, but instead a visitor's trail into abundant birdlife. The landscape forms a marked contrast with the agro-industrial greenhouse area.

The Amsterdamse Bos is now part of a broader policy approach: the Structural Vision Amsterdam 2040. The park is one of the 'green wedges' reaching from the outskirts of town right into the city center. It is essential for air purification but, unfortunately, with the very busy international airport (Schiphol) close by as well as a large motorway cutting through the forest, the forest actually has quite a poor air quality score!

A major complication is that the A9 motorway runs through the Amsterdamse Bos and connects the central part of the Netherlands with the airport. A decision to enlarge and broaden the motorway was taken by the National Government at the beginning of this century. The assignment is to "blend in" the motorway in order to reduce the damage as much as possible. This is to be done by establishing a "new landscape" years before the work on the road will be finished. To begin with, 1,500 trees have been planted and 1.2 hectares of new forest were added by transforming hard surfaces. Existing wild-life passages will be upgraded. This provides an opportunity to restore some aspects of the historical design, especially some linden trees and a lane bordered with oak trees.

Demands on green spaces have changed over time. Today, sustainability and coping with climate change is a high priority. The 85 years old Amsterdamse Bos is put to the test. Will the design be able to house 'city-agriculture', new sports and recreation concepts, events with masses of people, etc.? A monumental City forest may not be "trendy urban," and must adapt to or at least reflect upon these trends. The original design of the Amsterdamse Bos seems to manage to be flexible without losing its fundamental natural and cultural values. Adaptation and adjustment to new wishes can be executed if done in a detailed and respectful manner.

The emphasis on social functions of the park was directly inspired by the German example of the "Volkspark," especially the sports facilities.

CHORZÓW

Poznańska

Tadeusza Kościuszki

Katowicka

Parkowa

Siemianowicka

Heritage park

Scouting Center

Planetarium

Narrow Gauge Railway

Festival area

Mars Pond

Promenade Gen. J. Ziętka

SILESIA PARK

Stadium

Mars Field

Funicular Railway

Zoo

Big Meadow

Chorzowska

Japanese Garden

Rosarium

OSIEDLE TYSIĄCLECIA

Rawa

Staw Maroko

Legendia

Amusement Park

KATOWICE

WEŁNOWIEC-JÓZEFOWIEC

Telewizyjna

Bytkowska

DĄB

Józefa Pukowca

Kolońska

N

0   0,5   1 Kilometers
0   0,5 Mile

SILESIA REGION

Piekary Śląskie

Bytom

SILESIA PARK

CHORZÓW

Dąbrowa Górnicza

Zabrze

KATOWICE

SILESIA PARK, CHORZÓW, POLAND

SIZE: 6.2 square kilometers.

CREATED AND SUCCESSIVELY OPENED: 1950–1968.

OWNED AND MANAGED BY: The Gen. Jerzy Ziętek Voivodship Park of Culture and Recreation Inc.

WEB-SITE: www.parkslaski.pl

VOLUNTEER ORGANIZATION:: Silesia Park Foundation, Silesia Park Volunteers Academy.

HOW TO GET THERE: By tram lines 6, 7, 11, 19 and 23 or bus 6, 820, 830, 840; there are several parking places for cars.

BEATA FORTUNA-ANTOSZKIEWICZ,
JAN ŁUKASZKIEWICZ AND PIOTR WIŚNIEWSKI

Silesia Park, Chorzów, Poland

Silesia Park[1] in Chorzów is one of the largest urban parks in Poland (approximately 6 square kilometers). The park was established at the very heart of an expansive industrial agglomeration at the juncture of three large cities: Katowice (capital of the Upper Silesia region), Chorzów and Siemianowice Śląskie.[2] The park was founded in a heavily degraded area to improve the quality of life of local citizens. It was developed in a very difficult time when the country was being rebuilt after the devastation of World War II. Massive in scale, the park is composed of several large functional areas, and features a rich recreational program. Its final form is attributed primarily to Władysław Niemirski,[3] a professor of landscape architecture. Assisted by a team of professionals—including landscape architects from the SKTZ Department (Formation of Green Areas) at the Warsaw University of Life Sciences (SGGW), Poland's oldest school of landscape architecture—Niemirski developed the park between the years of 1950 and 1968. Now, years later, the park is the model of successful land reclamation in anthropogenic landscapes, and demonstrates a remarkable revival of a landscape heavily polluted and ruined by industry.

Since the very beginning, development of the park's vegetation was of paramount importance, as evidenced by large scale planting works. Silesia region residents, future users of the park, eagerly and spontaneously participated in planting trees in many park areas. On average, 500–1000 citizens participated daily in park development, with up to 4000 people reporting for duty on a single day!

Silesia Park is located in a heavily urbanized area—dense residential developments are located 0.2 to 0.5 kilometers away. Every year the park is visited by approximately 3 million people. Throughout the year the park houses numerous sports, educational, cultural, and artistic events. It is a very important place for all of the region's inhabitants.

Program elements that required a more assertive urbanistic and architectural treatment were merged into the valley beneath the park's highest hill and along the main streets surrounding the park. A key element of the composition is a promenade running parallel to the main communication route connecting Chorzów with Katowice (Chorzowska street). The design comprises areas for various activities. On the one hand, there is the intensive southwestern part with its rich functional

The park is the model of successful land reclamation in anthropogenic landscapes, and demonstrates a remarkable revival of a landscape heavily polluted and ruined by industry.

TOP: Industrial landscape of the 1950s. BOTTOM: Silesia Park vegetation after 60 years. The landmark of St. Mary Magdalene church in Chorzów is seen in both photos.

program, while in the eastern part, numerous passive recreational areas were designed, for example a forested area with clearings and meadows. The park's functional program kept evolving as the park was being built, and was subject to constant modifications to suit the needs of its users.

By 1950 development of the park had begun. The Festival area with a large dance arena and water sports center was created first (1951). Key facilities followed: The Planetarium and Observatory on the hill (1955), which came to dominate the park, along with the grand Silesian Stadium (1956). Other attractions followed, including: a Zoological Garden (1958); Amusement Park (1960); playgrounds; the Scouting Centre (1963); the "Wave," a complex of swimming pools (1966); the Rosarium (1968); Heritage Park (1975); as well as areas for seasonal gardening fairs and numerous compositional layouts with recreational facilities and characteristic vegetation.

To improve communication over such a large area, a 6-kilometer narrow-gauge railway line was built (1957), followed by a cable railway

TOP RIGHT: The main promenade—the starting point.
BOTTOM RIGHT: The Big Meadow—the main interior of the park.

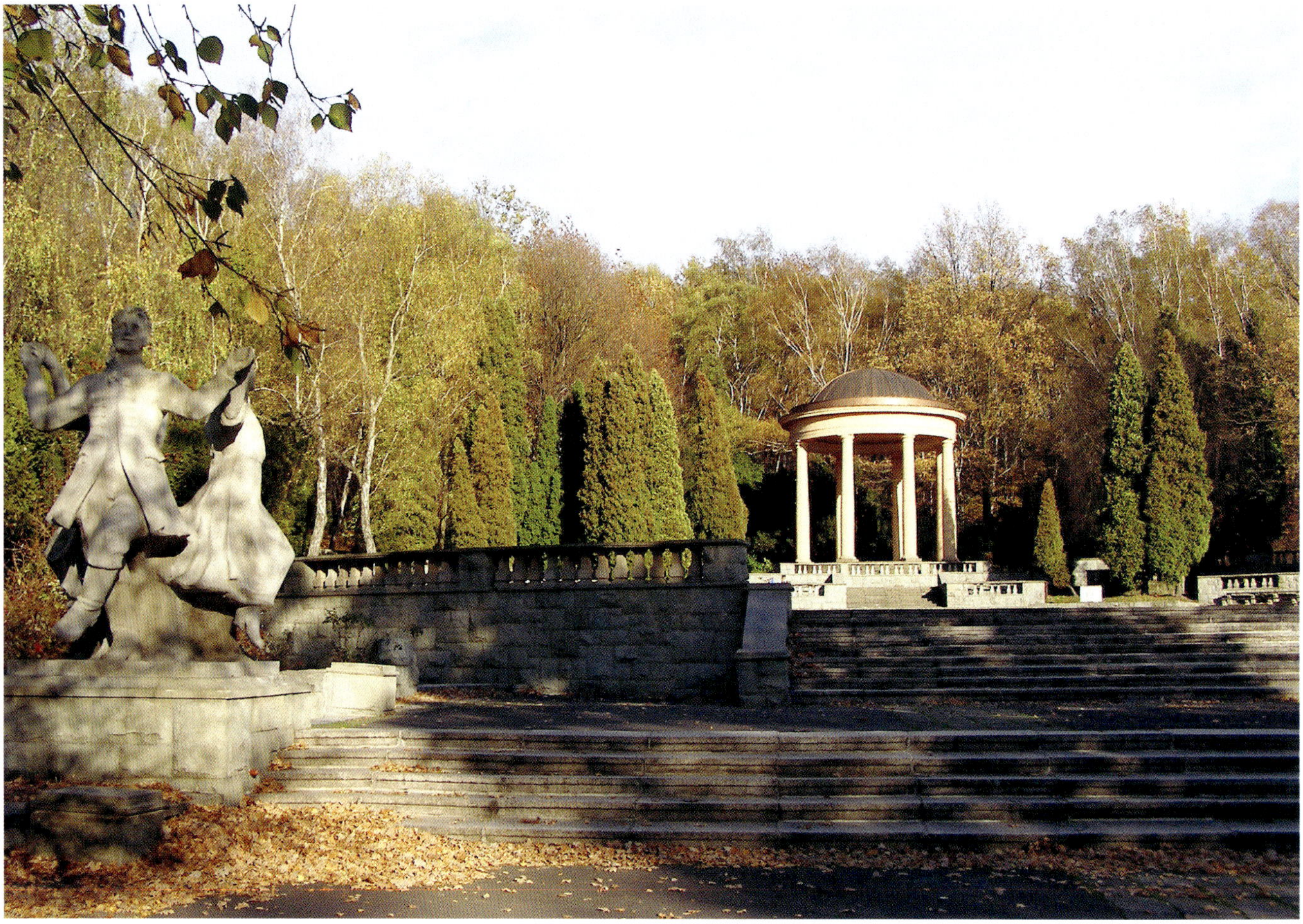

B. FORTUNA-ANTOSZKIEWICZ

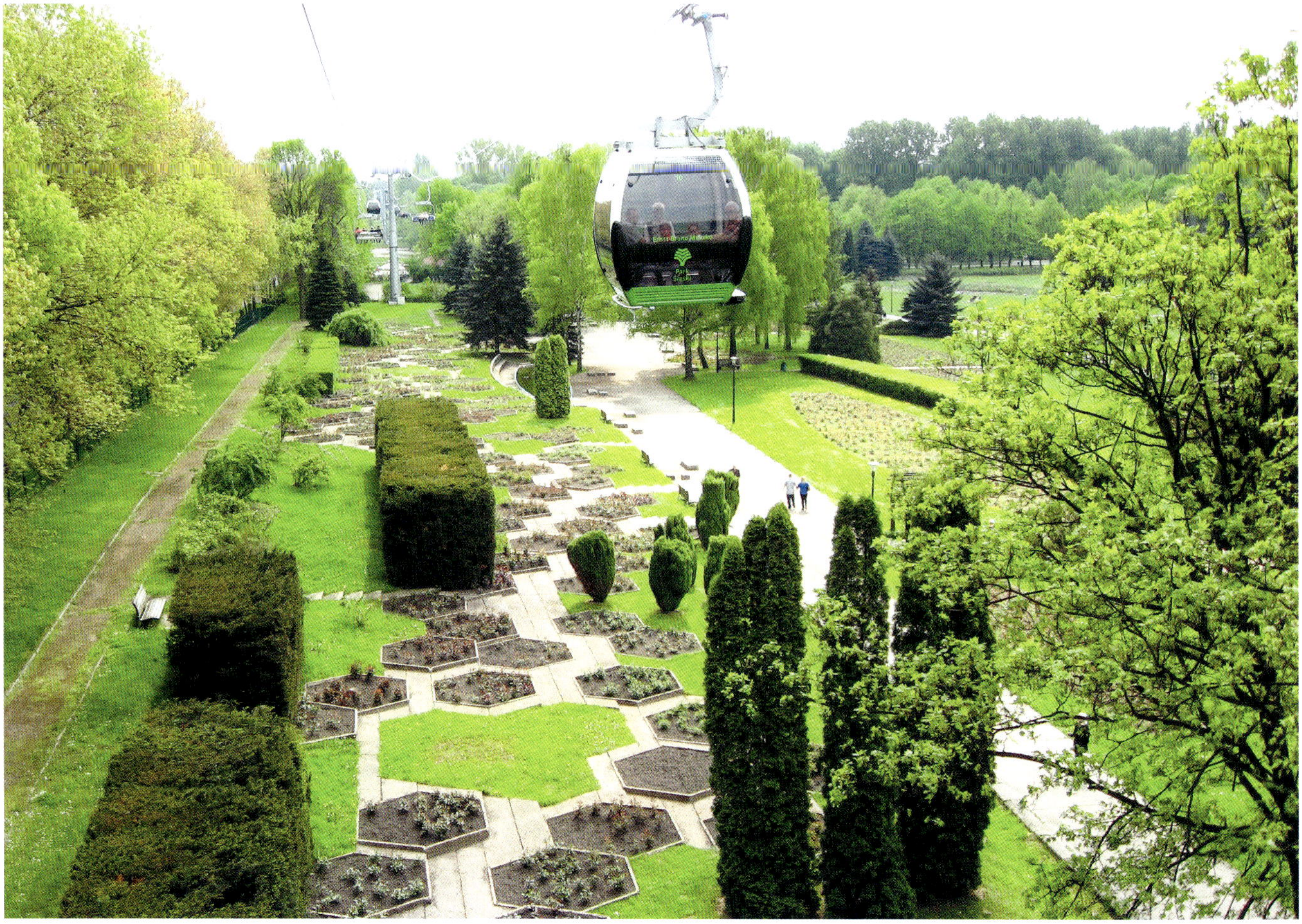

J. ŁUKASZKIEWICZ

line (1967). Together they form a triangular route through the park.

Silesia Park is one of the largest contemporary parks in Poland, which stands out with its original recreational and entertainment program. It also, despite being created in the spirit of socialist realism (imposed by communism), constitutes a leading example of a composition based on classical modernist[4] principles, which in their essence are strongly tied to the cityscape.

The design of Silesia Park, as was the case for model gardens of that era, centered around a fusion of landscape and architectural style. Into the geometrical layout of the area (accentuated in the western part of the park), some casual water and vegetation forms were introduced; rhythmical and asymmetrical layouts and contrasting designs were incorporated into the composition. The main features were positioned axially, avoiding the introduction of symmetrical patterns. The main theme was to counterbalance asymmetry with elements of a different character but similar in form (e.g. three-dimensional buildings in contrast with three-dimensional forms of the vegetation). Individual elements (architectural and natural) were interwoven to blur the boundary between architecture, vegetation, water, and land, creating harmonious spatial and visual patterns.

In some locations, there are thematic gardens (an alpine garden, a rose house, a perennial plant garden), ponds with aquatic plants, naturalistic adornments (boulders, picturesque tree silhouettes, flower compositions, vines climbing building walls, pergolas, trellises …). Modernism took inspiration from Japanese garden art, adopting its synthetic and symbolic form as well as sophisticated decorative vegetation compositions. It naturally followed that a Japanese garden was designed and is currently being completed. The designers extensively used natural materials - stone, wood, ceramic building materials, concrete, glass, and forged iron—exposing surfaces typical for each material.

Władysław Niemirski, whose skills and artistic sensitivity developed under the influence of modernist masters and philosophy, 'smuggled' the main precepts and principles of the movement (rejected in its entirety by the political establishment in Poland at the time) into the Silesia Park design. The exquisite quality of this spatial design is manifested in the perfect location of the main compositional axis and the very appealing manner in which the park's components have been handled, as well as in the superb functional quality of the park—even by today's standards.

TOP LEFT: Festival area with a large dance circle.
BOTTOM LEFT: Cable railway over Rosarium. Note the geometrical forms of rose beds introduced into an undulating landscape.

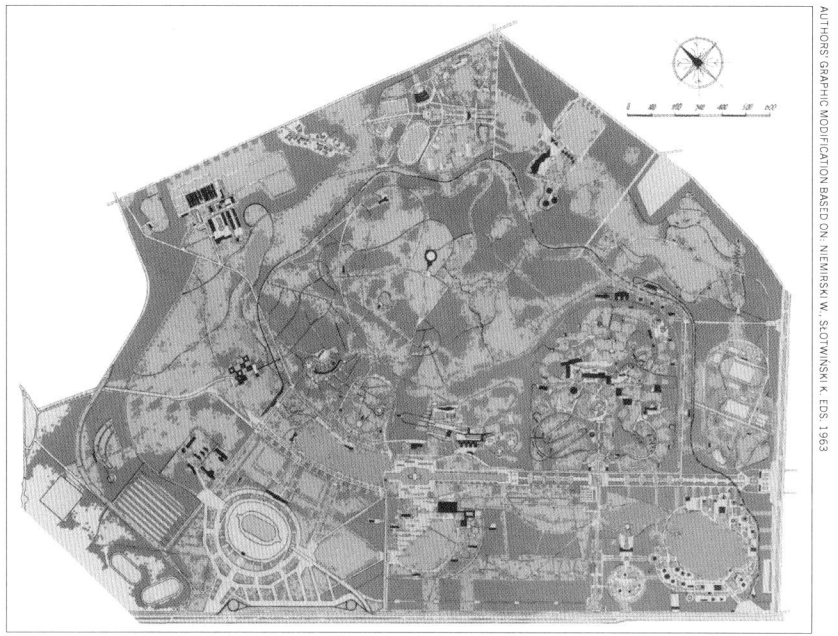

The Silesia Park master plan, 1962; main designer W. Niemirski.

Sixty years later, detailed research on spatial structure and vegetation development were carried out (Fortuna-Antoszkiewicz *et al.* 2013–2014, 2016, 2017). It was found that the park is divided into two distinct parts based on the way its space is utilized. In the southwestern "intensive" part, a rich functional program and vegetation are arranged into classical spatial forms (groves, groups, alleys, single trees, flowerbeds, etc.), delimiting clearly defined garden interiors. The most important functional elements are located here (Zoo, Amusement Park, Stadium, "Wave," ropes course) alongside thematic gardens (the Rosarium, the Japanese garden, the Perennial garden) which are clearly diversified in their compositional arrangement, constituting formal spaces with characteristic vegetation. All elements are in a sequential layout harmoniously linked with each other. The dominant feature of the park is the main promenade.

The eastern (hilltop) "extensive" part has a more limited functional program. The dominating feature of this area is a compact, dense tree stand—highly diversified and complex in terms of its structure (spatial, species, age), resulting from terrain variability and planting history (original plantings from the land reclamation period and compositional

TOP RIGHT: South-western part: an example of the park's interior with a single tree form (willow) as a main spatial element.
BOTTOM RIGHT: The main pond in the park's central valley, looking south.

B. FORTUNA-ANTOSZKIEWICZ

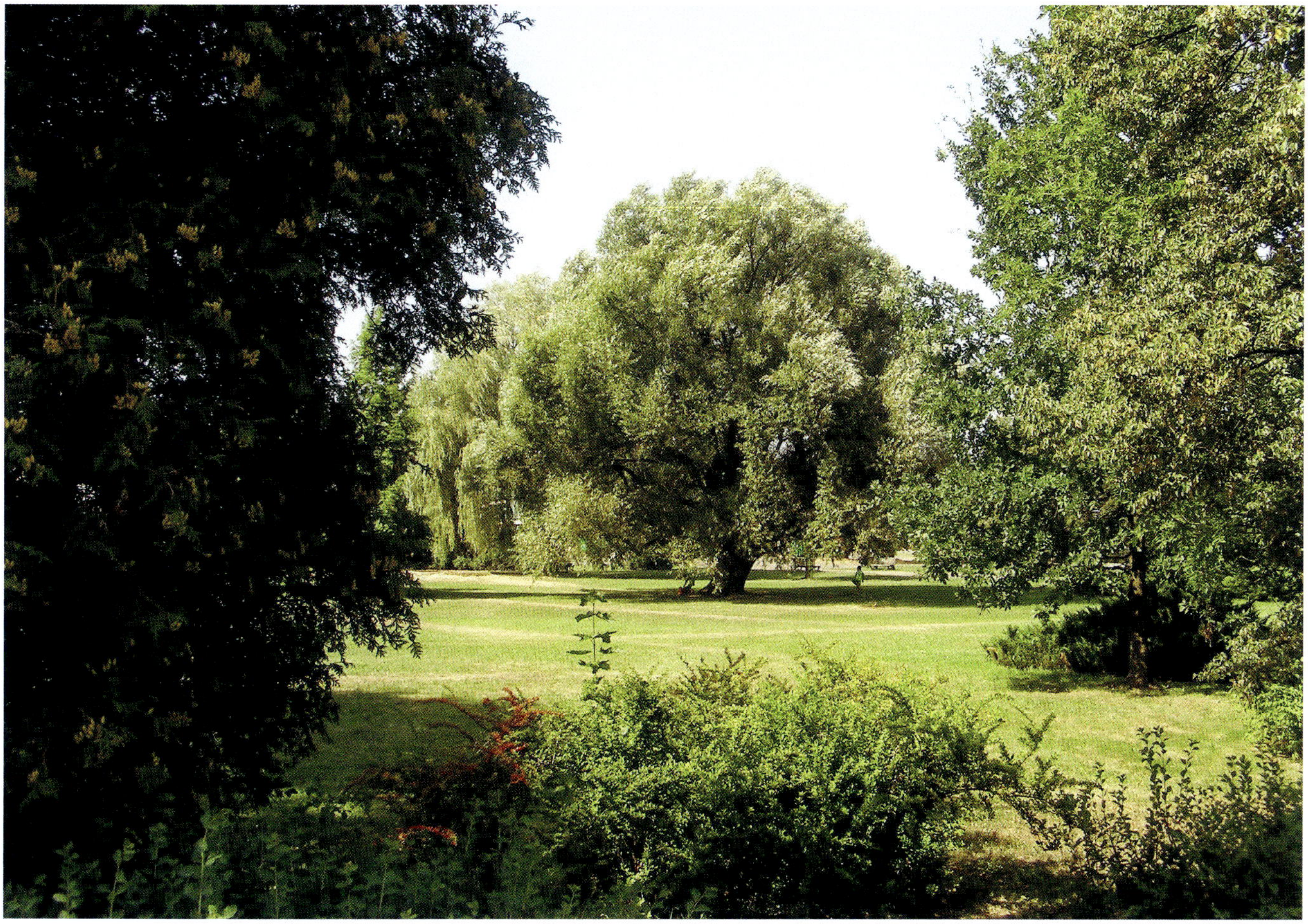

P. WIŚNIEWSKI

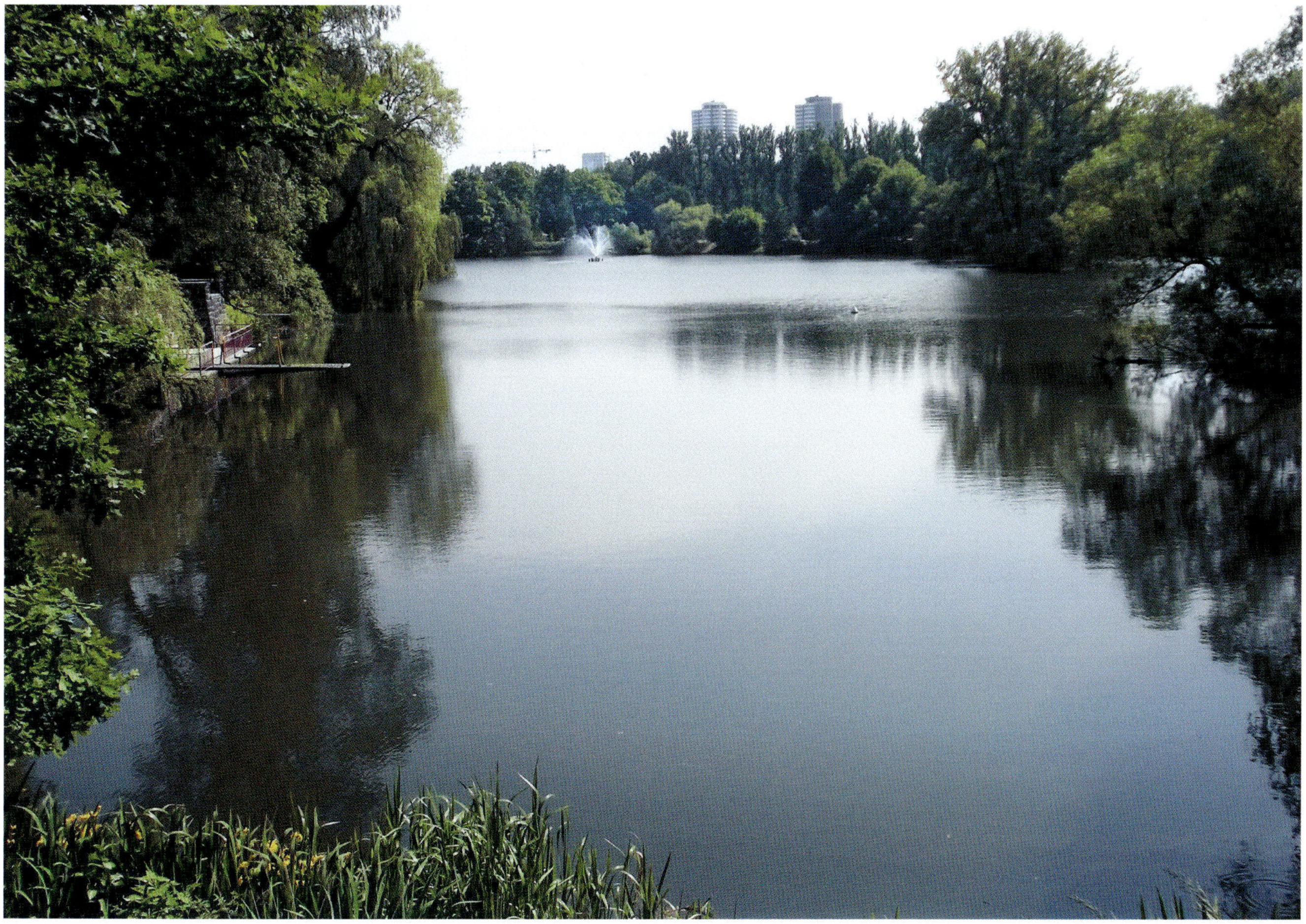

P WIŚNIEWSKI

P WIŚNIEWSKI

park plantings, such as a rock garden and Planetarium environs with a rich collection of conifer species and surrounded by water biotope).

Trees in the park are characterized by great diversification of species and varieties. They are distributed unevenly in the area - mainly a result of the division of the area into two formally distinct parts. Both native species (predominating in the extensive part) and introduced species, also rare and precious (mainly in the intensive part) have been identified, reflecting the original design and functional concepts.

Plant-communities of Silesia Park have reached a certain level of self-regulation and ecological stability. Due to successful soil reclamation (Łukaszkiewicz *et al.* 2017) and the formation of a specific plant microclimate, they now show signs of spontaneous biotic succession in which pioneering tree species (which were planted in the first stage due to their lower habitat requirement) now give way to more demanding ones[5].

The size of Silesia Park contributes to the green area's local and regional impact—only such a large green area with its compact spatial layout could have withstood all the harmful factors which have been present in its vicinity for many years (numerous steelworks and industrial plants in close proximity to the park emitted noxious substances in concentrations greatly exceeding norms; most of these factories are no longer in operation).

Thus, years after the park was created, its vegetation is well developed and in good general condition. This shows that the design premises (in terms of species selection) were ideally suited to spatial and habitat conditions. Time has proven the park to be a great success, even exceeding earlier expectations.

Only such a large green area with its compact spatial layout could have withstood all the harmful factors which have been present in its vicinity for many years.

[1] The full name is General Jerzy Ziętek Voivodship Park of Culture and Recreation. Ziętek (1901–1985) fought for Silesian independence from Germany after WWI and became Governor (Voivod) of Silesia and a member of the Polish Council of State. He took the initiative to create the park.

[2] The Upper Silesian conurbation is a unit of 19 bordering towns in the Silesia region in southern Poland, inhabited by 3,5 million people.

[3] Władysław Niemirski was a student and associate of prof. Franciszek Krzywda-Polkowski, the designer of the famous park in Żelazowa Wola, Poland (Fryderyk Chopin's birthplace).

[4] People's parks, reflecting at the time a trend in planning, were built in Poland even before WWII (e.g. a park in Łódź of 237 hectares, designed by Stefan Rogowicz, and built 1927–1939). It seems that in reality, Silesia Park's creators based their work on classical modernist principles and American parks such as Franklin Park and Jackson Park in Chicago, despite the fact that they were officially created according to the principles of socialist realism.

[5] In general, tree species typical for broadleaved forests, but also found in riparian forest and oak tree forest (mezzo- and eutrophic deciduous forests—*Querco-Fagetea* class) are increasing in number. Simultaneously, unfavourable changes are also visible—a succession of expansive and invasive (countrywide) species (e.g. *Quercus rubra, Prunus serotina, Reynoutria sp., Impatiens sp., Solidago canadensis*).

TOP LEFT: Horseback riding in the extensive eastern part of the Silesia Park.
BOTTOM LEFT: An extraordinary beech specimen in park's extensive part: 365 cm diameter at breast height (2013).

SOUTH MELBOURNE

Taylors Lakes

Airport

Bundoora

MELBOURNE

ALBERT PARK

Saint Kilda

Point Cook

Port Phillip Bay

Moorabbin

Airport

Ferrars Street

Albert Road

Albert Road Drive

Lakeside Stadium

Albert Sailing Club

The Point restaurant

Melbourne Sports and Aquatic Centre

Golf Driving Range

Aughtie Drive

Lakeside Drive

Albert Park Lake

Golf course

Albert Cricket Ground

St Kilda Road

Pasley Street

ALBERT PARK

ALBERT PARK

Mills Street

McGregor Street

Canterbury Road

Sports Centre

MIDDLE PARK

Queens Road

Wesley College Prahran

High Street

WINDSOR

Beaconsfield Parade

Melbourne Grand Prix Circuit

Princes Highway

Wellington Street

West Beach

Port Phillip Bay

ST KILDA WEST

N

0   250   500 Meters

0   0,3 Mile

© OpenStreetMap contributors

ALBERT PARK, MELBOURNE, AUSTRALIA

SIZE: 2.25 square kilometers.

LEGAL PROTECTION: Proclaimed a public park in 1864 and a reservation in 1876.

OWNED AND MANAGED BY: Statutory authority Parks Victoria.

HOW TO GET THERE: Yarra Trams Route 12 from Collins Street to stop 131 or Yarra Trams Route 96 from Bourke Street to stops 128, 129, 130, 131, 132 along the western boundary of the park.

FRAN HORSLEY

Albert Park, Melbourne, Australia

Albert Park, the playground of the people,
one of the few lungs of an ever-growing city …

The Age Newspaper 29 May 1939 [Bernard and Keating 1996]

Albert Park, "the playground of the people," as it was described in 1939, captures two exceptional values of large urban parks; as places of recreation and social interaction, and as "breathing spaces" supporting ecosystem services essential to the health of our growing urban populations. The story of Albert Park, just three kilometers from the center of Melbourne and hosting 6 million visits a year, exemplifies these benefits being enjoyed by multiple generations of Melbournians and visitors alike. Albert Park is, and has always been, a place that embodies the passions of people and the resilience of the landscape.

An area of 364 hectares was first proclaimed a public park in 1864 and named after Prince Albert, Queen Victoria's Consort. Permanent reservation of a diminished park size of 230 hectares did not occur until 1876. Since then it has reduced in size to 225 hectares, the most recent excision enabling creation of a much needed new primary school. It has, however, increased in its importance in providing nature-based recreation in a city with a population of nearly 5 million, projected to grow to 9 million by 2056.

Prior to European colonization, the traditional owners used and cared for the resources of the once swampy lagoon and wetlands area as part of a wider coastal landscape. The tangible and intangible values are part of the living history of the area. A large old red gum, the *Ngargee* tree, a Boon Wurrung[1] word for "a gathering, a celebration of community," remains as a tangible connection to their culture.

From humble beginnings, the now distinctive lake, fringed with palm trees and eucalypts against the backdrop of the city skyline, has been taken to the world through televisation of the Australian Formula One Grand Prix held in the park annually. Over time, the park has developed to accommodate a diverse range of activities including playing, walking, jogging, sailing, rowing, cycling, golf, picnicking, amateur and professional sports, and community and major events. It is also loved and valued as an urban oasis supporting birdlife, urban cooling and greening, and storm water management.

The fundamental importance of urban parks as places to connect city dwellers with nature for health and wellbeing was recognised at the time of Melbourne's establishment in the 1830s.

Albert Park, the playground of the people.

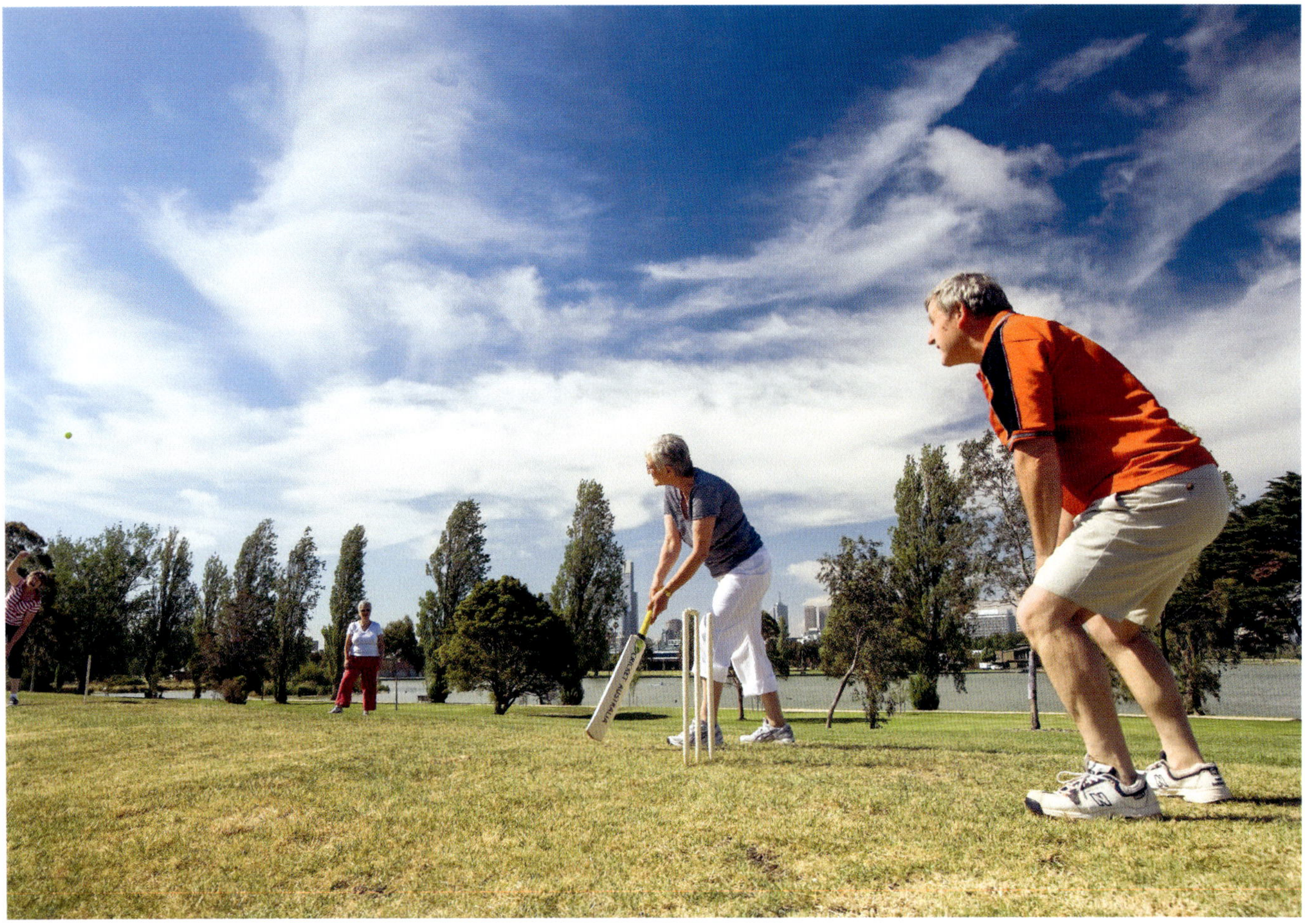

PARKS VICTORIA

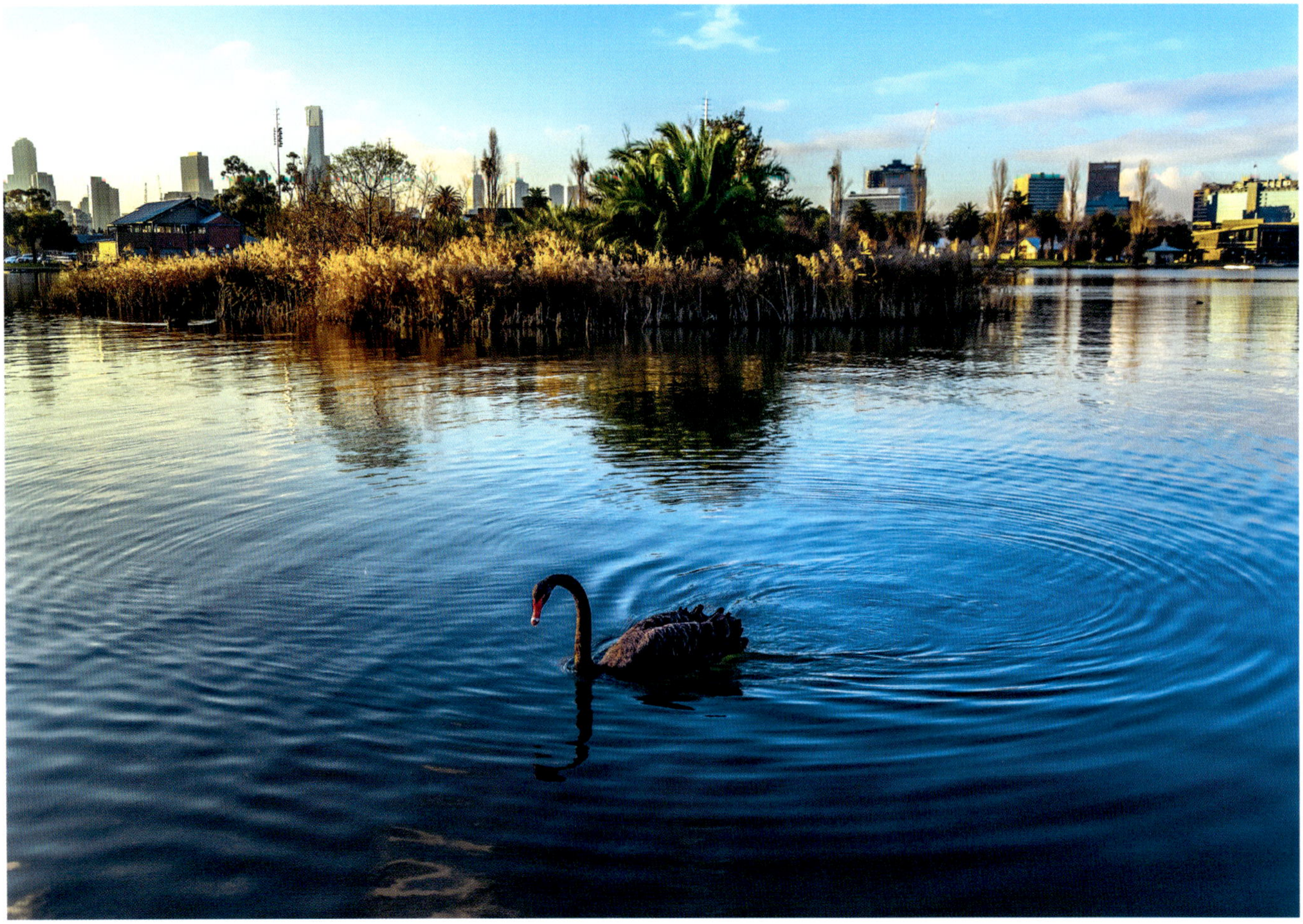

PRO DESIGN STUDIO.SHUTTERSTOCK

The fundamental importance of urban parks as places to connect city dwellers with nature for health and wellbeing was recognized at the time of Melbourne's establishment in the 1830s. As the young town grew, so too did the campaign for a "South Park," a breathing space for a population of tens of thousands. By 1853, the swampy land surrounding a natural lagoon and softened by scattered eucalypts was taking shape as the people's informal playground. This decade saw the area's use for amateur cricket begin, with archery, hunting for plentiful waterfowl, and cattle grazing also prominent.

With its declaration as a public park, and progressive development into a much-loved recreation reserve during the 1860s–1890s, came an inevitable increase in park regulations. The people's playground was at risk of becoming the people's battleground as interest groups appropriated areas, causing conflict amongst different user groups. A bemusing mix of social gatherings, school picnics, cattle and horse grazing, camping, hunting, cricket, Australian football, lacrosse, polo, archery, rubbish dumping, events, and informal horse racing graced, and disgraced, its grounds. The conflicts and demands on the park were further exacerbated over time with its reduction in size from the sale of land for residential development. Early plans for a cemetery were fortunately abandoned under the threat of it becoming a sinking field of watery graves. During World War II and beyond, a heavy army presence in the form of tents, barracks and use, including hosting American soldiers, dominated the south west area of the park.

By far the largest and most controversial use that has shaped design and purpose of the park in the modern era is the hosting of the Australian Formula One Grand Prix. The first Grand Prix in the park was enjoyed by over 100,000 visitors in 1956, only to be banned two years later following public uproar. The decision to revive the event in 1994 gave rise to one of the most sustained community protests that Melbourne has ever seen: the "Save Albert Park" campaign to stop the race. The group continues to be active to this day. Despite their efforts, a tired somewhat neglected park was transformed into a venue for an international car race, bringing benefits to park users through a major investment in redeveloped visitor facilities and park infrastructure. The event itself is an extraordinary example of coordination, involving 280,000 person hours over a four-month period each year. It requires assembling and dismantling infrastructure across two million square

TOP LEFT: Cricket in the Albert Park.
BOTTOM LEFT: There are black swans.

meters of parkland, including 3,400 square meters of marquees, 24 kilometers of temporary fencing, 3,324 concrete and safety barriers, and moving 6,000 tonnes of gravel, with an inevitable cycle of impact and reinstatement. In 2019, the Grand Prix attracted record crowds of over 324,000 visitors and was watched by tens of millions across the globe.

The size, shape, and design of the park mirror the ebb and flow of the city's development and the outcomes of debates between and within community and governments. But many elements have remained constant throughout its history. The park's shallow lake dotted with sailing boats, rowers, and wetland birds is the heart of its identity. Formalizing the once natural salt water lagoon into an artificial lake commenced in the 1870s with two clear objectives. Its primary value lay in its recreational purpose. However, Melbourne's urban growth had also disrupted natural water flows, and combined with upstream pollutants, the lagoon had become a health issue. The design solution gained much support, but funds were less abundant. The lake's depth continued to compromise its reliable use even as boathouses and clubs were being created with optimistic fervor. Works began in 1873 and continued in the 1880s and 1890s to clear aquatic vegetation, install a pump house, and create a promenade around the lake as the defining feature of the park. Managing aquatic vegetation and water levels remain ongoing park management challenges even today.

The lake's value as a critical part of a wider urban storm-water treatment system, not only provides the city with flood mitigation benefits but improves water quality before discharge into Port Phillip Bay.

In 2019 Parks Victoria released a new masterplan to guide the next 25 years of Albert Park's use and development. The plan reinforces the need for flexible, multi-purpose community use, allowing for programming of different activities, often under licenses and event permits. It targets park improvements to increase the resilience of the park rather than dramatically change the overall layout. Facilities and the environment will be better designed and managed to respond to, and mitigate, the impacts of climate change. Consistent with its long history, the community consultation period again raised many competing ideas and demands. The draft plan had included one option to reduce the 45-hectare, 18-hole golf course in size to create more casual open space for the growing population. After a strong campaign by golfing advocates, the idea was abandoned.

The park improvements are outlined under three key themes: Nature and environment; Community connection; Healthy and active. Proposals include a new wider path around the lake, a softened wetland

Consolidation of like activities and minimizing areas of exclusive use in favor of flexible spaces that allow for multiple use will help manage the varied demands on the park.

edge, more resilient sporting field surfaces to accommodate the need for more shared use, increased plantings for shade and habitat, and a clear objective to ensure the park becomes one of Australia's most accessible for people of all abilities.

Consolidation of like activities and minimizing areas of exclusive use in favor of flexible spaces that allow for multiple use will help manage the varied demands on the park. Supporting the dramatic growth in community events, such as fun runs and fundraising events, will require a balanced programming and scheduling approach. Time share management and alternative surfaces will extend the use of sporting ovals. With a nod to the desire for more informal and intimate community gathering spaces, mixed plantings providing shade and amenity, and improved design of the park's street frontages will help bring nature to the locals and locals into nature. Attention to "quality" is to the forefront of managing the complex demands on a finite space.

In response to increasing urban density and climate change, the masterplan is also focused on Albert Park's role in evolving to showcase a more environmentally sustainable ethos. Along with energy efficient lighting, storm water harvesting, and alternate sports ground surfaces, better facilitating sustainable transport and reducing congestion in the park is outlined. Walking and cycling links to and within the park will be safer and more legible, including trails linking to a major new Melbourne metropolitan multi modal transport interchange nearby.

Albert Park, as it has since its beginnings, will continue to evolve and change with equal measures of exhilaration and exasperation expressed by the community. But what is wonderfully enduring is that it remains a place where people simply gather in nature for sustenance of spirit and body and in doing so, create cultural and personal stories and memories. As Melbourne's population grows around it, the value of this large urban park as a breathing space for everyone will also continue to grow.

[1] A tribe of the Kulin Nation.

RIVIERENHOF

Main entrance

Riverenhof Castle

Sports ground

Sports ground

Sterckshof castle

Mirror Pond

Roeivijver

Open Air Theater

Visvijver

Schijn

Sterckshofler

Hooftvunderlei

RIVIERENHOF

Schelde

RIVIERENHOF

ANTWERPEN

Airport

DEURNE

MORKHOVEN

DRIEKONINGEN

BOTERLAAR – SILSBURG

KRIEKENHOF

Muggenberg Arena

Sint-Fredeganduskerkhof

Palinckstraat

Te Couwelaarlei

Leeuwlantstraat

Boshovestraat

Turnhoutsebaan

August Van de Wielelei

Ruggeveldlana

Herentalsebaan

Boterlaarbaan

E19

E34

E313

P

0   250   500 Meters

0   1000 Feet

N

© OpenStreetMap contributors

RIVIERENHOF, ANTWERP, BELGIUM

SIZE: 1.35 square kilometers.

OPENED in 1923.

OWNER AND MANAGER: Province of Antwerp. No special protection, "park" in the city zoning regulations.

HOW TO GET THERE: *Public transport:* from the city centre take tram 10 (stops Lunden, Venneborg and Schotensesteenweg are near the central part of the park).

PETER VERDYCK

Rivierenhof, Antwerp, Belgium

The Rivierenhof Park is the largest and most important park in the Belgian city of Antwerp. It has a surface of 135 hectares and is visited by an estimated one million people a year. It is managed by Antwerp Province and is home to a wide variety of activities. Undoubtedly, it exerts a large influence on the quality of life of many people from Antwerp and neighboring communities.

The current Rivierenhof Park originates from two summer residences whose antecedents are 15th century farmhouses that dated back to the early middle ages. In the 14th century, the first one, Hooftvunder, belonged to the powerful van Deurne noble family. At that time the second one, Ter Rivieren, was not named, but we know that from 1382 the grounds belonged to the well-known van der Biest tenant family and were used as farmland. During the following centuries several different families became owners of the respective properties.

The Rivierenhof is mentioned for the first time in 1514 when the Antwerp merchant Magnus van Bullestraete was owner of the castle. This family had it as a summer residence, just like many other upper-class families had summer houses in the countryside and by the sea. Hooftvunder castle became property of the banking family Sterck, which explains its current name Sterckshof. The next important fact is that the Jesuit-order bought Ter Rivieren in 1618 to make it a place of rest, study, and prayer. It was small, with a kitchen garden and a piece of woodland. It burned down and was rebuilt as a monastery. When the plague broke out again in the early 17th century, the Order opened the park so the inhabitants of Antwerp could walk there and get some fresh air. In 1693 they also obtained the Sterckshof castle.

Thus, in the 17th century both castles had become property of the Jesuits, who transformed the farmland into a park. After the suppression of the Jesuits by decree of Pope Clement XIV on 21 July 1773, they soon moved out of the park and castle.

In 1776, banker Jean-Baptiste Cogels bought the premises of Rivierenhof at public auction. Two years later, in 1778, he bought Sterckshof, and proceeded to join the two properties together. Subsequently, he built a new classical style castle in Ter Rivieren. Then, in 1780, the development of a 35-hectare park began. The park was developed by Charles De Wailly (architect of King Louis XVI of France) and included among other

The price of the Rivierenhof was high—costing 65 percent of the yearly provincial budget.

PUBLISHER G. CEULEMANS, PHOTOGRAPHY E. DESAIX, BRUXELLES

Entrance to the park. Postcard ca. 1925.

attractions: avenues, a labyrinth, statues, pavilions, a Chinese tower, an ice cellar, and a 1.5 hectare mirror pond.

In 1921 Rivierenhof Park was purchased by the province of Antwerp and became the first provincial park in Belgium. The price of the Rivierenhof was high—costing 65 percent of the yearly provincial budget (5,050,000 Belgian francs out of a total yearly provincial budget of 7,800,000 francs). The surface at the time of purchase was 87 hectares, of which 40 hectares were developed as a park.

Until the 19th century, parks like the Rivierenhof were a privilege of the aristocracy and the rich. Then the province chose to develop a new type of park, one that would not only be a place for the bourgeoisie to take strolls, but rather a park with sports and recreational facilities for everyone—a park for the people.

On 17 March 1922, an international competition was launched for the design/development of the park. An international jury received 43 proposals and the winner was Guillaume De Bosschere from Antwerp. He chose to expand the English landscape and to include sports infrastructures and other recreation facilities, an agricultural school, an arboretum, a tree nursery, a rose garden, and other park highlights. The further development of the park in an English landscape style took 10 years. As the park is situated on the banks of the river Schijn, the area was very swampy. Consequently, a system of ditches and two large ponds, one for rowing and one for fishing, were dug.

TOP RIGHT: Rivierenhof castle at the Mirror Pond.
BOTTOM RIGHT: The resting pavilion. In the background: Sterckshof castle.

PROVINCIAL PARKS REGION ANTWERP

PROVINCIAL PARKS REGION ANTWERP

PROVINCIAL PARKS REGION ANTWERP

PROVINCIAL PARKS REGION ANTWERP

The official opening of Rivierenhof Provincial Park took place on 7 May 1923. In the period 1923–1928 the park was extended with another 40 hectares by expropriations and purchase. The area of the park today is 135 hectares.

After many years of adding activities, there was little logic left in the spatial structure of the park. Valuable open spaces in the center of the park were taken by sport fields and could not be used by the general public. Furthermore, cars were parked in the large avenues. It was clear that the activities in the park had to be re-organized in a more logical way. Accordingly, the province created a new structural vision for the park, which was finally approved by ministerial decision in 2009. The process leading to the new plan involved about 15 different organizations giving advice about different aspects of the park (heritage, nature, sports, water management, etc.). After the identification of all activities and all values of the park, the area was divided into four main parts, each with a function of its own:

For such usage the event must add value to the park just as the park must add value to the event.

- Sports: two sport-cluster areas, one at the western and one at the eastern side of the park, close to the edges and to parking facilities.
- Heritage and gardens: the central part for soft recreational purposes, home to different themed gardens and most of the historical monuments.
- Nature: an area with a high ecological value close to the valley of the river Schijn.
- Management, administration, and education: an area where all the technical, staff, and educational buildings (school, children's farm) are present, including the main entrance area.

Put on hold was the issue of how to handle vehicle traffic through the park, north-south, in the western part of the park. Eventually, the road might be (partly) covered.

Once this vision for the future development of the Rivierenhof was worked out and approved, subsequent actions were undertaken to execute this vision. Realized projects include the renovation of several avenues, renovation of the arboretum, creation of several themed gardens, moving of sport infrastructure, etc. At the time of writing, several new projects are in progress, some are completed.

Currently, activities in the park are very diverse. They fall into five

TOP LEFT: Open-air theatre in the middle of the park.
BOTTOM LEFT: Art in the park. Maarten Schaubroek "Eagels."

PUBLISHER G. CEULEMANS; PHOTOGRAPHY E. DESAIX, BRUXELLES

Avenues in the park. Postcard ca. 1926.

categories: recreation, sports, botanical garden, nature conservation, and education.

Recreation has always been an important function. Many investments were made to create recreational infrastructures for both individuals and clubs. In addition to facilities for sport clubs, the park also hosts a club for miniature boat hobbyists and one for miniature steam train hobbyists. There are spaces available for petanque, miniature golf, a labyrinth, and several playgrounds. Picnicking, walking, and biking are encouraged. Food and drinks are available in the Rivierenhof Castle, the Sterckshof Castle, and at the miniature golf course.

During the summer, the open-air theater in the center of the park is a main attraction. Concerts are programmed from mid-June to mid-September and can host about 1700 people.

All year round, many events take place in the park. Several are organized by the Rivierenhof staff themselves, like the Rose festival, Ecodroom (an ecological market), and the Story Telling festival. Many more are organized by external parties that use the park for its décor. For such usage the event must add value to the park just as the park must add value to the event. An example of such an event is the medieval camp re-enactment "De Quaeye Werelt." Since its first edition in 1999 De Quaeye Werelt has become one of the most famous medieval events in the Benelux. Other examples are the yearly fireworks that commemorate the liberation of Antwerp after WWII, and the audio-visual spectacle "De Grote Schijn."

PUBLISHER NELS

The rowing pond, early 20th century.

The park also hosts different kinds of art projects. There is an art gallery in the western wing of Rivierenhof Castle that has new exhibitions every two weeks. A number of art pieces are found in the park and there are temporary open-air exhibitions with artists or school projects like "Art in the Garden."

Besides the individual sport possibilities in the park, there are several resident clubs that have their own infrastructure. Currently, there are three korfball, two soccer, and one tennis club. Other sports, for example, "Quidditch" (a game invented and described by J.K Rowling in Harry Potter), are played on the grasslands, but they have no specific infrastructure attached. The free and accessible infrastructure includes two running trails, calisthenics installation, and a small fenced field for basketball and soccer. Each year several sport tournaments (korfball, soccer, tennis) are organized in the park. There are two main running events: Valentine's Run and the Rivierenhof Run, but sometimes other running events pass through the park (e.g. the Antwerp Marathon).

Since the early days of the provincial park, botanical collections have been important in the park. The vision of Provincial Registrar J. Schobbens (who had persuaded the provincial politicians to buy the park in 1921) was that there should be an arboretum, providing an experimental area for the acclimatization of exotic tree species and the improvement of local plants. Accordingly, an arboretum was created. During the next decennia several valuable trees and shrubs were planted in several other parts of the park.

PROVINCIAL PARKS REGION ANTWERP

PROVINCIAL PARKS REGION ANTWERP

In 1937 the Royal Flemish Dendrological Society, the oldest dendrological society in Belgium, was created in the Rivierenhof. It is still active in the park, identifying the trees and organizing guided walks.

At the beginning of this century the arboretum had become neglected. In 2005 the overgrown arboretum was cleaned up and a selection of the valuable trees was made. New trees and shrubs were added.

The management of the botanical collection was recently professionalized, and an international seed exchange program started with 800 botanical gardens worldwide. Special plants are raised in the greenhouses. There are also training opportunities and workshops on different topics, e.g., rose garden guide, pruning, plant propagation, etc.

Besides a huge general collection, special attention was given to magnolias, rhododendrons, camellias, hazels, oak trees, roses, cornuses, ferns, edible garden plants, and plant varieties that belong to our living heritage. Special plant taxa can be found all over the park, but thematic plantings are present in the Arboretum (special trees and shrubs); the Rosarium (a rose collection with special attention for Belgian breeders); the Pinetum (a conifer collection); the Coryletum (a collection of hazels and other nuts); the Appalachian trail (trees and shrubs of Eastern North America); the Tree Horoscope (a collection of trees, each kind for a specific period of the year and with a special character); and the Azalea garden (a heritage collection of Hardy Ghent azaleas). At the Children's Farm there is a garden with edible small fruit crops, spices, and forgotten vegetables.

The rose garden has 340 different rose varieties of which many are Belgian breeds. One of them is the 'Rivierenhof' rose, a winner of several international prizes.

Although the park is surrounded by the city, it has several ecologically valuable locations, with interesting plants, insects, and birds. Special interest goes to the management of the old forest area, the wet forests, and the meadows. A few invasive species like *Fallopia japonica* (Donkey rhubarb) and *Impatiens glandulifera* (Policeman's Helmet) have entered the park, and special control programs have been started to keep them under control.

The park lends itself to educational activities. Each year over 20,000 school children are following one of its educational programs (for both infants and primary schools).

TOP LEFT: Fishing in one of the ponds.
BOTTOM LEFT: The fairy tale house.

GREATER LONDON

HATFIELD   HODDESDON   HARLOW

KINGS LANGLEY   CHIPPING ONGAR

WATFORD

Epping Forest

BRENTWOOD

Lordship Recreation Ground

EDGWARE

Hampstead Heath

Walthamstow Wetlands

ROMFORD

Colne Valley Regional Park

Lee Valley Park

WEMBLEY

Regent's Park

Saint James's Park

Kensington Gardens   Hyde Park

Red Cross Garden

The Thames

Gunnersbury Triangle Nature Reserve

Green Park

QEH Roof Garden

Rainham Marshes Nature Reserve

Battersea Park

Burgess Park

Greenwich Park

BEXLEYHEATH

Richmond Park

NORTHFLEET

THAMES

Bushy Park

WWT London Wetland Centre

Wimbledon and Putney Commons

BROMLEY

COBHAM   EPSOM

WESTERHAM

DORKING   REDHILL

SURREY

CRAWLEY

0   5   10 Kilometers
0   5 Miles
N

© OpenStreetMap contributors

LONDON NATIONAL PARK CITY, UNITED KINGDOM

49.5 percent of the area is green or blue

2 Special Protection Areas

4 UNESCO World Heritage Sites

7 RAMSAR Sites

37 Sites of Special Scientific Interest

147 Local Nature Reserves

300 Farms

673 Allotment sites

Over 3,000 parks and open spaces

INFORMATION FROM URBAN GOOD CIC MAP 2017

JUDY LING WONG

London: A National Park City, United Kingdom

Cities are extraordinarily dynamic places. London is exceptionally so, especially as the UK might have the most developed voluntary sector in the world. There are more than 120,000 voluntary organizations in London alone—a force of people used to grappling with issues on the ground, influencing policy through campaigning or celebrating different aspects of life. Beyond caring for and protecting nature, they act to create more nature. The unique selling point (USP) for the London National Park City initiative is people power. Londoners are aware that their health and the health of the environment are intrinsically linked. They are therefore motivated to work towards greener, wilder, and healthier lives.

In a poll of 1,000 Londoners for London National Park City in 2015, 85 percent said they wanted London to become a National Park City. The recognition that London already has 49.5 percent green and blue space with outstanding natural areas makes the concept of a National Park City across its entire fabric a possible dream. The model of a National Park City draws on the values of the UK's rural National Parks. It defines its urban context by paying equal attention to outstanding nature and the potential for wildness within the built environment.

London has 3.8 million gardens. Small individual actions within these can multiply to huge effects. With minimal effort, people can take up paving stones to transform front and back gardens, or use their garage rooftops and street spaces. Some people are claiming local streets as "their own," creating pavement gardens to form a lush urban oases – linking with efforts made by public bodies.

In various boroughs of London, green spaces set within social housing exceed the overall area of local public parks and gardens. However, many of these are often of the lowest quality with only mown grass and dog waste. Vast areas of concrete can also be transformed. Working to bring landlords on board and opening up participation by residents may result in wild nature and opportunities for outdoor activities, with pleasant spaces for leisure right outside the windows of some of our most disadvantaged communities. Partnership at different levels from local authorities, powerful civil society organizations, and national and local environmental and heritage organizations, right down to voluntary local community groups, will be needed for such a venture.

By 2019, over 250 organizations were supporting London National Park City. The idea is to regard all green infrastructure in Greater London as the core of a National Park City, providing a quality of experience on par with the 15 rural national parks in the UK. In early 2019, Daniel Raven-Ellison, one of the founders of the London National Park City Initiative said:

> This initiative is backed by our Mayor Sadiq Khan and London Assembly members representing all parties, and thanks to thousands of Londoners contacting their local ward councilors, a majority of local politicians. We now have the mandate to declare London the world's first National Park City.

Individual Londoners and organizations such as the Federation of City Farms and Community Gardens, Friends of the Earth, the Woodland Trust, the friends organizations of various parks, John Muir Trust, London Wildlife Trust, London Beekeepers, local environmental groups, international organizations such as World Urban Parks and IUCN Urban Alliance, and many companies with a green profile have signed the London National Park City Charter to pledge their support. The declaration was made on 22 July 2019 at a summit at London's City Hall, hosted by the Mayor.

Popular support for action is the driving force. A National Park City is not a legal concept or a designation. It is a concept owned by the people of London. It is defined as a "large urban area that is managed and semi-protected through both formal and informal means to enhance the natural capital of its living landscape." Any organization or individual signing the London Charter pledges to "enhance the natural capital of its living landscape." There is now an international working group using the National Park City Universal Charter, to work with other cities wishing to follow in London's footsteps.

The Mayor has since built London National Park City into the London Plan as an instrument to promote the green infrastructure of London. He has put in place a £12 million fund to support the people of London in greening the city in their own way, and is ambitiously working on access to nature by 100 percent of London's children. He also pledges to fund more trees, improve community green spaces,

TOP RIGHT: View towards The Millenium Dome in the center of London.
BOTTOM RIGHT: Queen Elizabeth Hall Roof Garden, South Bank.

LUKE MASSEY

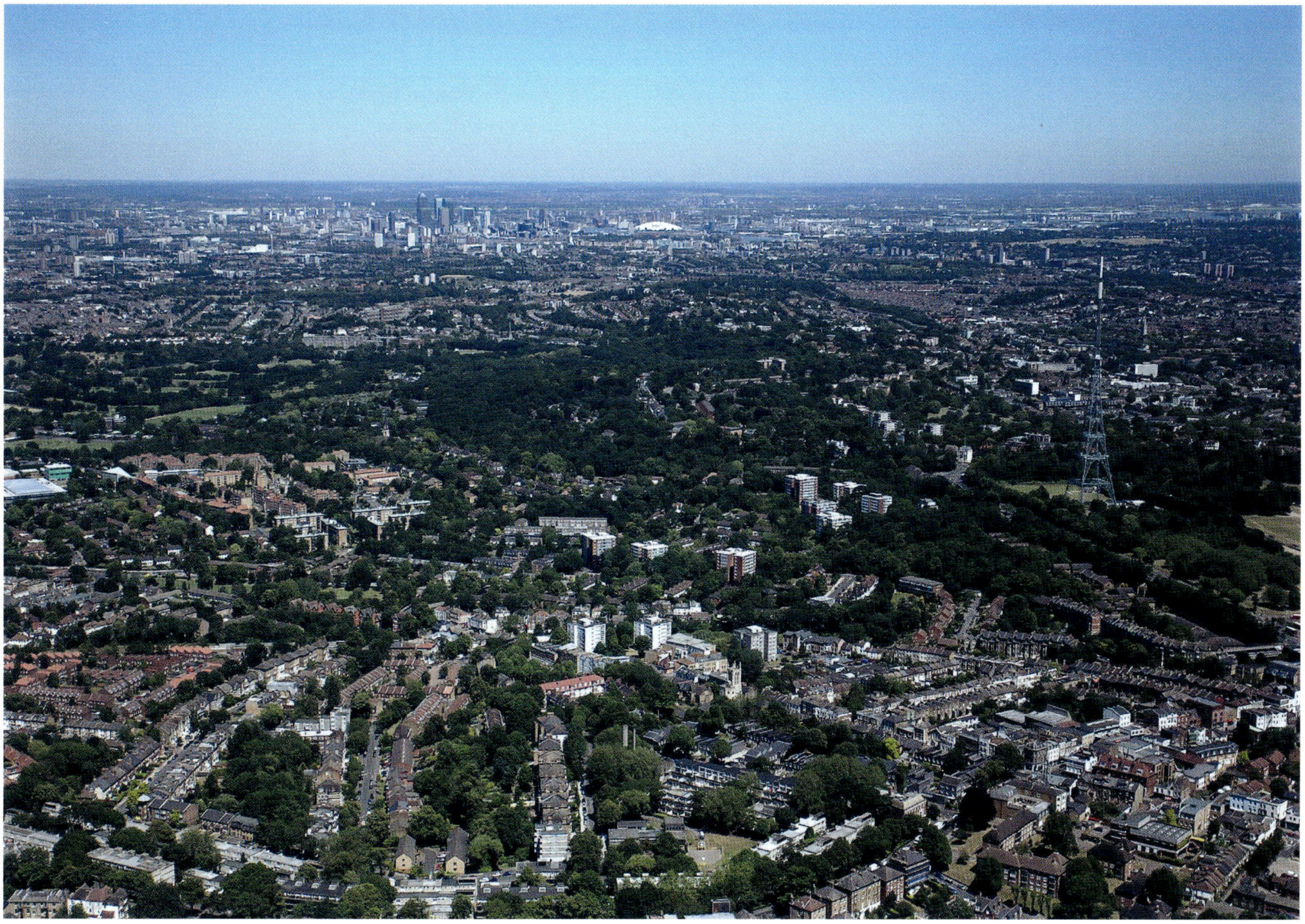

JUDY LING WONG

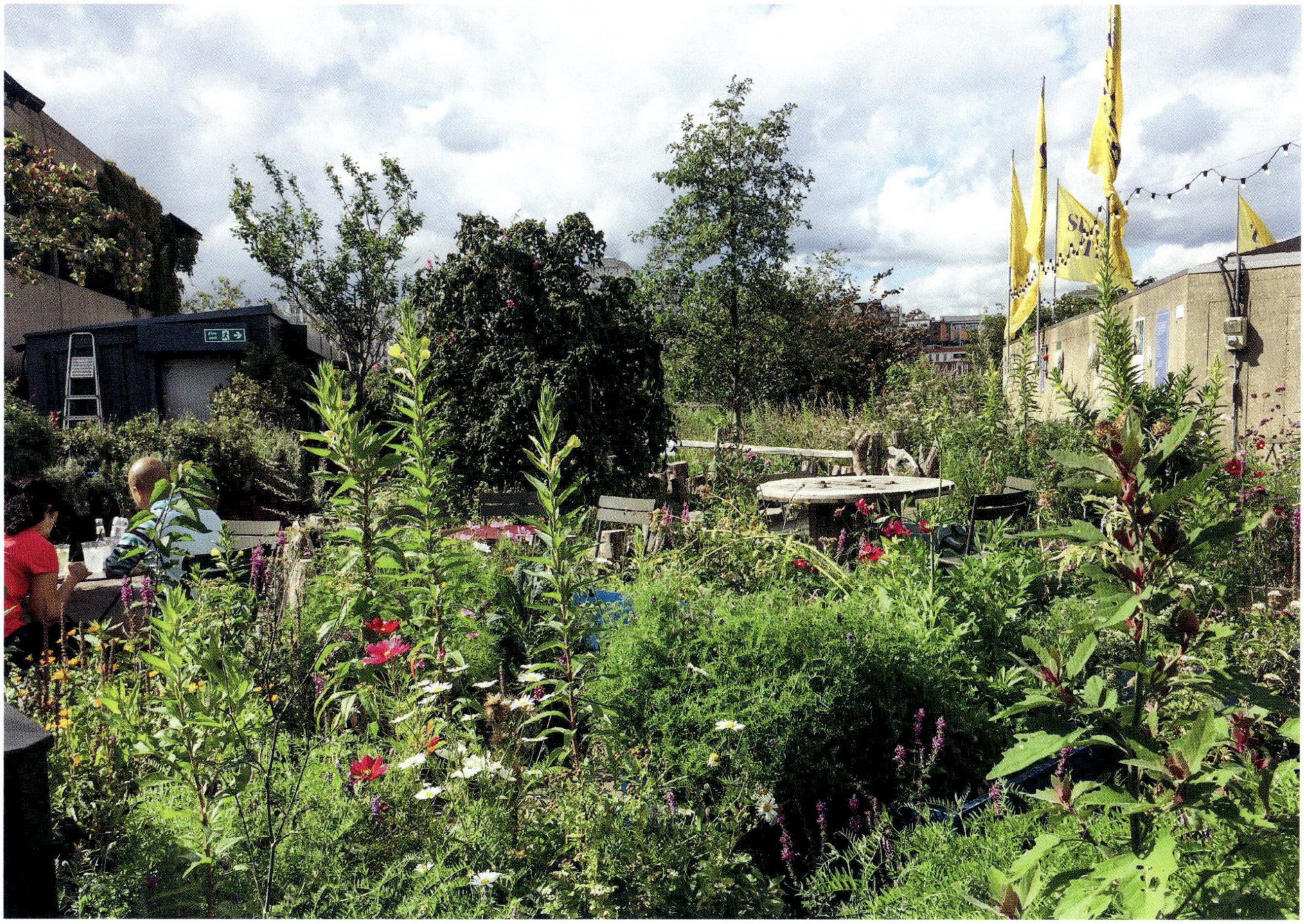

Deer in Richmond Park, London.

and support London's boroughs to invest in their much-loved parks.

Besides public parks and gardens, making up 5.8 percent of the total area, there are many other types of public areas that are green and blue: natural and semi-natural green spaces, pocket parks, gardens, outdoor sports facilities, reservoirs, rivers, canals, and so on. Public and private green and blue spaces amount to 49.5 percent of the area. London's Green Belt outside of the Greater London Authority area is not included in this account. Neither are green roofs and walls.

Ideas about protecting London open spaces emerged in the 1860s when legal protection was first given to Hampstead Heath, Wimbledon Common, and Epping Forest. In South London, Battersea Park was opened in 1860. In 1878, the Corporation of London acquired legal powers to bring areas into its ownership for the benefit of Londoners. With the introduction of the Crown Lands Act 1851, the Royal Parks were opened to the public: Bushy Park, Green Park, Greenwich Park, Hyde Park, Kensington Gardens, Regent's Park, Richmond Park, and St James´s Park. John Claudius Loudon published *Hints for Breathing Places for the Metropolis* in 1829, in which he proposed green wedges and a green belt for London, in order to provide the urban population with fresh air and to control city growth. A hundred years later, in 1935, the Greater London Regional Planning Committee proposed the Metropolitan Green Belt around London. In 1947, the Town and Country Planning Act transferred planning rights from private owners to the towns and boroughs, with the authority to set aside recreational green areas and to create green belts. In 1966, the 40 square kilometer Lee Valley Park to the east of London and the 128 square kilometer Colne Valley Regional Park (on the west periphery of Greater London) were created. In 2002, the Mayor Ken Livingstone launched the Biodiversity Strategy for London. This may be seen as the forerunner to the London National Park City campaign in 2013. The city is astoundingly biodiverse. 14,000 species of plants, animals and fungi have been recorded. Foxes are a feature, estimated to number 10,000. They are much loved.

To get a sense of the range and character of green and blue spaces in London City National Park, and to gain some insight into the role of Londoners and diverse organisations, a tour could start anywhere. In certain residential areas with very little traffic, locals have claimed the public realm to plant greenery of all kinds to transform their environment, creatively using pots and rubbish bins for flowering bushes or small trees, painting murals, or putting out decorated seats on pavements.

LEFT: Clapham common.

Cemeteries are an important part of the green infrastructure. Tower Hamlets Cemetery Park is one of London's greatest treasures of heritage and wildlife. Many cemeteries are specifically managed, involving volunteers, for wildlife.

Only 6 kilometers north of Trafalgar Square, you can experience gently rolling hills on Hampstead Heath, as if you were in the countryside. You can go swimming there in the summer, although Hampstead Ponds are deep and can be very cold. On Parliament Hill on a windy day, you will find all the kite flyers out, a real sight, and have great views right across London. This is where writers such as Keats, Shelley, and Byron hung out.

Quality green areas and access to nature are, however, not evenly spread throughout London. What Lordship Recreation Ground is like now is a supreme achievement by the local community. The grounds are situated in an area of deprivation in central Tottenham in North London. First opened as a public park in 1932, for years it suffered from poor quality, so that it was not used in an area of great need. In recent years, there has been unparalleled community engagement to turn this around. 20 user groups now collectively co-manage the park with the Council. They welcome visitors and run the new Passivehaus (low energy) Community Building with its café by the lake. Partnership and investment resulted in a Green Flag Award in 2013, the national award recognizing parks with excellent standards.

Friends of the gardens work with Back2Earth to maintain and support various activities. Local like-minded people can come, share skills and learn new ones. There is training provided in horticulture, food growing, garden and tool maintenance. Events have included music festivals. Many activities open up new areas of interest in people's lives as well as increasing their potential for employment.

Burgess Park in the Southwest is one of the 'young' parks, built between the 1950s and 1980s. It is Southwark's largest park. Recent investment transformed it into a superb multiuse park, serving the largest social housing estate in Europe. It has a very popular state of the art BMX track.

Bankside Open Spaces Trust (BOST) was set up in 2000 by local people to improve the area by making it a greener and more beautiful place. It maintains over 45 parks and gardens in the inner city borough. One of them, the historic Red Cross Garden, has been restored to its original Victorian layout. It was created by Octavia Hill, a social reformer who recognized the value of improving housing and contact with nature for poor workers. She was one of the founders of the

Quality green areas and access to nature are, however, not evenly spread throughout London.

National Trust. BOST involves residents, businesses, the borough council, health organizations, schools, and so on in its work. They have a strong program of public events, and a regular gardening group working to the highest standards. Many awards testify to their outstanding achievements.

Several wetlands and marshes have been created within Greater London. London Wetlands Centre is a world-class nature area reclaimed for wildlife, managed by the Wildfowl and Wetlands Trust in southwest London. It was restored from reservoirs serving London for many years. Rainham Marshes (411 hectares) on the fringes of the Thames estuary is a haven for all kinds of wildlife - birds, water voles, dragonflies, and more. It used to be a military firing range. Walthamstow Wetlands in north London is, at 211 hectares, Europe's largest urban wetland reserve. Based around ten operational reservoirs, the reserve is internationally important for migrating wildfowl, particularly overwintering species such as gadwall, shoveler, and tufted duck. It was opened in 2017, by virtue of a partnership between London Borough of Waltham Forest, London Wildlife Trust, and Thames Water. The reserve's majestic Engine House (1894) is now a visitor centre with a café and spaces for activities. Replacing an industrial chimney, demolished in the 1950s, is a 24-metre swift tower.

To get a complete sense of London National Park City, one must experience the extensive Royal Parks, a grand feature of London with over 20 square kilometers of high quality parkland. We have Hyde Park and Green Park right against Buckingham Palace. Regent's Park serves central London. Bushy Park and Richmond Park (designated National Nature Reserve) are in the west. Richmond Park features Isabella Gardens, with magnificent displays of azalea and rhododendra in Spring. Richmond Park is the largest of the Royal Parks, famous for Red and Fallow deer roaming free.

London National Park City, as a big idea, stems from an insight that Greater London, in spite of its concrete jungle of buildings, still boasts a phenomenal amount of green and blue, making up 49.5 percent of its area. This inspires us to dream that there can be more! The goal of the initiative is to make it even greener, healthier, and wilder, for people and for nature.

Greater London, in spite of its concrete jungle of buildings, still boasts a phenomenal amount of green and blue, making up 49.5 percent of its area.

City Meets Nature

The urban fringe tells a lot about how a city values nature. Some cities manifest an alienated relationship to nature, the periphery a wasteland of dumps, rusting factories, and polluted streams. Other cities invite nature into their very core. Many English cities used to have cattle grazing right behind the cathedral, and some still do, as a reminder of the fundamental conditions for life in cities.

URBAN EMANUELSSON
The City Meets Nature

"The city is the opposite of the countryside and nature," it is sometimes said, but without the countryside the essential conditions for the city would not exist. This can be clearly seen in historical studies where cities almost always grow up adjacent to major agricultural areas. If there is additionally a big river or an estuary with access to the sea, the conditions for a city are very good. If we look at the three metropolitan areas in Sweden this is obvious: Stockholm lies on the plains by Lake Mälaren and the province of Uppland, and is located at the mouth of a very short river called Strömmen; Gothenburg has Göta river and the plains of Västergötland; and Malmö has the plains of south-west Skåne and Öresund (the strait that functions like a river between the Baltic Sea and the ocean).

Although today we transport goods in all directions across the globe, it is a fact that humans and activities are drawn to places where big cities have existed for a long time. Competition between agricultural land and building sites is particularly noticeable there.

Many people in the world today have leisure time, which means that they want various activities which they find enjoyable or beneficial. Everyone today accepts that exercise and an outdoor lifestyle are good for body and soul alike. Modern sedentary office work and suchlike means that our need for the outdoors and for exercise is even more obvious. Many people in the world therefore want or need to get outdoors and get moving. A large share of these people lives in cities, and their numbers will surely increase.

The fact that large groups of people want or need to get outside and take exercise makes increasing demands on community planning. When in addition the use of cars must be limited in the future because of congestion and environmental considerations, it becomes increasingly difficult to solve the equation.

In some places community planning was quick to see the need for recreation for all city dwellers, but elsewhere the matter has had low priority. The demand for green areas is constantly challenged by the need for land for housing and other activities. Awareness that good arable land should preferably not be used for housing leads to greater development density inside cities, which often means that green areas are filled with buildings.

How have these problems been tackled? The answer is not simple, partly because 'green land' is scarcely an unambiguous term. 'Green land' can be used for a great many purposes: classical parks; landscape parks; areas for various kinds of sport, such as golf courses; small and unpretentious green spaces in residential areas; high-class nature of various kinds, such as old forests, natural pasture, and wetlands. The list could be very long, and in fortunate cases it is possible to combine different functions.

In certain countries such as Sweden, the right of public access is unquestioned; people are free, in principle, to move around anywhere except on developed land and where crops can be damaged. There is great variation in different parts of the world, from full rights of public access to modified rights, and to countries where people must confine themselves strictly to specific areas. The right of public access means that a privately owned forest near housing is an asset for the people living in that housing, as they can use it for strolls or to pick mushrooms, for instance. The right of public access thus means that there is recreational land quite close to where many Swedes live, without necessarily having been specifically arranged for the purpose or having any special status. This reduces the pressure on the authorities to engage in active planning of recreational areas. At the same time, if such land is close to a built-up area, it can easily be used for construction. The planning of recreational land and other 'green' areas must therefore be active if 'green' land is to survive. In countries with no or weak rights of public access there are greater demands on the authorities to ensure that there are 'green spaces' for the general public. A common way to manage this without much difficulty is the British system of public footpaths through areas which are otherwise, in principle, out of bounds to the public. It is mistaken to believe that in countries with the right of public access it is unnecessary to lay out footpaths actively. The more urban people there are, the more they need help to make their way through nature.

In many parts of the world, then, we see a growing need and a growing interest in creating large urban parks, whatever that term may mean.

This development is very interesting, and it is reflected in international organizations: the IUCN, which previously had only concerned itself with national parks far from cities, now has programs for green structure in cities; World Urban Parks, which formerly was only interested in city parks, now has Large Urban Parks as a section and has begun to take an interest in natural parks/landscape parks.

"Large urban parks" may be a way to meet the need for green areas close to where many people live, but we should not forget that a well-

Without the countryside the essential conditions for the city would not exist.

DIETMAR RABISH WIKIMEDIA COMMONS

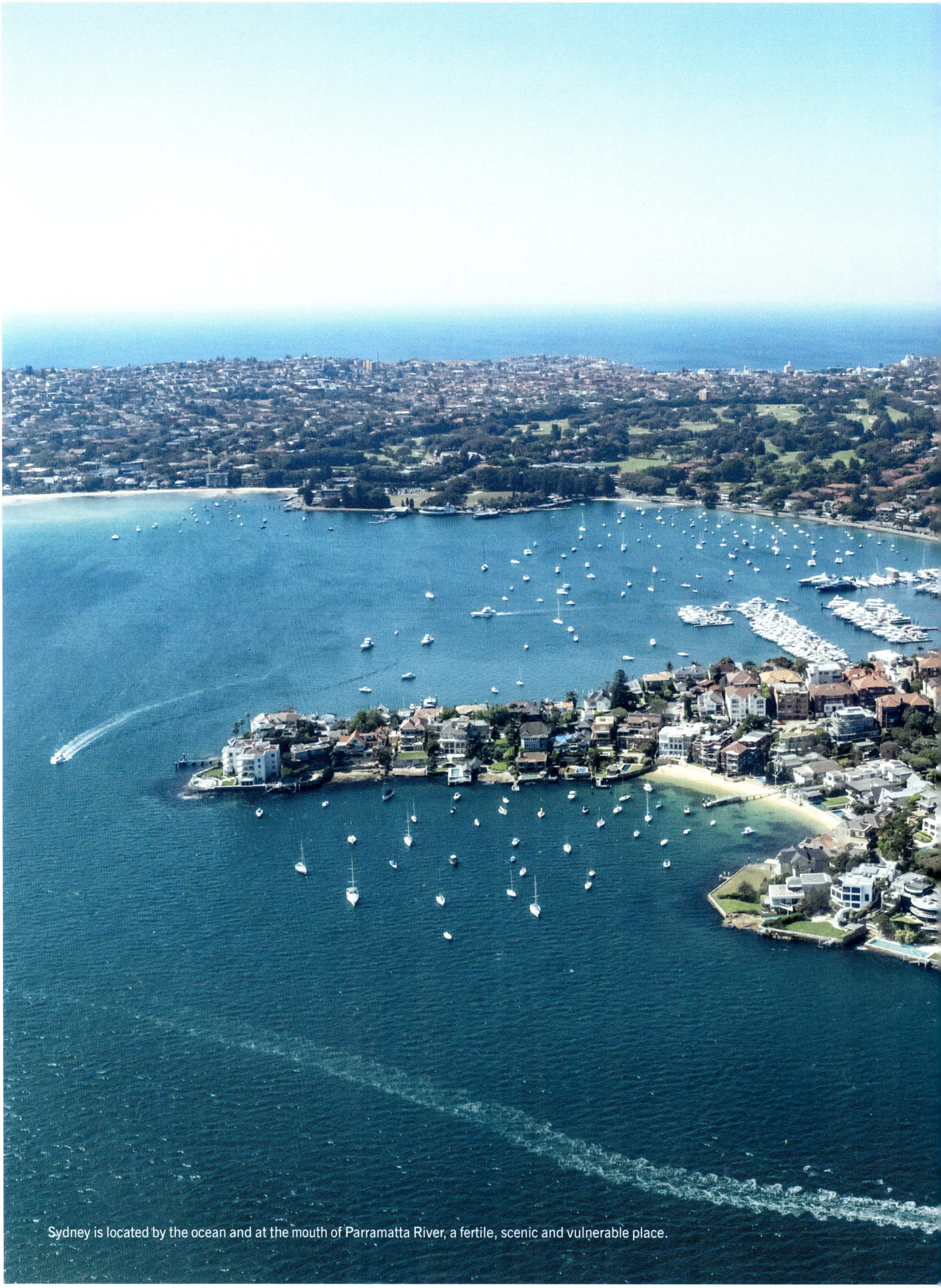

Sydney is located by the ocean and at the mouth of Parramatta River, a fertile, scenic and vulnerable place.

tended everyday landscape right beside an urban environment can be just as good a solution in many places. Landscape close to built-up areas has a very special role in these contexts. These are areas where very many people can enjoy recreation, but this is also landscape which is in danger when cities want to expand.

Several of the world's classical parks in cities have their origin in the desire of kings and emperors to have hunting parks close at hand. Djurgården in Stockholm, Tiergarten in Berlin, and Jægersborgsparken in Copenhagen came about in this way. Later the public were admitted, and in cities that were growing but had not possessed such parks, they began to be established during the nineteenth century and at the start of the twentieth century. In cities like Malmö, Sweden, which just lay there on a large arable plain, the parks really had to be created, with new ponds and planting of trees. If the growing towns were on undulating terrain, on the other hand, it was natural that the parks were actually just some kind of forest environment which had been tidied up and embellished.

In certain countries such as Sweden, the right of public access is unquestioned; people are free, in principle, to move around anywhere except on developed land and where crops can be damaged.

Certain cities realized early on that there was a growing need for recreation for the increasing population. Stockholm was one such city, buying up forests in neighboring municipalities. At the start of the twenty-first century, however, the center-right majority that then governed Stockholm realized that owning forest was not something a municipality should engage in. Much of the forest was then sold off. Sometimes the buyers were private citizens and companies, but some of the municipalities in which these forests were located wanted to continue providing opportunities for recreation, although now perhaps chiefly for their own inhabitants. Botkyrka is one such municipality. It bought a large area of forest from Stockholm, partly with some financial help from the Environment Protection Agency. The result was the Lida nature reserve, which has a distinct recreational character. Inside the reserve, for example, there is the Lida outdoor center, a starting point for walks, and for skiing and skating in the winter. What is interesting about the Lida nature reserve is not just that a large recreational forest near a city was "rescued"; there were also higher ambitions for its management. When Stockholm had owned the area, fairly rational forestry had been practiced, albeit with a gentle hand. In the new nature reserve, more thought was given to conservation. Various measures have been taken to increase the biological diversity and simultaneously make the area more attractive for walking. Controlled fires were started, dismal spruce plantations were felled, and grazing has been introduced in a semi-open woodland environment in a part of the reserve. The Lida reserve is an example of the continuation of the tradition of public ownership of strategically important recreation areas while simultaneously bringing classical recreational forest and conservation-driven forest management much closer to each other. There are great advantages in doing so elsewhere, in places where we need both recreational land and areas where a rich flora and fauna can continue to exist.

Close to London there is also a fine example of a municipally purchased forest which has become important for both recreation and nature conservation. This is Burnham Beeches west of London, which is both a large recreation area and highly valuable as a natural area. Somewhat similar is Epping Forrest north of London. It was Queen Victoria who donated the royal property to the public. Epping Forest, a long, narrow strip, is likewise very important for both nature conservation and recreation. What is interesting about the management of Burnham Beeches and Epping Forest is that close attention is paid to looking after the cultural history. The beeches in both forests have been pollarded for a very long time, and this has been continued, as has the use of grazing animals in certain parts.

When discussing 'green areas' in and around cities, it is important to bear in mind that there are different ways of using such places. These uses can very well exist side by side, but it is important to see the differences so that planning can take them into consideration.

Inside old cities there is often some form of well-tended park. These fine parks can have a background in palace parks, hunting parks, and the like, but many such parks were created, especially in the late nineteenth and early twentieth centuries. Today these parks often have large, mature trees, lawns, flowerbeds, and some form of water installation. From having been parks for strolling and playing, or for "déjeuner sur l'herbe," these parks have become increasingly important for various kinds of more "active" exercise, particularly jogging. Such parks have sometimes become more valuable for their biological diversity, especially if the management has succeeded in preserving old trees, not just chopping them down when they have become frail.

Since the end of the twentieth century, new parks of a certain kind have been created, often on old rubbish dumps or as a way to get rid of excavated rubble. These parks, often quite big, typically have hills, open grassy areas, and lots of bushes. They have a modernistic character showing a relationship to nearby high-rise buildings. In these parks there is usually little ambition to maintain flowerbeds, whereas areas of grass are extensive. These parks are suitable for large-scale sports events such as races. Mostly they are poor in wildlife and plants and thus not particularly attractive for enjoying or studying nature.

The green wedges nevertheless have required some protection, and the pressure of development is felt.

A type of park that has often grown up, almost unplanned, is the area close to a city that has become attractive for its natural qualities and has perhaps resisted development because of opposition by nature and culture conservationists. Such places are very often located on sites that are difficult to develop, such as rocky hills with limited topsoil, which has also helped to preserve them. Ramberget in Gothenburg, Sweden, is a typical example. As with Ramberget, several of these 'natural parks' have also come to include grassy areas for sport and meetings, and also flowerbeds. Here we see areas that combine several different uses. The Royal National City Park in Stockholm also has this character, although it has its origin in an older tradition of a royal deer park.

Such mixed parks can be useful in many ways if they extend from the city center out towards purely rural areas. Thanks in large measure to its physical geography, Stockholm is gifted "almost undeservedly" with some such wedges, for example, the Nacka wedge where virtually "wild" natural woodland extends from the rural forest landscape almost the whole way to the city center.

The thought of having wedges of 'green land' from the city center out to the surrounding landscape—an excellent geometrical construction! —is a fairly natural idea if you want to have large urban areas, i.e. a "big" city, and simultaneously give many inhabitants access to different forms of "green land." But how has this way of planning functioned in reality? It was actually easy for Stockholm when it started planning its green wedges, as water and mountains had prevented massive building, and the military exercise ground of Järvafältet had also contributed. The green wedges nevertheless have required some protection, and the pressure of development is felt. Another city of Stockholm's size, Copenhagen, is located on what is, in principle, purely arable soil. To achieve the wedge ideology a "finger plan" was introduced at an early stage, that is, green wedges and communication lines would radiate from the center. But if you look at a map of Copenhagen today you see that, while the communication lines follow the finger plan, it is very hard to see any clear green fingers. The example of Stockholm-Copenhagen illustrates well how important the underlying physical geography is for achieving true green wedges. A survey of several big European cities in decidedly arable areas shows how almost impossible it has been to create green wedges. In contrast, where there is broken topography and complex water systems, there are often green/blue

TOP RIGHT: In the United Kingdom public access is restricted to public footpaths.
BOTTOM RIGHT: Pildammsparken in Malmö. Sweden, laid out in the 1920s.

R. MOORE/SHUTTERSTOCK

JORCHR/WIKIMEDIA COMMONS

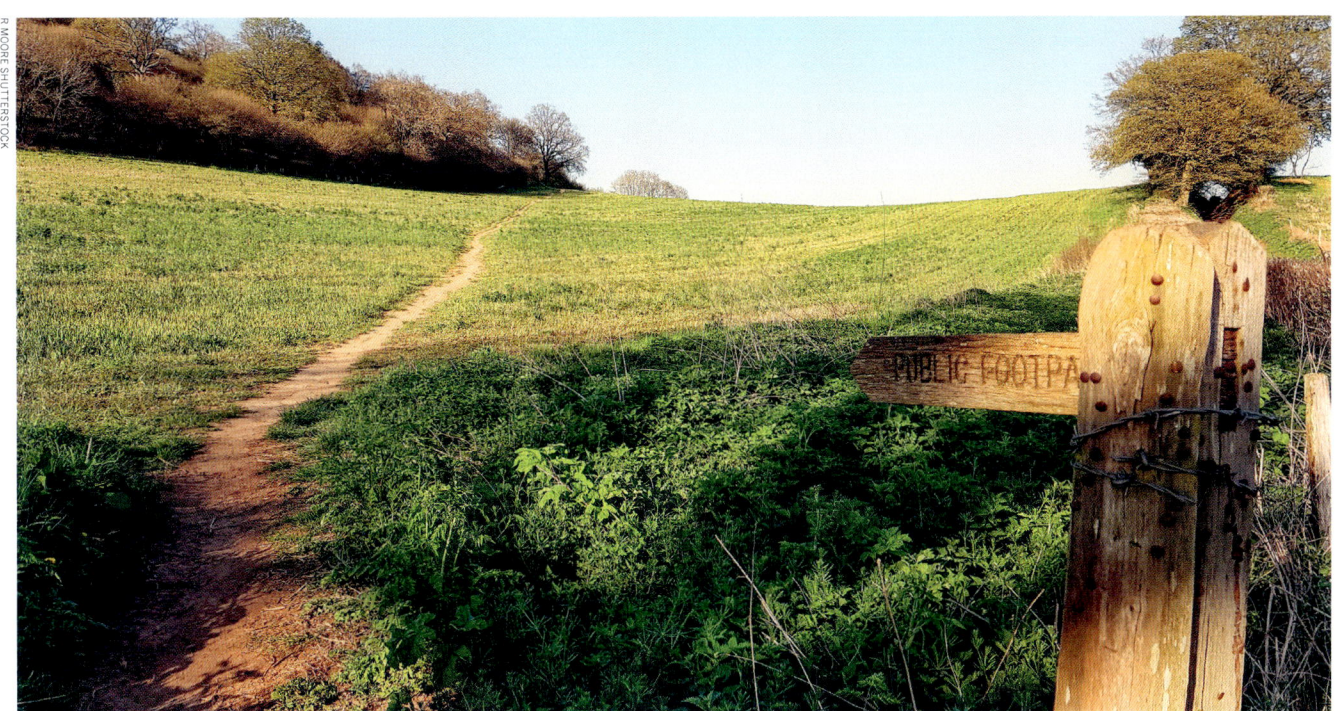

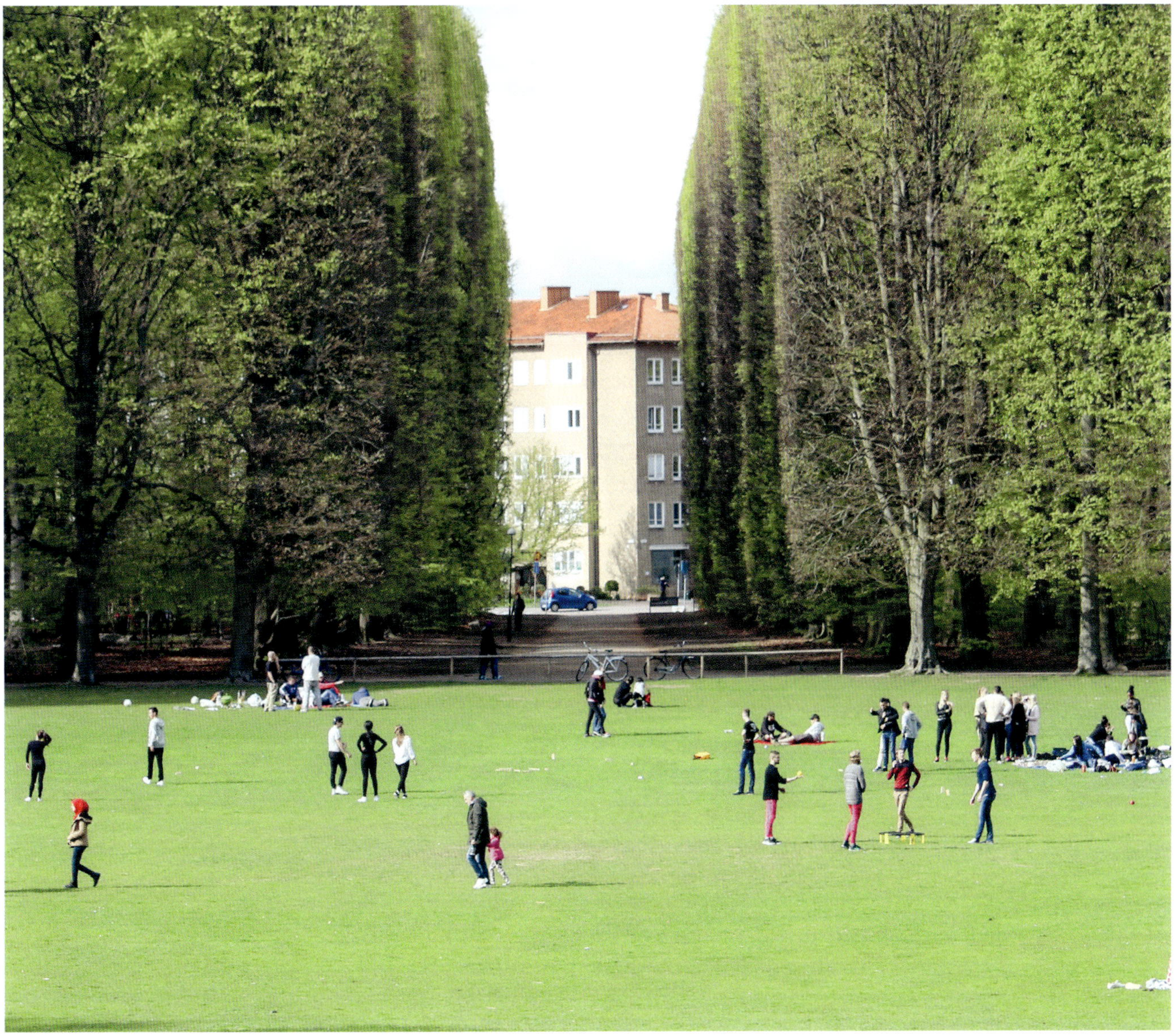

In contrast, where there is broken topography and complex water systems, there are often green/blue wedges.

wedges. In Madrid, a large natural area almost forces itself in from the north into the city center. Oslo is an example of how rocky terrain and a complex coast can help to create green/blue wedges.

Does this mean that there are no ambitions in big cities located on plains to create green wedges despite the difficulty? If we study cities with an express ambition to create "green land" we find that the result is not so much wedges as islands. Malmö is an example. Since the start of the twentieth century the planners have tried to remedy the lack of recreational land by establishing wholly artificial parks. The most recent is the Lindängelund recreation area, covering about 100 hectares on the southern outskirts of Malmö. Broken topography is built here with rubble excavated elsewhere, and trees are being planted, including mammoth trees from California! The park becomes an island without any contact with the landscape outside. This is not so strange, in fact, as the park is bound to the south by the worst enemy of green wedges, a major ring road. When you look at big European cities and their large parks in peripheral locations, you very often find that they are demarcated by major communication lines which make proper green wedges impossible. At the same time, perhaps the island parks in Malmö are in fact an adaptation to the fact that a hypothetical green wedge out of Malmö would perhaps just end up in pure arable landscape with little biological diversity and few spaces for people to take exercise.

If you really want to create functioning green wedges in cities on arable plains, you must have very long-term planning and include the construction of attractive landscapes in what are today monotonous expanses of industrial farming. Against this, one could argue that we need all the arable land to provide our food, and to withdraw such land from food provision may seem odd when we try to impose restrictions on building on such land. This equation, however, need not be nearly as difficult as it looks.

Many studies and practical experiments show that a reduction of wholly arable land by 3–10 percent in highly industrialized agricultural landscapes, replaced with, say, recreated wetlands, restored natural pasture, herb-rich border zones, lines of trees, and clumps of bushes and trees, can lead to a considerable rise in production on the remaining 90–97 percent arable land. If these new or restored biotopes are located in smart patterns and stretches, one can create attractive landscapes both for biological diversity and for people. In this way green lines inside cities could continue out into an attractive agricultural landscape. This, however, would require forward-looking long-term planning (Environmental Protection Agency, 2018).

It should also be stated that there are plenty of fairly small conurbations all over the world which, despite the fact that it may be only a few hundred meters from the center to the periphery, have limited opportunities for recreation in the surrounding landscape. Transforming surroundings like these according to the model above would be a major contribution to greater biological diversity and better public health in many places.[1]

Another category of community which could quickly benefit from such a transformation of the nearby landscape is the kind that is built up around sun-and-sea tourism, for example, in the Mediterranean and the Canary Islands. Usually these places are beside the sea for natural reasons. Sometimes it has not even been possible to protect the beach zone, with hotels being built right by the water. Sometimes the hotels are 50 meters back from the water, and sometimes there may be a more generous coastal zone with a footpath. The planning of beach zones really needs to be strengthened in many places, but behind the hotel areas, which often form just a narrow strip along the coast, one can often find land that has been badly treated and left unplanned. Many areas are just waiting for the possibility of development, and if there is functioning agriculture it is often rather rational and does not welcome walkers. Improving such environments from the point of view of accessibility and for the sake of biodiversity and culture would surely pay off well in the future. Already today, many sun-and-sea tourists are keep-fit enthusiasts, and good coastal paths are already becoming an important selling point. A better landscape behind the hotels would no doubt also reward the investment, while agricultural production here could simultaneously become more nature-friendly.

Although I have fairly broad international experience of different landscapes, it is in Sweden that I have the best understanding of the different processes for strengthening 'green areas' inside cities and on their periphery, bordering the surrounding landscape. When I cite some examples of how such work has been successfully done, they are therefore mainly from Sweden.

First, I shall take the case of Linköping, mid Sweden. This is not actually an example of how nature has been shaped, but rather an example of how a unique opportunity was seized and perhaps the best green wedge in Sweden was created, leading from beautiful landscape outside the city almost all the way to the center. South of Linköping there was a military exercise ground until a few decades ago. This had been created by transforming a number of agricultural properties with some forest and pasture into an area with extensive grazing. The army

If you really want to create functioning green wedges in cities on arable plains, you must have very long-term planning.

313

LUCA GRIONI ARCHIVIO PARCO AGRICOLO SUD MILANO

URBAN EMANUELSSON

had thus, almost unconsciously, prepared the shaping of a fine area for recreation and nature. When the municipality then took over the area, it developed both its recreation potential and the natural landscape. The nature reserve is called Tinnerö Oak Landscape Culture and Nature and it covers an area of 687 hectares, starting in the oak landscape of Östergötland to the south and ending just two kilometers from the central square, Stora Torget. Linköping Municipality saw possibilities in a way that has not happened anywhere else.

In Örebro, also mid Sweden, they had to work even harder to create a fantastic green/blue wedge in towards the city center. In the eastern part of the city, towards Lake Hjälmaren, there was an area that was not a pretty sight, with various storage places, harbor installations, dumps, and the like, in a badly damaged delta. Örebro, however, has an old tradition of nature conservation: one of Sweden's foremost ornithologists and conservationists lived in the Oset area south of the Rynningeviken bay and wrote many popular books about birds and the environment. This gave the inspiration for beginning in the 1990s to undertake a complete renovation of Rynningeviken. Rubbish and pollutants were removed and a lovely wetland was created with plenty of room for different kinds of recreation, not least skating in the winter. Bird life returned and is now flourishing in the Oset and Rynningeviken nature reserve. The people of Örebro now have a wonderful blue/green wedge leading towards Lake Hjälmaren.

When the author of this chapter was a field biologist at the end of the 1960s, a group of nature enthusiasts cycled from Lund, in the south of Sweden, to the coastal town of Lomma to look at birds. We were particularly interested in the old brickworks' ponds scattered here and there in the town. Masses of clay had been extracted here for the brick industry, and the ponds were regarded by the politicians of the time as worthless. Waste management in south-west Skåne (Scania) realized the potential of the ponds in Lomma and began to fill them with waste that had to be disposed of somehow. This was before the days of recycling and rational power and heating plants burning waste. We field biologists protested and demonstrated, but to no avail. Time passed and more and more ponds disappeared, but the waste management got better. Some time in the 1990s the scales had begun to tip, and soon some of the ponds became nature reserves, but there was still some landfilling,

TOP LEFT: Parco Argricolo Sud Milano borders directly to the city. The park is huge, agriculture is ecological and visitors are welcomed.
BOTTOM LEFT: Tinnerö green wedge stretching to the center of Linköping, Sweden.

because Lomma now wanted attractive land for new housing near the coast. Today a kind of balance has been achieved: Lomma has acquired fine wetland nature reserves with good paths for rambling, and it has also acquired splendid new housing. Some of the ponds, however, are not reserves, and it is not entirely certain that no more of them will be filled in, but one may hope that they will continue to contribute to the species-rich man-made environments in Lomma. These places came very close to being eradicated, but today the ponds and the surrounding land are a great asset.

All three places—Tinnerö Oak Landscape, Rynningeviken, and the Lomma ponds—are examples of areas that are scarcely traditional parks, nor are they traditional nature reserves; they combine several functions, besides which they extend from inner urban areas out towards the countryside.

With or without water, that is the question. A city by the sea or beside a large lake, or even just beside a big river, appears to have much better natural conditions for creating or preserving good recreation opportunities and nature than a city without water. Lund, the city where I grew up, is a deterrent example for me! Almost without water and surrounded by industrial-scale agriculture—could there be a worse starting point? By good luck, Lund, being an old university with medieval roots, has other, cultural, advantages! Nearby cities like Malmö and Landskrona, surrounded by the same intensive agriculture, have drawn on the benefits of their coastal environment. Malmö, though, wasted large, highly interesting areas along the coast as industrial land many years ago. If we could go back a hundred years, knowing what we now know, we could probably do a better job!

One difficulty with flat coasts like the one at Malmö is that many of the biologically most interesting areas here do not seem to be in any way striking at first sight: no high cliffs, very few sandy beaches. It can be difficult to justify the preservation of beach meadows and very shallow bays, but today most people appear to have seen the value of such areas.

Cities beside inland water can also benefit greatly from their shores and lakes, but the struggle against insensitive development has been waged for a long time. Växjö, another Swedish town in a more wooded area a bit north-east of Lund, has beautiful blue wedges leading right into the center, but had mistreated these for a long time. The value of these has since become obvious, and major restoration projects have been undertaken, including fine paths for walkers.

You can travel to Uppsala, just north of Stockholm, if you want to see the difference that water and high ridges can make for excellent recreation

It is mostly just where nature or chance has helped the authorities that it has been possible to shape nice transitions between town and country.

environments extending in towards built-up areas. But you have to go to southern Uppsala; you will not see anything like this in the north-eastern parts, where the apartment buildings look straight out on agribusiness fields just as in, say, Lund. In the south, the river Fyrisån runs right through the city, creating a rich natural landscape that starts at the center of the town.[2] Uppsala needs to preserve its green strips in the south but must also try to create something similar to the north-east.

In Sweden, as in several other countries, we have seen coast protection, which means in principle a ban on new buildings in a coastal zone which can vary in width, sometimes up to 300 meters wide. This legislation, when it works, is one of the best steering instruments for retaining recreation land close to built-up areas. Some municipalities in Sweden have been scandalously lax: for example in Värmdö, a suburb of Stock-holm, it is very difficult to follow the shoreline on foot. In many coun-tries, for instance in Southern Europe, it is like Värmdö, where a very important environment for biological diversity and recreation has been eradicated by housing construction. It is not always necessary to have green-structure planning to achieve green areas; sometimes it would be enough just to enforce the existing law.

It is not easy for me in a chapter like this to say very much about the planning of green areas in cities outside Europe. I have travelled widely and worked in different parts of the world, but to understand why things have turned out as they have and what ambitions there are would require a great deal of background research. What I can share here are some observations, at least.

Australia and Brazil both built new capital cities in the twentieth century. I have visited both and can see certain similarities but also major differences. Brasilia is an almost frightening city with colossal modernistic structures and very little greenery. What Brasilia concen-trated on for recreation was a central artificial lake where people can bathe and enjoy beach life in the hot climate. That part appears to have worked. Canberrra also has en central lake, but has, contrary to Brazilia, preserved the original landscape and its greenery to a great extent, even added greenspace, and with care built the city around the lake. Here I felt a completely different sense of well-being, besides which it was possible to carry out many interesting nature studies inside the city.

China has a well-documented attitude to nature, and one which seems very strange to us. Nature has to be experienced from paths, steps, and preferably in enclosed areas where you pay for admission. Being outdoors for an urban Chinese person is something like what a visit to a fairground or a museum is for us. Strolling freely in nature would be virtually inconceivable. In a large coniferous forest reserve in China I once met a Chinese woman who spoke English and described her total fascination with having walked several kilometers on small paths through unsigned forest. She thought it was fantastic.

Chinese people, however, do want to get outside, and the tradition of exercise, for instance in the form of shadow boxing, is strong. One solution is to have nature in an urban environment. There is an old tradition in China of having embankments along lakes and wetlands where it is easy for city dwellers to get a view of some form of nature.

Cape Town is in large measure a result of apartheid. Towards the sea to the west and north live the well-off people, nowadays of all races. In the east, on the other side of Table Mountain, are the poorer residential areas. In the west, above all, they have splendid opportunities to enjoy the sea and nature at Table Mountain. In the east people often live a long way from the sea and the mountain, although there are shanty towns here that extend down to the sea. It would have been interesting to examine how it might be possible to improve the access of the eastern parts to recreation areas. Perhaps it is the same as in poorer parts of the world: for ordinary people, thinking about recreation areas is far down on the list of priorities. The rich have access to vehicles and can drive to excellent nature reserves when they want to.

Golf courses are often given priority, not only in rich countries but also in countries with more limited resources. Here the better-off people can have their outdoor recreation. In China it can be the case that it is difficult to get out into other recreation areas, for instance near Peking, but for those who can afford it, it is perhaps both safe and a status symbol to play golf. It is also very clear in many places for sun tourism that golf courses take high priority. It is easy to get there, but finding your way out into more everyday nature can sometimes be very difficult; all that remains, at best, is coastal paths. My worst experience of this has been in the Gulf States, where it is practically impossible to find desert scenery that is not disfigured by rubbish or spoiled by vehicles, where one can stroll close to the tourist resorts.

After this short descriptive tour of the world, and with more experience from Europe, I can observe that it is mostly just where nature or chance has helped the authorities that it has been possible to shape nice transitions between town and country. Much work needs to be done,

TOP RIGHT: Rynningeviken, outside of Örebro, Sweden, a formerly badly damaged area, now restored.
BOTTOM RIGHT: Brick clay ponds turned into nature reserves, Lomma, Sweden.

URBAN EMANUELSSON

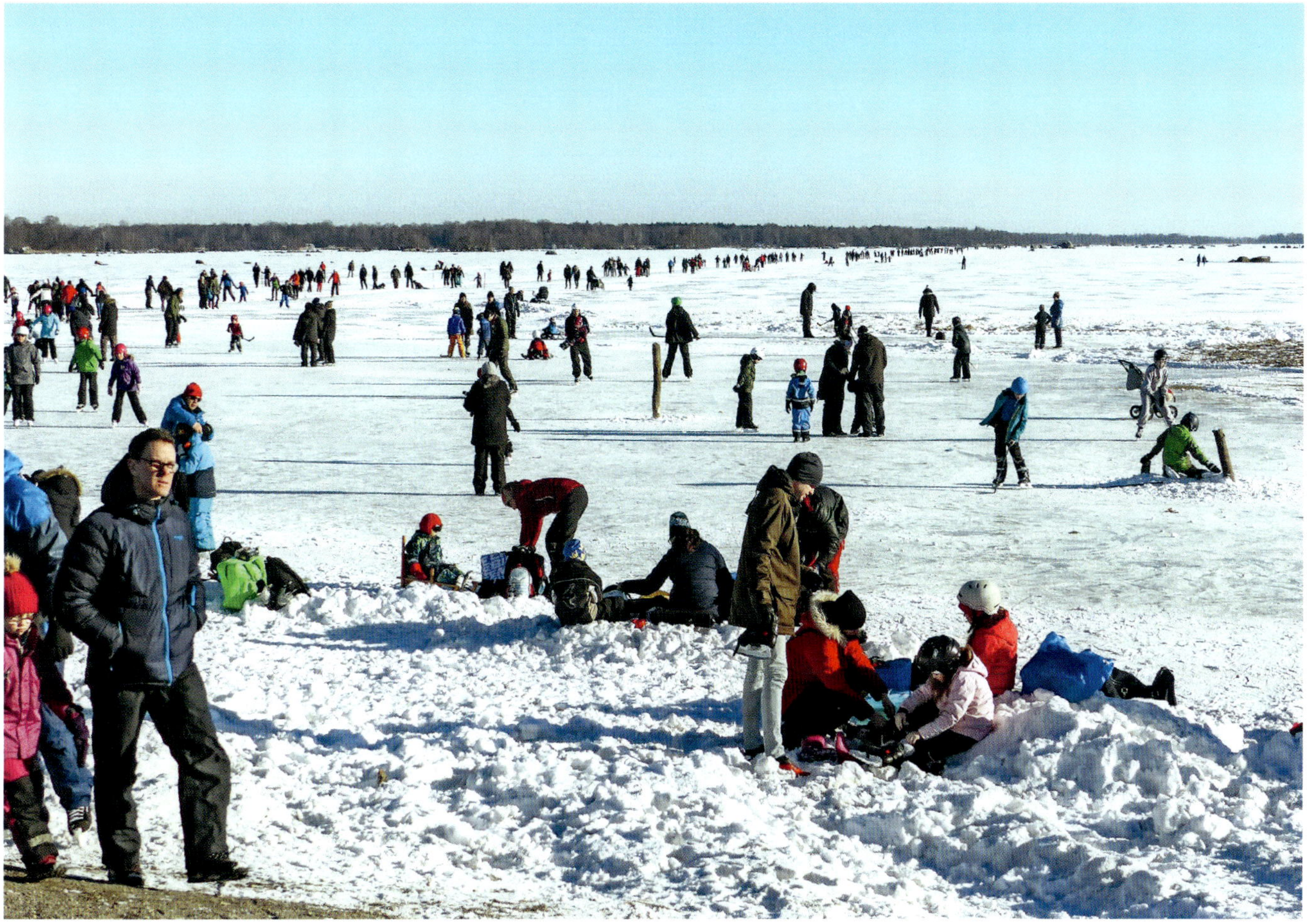

URBAN EMANUELSSON

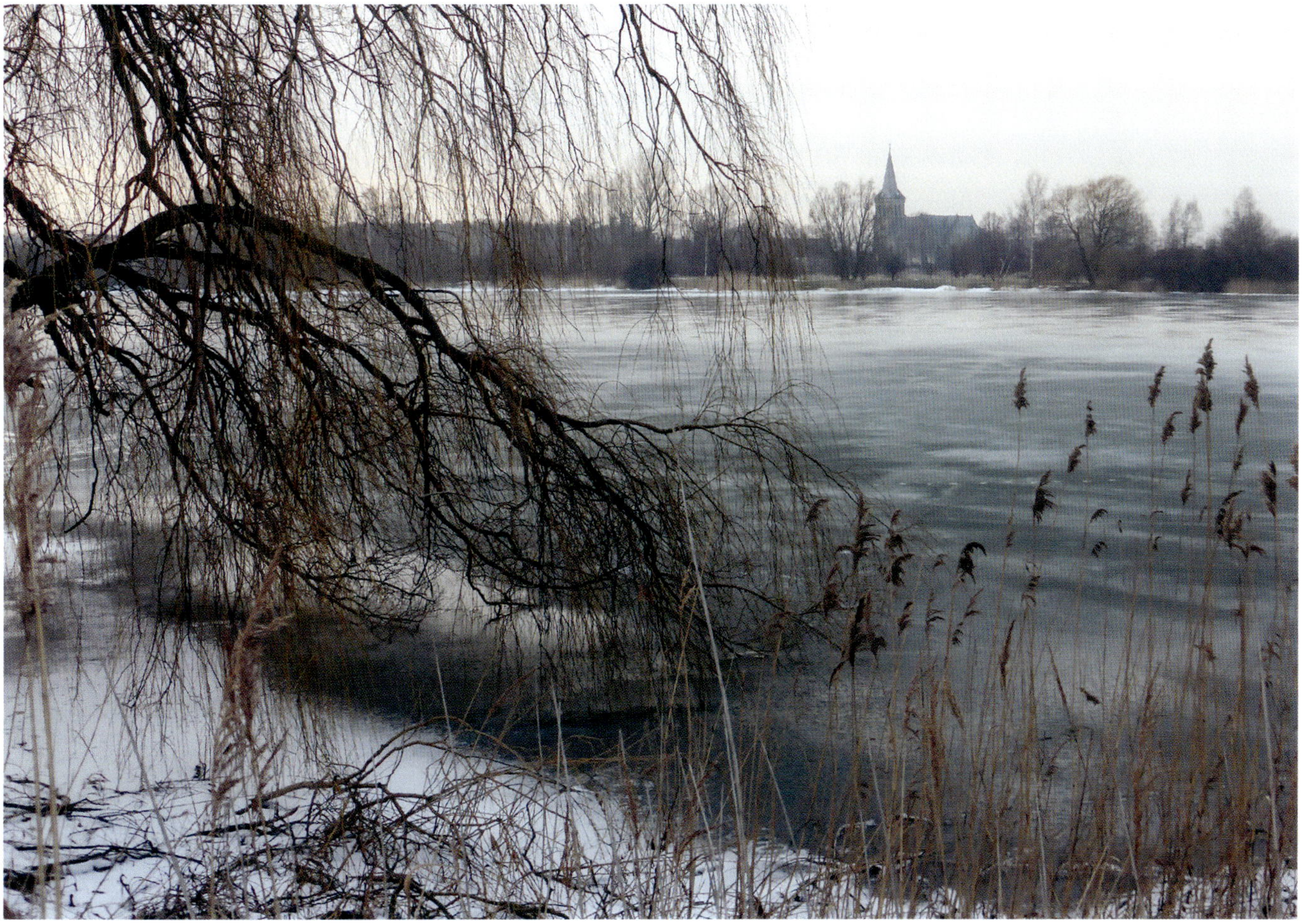

Tram leading to a major nature reserve and skiing area, just north of Oslo, called Nordmarka. The tram is now heading back to the parliament ("Stortinget") in central Oslo.

and in many places it seems as if people have not even bothered to take simple, cost-effective measures.

Working with green areas near cities and green wedges is not only interesting from the point of view of recreation. Classical nature conservation has unexpectedly good opportunities to be successful here. A simple fact that is often pointed out is that in quite a few places there are more old, valuable trees inside and around cities than further out in the countryside proper. This is because rational forestry has not been able to flourish here. Large old gardens and stately homes have also preserved many old trees.

But there is another thing that is not very well known: if the land is, for example, municipally owned, and if one wants to invest money in special conservation and development projects, it is often easier to do so in places close to cities. Just one example: "The Bee Paradise" close to the city of Växjö is an area where special microbiotopes have been shaped, often with sand and water, and where selected plants have been planted for various rare insects. The area has become a special attraction even for people who know nothing about insects. At the same time, it helps to strengthen several endangered populations of species such as wild bees.

Let us end by leaving the discussion of the physical planning of 'green areas' and look at what needs to be done on the ground to create fine environments. The truth is, where you want to create green wedges or strengthen a stretch of landscape, it is mostly not enough to just wait and let nature herself arrange it. We need techniques to go from boring industrial fields, spruce plantations, and in more southerly countries dry wasteland, to rich greenery and biological diversity. A common question that I have been asked in recent years has been: How do you do it and how long does it take to turn an arable field into natural pasture or a species-rich hay meadow? I have some answers, but a great deal depends on the soil, the surroundings, and how much it is allowed to cost. Then it has to be admitted too that our knowledge of such matters is limited. Research is needed here.

Another question is how to go from boring production forests to a more species-rich and fascinating woodland. Some techniques are being started, such as 'veteranization', whereby you inflict moderate damage on young trees so that rare insects and fungi have a chance when almost all the trees in the locality which are old and provide a home to these organisms are disappearing. A similar question, but perhaps not so difficult, is how to take a uniform stand of trees and make a varied forest with gaps and dense patches. A chance to do this exists in the nature reserve and recreation area of Skrylleskogen about 10 kilometers east of Lund. Here they are thinning the young deciduous trees that were planted 10–20 years ago after the spruce trees were blown down by storms. The thinning seems to be following normal principles of forestry, with even distances between the surviving trees. There would be room here for new thinking.

Finally, a challenge to biologists, landscape architects, landscape researchers, recreation researchers, farmers, and cultural historians: How do we recreate a more attractive agricultural landscape without losing more than 3–10 percent of the area? We know that production can be maintained, but what can be done in more concrete terms, for example, to follow a historical tradition, to prioritize biotopes, to give good walking paths, and simultaneously get the landowners to cooperate in this?

[1] The Finnish national urban parks have precisely this function. Around relatively small cities they create protection against development and preserve a cultural landscape.

[2] This area in Uppsala was a candidate to be designated a National Urban Park by the Government, but the city rejected the idea.

TERESA PASTOR
Peri-Urban Parks

As development intensifies, peri-urban parks may eventually end up surrounded by the urban fabric.

This is a very personal account. For many years I worked with an organization called Fedenatur, an association of 31 peri-urban parks, which are parks found on the periphery of cities mainly in southern Europe. Set up in 1995, it was inspired by the "sustainable city objective" set out in Agenda 21, adopted during the Earth Summit in Rio de Janeiro in 1992. At the time I was based in Barcelona, which meant close contacts and work with Collserola Natural Park, the large peri-urban park in that city. Since a couple of years I work with Europarc, the organization with which Fedenatur merged in 2017. The merger was motivated by the will to strengthen the management and development of Europe's nature reserves and parks and enhance their protection. Europarc had been set up in 1973—one year after United Nations Conference on the Human Environment in Stockholm—and now organizes parks and nature reserves in 40 European countries. This includes 40 percent of all Natura 2000 areas in Europe. The growing concern for urban ecology and green infrastructure in cities is manifested by the International Union for Conservation of Nature (IUCN) that in 2008 adopted a recommendation for "setting up networks of protected urban and peri-urban natural areas." IUCN will henceforth not only deal with and care for wild nature, but also urban nature.

Here I give an overview of the situation regarding peri-urban parks, their history, and the particular needs to be met in designing such parks and dealing with their management challenges.

Many cities have some kind of protected natural area located on their outskirts. The degree of protection varies, depending on the state, region or local legislation, as does the category of protection and assigned designation, such as urban protected area, urban national park, green belt, agricultural park, natural park, or recreational park.

These protected areas are globally referred to as peri-urban parks (peri-urban means being located on the periphery of an urban settlement). This designation refers to the location of the parks and not to the level of protection. As development intensifies, peri-urban parks may eventually end up surrounded by the urban fabric. When this is the case, peri-urban parks are termed metropolitan parks as they are enclosed within a large metropolitan area. In fact, one of the main functions of peri-urban parks is to contain urban sprawl by constituting a natural

The Green Heart of Holland. The concept of the Green Heart was presented in 1956. Regulation was not required because there was consensus between the national planning authority and the provinces about the importance of preventing urban sprawl in the area. Today there is open conflict about the future of the area.

and legal barrier to urban development. Peri-urban parks are important non built-up spaces that act as "territorial separators" to avoid the fusion of neighbouring cities into one single mega-city, thus preserving their respective identities. This is the case, for example, of Collserola Natural Park, which is surrounded by 9 different cities, Barcelona being one of them. Similarly, the Green Heart of Holland, an area of wet meadows and marshland suitable for agriculture but not for building, is surrounded by the Randstad ring of the four largest Dutch cities (Amsterdam, Rotterdam, The Hague, and Utrecht).

Peri-urban parks have specific features and roles that distinguish them from both urban parks and remote natural parks (Fedenatur 2004). Peri-urban parks are wilder, more biodiverse, and substantially larger than most common urban parks. That is why they can deliver numerous ecosystem services to the city, such as provisioning services (fresh water, food, and timber production); regulating services (clean air and control of temperatures); and supporting services (soil and habitats). Furthermore, they offer recreation opportunities to the city population

JACE48 WIKIMEDIA COMMONS

SPAY—UNION FOR THE PROTECTION AND DEVELOPMENT OF HYMETTUS

in a healthy natural environment—a cultural service very much needed in highly dense compact cities. It is quite common, in many peri-urban parks, to come across hikers, runners, cyclists, and horseback riders. Moreover, peri-urban parks are crucial in the struggle of towns against climate change by cooling down temperatures and removing greenhouse gases.

Most peri-urban parks host high levels of biodiversity, one reason being that they tend to have a heterogeneous structure (due to their location at the crossroads between urban and rural landscapes), which in turn gives rise to a highly diversified mosaic of biotopes that can support a wide range of different species. As compared with remote natural parks, peri-urban parks usually contain fewer sensitive species, such as large carnivores (Santa Monica Mountains National Recreation Area in California being a notorious exception), but tend to contain a higher number of exotic species that have been introduced from the city.

Peri-urban parks present different typologies. They might be quite natural if they are the result of natural habitats that have been preserved around a city for different reasons (mountainous territory, marshlands, hunting area, etc.); semi-natural (agricultural land, etc.); or completely man-made (restored brownfields, reforested areas, etc.).

Peri-urban parks can be fully or partially devoted to agriculture, offering food to consumers from the nearby towns and thus contributing, to some extent, to food security and food sovereignty. Parks substantially dedicated to agriculture are known as rural or agricultural parks. Peri-urban agriculture plays a role of primary importance, not only for soil conservation but also for food production; environmental education about food security and safety, and food production chains.

Furthermore, some peri-urban parks, despite not being agricultural parks, promote agricultural activities for environmental and/or social reasons. Harvesting some agricultural land within a forest park, or on its edges, is a means of diversifying habitats and thereby increasing biodiversity. In fact, many forest species depend on open spaces for food. On the other hand, agriculture is a potent tool used to promote social cohesion. Some peri-urban parks, such as the Parco Nord Milano, harbour communal gardens in which citizens share organic farming practices while also learning about sustainable lifestyles.

Notwithstanding the degree of naturalness of a peri-urban park in its origin, the way it is conceived, designed and/or managed will greatly

The way it is conceived, designed and/or managed will greatly influence biodiversity.

TOP LEFT: Kinderdijk windmills in the Green Heart of Holland.
BOTTOM LEFT: View from Mount Hymettus over Athens.

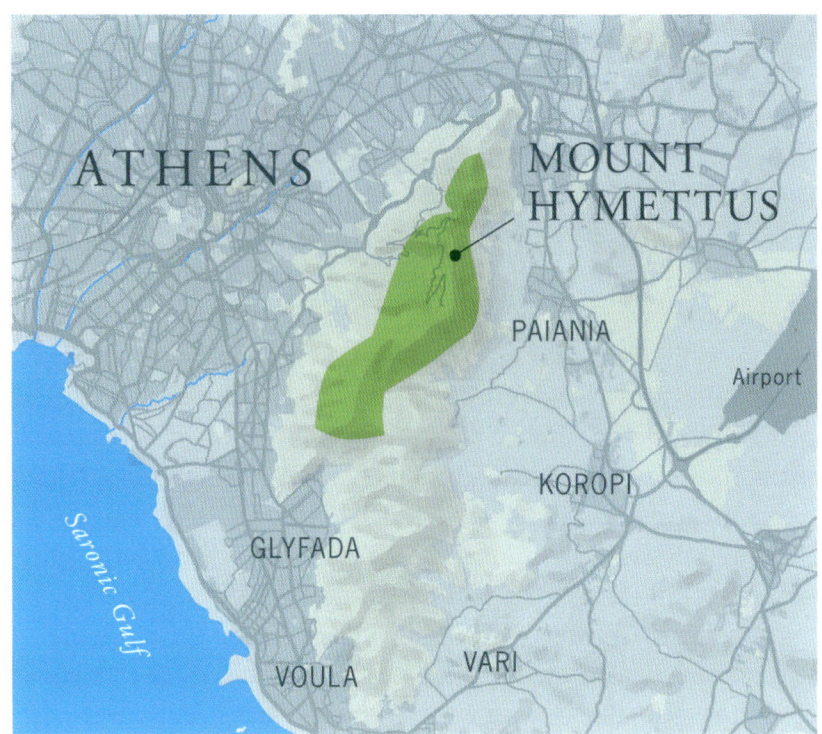

Mount Hymettus (81 square kilometers) in Athens, Greece.

Parks should be connected to other natural areas in order to avoid isolation.

Grand Parc Miribel Jonage (22 square kilometers) in Lyon, France.

Milano, Italy, is enclosed by large parks: Parco Agricolo Sud Milano (470 square kilometers); Parco Nord Milano (6 square kilometers); and Parco delle Groane (38 square kilometers).

influence biodiversity, the type and quality of habitats, and the presence or absence of sensitive species.

Because of their situation, peri-urban parks have the double task of preserving biodiversity while also being prepared to host a large number of visitors. These visits occur at a high frequency with peaks of intensity during weekends and public holidays. This is highly challenging from a conservation point of view, since an excess of visitors can badly affect biodiversity. Therefore, planners need to organize the park by devising different zones, i.e., core and buffer areas, which are large enough to host viable ecosystems but at the same time allow visitors to enjoy some parts of the park.

Zoning refers to dividing a park into several different zones for the purpose of distributing different types of use or protection in the most appropriate places. Each zone has its own set of rules and regulations for activities carried out within its boundaries. Peri-urban parks should contain large buffer zones dedicated primarily for public use and core zones, with more restrictive public access devoted to biodiversity protection.

Collserola natural park (80.7 square kilometers) and the peri-urban parks Parc del Garraf and Parc de la Serralada outside of Barcelona, Spain.

As an example of regulation, the federal government of Switzerland has established a category labelled "nature discovery parks" that consists of a core natural area of at least 4 square kilometers, with restricted use; and a transitional area that is used as a buffer and for people to enjoy nature. Examples of such parks are the Wildnispark in Zurich and the Jorat peri-urban natural park in Lausanne.

Quite too often, urban fringes—where peri-urban parks are located— have been disregarded and have not been the object of good planning. Instead, the borders of cities have been prone to degradation. Landscape restoration of degraded areas and installation of equipment to host visitors is vital in order to transform the degraded areas into attractive areas.

Furthermore, parks should be connected to other natural areas in order to avoid isolation, which would be harmful in the long term for persistence of biodiversity. Maintaining connection between different

TOP RIGHT: Peri-urban parks that are close to major cities are often cut through by heavily trafficked roads. Above is a good example of an ecoduct built in the National Park Hoge Weluwe outside of Amsterdam.
BOTTOM RIGHT: Parco Agricolo Sud Milano, Italy. The ecologically managed farms border directly onto the city.

APDENCY WIKIMEDIA COMMONS

TONI NICOLINI, ARCHIVIO PARCO AGRICOLO SUD MILANO

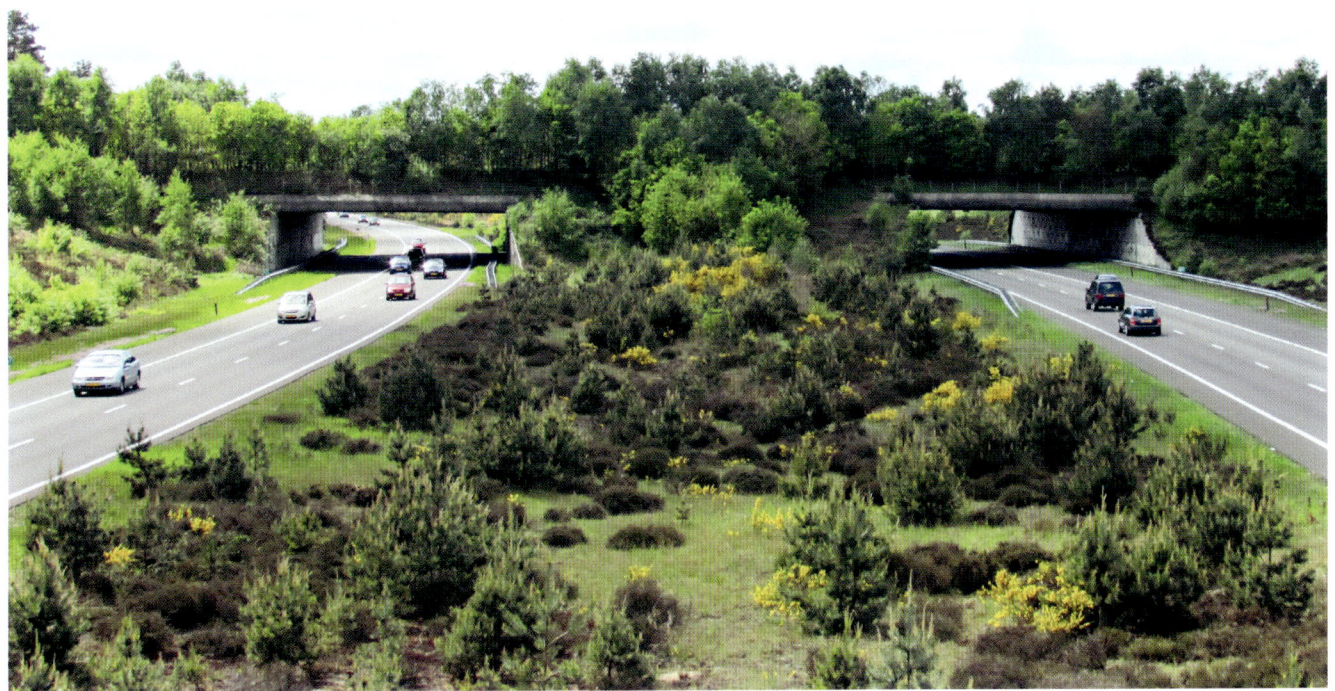

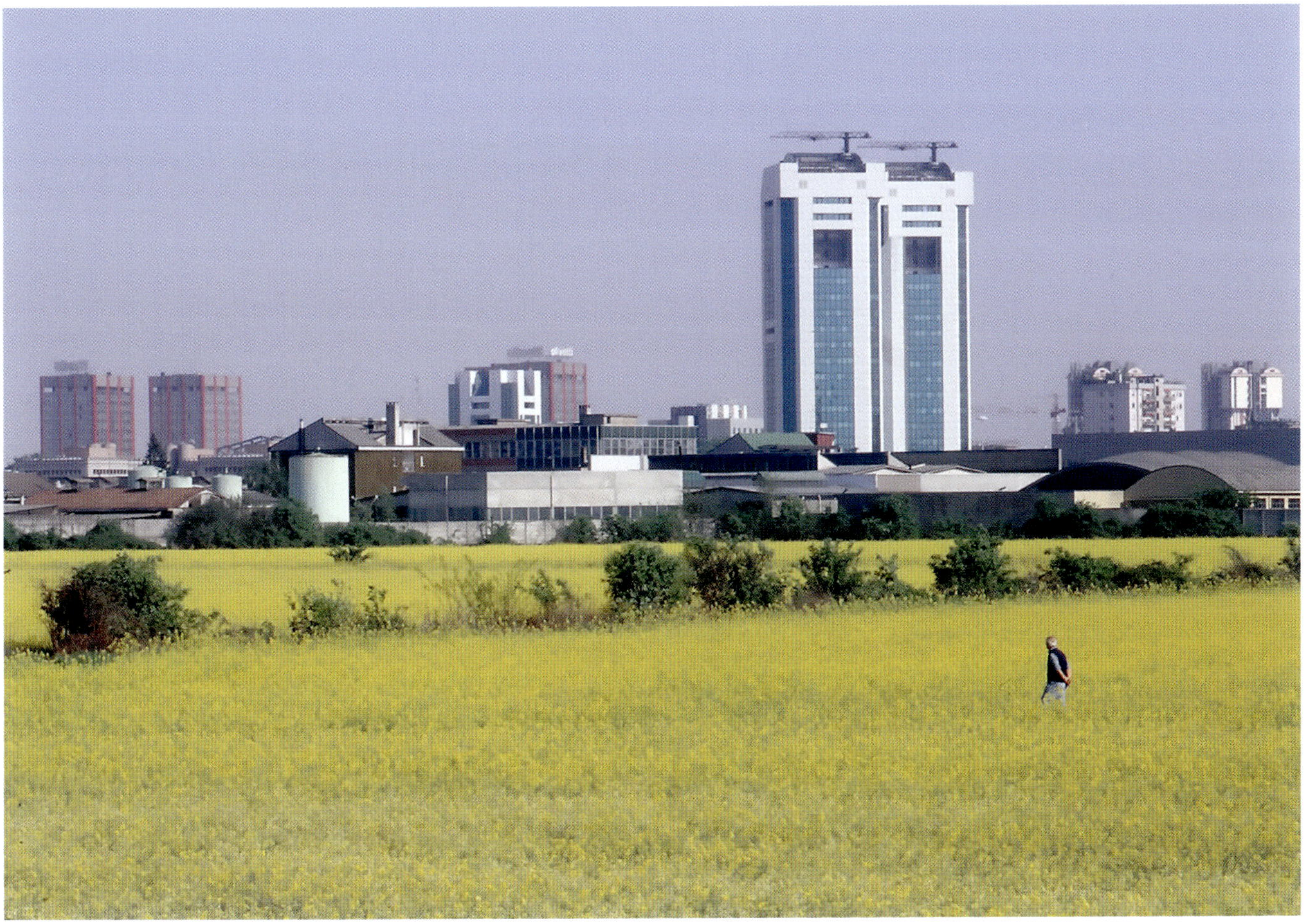

CONSORCI PARC NATURAL DE LA SERRA DE COLLSEROLA

DONA REISKOFFER WIKIMEDIA COMMONS

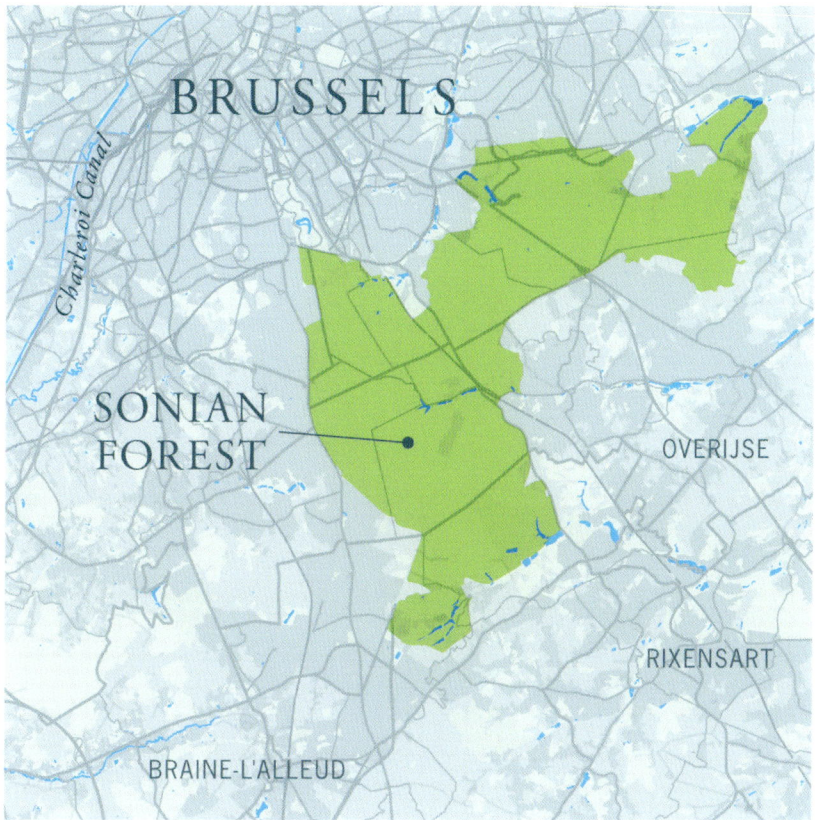

Sonian Forest (43.8 square kilometers) southeast of Brussels, Belgium.

natural spaces and natural elements in an urban context is one of the main challenges with which these parks are confronted.

Given their location on the outskirts of towns, peri-urban parks tend to be crossed or surrounded by heavy infrastructure (highways, roads, railways, etc.) causing fragmentation of the landscape. Therefore, strong investments such as viaducts or underpass tunnels are often needed to restore connectivity with other natural areas. Ecological connectivity was not considered 30-50 years ago when many of the current peri-urban large parks were established.

One of the main features of peri-urban parks is that they should be easily accessible by all citizens in order to allow people frequent contact with nature.

TOP LEFT: Public welcome structure on the border between the city and the Collserola natural park, Barcelona, Spain.
BOTTOM LEFT: Sonian forest is a World Heritage Site inscribed by UNESCO as an "ancient and primeval beech forest." Unfortunately, the forest is cut through by a major road and thus an ecoduct has been built to connect the two halves.

L'Arche de la Nature (4.5 square kilometers), Le Mans, France.

This access should be as universal as possible, taking into account those persons with special needs. The park should be designed in a way that promotes access on foot, bike, and public transport, whereas the use of the car should be restricted inside the park. Some design techniques could be employed to discourage people from parking outside the designated parking facilities. There should be reserved parking places for less mobile persons.

Due to their proximity to the city, peri-urban large parks play an important social and educational role. They receive a large number of visitors, with very different interests. Peri-urban parks thus need to be prepared to attend to people with special needs; families; the elderly; and sport participants. Furthermore, hosting visitors requires that the available equipment should respect safety conditions, even if it is in a natural environment. Such equipment needs constant maintenance and replacement because it is submitted to high levels of use.

In addition to the already-existing cultural heritage in the form of buildings, peri-urban parks are usually equipped with visitors' and environmental education centres.

Other pieces of equipment that are usually found in large parks are tables and benches, even barbecues, in delimited picnic areas. Bars, restaurants, and even accommodation facilities are common features of peri-urban large parks. Some parks offer regulated orchard lots to give the local population a chance to grow their own vegetables.

Peri-urban large parks have the challenge of looking as natural as possible while at the same time offering a certain level of safety for both users and inhabitants. Due to their proximity to towns and their high number of visitors, natural hazards are potentially more dangerous in peri-urban large parks than in wild nature elsewhere. The potential sources of natural threats may include forest fires, floods, landslides, falling trees, etc., which all need to be taken into account both in planning and management.

Though there are many peri-urban parks found across Europe they do not share a common history. Nevertheless, there are common factors that can explain their existence in many cities, such as the geographic and topographic conditions; the involvement of local administration; and social demand. Geographic conditions constitute a particularly important parameter, setting the types of natural environment as well as the type of human activities carried out there. Many peri-urban parks are the consequence of the existence of strong physical constraints, such as wetlands and flood plains (e.g., Grand Parc de Miribel Jonage in Lyon, France, and the Green ring of Vitoria-Gasteiz in Spain); rough faults, mountains, and hills (e.g., Collserola Nature Park in Barcelona, Spain, and Buda Hills in Budapest, Hungary), or a mountainous coast (e.g., Parc périurbain des Calanques outside Marseille, France, and Parco di Portofino outside Santa Margarita di Liguria in Italy) that have hampered urbanization and thus preserved natural heritage. Other peri-urban parks were former royal hunting areas (Monte El Pardo in Madrid, Spain) or areas devoted to agriculture (Parco Agricolo Sud Milano, Italy, or Parc del Baix Llobregat outside Barcelona, Spain). Yet many peri-urban parks are the result of important restoration and reforestation works of former industrial sites and brownfields (Parco Nord Milano in Italy and Espace Nature in Lille, France).

Another commonality of many peri-urban parks is that their protection stems from the local administration, quite often as the consequence of social demand. In France, some peri-urban parks (e.g., Base de loisirs Saint Quentin en Yvelines just west of Paris) were first established to offer a leisure site in an open-door environment for low-income families that could not afford to travel away for holidays. Nowadays, after global economic crises, rapidly growing cities, new life-styles and new holiday patterns, both urban parks and peri-urban parks have seen a steady increase in the number of visits both on week-ends and in the summer months. That is another good reason to pay attention to peri-urban parks.

Peri-urban large parks have the challenge of looking as natural as possible while at the same time offering a certain level of safety for both users and inhabitants.

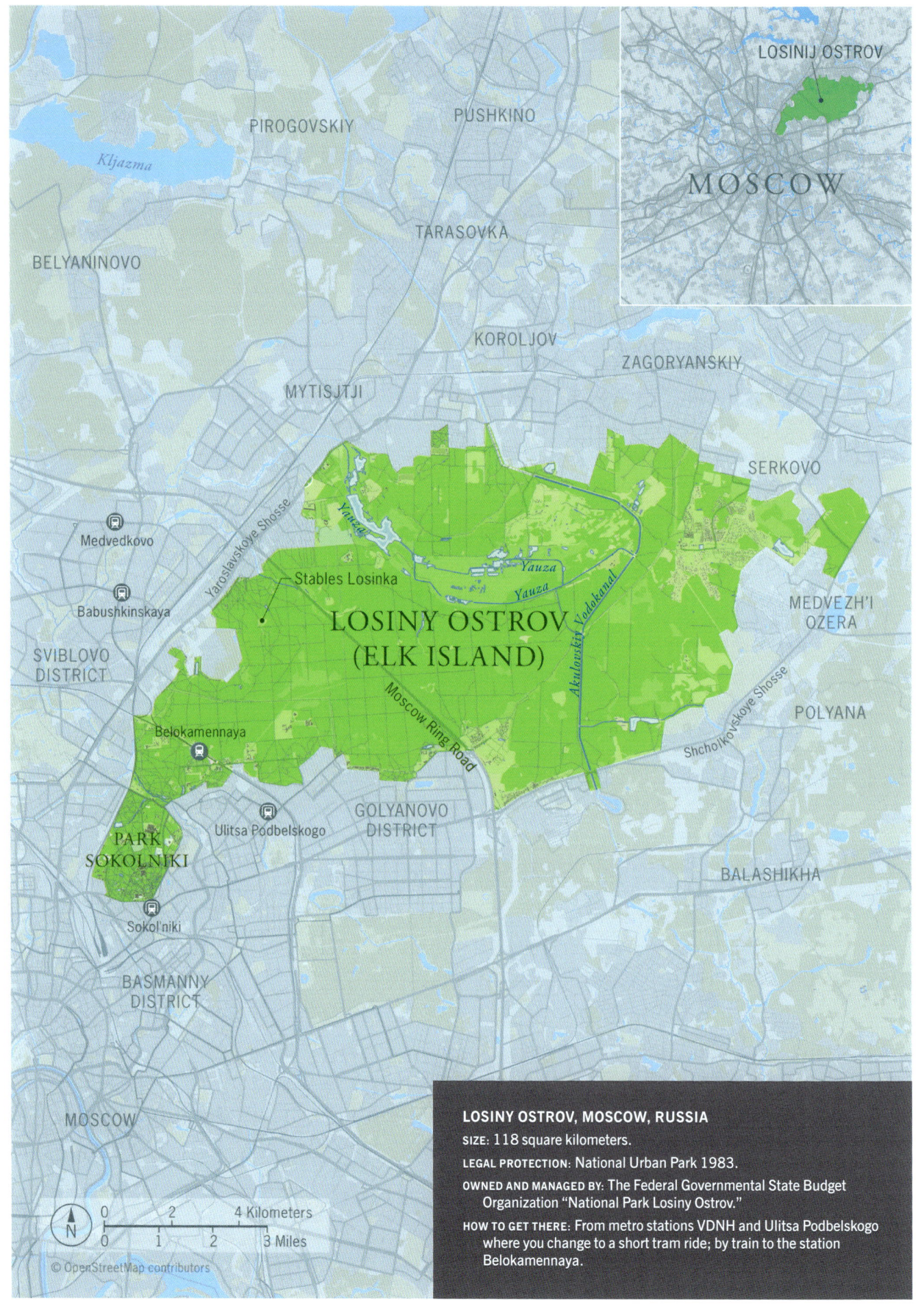

LOSINIJ OSTROV

MOSCOW

PIROGOVSKIY

PUSHKINO

Kljazma

TARASOVKA

BELYANINOVO

KOROLJOV

ZAGORYANSKIY

MYTISJTJI

SERKOVO

Medvedkovo

Yaroslavskoye Shosse

Stables Losinka

Yauza

Yauza

Yauza

Babushkinskaya

**LOSINY OSTROV
(ELK ISLAND)**

MEDVEZH'I
OZERA

SVIBLOVO
DISTRICT

Moscow Ring Road

Akulovskij Vodokanal

POLYANA

Belokamennaya

Shcholkovskoye Shosse

Ulitsa Podbelskogo

GOLYANOVO
DISTRICT

PARK
SOKOLNIKI

BALASHIKHA

Sokol'niki

BASMANNY
DISTRICT

MOSCOW

N

| 0 | | 2 | 4 Kilometers |
| 0 | 1 | 2 | 3 Miles |

© OpenStreetMap contributors

LOSINY OSTROV, MOSCOW, RUSSIA

SIZE: 118 square kilometers.

LEGAL PROTECTION: National Urban Park 1983.

OWNED AND MANAGED BY: The Federal Governmental State Budget
Organization "National Park Losiny Ostrov."

HOW TO GET THERE: From metro stations VDNH and Ulitsa Podbelskogo
where you change to a short tram ride; by train to the station
Belokamennaya.

DMITRY KAVTARADZE
Losiny Ostrov, Moscow, Russia

Losiny Ostrov National Park (Elk (Moose) Island) is one of the few national parks of its size in the world to be situated within a megalopolis. Located in Moscow and Moscow Oblast it is said to be the third largest forest in a city of comparable size, after Table Mountain National Park in Cape Town, South Africa and Pedra Branca State Park in Rio de Janeiro, Brazil.

Losiny Ostrov is located to the north east of Moscow, one-third of it within the city boundaries of Moscow, two-thirds within the city of Korolev. From Red Square (the heart of Moscow City) it is 16 kilometers and 20 minutes by train. Losiny Ostrov is one of the first national parks in Russia. It is also the biggest among forests (with remnants of native forests) located within Moscow boundaries. There are over 40 forests in the City that cover 12 percent of the area.

There are many forest-parks in the Moscow metropolitan area, such as Bitzevsky and Kuzminsky parks. Forest-park is a special category of green areas in Russian classification and refers to a partially planted or modified natural forest within the city limits that has design and planning elements and a maintenance regime.

There are other large parks within the Moscow administrative boundary. Quite a few of them are historical and as such are under government protection deemed as "garden and park art heritage," such as Ostankino Park and Tsaritsyno Park. In the central part of the city there is an important series of parks which are located next to the Moskva River, for example Neskuchny Garden. Neskuchny Garden was designed in the 19th century and is now the biggest historical park in the central part of Moscow. It is now part of the Gorky Park, the famous Park of Recreation and Culture named after Maxim Gorky. The Gorky Park opened in 1928 and is considered to be the "father" of the generation of Soviet "culture and recreation parks."

The most recent park in the centre of Moscow is Zaryadye Park (13 hectares), opened in 2017. It was built on the site of the demolished Rossiya Hotel, just next to the Red Square.

Losiny Ostrov covers 118 square kilometers. It is surrounded by urban areas, but still contains a rich array of wildlife. Forest occupies 83 percent of the area, 2 percent is water, and 5 percent is swamp. An additional 66.45 square kilometers is reserved for expansion of the park,

Losiny Ostrov National Park is one of the few national parks of its size in the world to be situated within a megalopolis.

Big parks in and around Moscow.

The Gorky Park opened in 1928 and is considered to be the "father" of the generation of Soviet "culture and recreation parks."

mostly towards the north-east. Nowadays there are 45 mammal, 185 bird, 4 reptile, 9 amphibian, and around 20 fish species, as well as over 500 species of the higher vascular plants. The territory of Losiny Ostrov is in a sub-zone of mixed coniferous-broad-leaf forests.

Since ancient times Losiny Ostrov has served as the strictly guarded hunting area of Russian grand princes and tsars. Its territory was declared as a reserve already in 1799 and the first forest management was established here in 1842. The idea of the creation of the national park was expressed as early as 1909.

After the transfer of the capital from Moscow to Saint Petersburg in 1712, this area lost its value as a tsarist hunting ground, but the government property continued to be guarded by the imperial edicts. Around this time the area finally received the name Losiny Ostrov. In 1798, these forests passed to the management of the newly formed Forest

TOP RIGHT: Zaryadye Park, 100 meters from the Red Square in Moscow.
BOTTOM RIGHT: Birch forest, Losiny Ostrov.

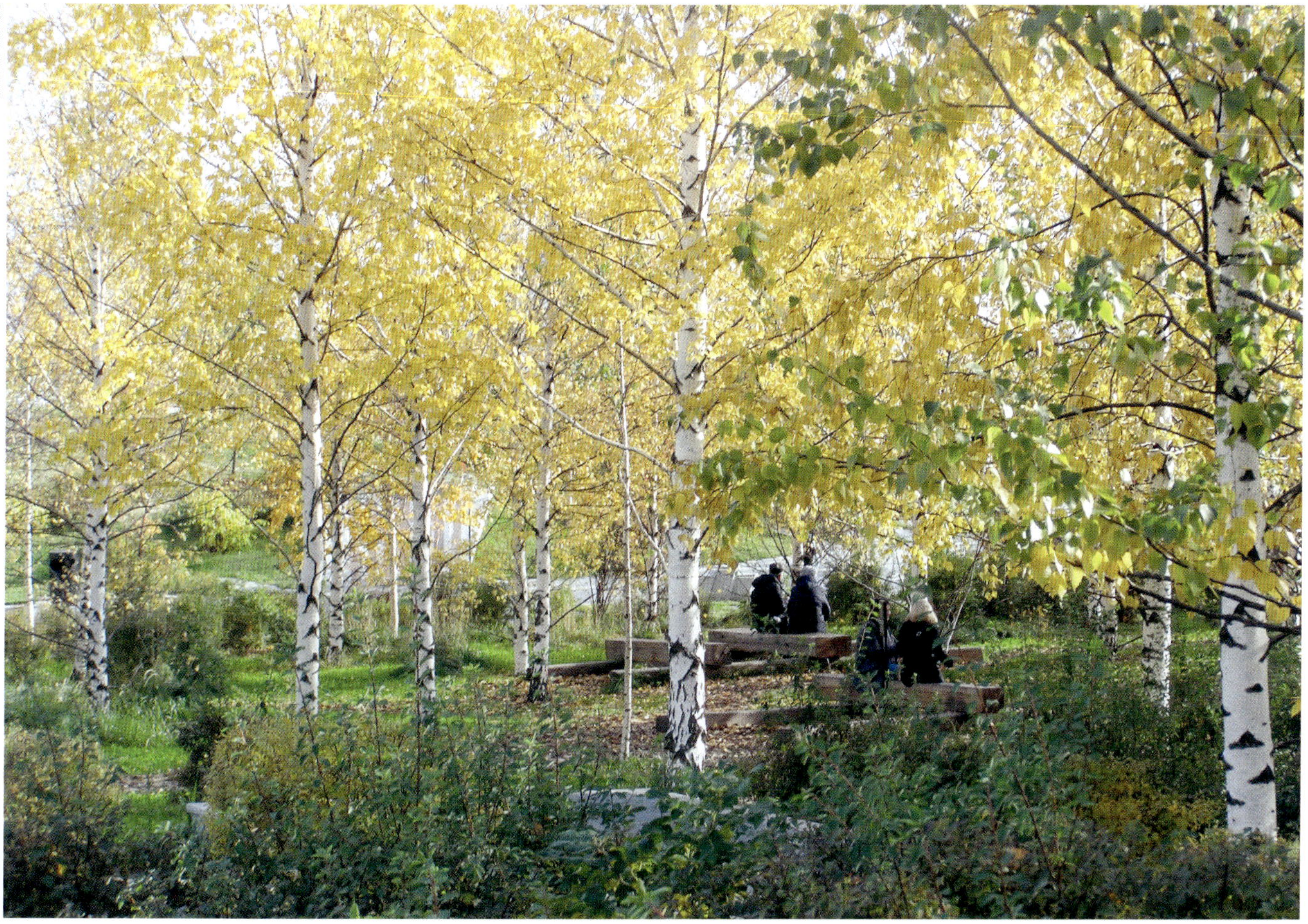

RICHARD MURRAY

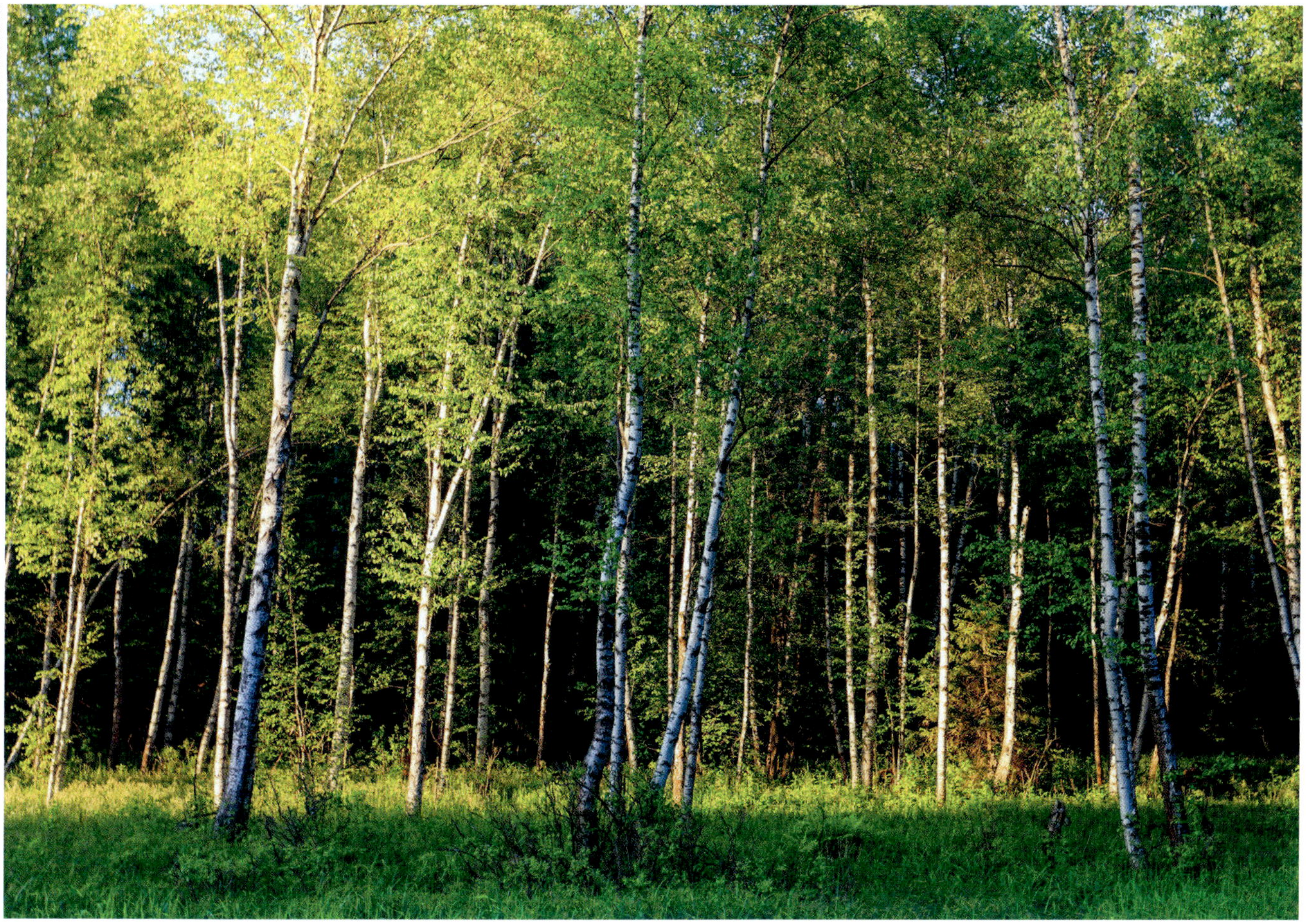

SERGEY GORDIENKO SHUTTERSTOCK

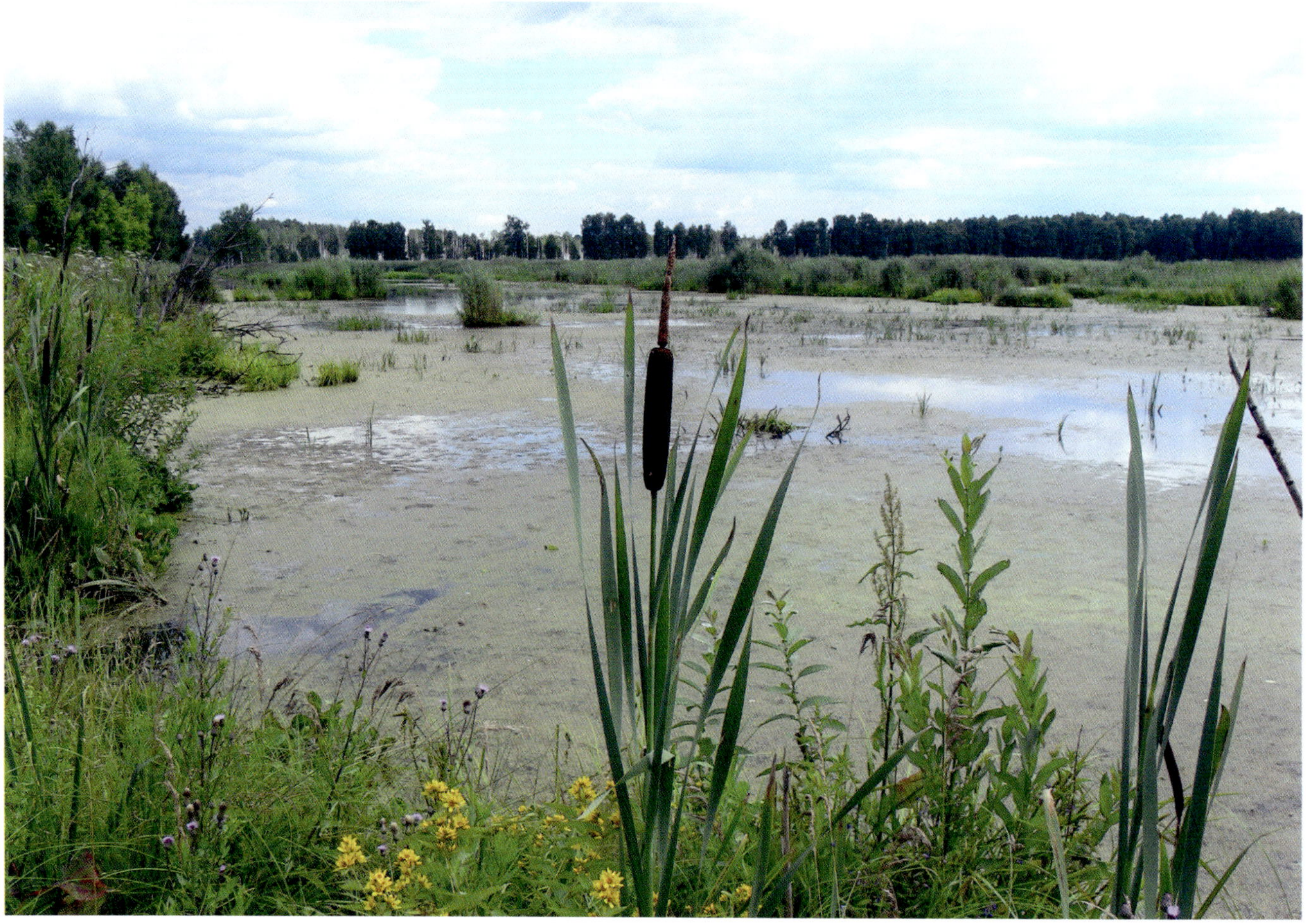

VICTOR SOLODUSHKIN

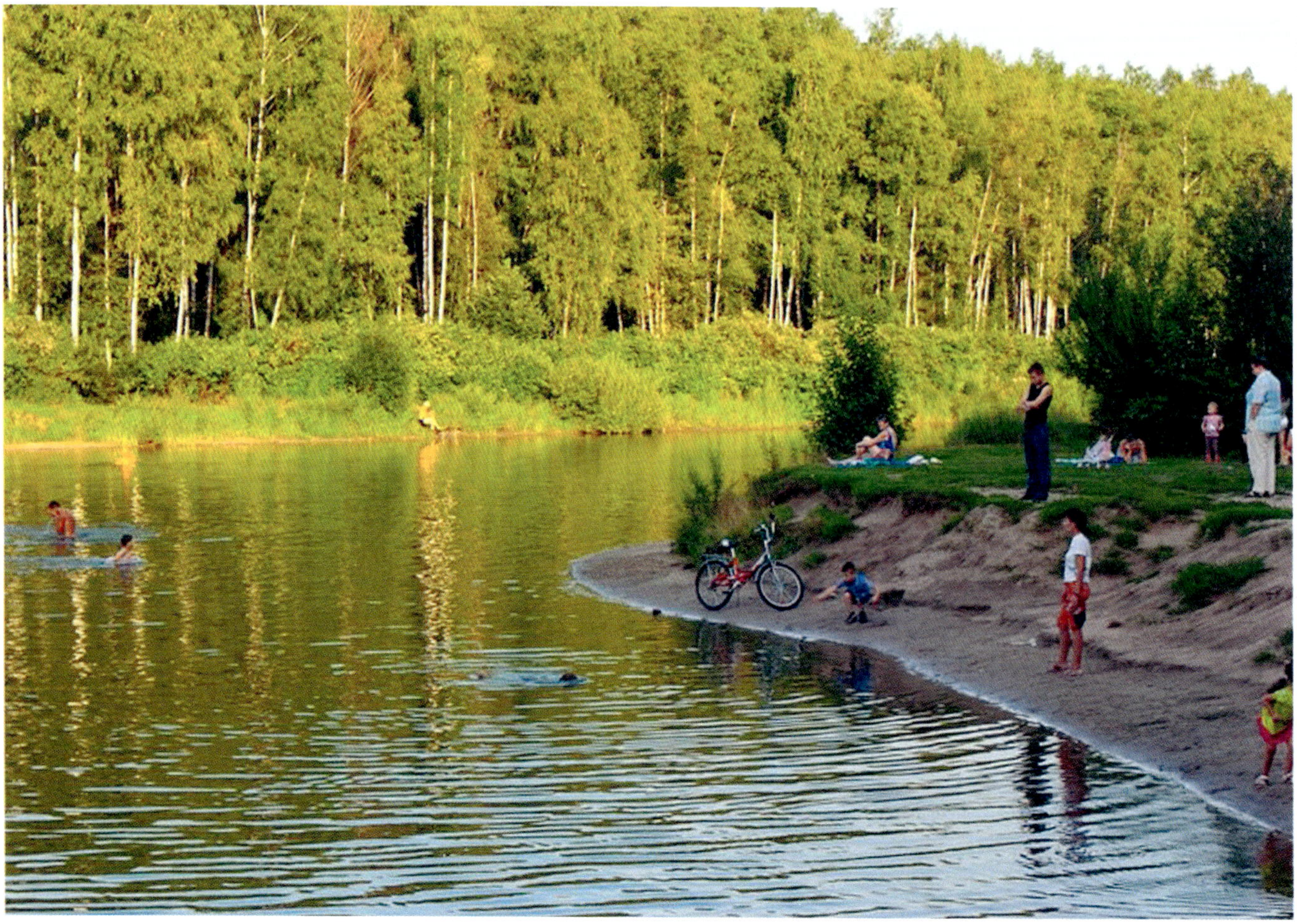

OPEN SOURCE

Department. In the middle of the 19th century the Losinoostrovskaya Lesnaya Dacha (forest mansion) was opened, and the period of systematic forestry activity began. In 1934, the Losiny Ostrov was included in the 50 kilometer greenbelt zone of Moscow.

A large area of the forest was cut down during the Great Patriotic War 1941-1945 (World War II). Since that time, Losiny Ostrov has suffered from arbitrary seizures of the land for vegetable gardens, intense cattle pastures, and even illegal cuttings. In the late 1950s, construction of the Moscow Ring Road split the forest into the inner and the outer sectors.

In 1979, the united resolution of the Moscow Urban and Provincial Soviets of People's Deputies declared the Losiny Ostrov the status of a natural park. In 1983 the decision of the Council of Ministers of the Soviet Union established Losiny Ostrov National Park. According to Moscow Law No. 48, "On Specially Protected Natural Areas in Moscow," passed on 26 September 2001:

> The National Park is a specially protected natural area of federal importance, located in the city of Moscow and has a special environmental, educational and recreational value as a unique natural complex, over 500 hectares, with high natural diversity and the presence of rare or well-preserved typical natural ecosystems, rare and vulnerable species of plants and animals. The use of the national park is permitted for conservation, educational, scientific purposes and for regulated recreation of the population in designated places.

The climate of Losiny Ostrov is characterized by moderately cold winters and warm summers. The coldest month is January with average temperatures from −9 to −11 degrees Celsius and the hottest is July with average temperatures from +19 to +20 degrees.

The Yuaza river basin is part of the water supply system of Moscow.

Losiny Ostrov is important for the preservation of rare plants and animal species given that there are no other protected areas nearby. But the park is under pressure from surrounding urban areas, which threaten to fragment the landscape even further. The city of Korolev is a research-oriented settlement with numerous scientific institutions and related facilities and will need land for new housing. Several other urban settlements border the national park and thus, affect the park's unique ecosystems.

The Moscow part of the Losiny Ostrov is also crossed by the Moscow

Establishing green corridors from Losiny Ostrov to the surrounding landscape has been an essential goal of local and regional planning strategy for several decades.

TOP LEFT: Upper Yauza Swamps, Losiny Ostrov, 16 km from the Red Square in Moscow.
BOTTOM LEFT: Swimming in one of the rivers in Losiny Ostrov.

Little Ring Railway. Built in 1902, this railway was used mostly for industrial train operations. In 2010 it was converted to passenger transportation. The Belokamennaya Station of the Moscow Central Railway Ring is located inside the Park. The railway offers a quick route to Losiny Ostrov for the people of Moscow.

Establishing green corridors from Losiny Ostrov to the surrounding landscape has been an essential goal of local and regional planning strategy for several decades.

The park is tremendously important for the thirteen million inhabitants of the Moscow metropolis.

The National Park Losiny Ostrov is divided into three zones:

- 53.94 square kilometers is specially protected and closed to the public.
- 31.30 square kilometers is for training and excursions, open for restricted visitors and only along the park's established routes.
- 29.81 square kilometers is open for recreation.

With 8 million visitors per year it is important to monitor activities going on in the park and provide appropriate maintenance, as well as to enact the protection and educational policies.

TOP RIGHT: In winter it is hard to get a drink in Losiny Ostrov.
BOTTOM RIGHT: Forest man with a group of children.

VICTOR SOLODUSHKIN

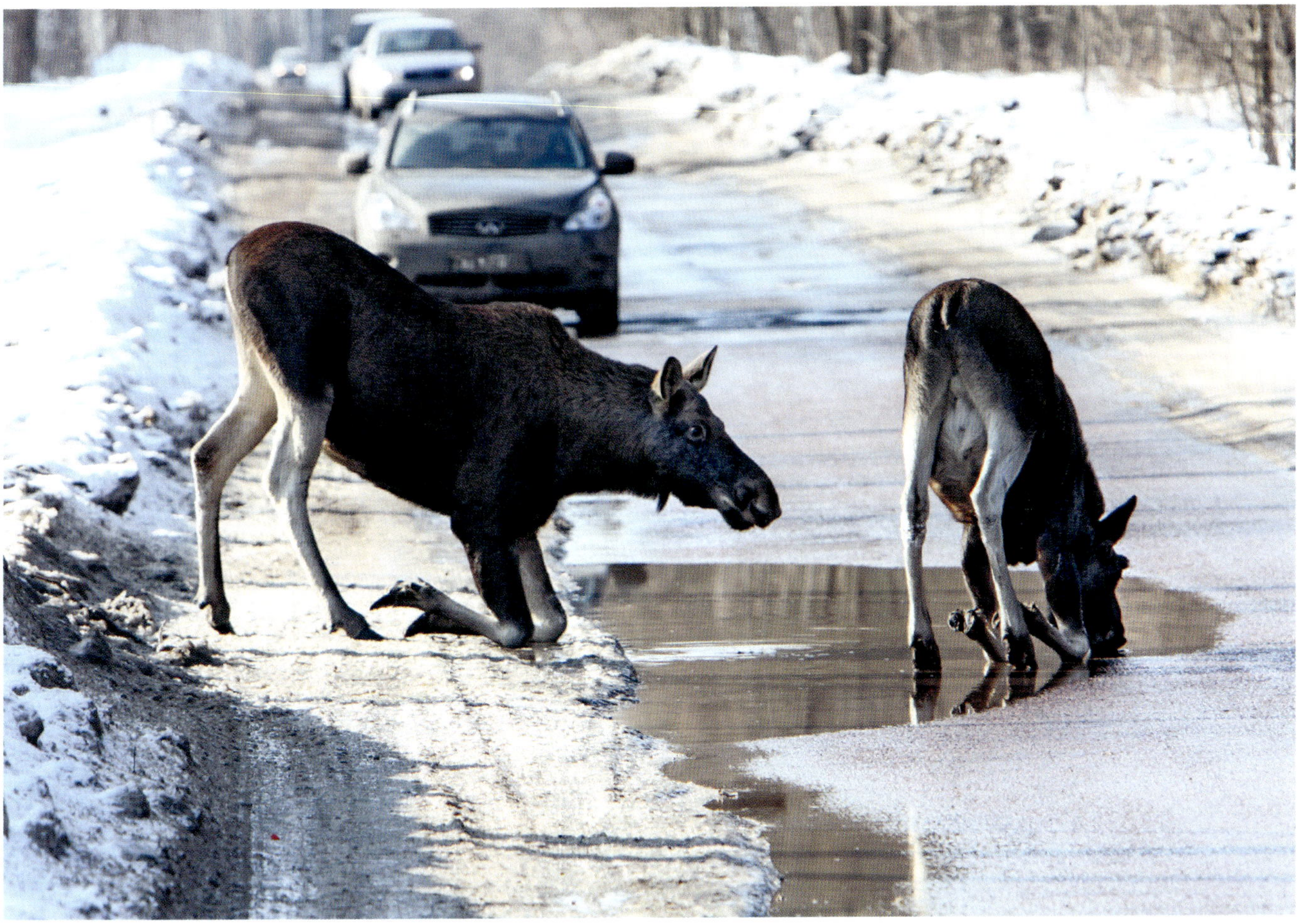

OPEN SOURCE

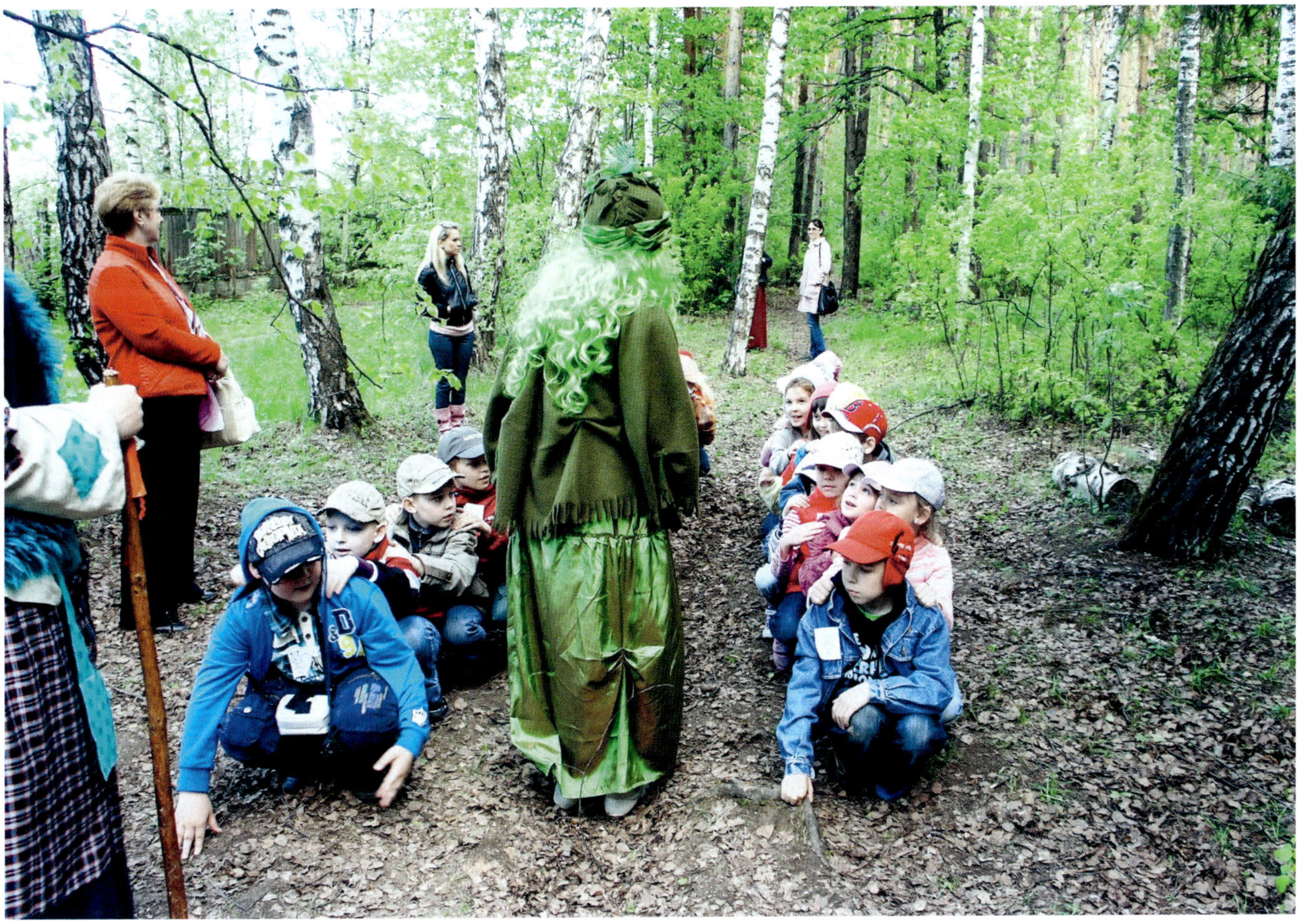

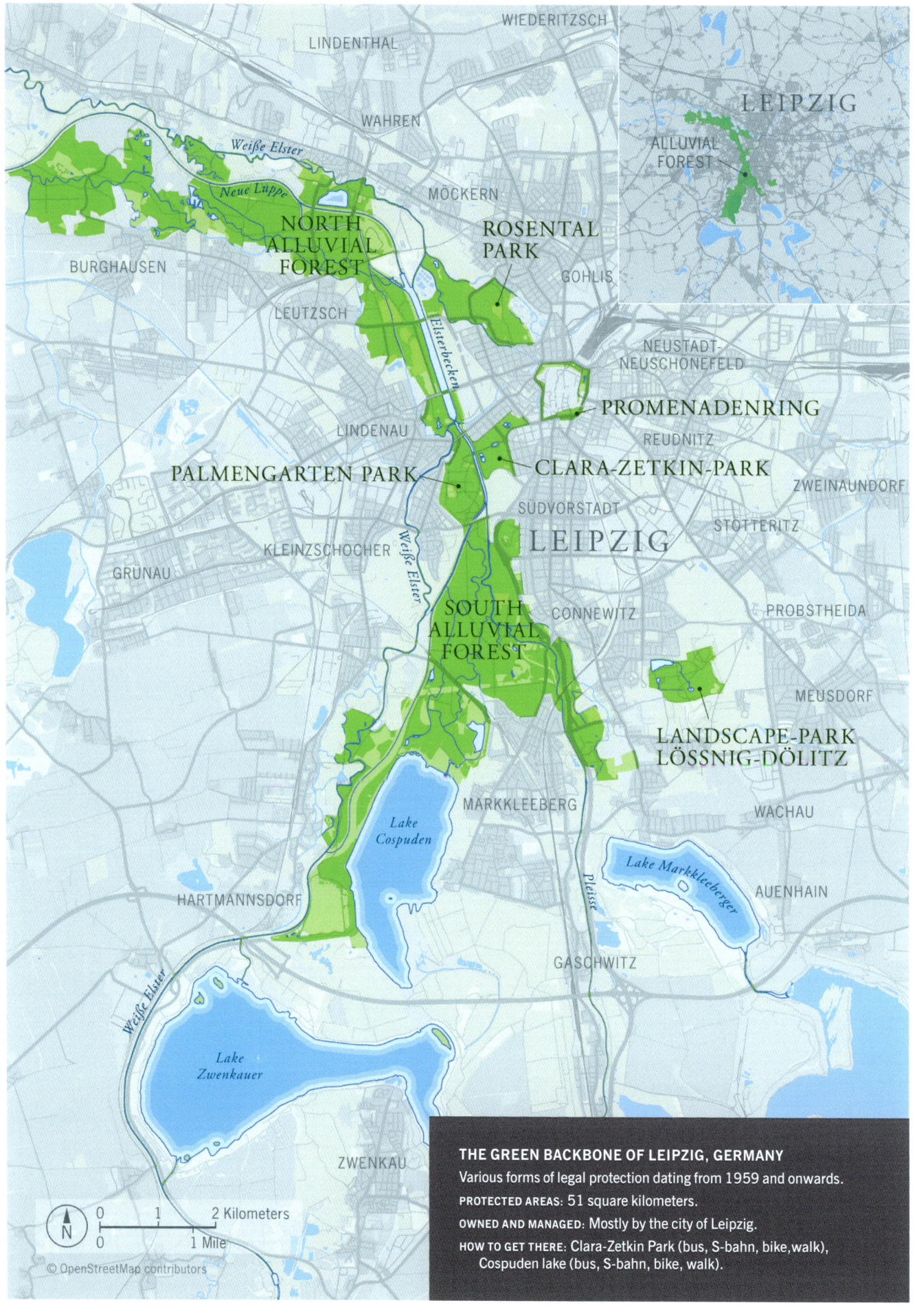

LINDENTHAL

WIEDERITZSCH

LEIPZIG

ALLUVIAL
FOREST

WAHREN

MÖCKERN

Weiße Elster

Neue Luppe

NORTH
ALLUVIAL
FOREST

ROSENTAL
PARK

GOHLIS

BURGHAUSEN

LEUTZSCH

Elsterbecken

NEUSTADT-
NEUSCHÖNEFELD

PROMENADENRING

REUDNITZ

LINDENAU

PALMENGARTEN PARK

CLARA-ZETKIN-PARK

ZWEINAUNDORF

Weiße Elster

SÜDVORSTADT

LEIPZIG

STÖTTERITZ

KLEINZSCHOCHER

GRÜNAU

SOUTH
ALLUVIAL
FOREST

CONNEWITZ

PROBSTHEIDA

MEUSDORF

LANDSCAPE-PARK
LÖSSNIG-DÖLITZ

Lake Cospuden

MARKKLEEBERG

WACHAU

Lake Markkleeberger

AUENHAIN

HARTMANNSDORF

Plelsse

Weiße Elster

GASCHWITZ

Lake Zwenkauer

ZWENKAU

N

0   1   2 Kilometers

0   1 Mile

© OpenStreetMap contributors

THE GREEN BACKBONE OF LEIPZIG, GERMANY
Various forms of legal protection dating from 1959 and onwards.

PROTECTED AREAS: 51 square kilometers.

OWNED AND MANAGED: Mostly by the city of Leipzig.

HOW TO GET THERE: Clara-Zetkin Park (bus, S-bahn, bike,walk),
Cospuden lake (bus, S-bahn, bike, walk).

TORSTEN WILKE

The Green Backbone of Leipzig, Germany

Leipzig is blessed with one of the biggest floodplain (alluvial) forests in Europe, the Leipziger Auwald, located in the heart of the city. This unique biotope stretches through the city from north to south in a wide band, including several rivers. The Leipziger Auwald is the Green Backbone of Leipzig—very important for the climate, environmental and flood protection, and the enhancement of the urban residential environment. In and around the city and connecting with the forest there are numerous parks, other kinds of woodlands, meadows, marshes, and green "active axis," which can be used to relax, for sightseeing, leisure activities and sports like hiking, cycling, horseback riding, canoeing, and other water sports.

In some surveys, Leipzig is rated as the most livable city in Germany.

When the wall fell in 1989 and East and West Germany were united a great exodus of young people from the east to the west started. From 1990 to 1998 the city lost 70,000 of its 510,000 inhabitants.

The first priority after unification was housing and efforts to stop the shrinking of the city, by means of a policy to create jobs. One of the strategies to attract people back to the city was to create a livable city, full of parks and forests. A series of centrally located city parks were created, such as district park Plagwitz (opened 2001 on a former expo site), Lene-Voigt Park (built on a former railway station site, opened in 2001), and Henriettenpark (opened in 2005, built on a former factory site). Also the Promenadenring (a string of parks around the Medieval city, created for strolling in 1777 when the defense wall around the city was torn down), was renovated. There was strong citizen engagement for these projects and a deliberate policy for citizens' participation.

But this was just the start. Already in 1993 a landscape plan suggested a Green-Ring-Radial-System of urban green. It has since then been improved twice. Today Leipzig has changed from a shrinking to a growing city. In some surveys, Leipzig is rated as the most livable city in Germany. And its growth has been the fastest of all German cities for several years.

Leipzig is today a city of 600,000 inhabitants, 90,000 more than before the unification.

The city dates back to the early Middle Ages. It is located about 160 kilometers southwest of Berlin at the confluence of the White Elster, Pleiße, and Parthe rivers at the southern end of the North German

Plain. It was first developed to the east of the river White Elster. Stretching along the river are the floodplain forests. When the city started to grow also on the west side of the river, a wide green and blue area was left untouched. This is still the case and is what makes up the Green Backbone.

Leipzig covers a total area of about 298 square kilometers. 17 percent of this area is urban green (10 percent recreation area and 7 percent forest, in all 51 square kilometers) and almost 3 percent is water. In addition, there are 61,000 street trees.

The challenge today is to preserve this green infrastructure when the city is growing rapidly. The east and west urban parts of Leipzig gravitate to close the gap between them. There have for many years been roads and bridges connecting them. Centrality attracts builders.

Legal protection is a patchwork. A 25 square kilometer large section of the Leipziger Auwald was granted landscape protection already in 1959, when this was East Germany. This protection was enlarged to 59 square kilometers in 1996 and encompasses the forested parts of the floodplains that run through the east and west parts of the city. Since then, other forms of protected area have been added: a 9.8 square kilometer nature conservation area; 28 square kilometers of biotope areas (Natura 2000); and 29 square kilometers of bird sanctuaries. They overlap to a great extent, thus making the total protected area 51 square kilometers.

The landscape protection is fairly loose, with many encroachments, while other forms of protection are stricter.

The Leipziger Auwald is at risk because of a falling water table and the lost dynamics of the rivers. Traditional kinds of trees like Ash, Elm, and German Oak are becoming rare because of that. In order to save the forest, the water table has to be raised and the rivers given back their dynamics as far as possible. The reason for the falling water table is the river Neue Luppe, built as a canal through the forest in the 1930s in order to cope with flooding. Now, eighty years later, the project Lebendig Luppe aims to raise the water table by redirecting flow in various segments of the forest river system back to more natural flood-plain-dynamics.

Another big landscape transformation takes place in the former open-cast-mining-landscape around the city. Since 1924 lignite (brown coal) had been excavated south of Leipzig. In 1937 another field opened up. Together these two fields covered an area as large as the whole city

When the city started to grow also on the west side of the river, a wide green and blue area was left untouched.

TOP RIGHT: Equestrian race course in Clara-Zetkin-Park, the largest park, one kilometer from the very center of Leipzig. To the south the Green Backbone of Leipzig stretches further along the river White Elster to the recently created lakes, where brown coal was mined until 1996.
BOTTOM RIGHT: Leipzig center looking towards the northern part of the alluvial forest.

GRÜNER RING LEIPZIG

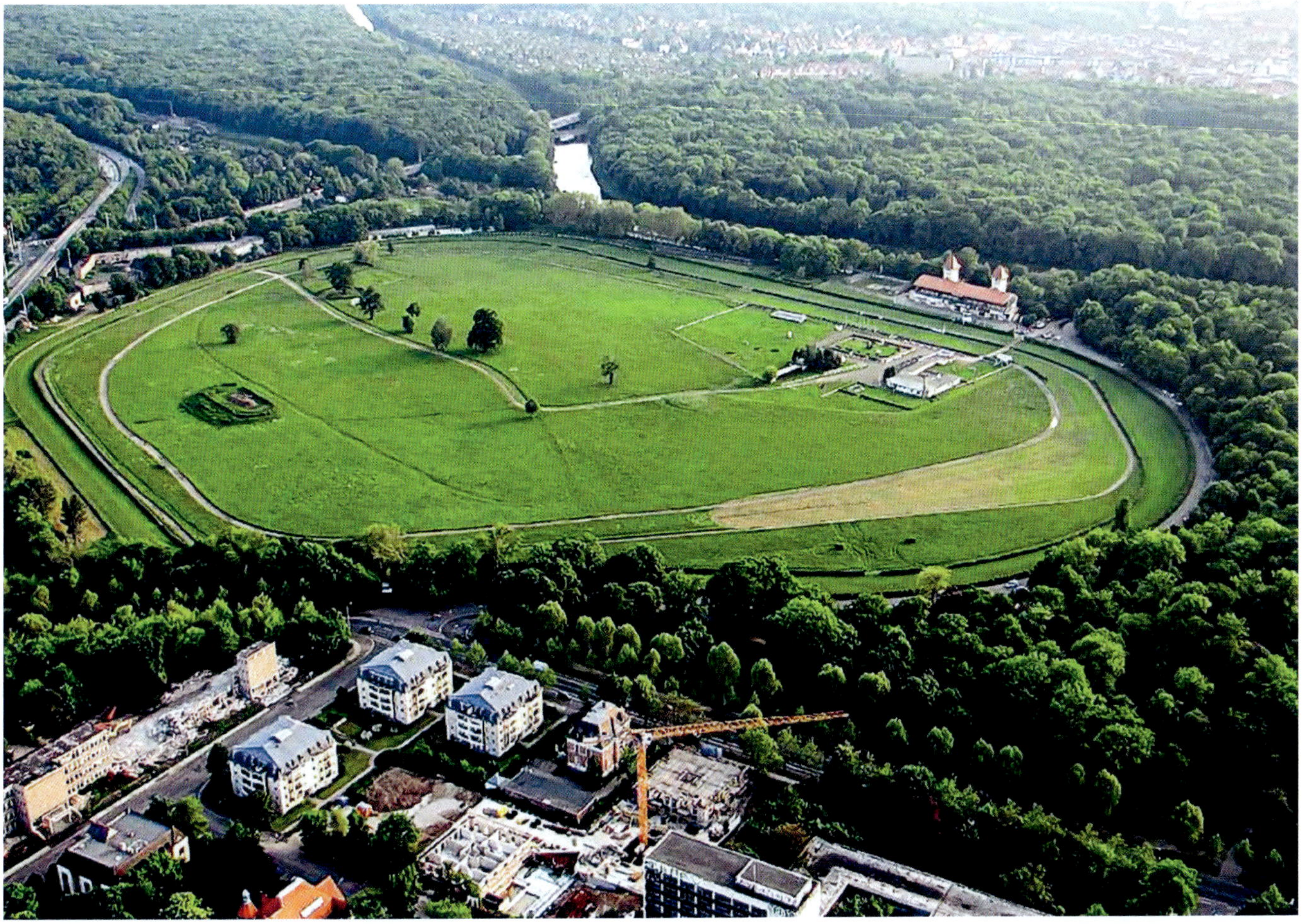

FOTO-SELECT SHUTTERSTOCK

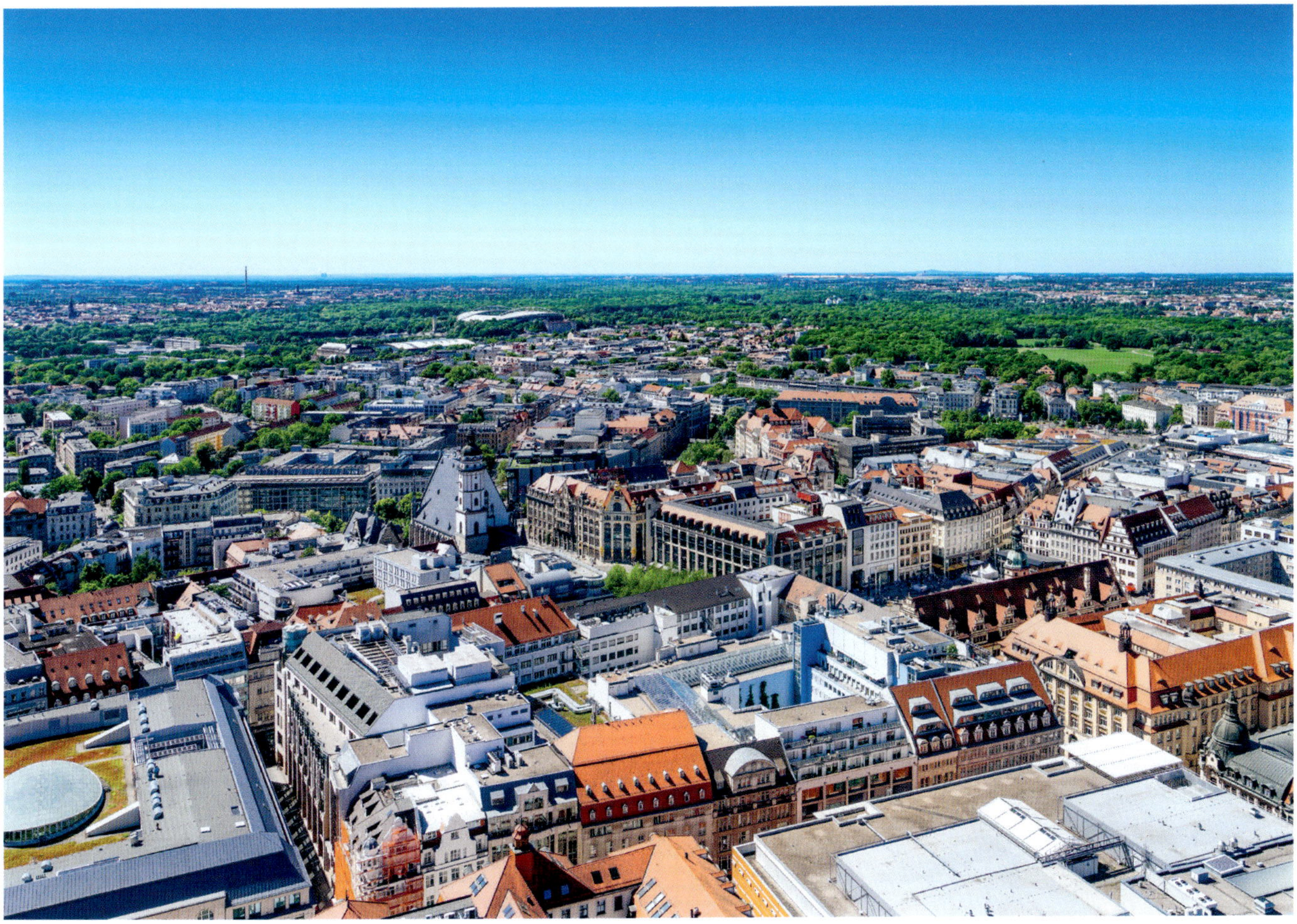

Center of Leipzig with City Hall.

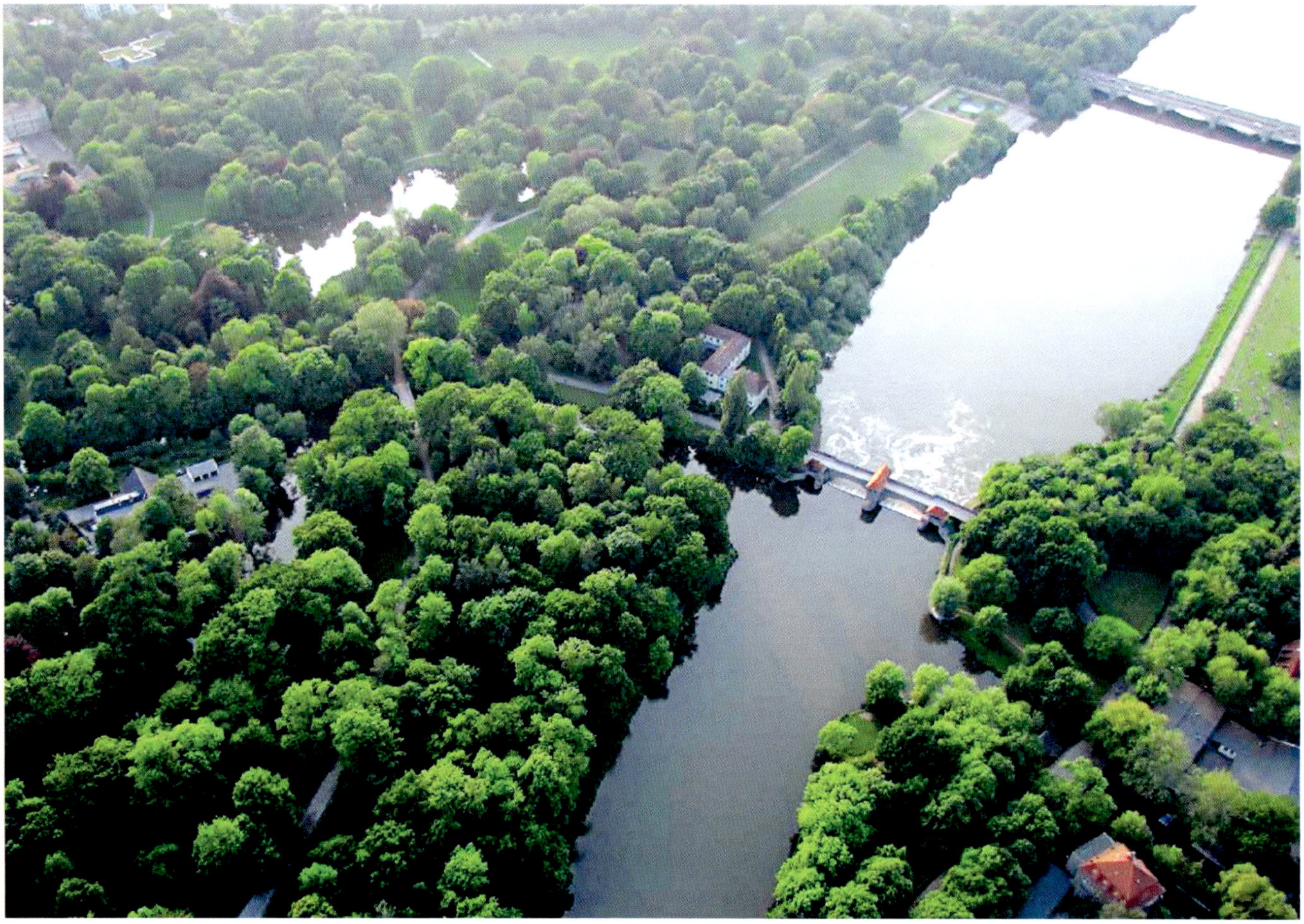

GRUNER RING LEIPZIG

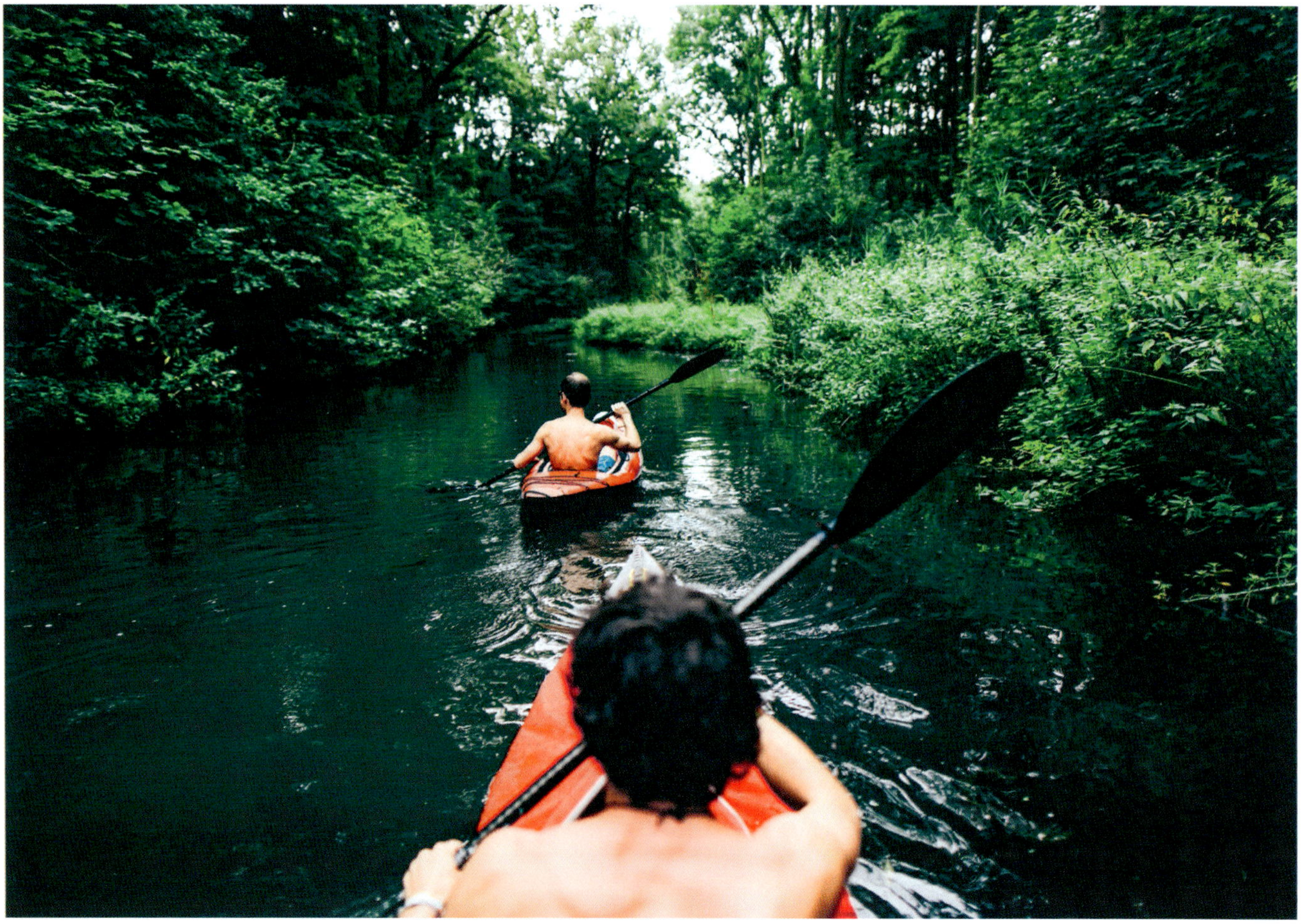

SHAKISHAN SHUTTERSTOCK

and bordered it closely to the south. In the seventies and eighties two more fields opened up, even closer to the city. These coal fields eradicated most of the southern part of the Leipzig floodplain forest.

Over a large part of the two Federal states of Saxony and Brandenburg, mining for for lignite in vast open-cast mines has been going on for decades. However, even during the time of the German Democratic Republic (GDR, East Germany), efforts were being made to rehabilitate the damaged landscape. A landscape planner by the name of Otto Rindt suggested that a pit close to Dresden should be flooded. It was turned into a lake—Lake Senftenberg, nicknamed "Dresden's bathtub"—and became a model for regeneration. Following reunification, most of the remaining mines were either taken over and cleaned up by the Swedish company Vattenfall[1], or handed over to the federally owned Lausitzer und Mitteldeutsche Bergbau and Verwaltungsgeschellschaft bmH (LMBV, founded in 1994) for transformation into recreational areas, reborn forests, or agricultural fields. The federal government of Germany allocated € 10 billion for this purpose.

Neuseenland (New Lakeland) is an area south of Leipzig where old open-cast mines are being converted into a huge lake district. It is planned to be finished in 2060—it takes time to fill up the pits in order to make them into lakes! Work had started already in the GDR era, but it took a great leap forward after the unification.

Because of the present-day growth of the city, a visionary program for sustainable city development was created. It says, in a vision for 2030:

The creation of new parks has still proved its worth as an instrument of the city development in Leipzig. Both to bring new qualities in existing city quarters, to fit new ones and to extend the green network of Leipzig.

Every year since 1990 just about 1,000 hectares of green public space has been added in the city. Present work includes creating green corridors through the cityscape so as to improve accessibility to parks and green spaces throughout the city. It is seen as especially important to connect the suburbs with green areas. Redevelopment of the Bavarian railway station into a park will lead people on to the Lakeland in the south of Leipzig.

Another important project is the renovation of the natural and cultural values of the landscape along the river Parthe. The river runs from

Neuseenland is an area south of Leipzig where old open-cast mines are being converted into a huge lake district.

TOP LEFT: Elsterbecken, the narrowest part of the Backbone of Leipzig, in the very center of the city. The river banks here are surrounded by sports arenas, exhibition areas, and parks. This is the weak point of the Green Backbone and at the same time very important for the connection between the floodplain-biotope-complexes south and north of this location.
BOTTOM LEFT: Canoeing in the midst of Leipziger Auwald.

Citizen protest in 1990 against the Cospuden coal field that had opened in 1981, just south of Leipzig.

pure countryside into the very center of town, before joining White Elster. This will enhance a green wedge.

The Green Backbone is very important for the climate, environmental and flood protection, and the enhancement of the urban residential environment. This at the same time creates an interesting landscape to visit and promotes biodiversity. Parks also seem to play an important role in attracting business.

[1] Vattenfall sold several lignite mines (Jänschwalde, Nochten, Welzow-Süd, Reichwalde, and Cottbus Nord) to a Czech company in 2016, despitet protests both in Germany and in Sweden.

TOP RIGHT: The former coal mining landscape today, after intensive landscaping. Cospuden lake was finished and filled up in the year 2000.
BOTTOM RIGHT: In the midst of the Leipziger Auwald August the Strong (1670–1733) had a field cleared to build a palace. The palace was never built. That is how the Rosental park, now in the center of the city, came into being.

LMBV PETER RADKE

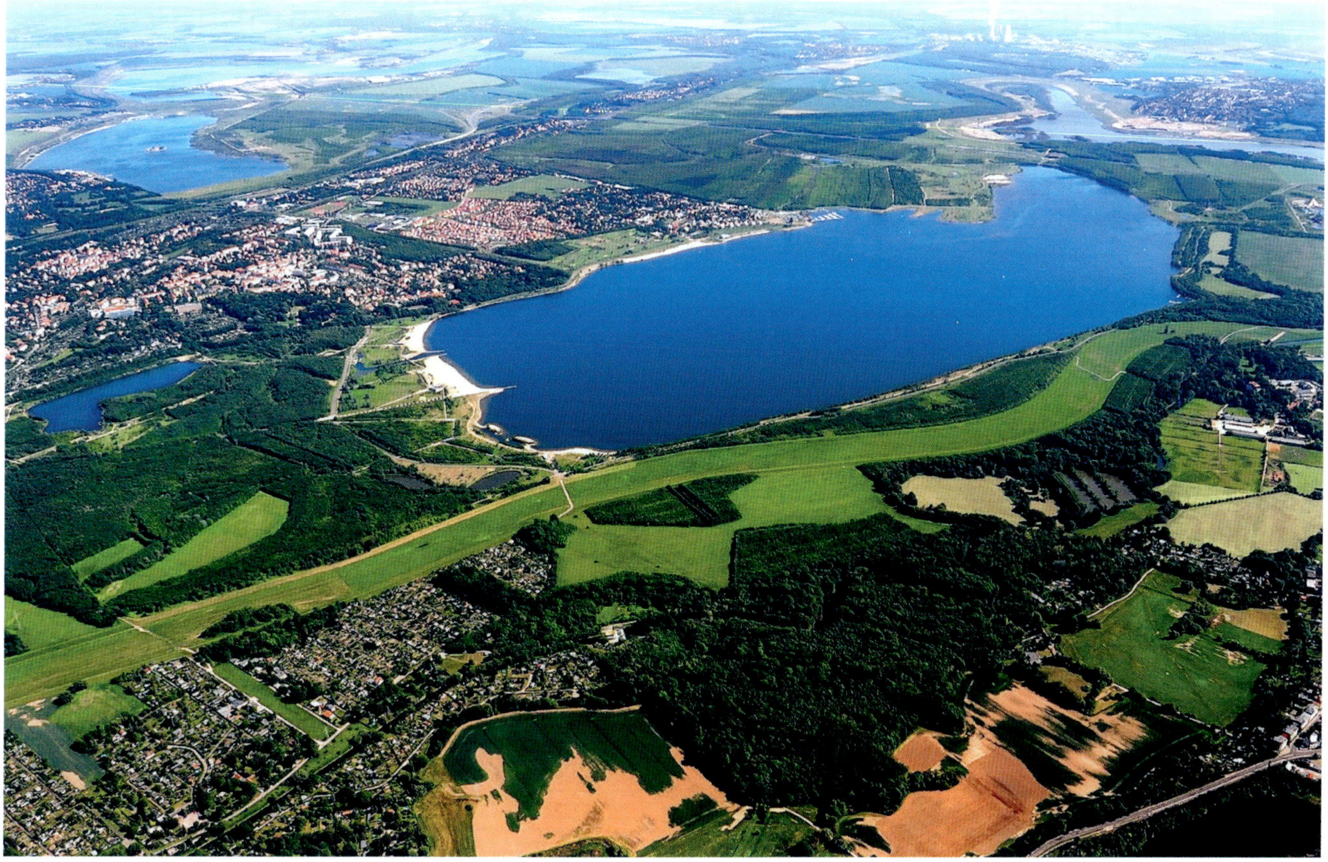

1362239105 SHUTTERSTOCK

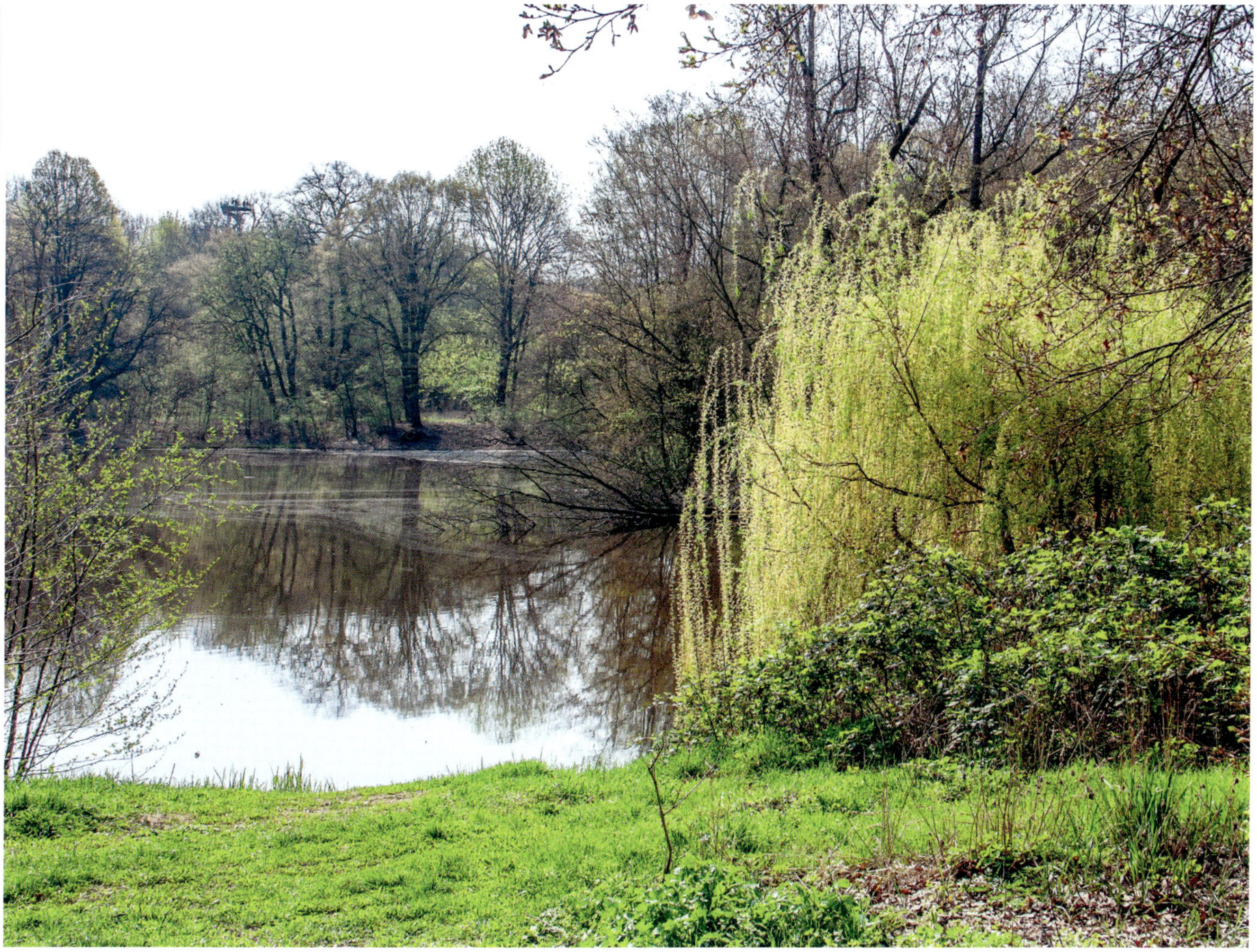

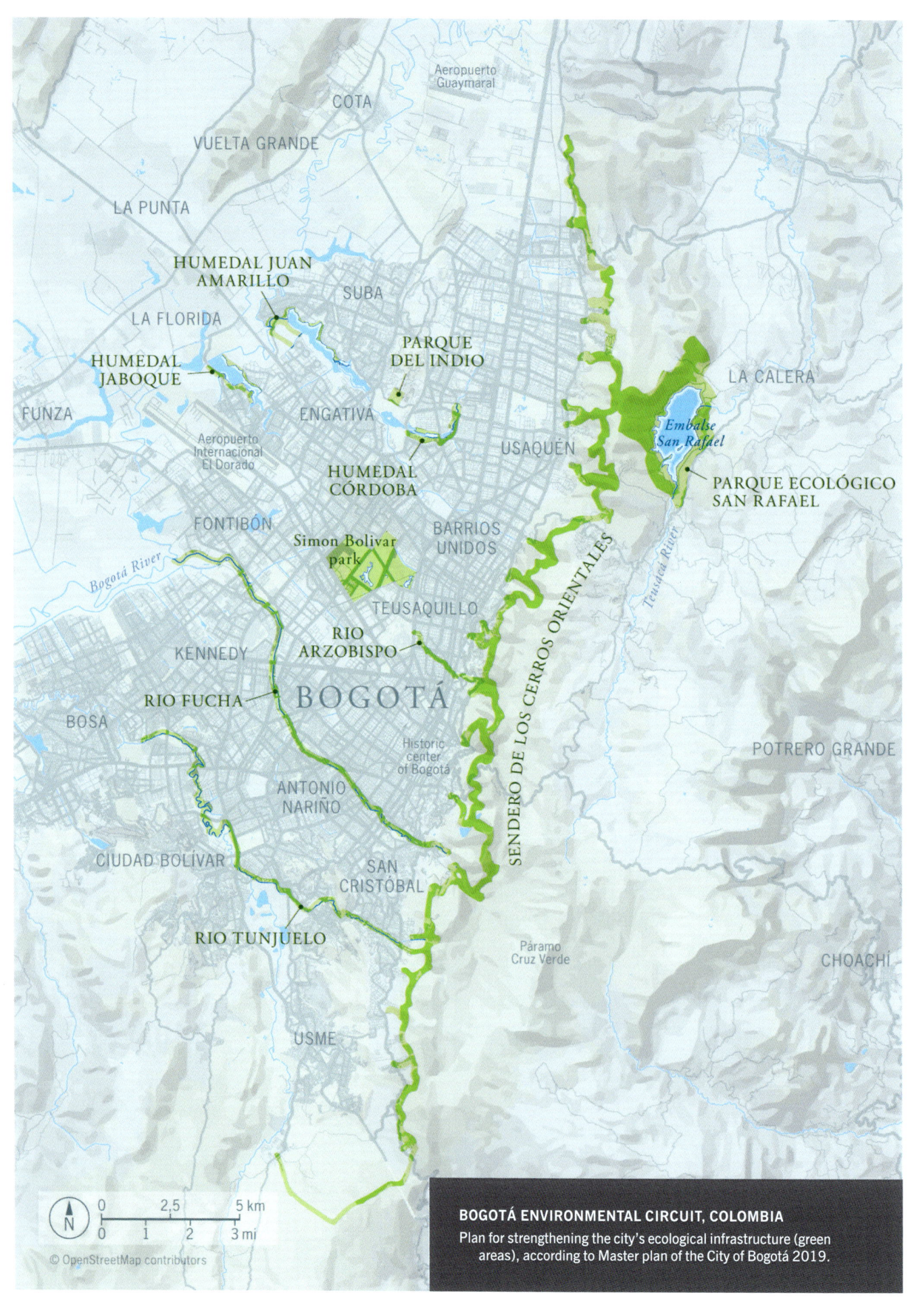

COTA

Aeropuerto
Guaymaral

VUELTA GRANDE

LA PUNTA

HUMEDAL JUAN
AMARILLO

SUBA

LA FLORIDA

PARQUE
DEL INDIO

HUMEDAL
JABOQUE

LA CALERA

FUNZA

ENGATIVÁ

Embalse
San Rafael

Aeropuerto
Internacional
El Dorado

USAQUÉN

PARQUE ECOLÓGICO
SAN RAFAEL

HUMEDAL
CÓRDOBA

FONTIBÓN

BARRIOS
UNIDOS

Simon Bolivar
park

Teusacá River

TEUSAQUILLO

RIO
ARZOBISPO

BOGOTÁ

SENDERO DE LOS CERROS ORIENTALES

KENNEDY

RIO FUCHA

POTRERO GRANDE

BOSA

Historic
center
of Bogotá

Bogotá River

ANTONIO
NARIÑO

CIUDAD BOLÍVAR

SAN
CRISTÓBAL

RIO TUNJUELO

Páramo
Cruz Verde

CHOACHÍ

USME

0   2,5   5 km
N 0   1   2   3 mi

© OpenStreetMap contributors

BOGOTÁ ENVIRONMENTAL CIRCUIT, COLOMBIA
Plan for strengthening the city's ecological infrastructure (green
areas), according to Master plan of the City of Bogotá 2019.

MARÍA CLAUDIA LÓPEZ

Bogotá's Environmental Circuit, Colombia

This is the story of the introduction of green infrastructure into city planning. The city of Bogotá has made great leaps forward in improving living conditions for the general public through promoting and instituting public transport, housing, schools, bike lanes, public spaces and parks, but until now there has been limited understanding of the value of planning in accordance with ecological foundations. A twelve-year master plan is now starting to be realized. In this article you will learn about the green vision of a megacity in Colombia.[1]

Bogotá has undergone important urban transformations in the past. Between 1998 and 2001, the BRT (Bus Rapid Transit) system made significant improvements to the transport infrastructure. Indeed, Transmilenio became a symbol of good practice all over the world as it showed how organized transportation could be quickly developed in a city with, at that time, over 7 million people. The system nowadays moves more passengers per hour and kilometer than almost any international subway, doing so at similar speeds but just 10 percent of the cost.

In a very short time hundreds of kilometers of new sidewalks, linear parks, plazas, and bikeways were created. Bogotá became the Latin American capital for bike use and cars were banned from occupying sidewalks. Public city life has flourished.

Informal settlements that had long been neglected were legalized and basic services and public buildings were provided to them. In these neighborhoods the city government built 123 traditional children's nurseries and 12 mega-nurseries. There were also important investments in culture and education; 3 large libraries were designed and built along with 12 smaller ones plus 24 new schools—of a quality unseen before in low income neighborhoods. Furthermore, an additional 25 were rebuilt because of their state of decay. In the span of 3 years the city's physical infrastructure improved dramatically and a sense of pride and belonging flourished among its inhabitants.

However, Bogotá has kept growing at an ever-increasing rate and its planning and implementation mechanisms have not been able to keep up the pace. The city has been the main destination for internal migration in Colombia. Internal displacements, due to armed conflict that the central government has faced with illegal groups, and the lack

The city of Bogotá has made great leaps forward in improving living conditions for the general public ... but until now there has been limited understanding of the value of planning in accordance with ecological foundations.

of opportunities in the countryside, have led people from all around the country to seek refuge in Bogotá. This has put a strain on the city's physical infrastructure and nature. The lack of a proper development strategy that recognizes the complexities of the situation has only worsened the predicament in the last decade.

In order to set a roadmap for further development the city needed to understand the delicate relation between the man-made city and its natural ecosystems. Bogotá still has a rich and unique ecological structure although it has been severely damaged by the continuous and rapid growth of the city. Bogotá was founded in 1538, close to the eastern mountains, where water and wood was plentiful. By the late 1960s, the mountains had been exploited to such an extent that there were barely any trees left and its rivers started to dry out. The main rivers that flowed from the mountains to the west faced a similar fate. The continuous expansion of the city fabric meant that these rivers became sewers and carried polluted water and waste from people's homes to the Bogotá River.

... the ambition to give the city's current 8 million citizens access to their own rivers, mountains, and marshes.

It took decades of this decay of our primary ecosystems before the government took appropriate measures to improve the situation. The publicly-owned water company started by purchasing big watershed areas in order to protect them. It also began planting exotic fast-growing vegetation in the mountains. Lastly, it built a proper sewer system that separated rainwater from sewage. These efforts meant that our ecosystems recovered partly. However, it did not stop further damage. Preservation efforts have focused mostly on secluding these areas from the public. This has led to their neglect. But, finally, a sense of urgency to save them has developed. It is strengthened by the ambition to give the city's current 8 million citizens access to their own rivers, mountains, and marshes.

A new revolutionary way of thinking is now taking place, namely the Environmental Circuit of Bogotá. It will forever change the relation of the city's inhabitants to the natural ecosystems that are part of its territory. The approach is based on a new development strategy for the city that properly integrates the ecological structure and the built environment so that they work as one. The Environmental Circuit will be an integral part of the new POT (the city's central masterplan) for the next 12 years. It has four main strategies. The first consists of cleaning rivers, mountains, and marshes. The second consists of planting to replace exotic invasive vegetation that is threatening existing eco-

TOP RIGHT: Bogotá from Mount Monserrate.
BOTTOM RIGHT: Periodista Square in the historic part of Bogotá as BRT passes through.

CITY OF BOGOTÁ 2019

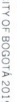

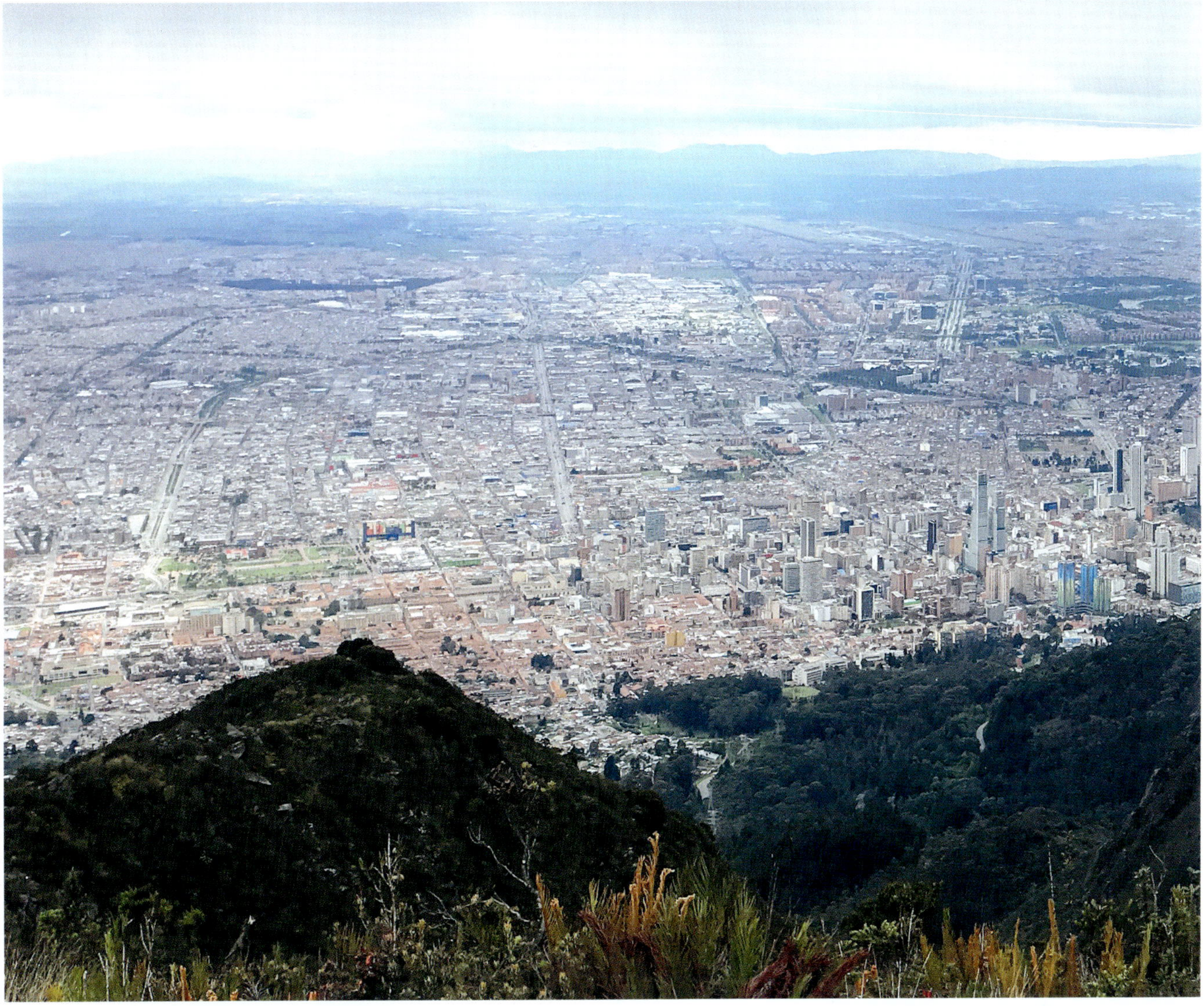

MARK PITT IMAGES SHUTTERSTOCK

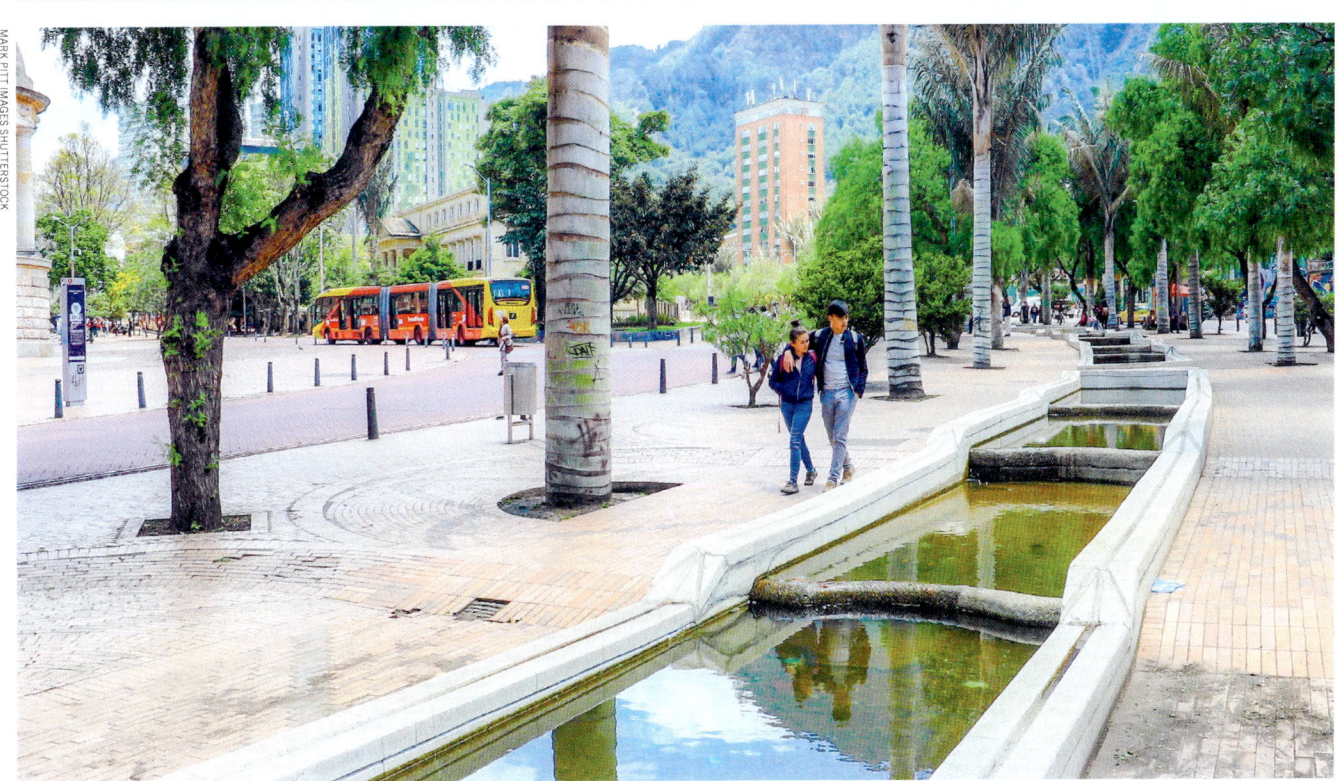

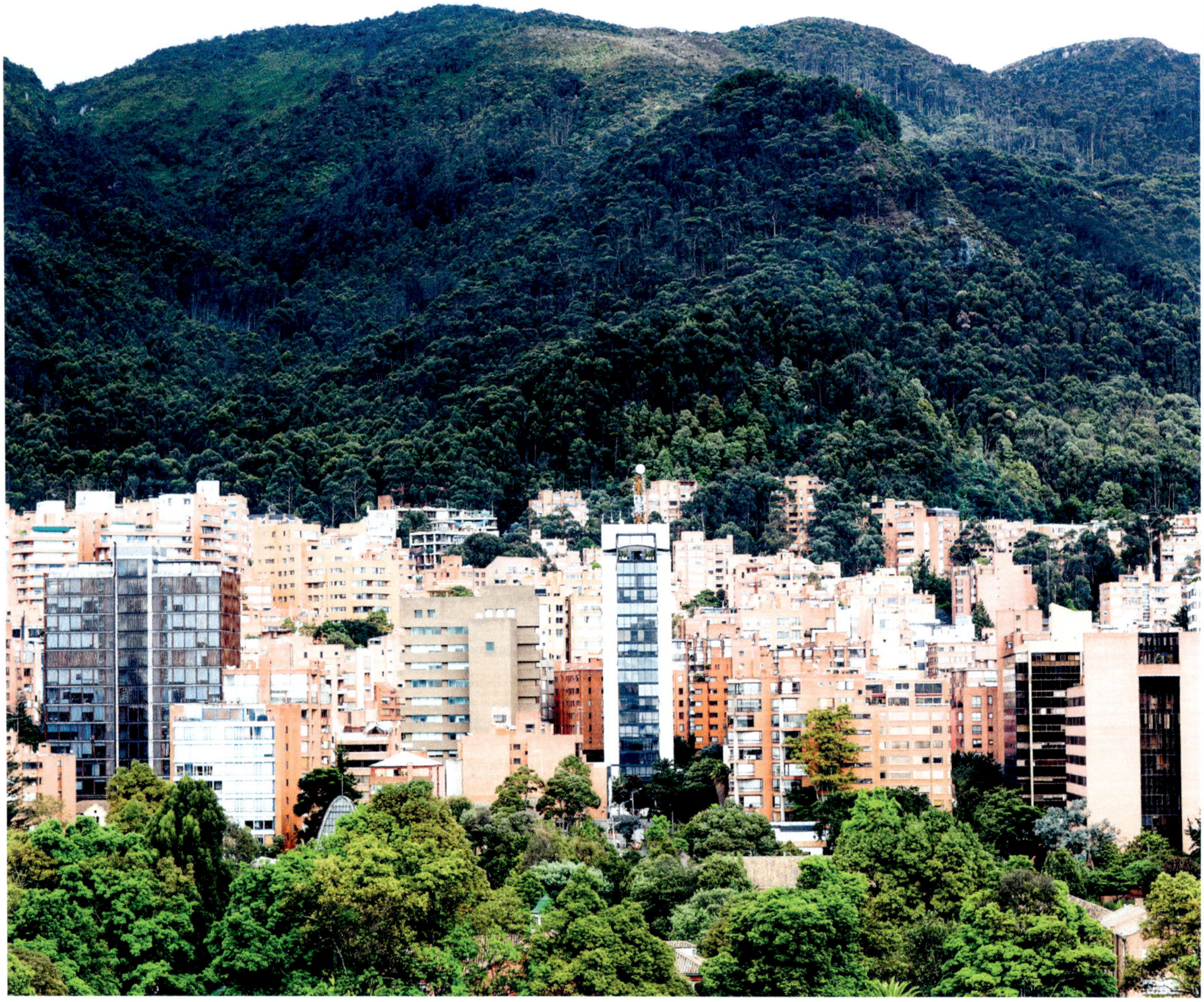
ESB PROFESSIONAL SHUTTERSTOCK

CITY OF BOGOTA 2019

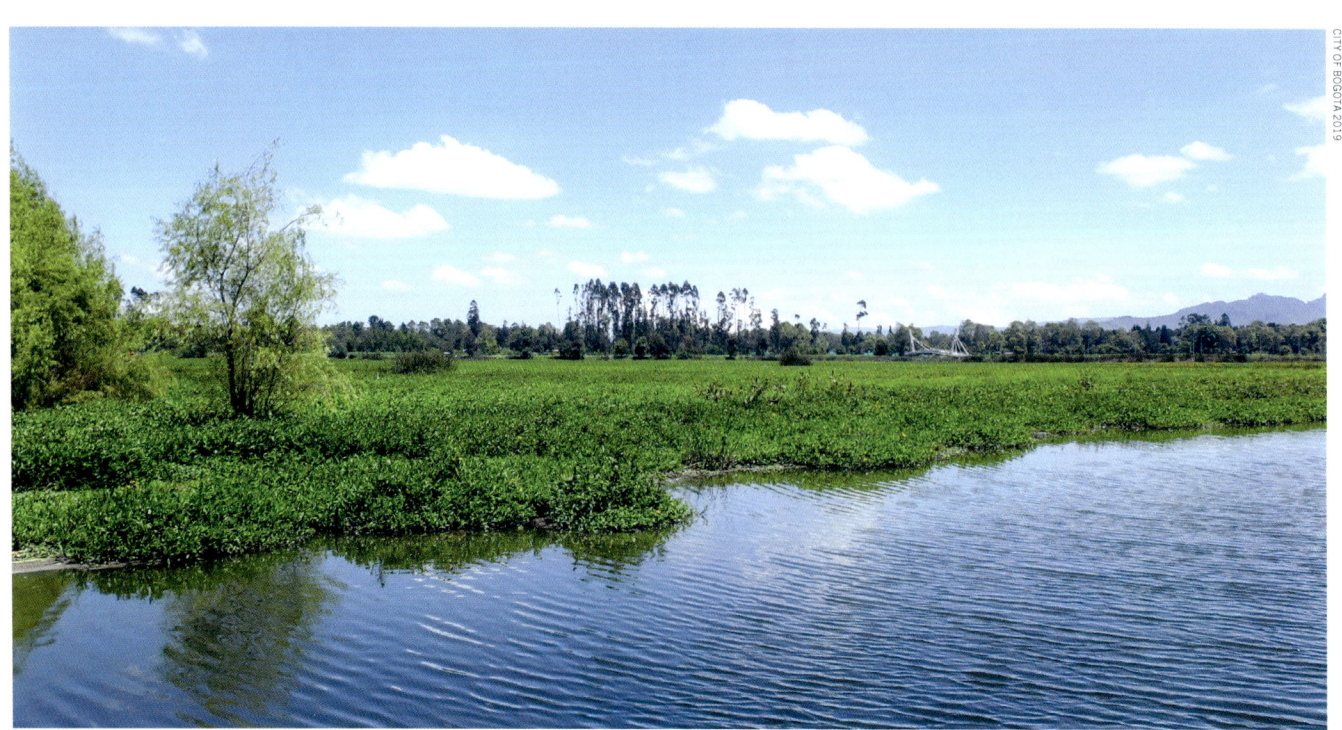

systems. The third consists of allowing the use of the existing ecological structure for recreation. The fourth is a series of strategies for future planning and development that regulates both the private and the public sector.

The first two strategies entail a joint effort from different government entities to help improve the existing conditions of our ecosystems and to educate the city's inhabitants in their proper use.

The third strategy aims at turning the existing ecological structure of the city into areas for the enjoyment of its citizens. This will recognize and follow the flow of water from the moorlands in the east across the mountains and to the rivers and marshes within the city that finally connect with the Bogotá River on the west. By implementing the different projects of the Environmental Circuit, the ecological structure will cease being a scattered "sewer and drainage system" and again become natural corridors for people, fauna, and flora. Nature will recover its rightful place as the underlying structure for all future city development.

In order for it to work properly, the Environmental Circuit also includes areas outside the city limits. The ecological structure of Bogotá extends far beyond the city's political boundaries and therefore needs to be articulated at a regional level. Both the water and energy companies of Bogotá have significant areas of land outside the city that provide ecological services to the city and nearby municipalities. Articulating the Environmental Circuit regionally will be one of the main challenges.

There are four project types within the Environmental Circuit based on the four interlinked ecosystems: mountains, rivers, wetlands, and water reservoirs.

The eastern part of the city is composed of an intricate mountain range where an ecological trail will serve as a safe public amenity. The trail will extend more than 70 kilometers along the mountains and connects neighborhoods from north to south. It will have 26 entrances and the trails will be a total of 107 kilometers in length. This will allow people to enjoy forests and panoramic views towards the city and the Andes while being educated about the rich variety of flora and fauna in the mountains. The Sendero de las Mariposas will serve as the backbone of the Environmental Circuit and will be a destination for every visitor to the city.

In the western part of the city is the Bogotá River. The river is undergoing an important revitalization process that will clean it and bring

A new development strategy for the city that properly integrates the ecological structure and the built environment so that they work as one.

TOP LEFT: To the east is Mount Monserrate.
BOTTOM LEFT: To the west marshes eventually meet the Bogotá River.

In the western part of the city is the Bogotá River. The river is undergoing an important revitalization process that will clean it and bring back its fauna and flora.

back its fauna and flora. A linear park along its banks will allow pedestrians and cyclists to follow its sinuous path. Ciudad Rio, as the project is called, also includes new developments along its banks that will showcase contemporary urbanism and sustainable city development. The river will also be a connection point to the surrounding municipalities to the west.

Between the eastern mountains and the Bogotá River, other important rivers and marshes connect the flow of water from east to west. Linear parks along the rivers will be intimately linked to the development of housing, schools, and public amenities. They will become important mobility corridors for alternative forms of transportation. The Fucha, Arzobispo, and Tunjuelo rivers will add more than 25 kilometers of linear ecological parks to the city. On the other hand, the Juan Amarillo, Córdoba, and Jaboque wetlands are being restored, allowing for contemplation of wildlife, relaxation, and other types of recreation.

On the outskirts of the city, the Tomine and San Rafael reservoirs will become two of Colombia's largest natural parks and will complement the need for recreational areas in the metropolitan area and neighboring municipalities. It will add more than 60 square kilometers of green public space to Bogotá and the region.

The San Rafael reservoir is located in the adjacent municipality of La Calera. Its area is 12.5 square kilometers. The existing technical infrastructure of the reservoir will be used as public amenities for the park's visitors. Around 95 percent of its territory will remain as a natural reserve for citizens of both Bogotá and the region to enjoy.

The Tomine reservoir is 49 square kilometers with a circumference of almost 65 kilometers. The park will be a series of smaller parks with themes that relate to the indigenous past of the territories and nearby municipalities and cultures. As a showcase for sustainable energy systems it will set an example for energy generation and consumption in the area.

Regarding the fourth strategy, the Environmental Circuit will integrate future private and public spaces into the main ecological structure of the city by establishing a series of strategies for water management and for nature corridors relating to new developments. The city has been divided into ecological units that have distinct

TOP RIGHT: A vision of the trail along the eastern mountains, given the poetic name "Sendero de los Mariposas" (Butterfly Trail).
MIDDLE RIGHT: A vision of a linear park along the Tunjuelo River.
BOTTOM RIGHT: A vision of a linear park along one of the canals.

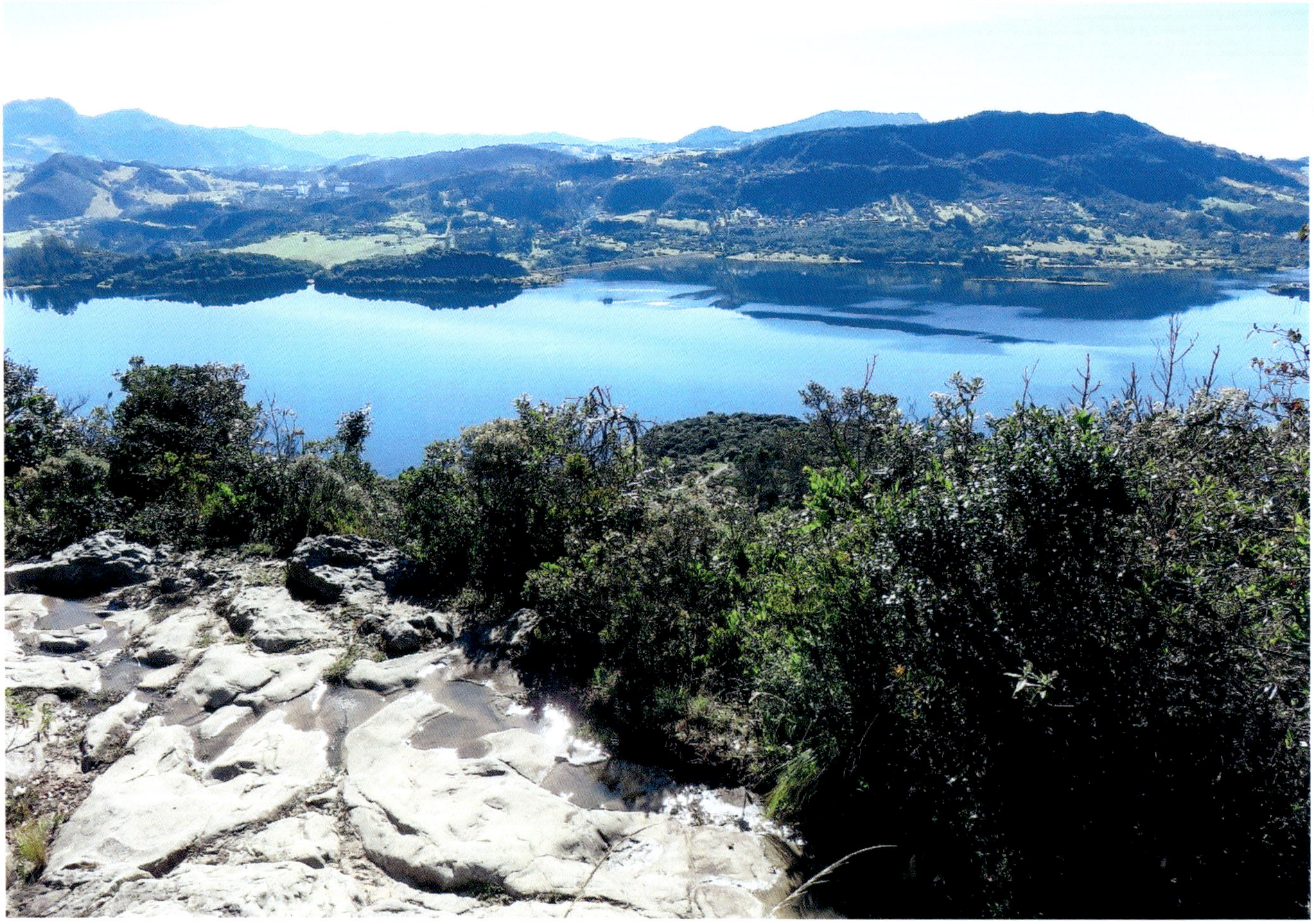

CITY OF BOGOTÁ 2019

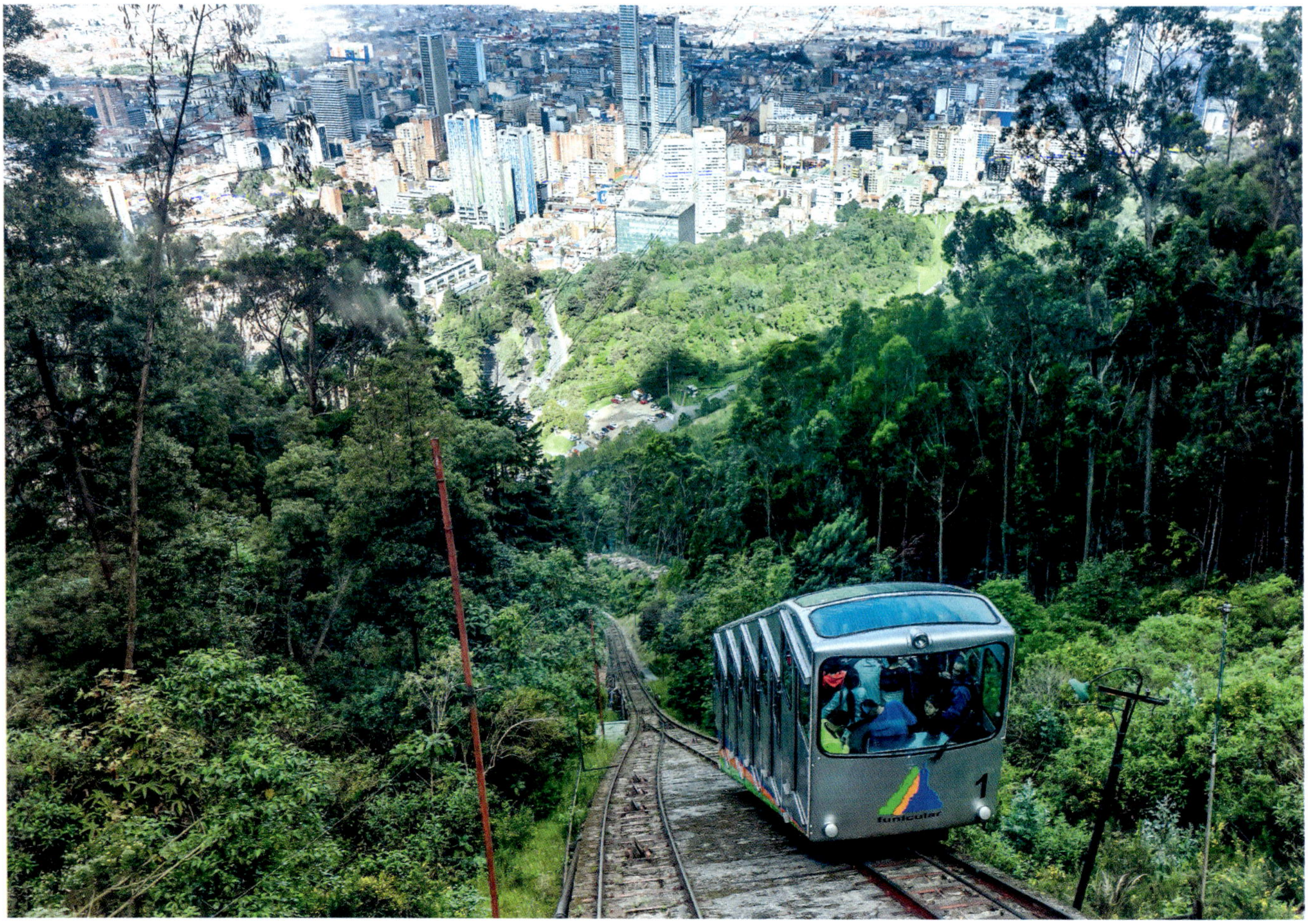

SDARUKO/SHUTTERSTOCK

emphases depending on the current buildings/constructions and natural conditions of the area.

The plan aims at improving the citizens' access to clean air and water. Substantial health benefits are expected. The plan anticipates the growing demand for 'nature recreation' and exercise and this will ultimately generate a decrease in health expenditure in the long term. Indeed, the mountain path alone will directly benefit 57 different neighbourhoods that border the eastern mountains in Bogotá.

The first stage of the Environmental Circuit is currently underway. The first nine connecting projects are currently being designed and are set to be built in the next three years. The investment is approximately 1.07 billion pesos (around 400 million USD). Although it is a very ambitious endeavour it is imperative for the city to undertake it.

There are still many challenges ahead, but the administration of the city understands the importance of a paradigm shift regarding the city's development, particularly urgent in the age of climate change. It is hoped that the Environmental Circuit will be a turning point for the city and that it will be a first step toward a more sustainable approach to city design. Perhaps it will serve as an inspiration to other cities worldwide.

[1] The text refers to the situation in December 2019.

Substantial health benefits are expected.

TOP LEFT: The San Rafael reservoir.
BOTTOM LEFT: Funicular going up Mount Montserrate.

Heritage and Identity

A large park in a city can tell the story of centuries of local human settlement, serving as a history book for the urbanites. Though asphalt and concrete add a lot to that story, they also cover up the landscape and essential parts of history. That those presently living in that city have a right to its—their!—history ought to be universally recognized. It is therefore a human right that citizens have a say regarding the future of their environment, whether built or green.

GRAHAM FAIRCLOUGH

Florence and Faro, Two Cities and their Conventions: Parks and Everyday Life in Major Cities

Parks perform necessary urban functions, and are one of the defining features of an urban place, being as essential as its streets or marketplaces, its houses, workshops, offices, or shops.

Can a place be called a city if it has no park (or square)? The coming-together of people in such places symbolizes the very purpose of the city in human society. Urban parks have value and meaning not only as greenspace and designed landscapes but also as essential, functioning parts of the "city machine." They may sometimes seem to be apart from the city, or even as left behind relics of a pre-urban world, but they are *not apart from* the city, nor outside the urban mainstream. Rather they are an integral part of the city and its landscape.

Two treaties of the Council of Europe have particular relevance to the future of urban parks: the European Landscape Convention (the Florence Convention, 2000) and the Convention on the Value of Cultural Heritage for Society (the Faro Convention, 2005). Both recognize the inter-connectedness and one-ness of culture and society, of human life and nature, and of people and place. The very first biennial Landscape Award, supported by the Council of Europe's Florence Convention, was awarded to a new urban park, the Parc de la Deûle in Lille, France.

Parks perform necessary urban functions, and are one of the defining features of an urban place, being as essential as its streets or marketplaces, its houses, workshops, offices, or shops. They are often amongst the *foundational* features of a town or city, around which the urban topo-graphy has accrued. Consider their place in culture and, for example, how often parks feature in literature as play areas for children and young people; trysting places in romantic novels; a source of, or locus for, fear and danger in certain types of police thriller; the location in films where spies meet and plot; a place for debate and demonstration. And not to forget music: what self-respecting park did not once contain a band-stand? Parks are often where communities place their sculptures, and where all manner of social or artistic performance takes place. Unfortu-nately, perhaps paradoxically, it seems all too easy for parks in cities to be taken for granted, to be treated as if they are "just there." By virtue of their familiarity, parks are so much a natural part of everyday life that they can become almost invisible to us. It is easy to forget that they are not naturally self-sustaining places, but rather cultural, artificial con-

CHRISTIAN CABLE

The park … is just there! Williamson Park, Lancaster, United Kingdom, created from circa 1880 by millionaire James Williamson (Baron Ashton). Its focal point is the Ashton Memorial, dedicated to the memory of Williamson's wife (but still, nearly 100 years later, locally known only as "The Structure"). The park itself, built partly in romantically redesigned stone quarries, is now in continual use as a recreation ground, tourist attraction, and theater in summer evenings, among many other uses.

structions utilizing nature as part of culture to support city life ; and that parks themselves require continuing care and management. In the centuries-long lifespan of a city, the uses and purposes of parks can change, for example from royal or aristocratic private places to the haunts of "respectable society," or to the living places of the homeless and the marginal in society. In the shorter human life span, we may use them only at particular stages of life, first of all in childhood, for instance. In adult life, if we find ourselves too busy to enjoy them, they are still with us in memory, and we assume they will still be there when we need them again. Imagine taking your grandchildren to the park where you played as a child and finding it no longer exists! Parks are indeed just there, but that is why they are important.

The collective and communal uses, purpose, and potential of parks as places where people meet and act in concert, and in which communities are created and sustained, is central to the value of parks. Many

But this requires forward looking planning in terms not just of park creation but of urban design and territorial planning.

urban parks indeed began life as, and often remain, common land. This aspect of parks calls to mind the early medieval sense of landscape (often called "Nordic") as a territory both real and imagined that contains and nourishes a community. Parks—and the cities that contain them—are all undoubtedly landscapes within the presently-dominant definition of the Florence Convention, the European Landscape Convention. This definition—"an area, as perceived by people, whose character is the result of the action and interaction of natural and/or human factors"— makes no distinction between rural and urban landscapes (or indeed seascapes). Indeed, urban parks are becoming more important, since for more than half of the world's population (indeed for almost the whole population in many countries) the landscape of their everyday life is first and foremost urban or civic.

The words "urban landscape" should not only conjure up in our mind's eye and beneath our feet the "green bits" of the city, its trees, parks, and gardens. The European Landscape Convention invites us to see the whole of the city, whether grass, stone, or concrete, as a part of the landscape, and it is probably essential to see it in that way if any sort of culturally-framed sustainability is to be achieved. Parks in cities must be treated as part of their whole urban landscape in order to make cultural and social sense. If urban parks do not make cultural and social sense there is a risk that they will come to be seen as merely 'land'—and land too readily becomes little more than a raw material for commercial exploitation, 'development', and 'reclamation'. To be safe, city parks need to be culturally used.

Cities through the centuries - and in some cases millennia—have been forward-looking centers of creativity and innovation. They have been places to which people moved to find greater liberty and free-dom; places of resistance, rebellion, and resilience; places to meet and trade. Urban parks have many different origins—some evolved from "working" medieval urban commons such as the Town Moor in New-castle, and others were once the private hunting or pleasure grounds of royalty. Many cities, as they grew, incorporated the parklands of aristocratic houses that once surrounded the town, opening them to a wider public. And in the 19th century, new urban parks were created,

TOP RIGHT: The park as urban commons. The Town Moor, Newcastle, England, is 400 hectares of open space providing a green heart to Newcastle's city center, with a history as commonly shared pasture land dating back to the 12th century, with land tenure and use regulated by its own Act of Parliament. The Fair (the "Hoppings") is still held annually, every June. Painting *Fair at the Town Moor* (c.1810) by Jack Wilson (1774–1855).
BOTTOM RIGHT: The Town Moor of Newcastle on a quiet day nowadays.

LAING ART GALLERY/BRIDGMAN IMAGES

HARRY GREEN SHUTTERSTOCK

Boston Commons, United States.

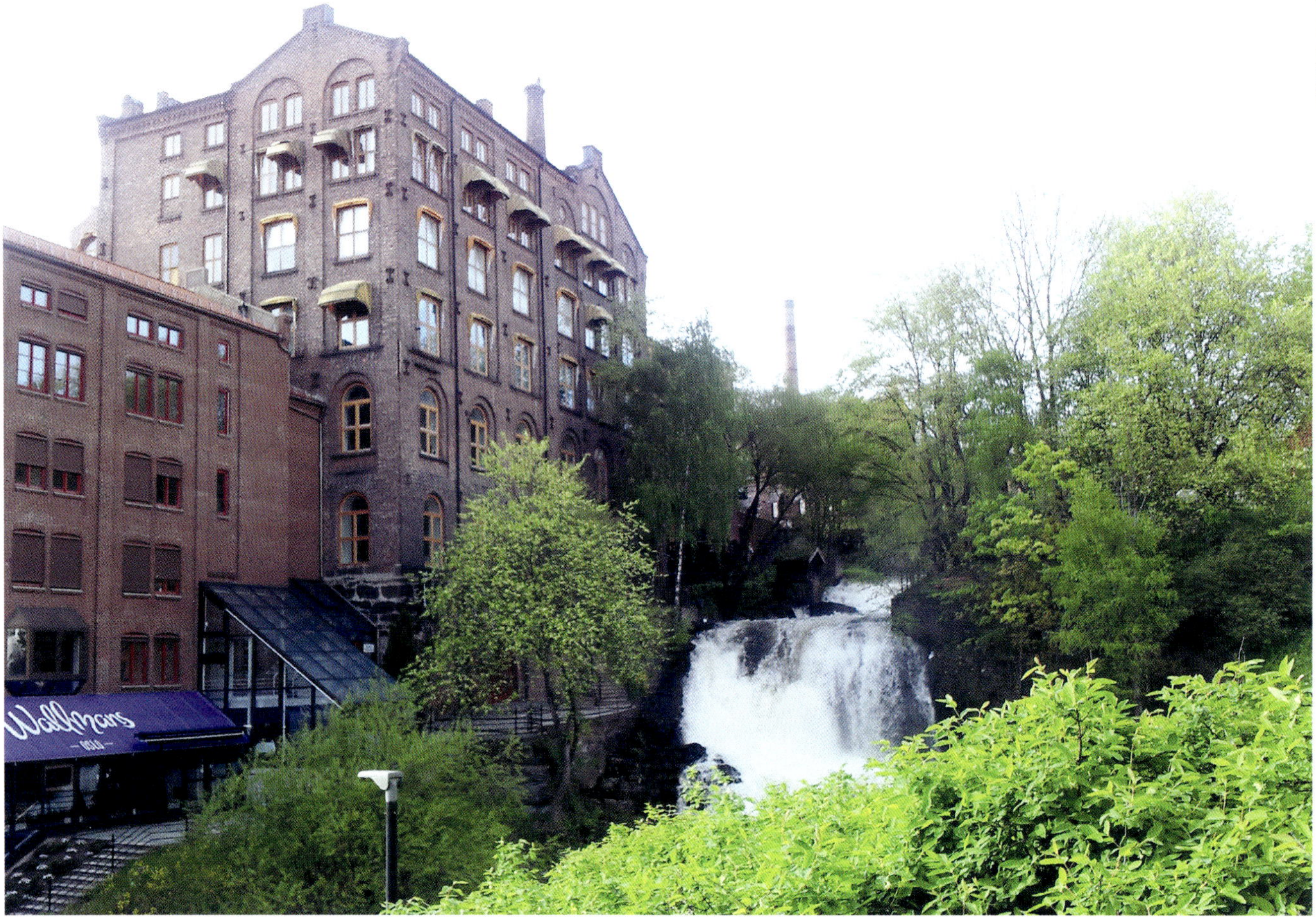

GRAHAM FAIRCLOUGH

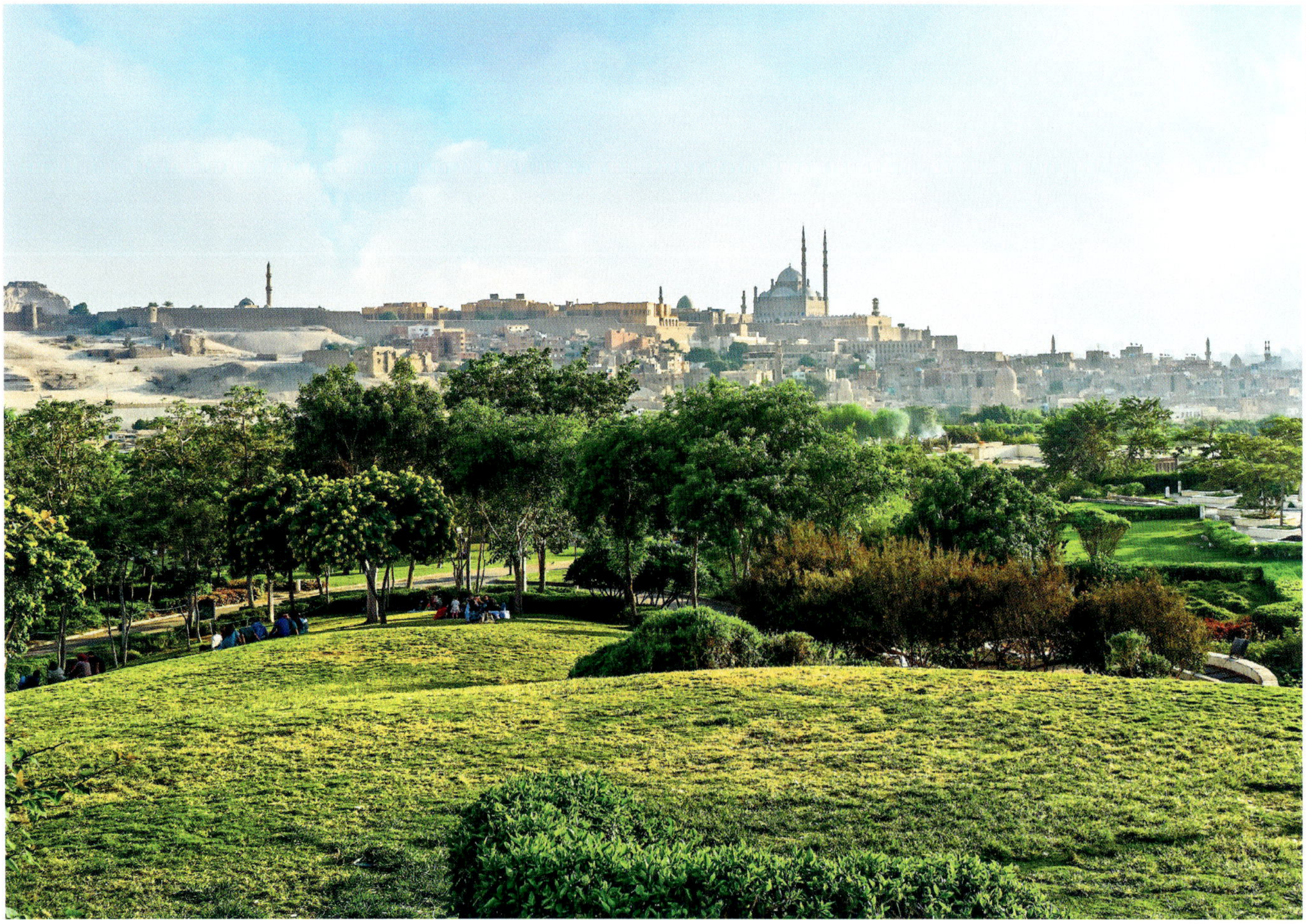

EF ESENKO SHUTTERSTOCK

sometimes explicitly as "Peoples' Parks," most famously Birkenhead Park in Liverpool, England. When they were located at the urban edge, new parks would in turn become embedded in the enlarging city (as, for example, with Liverpool's Newsham (1868), Stanley (1870), and Sefton (1872) parks). In other cities, linear parks grew from the abandonment of industries (often water-powered industry in valleys) creating some of the most dramatic city parks, such as Akersleva Park in Oslo, Norway. Others again—even New York's Central Park (United States)—owe their origin to protective zones around water supply reservoirs.

Very few parks are "left-over" fragments of pure Nature. Very many, in contrast, have entirely cultural origins, created as part of their city's own natural evolution. How can we ensure that cities are able to continue to create parks as they grow? For new parks, substantial investment is required, as in the case of the inner-city Al-Hazar Park in Cairo, created by the Aga Khan Trust for Culture as "a catalyst for urban renewal." More commonly, parks are created where the city meets the country-side, at the edge of the built-up area, as in the award-winning parks created in Lille. Establishing new parks at the current urban fringe is potentially a future-oriented sustainable approach: as cities expand (as almost all current predictions say they will) the future greater city "inherits" parks within its boundary, as Liverpool did in the 19th century. But this requires forward looking planning in terms not just of park creation but of urban design and territorial planning.

The Florence and Faro conventions are relevant in all this because they both insist on the relationship between places and people, and between past and future; moreover, they insist on the social, democratic, and cultural vitality and virtues of landscape and heritage. While Florence is about landscape and Faro is about heritage, they share many attitudes, concerns, and hopes about the ways in which people see and relate to the world around them; the ways in which they live in the world, reconciling past and future; and how they manage the daily interaction between themselves and "others." We simply use different words depending on our cultural, professional, or academic backgrounds, and on which aspects of life we wish to place in the foreground. Some people call it landscape, others heritage. Others again use the language of ecosystems, environmentalism, or politicall ecology. But behind these various

The actors in the heritage process are all of us—not only heritage experts or specialists, not only owners and governments, but everyone—whether individually or collectively.

TOP LEFT: Akerselva Park, Oslo, Norway, is part of the town, a former industrial zone using water power from the Akerselva River, now a much-used linear park threading its way through the city and—ultimately—connecting to the surrounding countryside to the east and to the much larger hilltop urban park of Ekebergparken.
BOTTOM LEFT: Al-Hazhar Park in Cairo, Egypt.

specialist terms there are similarities verging on congruence, and all of them denote some form of social and environmental common good that contributes to (or challenges, if our stewardship and decision-making are poor) quality of life, democracy and equality, and the meeting of inter-generational responsibilities.

Whilst the term 'landscape' in popular usage still often automatically prompts images of a scenic rural view, the Florence Convention promotes a much more ambitious view of what landscape is. It is urban as often as it is rural, as mentioned previously, but it is also much wider than the definitions of landscape bequeathed to us by Renaissance and early modern thinking. Scenery, picturesque and romanticized nature, and the sublime are only part of this new understanding of landscape. Perceiving landscape potentially involves all our senses, not only sight; just as important are the knowledge, cognition, memories, symbolism, historical associations, and imagination that we bring to our perceptions of landscape. Landscape also concerns experiential, physical, and embodied aspects of being in and of the world, of moving and acting, and of sharing—for example, playing, walking, running, taking dogs for their walks (or *vice versa*), sometimes foraging for food (from trees or food-vans), meeting friends, reflecting on life. It is those additional human perceptions, and the overlay of culture, that changes 'land' into 'landscape'; or (in other terminologies) space into place, nature into ecosystem, or property into community.

Both landscape and heritage are continuously being re-made, re-conceived, and re-configured. Protection or preservation is important, but management and enhancement, and the continued construction of new heritage, and the design and creation of new landscapes, as the European Landscape Convention emphasizes, are more so. Urban parks have changed through history and they continue to respond to new social needs, as "audiences" change and as new communities bring different cultural attitudes to nature and the outdoors. Parks will somehow adapt to those new demands, and to this end the Florence Landscape Convention offers the concept of "Landscape Quality Objectives" (the aspirations of the public for their future landscapes) and an insistence on broad levels of public participation in making decisions.

The focus on participative democracy is a key link between the Florence Landscape Convention and the Faro Convention on the Value of Cultural Heritage to Society. In urban landscapes we continually see the remains, traces, and reminders of the past: from long-standing street patterns (the skeleton) to more transitory, oft-replaced buildings (the clothes). From the perspective of the Faro Convention, heritage is

An urban park's first importance is its cultural significance from which all else flows.

**FLORENCE
– THE EUROPEAN LANDSCAPE CONVENTION**

ARTICLE 5

Each Party undertakes:

a) to recognise landscapes in law as an essential component of people's surroundings, an expression of the diversity of their shared cultural and natural heritage, and a foundation of their identity;

The European Landscape Convention, adopted by the Committee of Ministers of the European Council of Europe on the 19th of July 2000 (Florence convention). It underlines the importance for people's identity of the landscape as an expression of their shared cultural and natural heritage.

**FARO
– THE CONVENTION ON THE VALUE OF CULTURAL HERITAGE FOR SOCIETY**

ARTICLE 1—AIMS OF THE CONVENTION

The Parties to this Convention agree to:

a) recognise that rights relating to cultural heritage are inherent in the right to participate in cultural life, as defined in the Universal Declaration of Human Rights;

b) recognise individual and collective responsibility towards cultural heritage;

c) emphasise that the conservation of cultural heritage and its sustainable use have human development and quality of life as their goal.

ARTICLE 4—RIGHTS AND RESPONSIBILITIES RELATING TO CULTURAL HERITAGE

The Parties recognise that:

a) everyone, alone or collectively, has the right to benefit from the cultural heritage and to contribute towards its enrichment;

b) everyone, alone or collectively, has the responsibility to respect the cultural heritage of others as much as their own heritage, and consequently the common heritage of Europe.

The European Framework Convention on the Value of Cultural Heritage for Society (Faro convention), adopted 2005-10-27. It underlines a broad definition of cultural heritage ("all aspects of the environment") and its importance as an expression of people's beliefs, values, knowledge, and traditions.

… both have begun to frame themselves within democratic/ human rights discussions.

an everyday thing, which begins on your doorstep and surrounds you during the day. It is everywhere, like landscape, and most importantly it is a continuing process, the action of inheriting, using, and bequeathing (passing on). The actors in the heritage process are all of us—not only heritage experts or specialists, not only owners and governments, but everyone—whether individually or collectively (through "communities of heritage" as the Faro Convention calls them, or through "communities of place"), again just like landscape. These new ideas of heritage have been rising up for the past two or three decades before being encapsulated in Faro to represent heritage as a process that offers opportunities and imposes burdens, and that never ends. It is a way of seeing heritage that is concerned more with people than with objects; democratic rather than canonical; and about the future as well as the past. UNESCO, through its Historic Urban Landscape declaration (2011), has also begun to adopt a view of heritage that is about change as well as preservation. These ideas surely have especial applicability to parks, inherently shared places.

This simple idea is however complicated by the plurality of heritage and of landscape. There can be as many opinions of what is important as there are people; heritage and landscape are about compromise as much as consensus. The Faro convention argues for a human right to heritage (and such a claim is implicit for landscape in the Florence convention), but like all rights it brings responsibilities, in this case respect for another community's heritage or landscape (and of course intergenerational duties). Such debates are necessary, and we thus come full circle again to the idea of landscape as meeting or assembly place—*forum, agora,* and *thing*—and as a place of community. At the risk of overstating the point, what are urban parks but first and last, community and meeting places? In between, of course, they are many other things as well—nature reserves, biosphere resources, green lungs, the mitigation of 'urban heat islands' —but an urban park's first importance is its cultural significance from which all else flows.

For the future, civic society will decide what its cities and their parks should look like. The ideas of heritage and landscape have converged and matured into awareness of the overriding importance of the whole —coherence, setting, townscape—and both have begun to frame themselves within democratic / human rights discussions. Bringing them together is particularly important—and simple—in the context of large urban parks. Urban parks have not evolved naturally, and their development will continue to be guided by a billion small decisions that reflect broader societal attitudes. We should continue to work hard to keep and

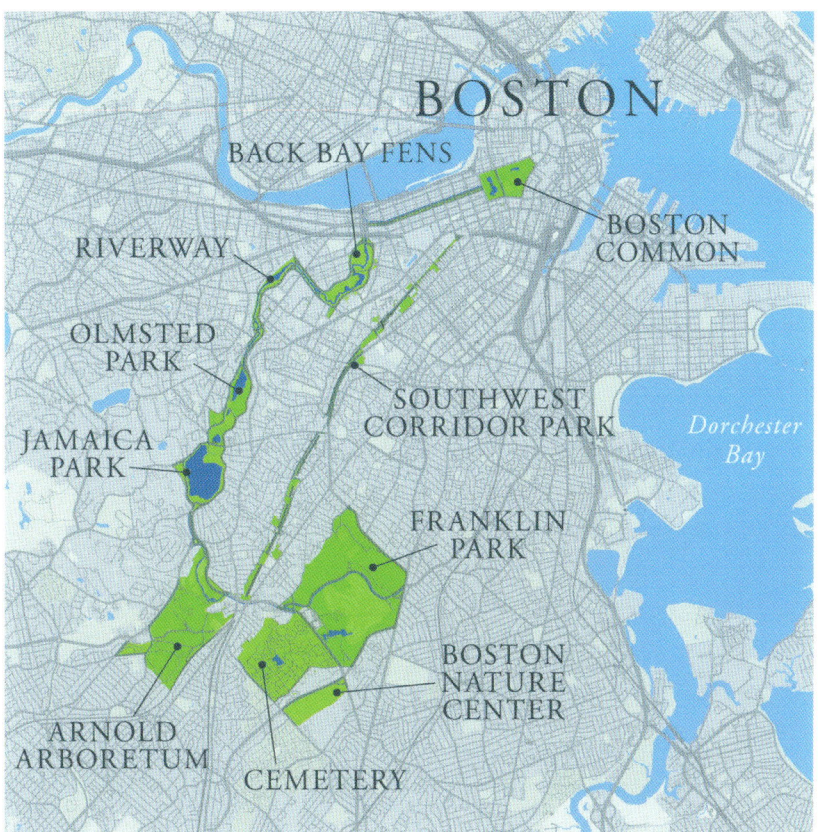

Forward looking. Boston. The Emerald Necklace, initially conceived in 1878 by Fredrick Law Olmsted Sr., Landscape Architect, creating a linear park along a water course to improve sanitary health, now sits in the heart of a large city. Since the 1980s, this unmatched urban legacy has been renewed and expanded (for example by adding Boston Nature Center and Boston Common), and is currently supported by Boston, Brookline and the Emerald Necklace Conservancy, landscape architects, staff and eager volunteers. This revitalization respects the original design intent and addresses contemporary sustainability practices in environment, society and economy.

improve the parks we have inherited, but new parks should be created as cities grow, for example at the current city edge so that in a generation or two they will be embedded within the enlarged city. We can aim to "green" the suburbs, if economic "imperatives" can be coaxed into a civic-minded context, but we can also "park-ify" the city as whole, and not primarily for the benefit of nature or for "green" motivations, but for any one of a number of practical reasons grounded in human, cultural and social needs, whether concerned with aesthetics, recreation and amenity, air quality, the cooling of urban heat-islands, politics … or simply, in one word, civic life.

DANDERYD

DJURSHOLM

ROYAL NATIONAL CITY PARK

Hässelby

Vaxholm

Baltic Sea

STOCKHOLM

Solna

Ekerö

Haninge

ULRIKSDAL

Ulriksdal Palace

SOLNA

Bergshamra

STOCKSUND

Ulriksdal

E18

E4

Brunnsviken

Roslagsvägen

Solna station

TIVOLI

Bergius Botanic Garden

FRESCATI

NORTHERN DJURGÅRDEN

ROYAL NATIONAL CITY PARK

LIDINGÖ

Haga Palace and Pavillion

Universitetet

HAGA

Stockholm University

Solna centrum

Uppsalavägen

ALBANO

Ropsten

SOLNA

HJORTHAGEN

Lilla Värtan

VASASTADEN

Tekniska högskolan

Olympic Stadium 1912

Gärdet

FRIHAMNEN

Stadion

The Kaknäs tower

Karlaplan

ÖSTERMALM

KUNGSHOLMEN

STOCKHOLM CITY

Vasa-Museum

The Maritime Museum

Rosendal Palace

FJÄDER-HOLMARNA

Royal Palace

Riddarfjärden

OLD TOWN

SOUTHERN DJURGÅRDEN

Saltsjön

STORA ESSINGEN

Museum of Modern Art

Skansen open air museum

NACKA

SÖDERMALM

LILJEHOLMEN

VÄSTBERGA

ÅRSTA

N

0   1   2 Kilometers
0   1 Mile

© OpenStreetMap contributors

THE ROYAL NATIONAL CITY PARK, STOCKHOLM, SWEDEN

SIZE: 27 square kilometers.

LEGAL PROTECTION: National Urban Park 1995.

OWNED BY: Three municipalities, the national government, and private owners.

MANAGED: 60 percent by the Royal Court, the rest by municipalities, private owners, and the National Property Board. Surveillance of the park is conducted by the county administration (www.nationalstadsparken.se).

VOLUNTEER ORGANIZATIONS: Ekoparken Association, The Committee for the Gustavian Park and WWF Sweden.

HOW TO GET THERE: *Subway stations:* Bergshamra, Gärdet, Tekniska Högskolan, Universitetet. *Train stations:* Universitetet. *Tram:* nr 7. *Ferry. Busses:* 1, 6, 50, 54, 55, 65, 67, 69, 76, 77, 176, 177, 503, 526, 540. By bike or on foot from the city center.

RICHARD MURRAY

The Royal National City Park, Stockholm, Sweden

The Royal National City Park stretches right into the very center of the Stockholm metropolitan area with its two million inhabitants (2018). Commonly and affectionally it is called Ekoparken[1]. Its diversified landscape contains 27 square kilometers, of which 19 kilometers are comprised of land, the rest of water. The story behind the park is rich and varied and entails stories of Swedish kings, military exercise fields, early science-city development, political history, and much more. In effect, it tells important parts of the history of Sweden. In early medieval times the area served as a source of food for some large monasteries and later for the royal court from gardening, agriculture, livestock, and fishing activities. Kings and noblemen used it for hunting deer, fowl, and wolves in the area, most of which, up until the 1820s, was fenced in. Archaeological investigations have revealed evidence of early settlements in this part of Sweden, including 1500 BC sites. Even though, today, some 15 million visitors yearly stroll, jog, bird-watch, canoe, kite-fly, go to museums, picnic, and so forth, around 75 percent of the flora and fauna species of mid-east Sweden is to be found in the park.

In effect, it tells important parts of the history of Sweden.

The park is made up of five interconnected parts: Ulriksdal, Haga-Brunnsviken, Northern Djurgården, Southern Djurgården, and Fjäderholmarna. This is the very center of the metropolitan region of Stockholm. A unique area is the inner harbor, parts of which are visible in the photo below, where shipyards, the navy, and cargo ships formerly predominated—it now constitutes a marine heritage site.

The park in its entirety received legal protection in 1995 as an area of national concern ("riksintresse") and was designated as a national urban park ("nationalstadspark"). It is made up of a series of smaller parks and green areas and a considerable amount of water bodies. These areas had been recognized in earlier years, but with no legal protection—the uniqueness in the creation of the national city park in 1995 was in connecting these areas into one whole.

The northern part of the park is occupied by the Royal Palace Ulriksdal, with its extensive grounds that used to be a baroque garden. The park has been restored but important baroque elements have been kept or mimicked. Through the park runs a stream—Igelbäcken—with a rare fish species, stone loach (*Barbatula barbatula)*. Work has been undertaken to preserve the fish and it seems to have been successful. A series of nature reserves protects the stream almost all the way to Lake Säbysjön, which lies outside the park, 15 kilometers north. The palace dates from the 1640s and has been modernized. A horse riding stable was in 1753 turned into a theatre for the royalties and is now, restored, used for opera productions.

King Gustavus III in 1780 bought Haga, a piece of land south of Ulriksdal and north of Stockholm. He wanted, like many artists, philosophers, and rulers in that time, a secluded retreat where he could live as an ordinary man. Only five years later he added another large tract of land. This was after he had visited Italy and seen the famous places outside of Rome. He now wanted to create a magnificent landscape park and to build a large palace there, overlooking Lake Brunnsviken. The areas on the other side and in view from Haga, he named Tivoli, Frescati, and Albano, with reference to Italy. This part of the Royal National City Park now makes up one of the most magnificent English landscape parks in Europe. The king liked theatre and play and had a Roman tent, a Chinese pavilion, and a Turkish kiosk built in the park. He moved his residence to a pavilion built in a special Swedish style of the late 18th century—"Gustavian"—Rococo on the way to Empire. Unfortunately—or fortunately—the planned big palace was never built. Only the foundations, the cellars, were completed when construction was stopped in 1792 upon the assassination of the king at a masquerade ball at the Stockholm Opera (Verdi later composed an opera about that event).[2]

As a tribute to the bucolic landscape, three distinguished individuals have chosen to be buried at the shores of Lake Brunnsviken—the German-born composer Joseph Martin Kraus (1756–1792); the poet and pioneer in physical education Pehr Henrik Ling (1776–1839); and the Crown Princess Margareta (1882–1920). The latter grave site has become the Royal Burial Ground.

To the east of Lake Brunnsviken extends a large tract of land, Northern Djurgården[3], with deep forests. King Charles XI and his son—eventually to become King Charles XII—hunted wolves there in the late 1680s. The large area, bounded by Lake Lilla Värtan in the east, was completely enclosed by a fence, that stretched southward to also include Southern Djurgården. Unlike royal hunting grounds in many other other countries, the general public was allowed to enter the area but had to pay a fee

Around 75 percent of the flora and fauna species of mid-east Sweden is found in the park.

TOP RIGHT: The Royal Palace of Ulriksdal.
BOTTOM RIGHT: Echo Temple (inside you can hear a whisper from one end to the other), originally the "Green Dining Room" for Gustav III, who loved eating outside.

EDDIE GRANLUND

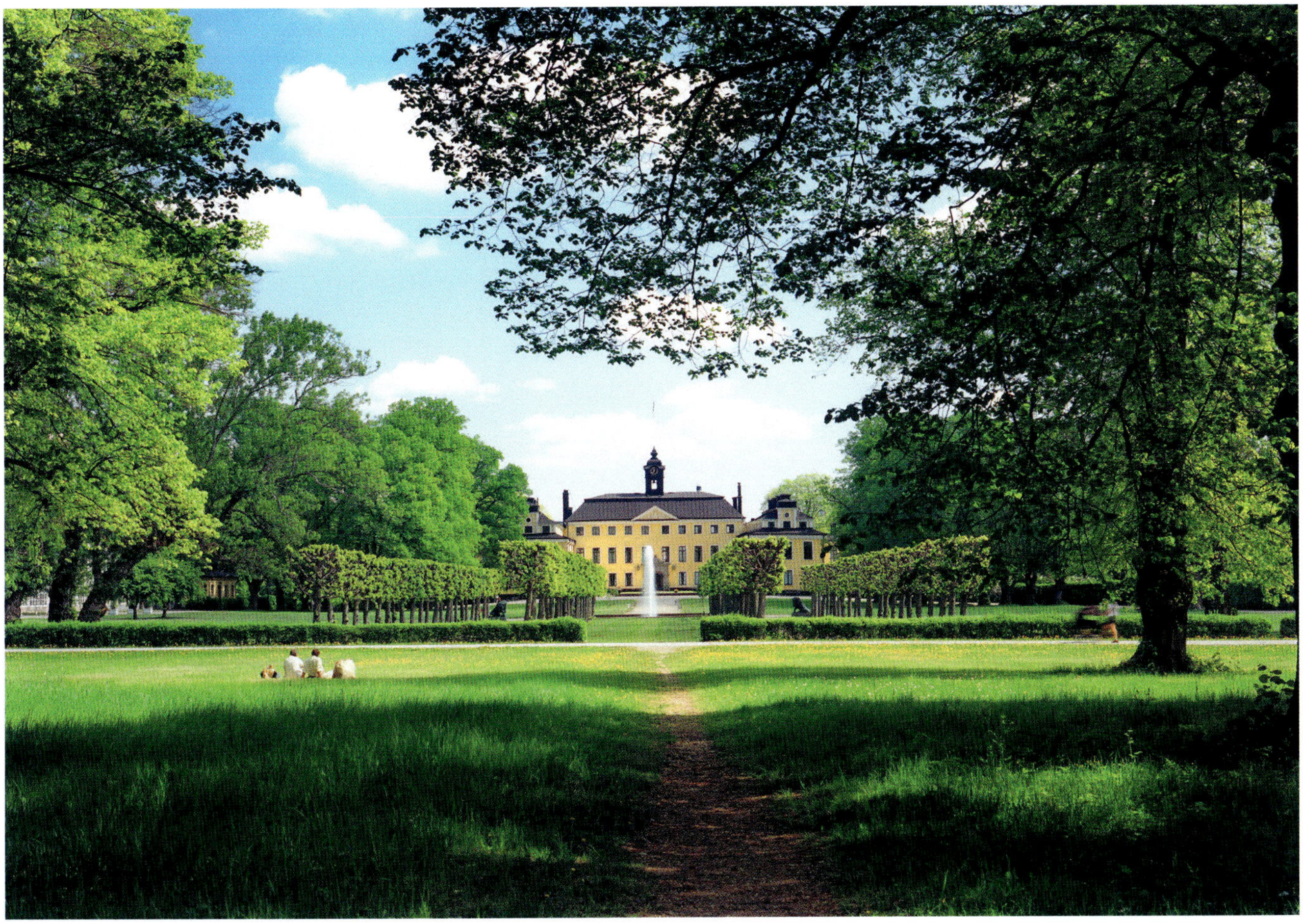

LSF_3STEPS SHUTTERSTOCK

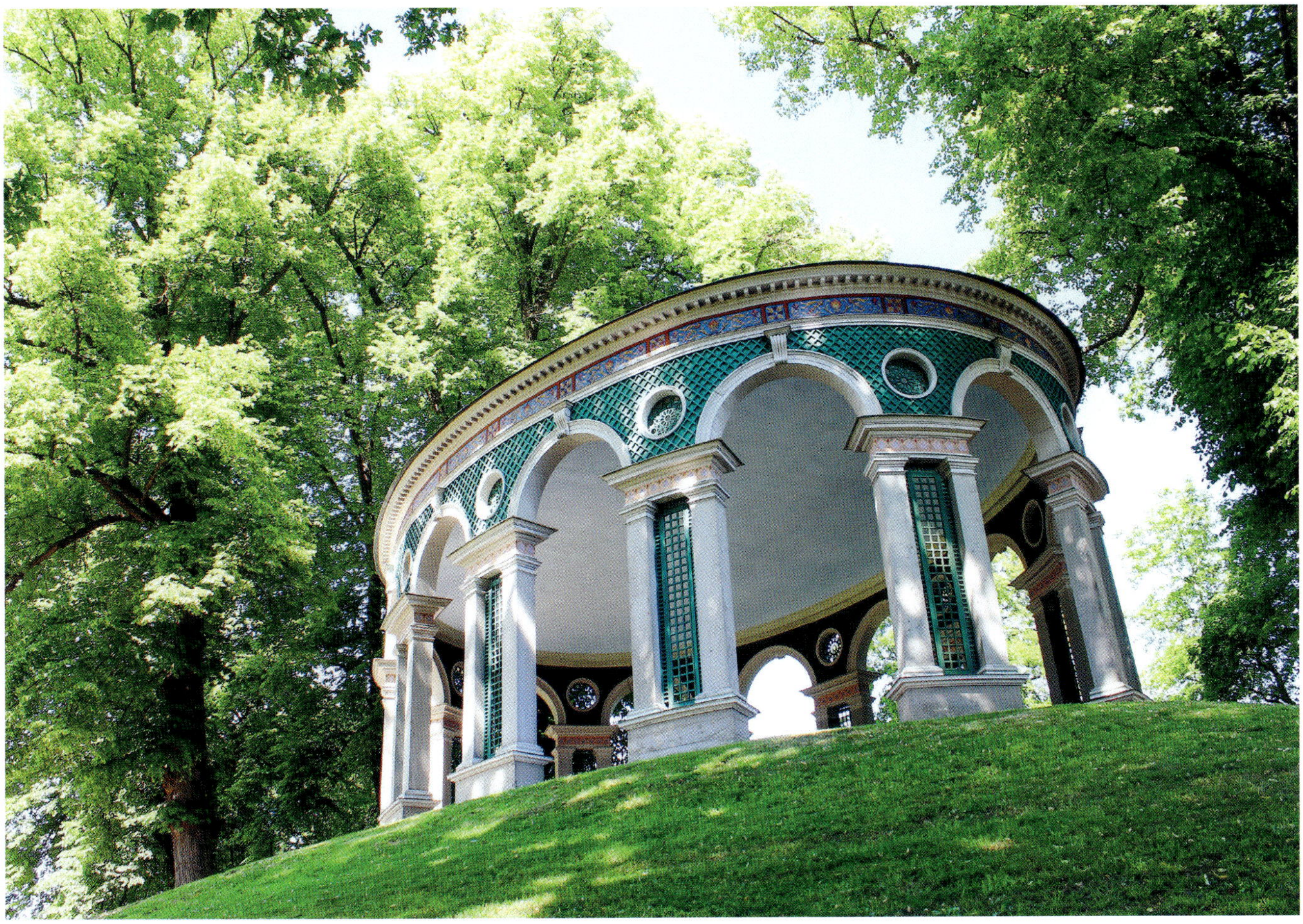

The Royal National City Park borders the Old Town and the very center of Stockholm.

in order to gather wood, pick berries, hike, or have cattle grazing. They were not allowed to hunt. That lasted until 1860/70 when fees were removed. The fence constituted valuable material and was often stolen, which let deer out and wolves in. The area is still rich in animals and flowers and weeds. There are owls, deer, foxes, badgers, herons, and seven kinds of bat. Also found there is a collection of giant oaks, the largest number in northern Europe, which hosts a very special set of birds, insects, and mushrooms. Wetlands flourish with salamanders. Lately, however, frogs have become sparse, which is cause for attention.

Northern Djurgården has had many usages. Fishing and hunting were dominant in the days before the early 19th century when modern methods of agriculture and gardening started to attract interest. Then the Academy of Agriculture was started, and fields of experimentation were created in this area. Also, a School of Forestry was located in the area. In the late 19th century, part of Northern Djurgården was labeled Science City ("Vetenskapsstaden"). The Royal Academy of Science and the Museum of Natural History were placed here in the early 20th century, and Stockholm University moved there in the 1960s.

There is also a rich history of sports. The stadium for the Olympic games of 1912 is located here. Nordic winter games—long-distance skiing, ski-jumping, and skating—took place in Northern Djurgården in the early 20th century. The ski jump is still there, now a conference center.

Connecting Northern and Southern Djurgården is a park with distinct natural qualities, the Tessin Park, named after a French father and son who came to Stockholm in the early 17th century and worked as architects in Stockholm up until the early 18th century. The son was the architect of the Royal Palace (see photo above). The park is a modernistic —functionalist—development. Smooth rock grounds have been preserved and incorporated, as an important element, into the set of pastures and greens in which playgrounds are inserted. Large trees and lots of bushes strengthen ecology. This is a needle-hole for the spread of animals and vegetation to the Southern Djurgården.

The park extends into a large grassy field, Ladugårdsgärde, which hosts kite-flyers, horseback-riders, rugby-players, picnickers, model airplanes enthusiasts, horse-riders, dog-walkers, and others. The field is used for large events like horse-races, vintage airplane shows, concerts, and much more. In the winter, if there is enough snow, tracks are prepared for long-distance skiing.

On the other side of the field, a set of museums are located: The Ethno-

---

LEFT: Forest known as "The Big Wolf Hunt" at Northern Djurgården.

graphic Museum, the Technology Museum, the Maritime Museum, just to mention the most important. They are located on the shore of a bay, which is part of the Baltic Sea. The bay leads into a channel that creates the island of Southern Djurgården.

The big, grassy field was once used for military exercises. Many regiments were located around the field in the late 19th century. When the regiments moved out, there was a lively discussion about extending the eastern part of Stockholm this way. This was at a time when the interest in nature and in preserving the landscape and its biodiversity had sprung up (in the US national parks were being created) in the wake of rapid industrialization and urbanization. The natural landscape, cultivated through centuries, was suddenly threatened. In Sweden this was the impetus for the preservation of an area that was deemed outstanding in both beauty and richness. However, only after a long and winding battle did legislation—in 1995—come about.

Southern Djurgården is an island with very different characteristics. In the western end there are museums, among others the Wasa Ship (an almost intact warship from 1628), Skansen (the first open-air museum in the world, holding a zoo with Nordic animals), and an amusement park, theatres, and restaurants. The rest of the island is a very pastoral landscape, in which a set of villas are located. King Oscar II sold or gave away plots of land to noblemen and friends in the late 19th century. They built grand villas, designed by the best architects. The collection provides an interesting overview of architectural styles at the time. Some villas have become art museums. This privatization of the land caused uproar because it infringed upon the picnic grounds of the general public that for centuries had allowed them to flee from the dirt, stench, and smoke of the city. This resulted in the parliament stopping the king from further bestowing such land deeds. Thus, the idea of preserving the area for the benefit of the general population of Stockholm took hold.

In fact, King Karl XIV Johan[4] had already called for the restoration of this part of the park for the pleasure of the general public. The beautiful, curved canal that separated Southern Djurgården as an island was dug on his orders. He and his entourage, along with masses of Stockholmers, used to ride and walk there on May 1st to welcome Spring.

Southern Djurgården houses a bird-watcher's paradise, Isbladskärret. Many of the 100 species of bird that breed in the park are found here.

This privatization of the land caused uproar because it infringed upon the picnic grounds of the general public that for centuries had allowed them to flee from the dirt, stench, and smoke of the city.

TOP RIGHT: White-tailed sea eagles (*Haliaeetus albicilla*) at Fjäderholmarna.
BOTTOM RIGHT: Some heavily trafficked roads border the park, some even cut through.

HENRIK EKMAN

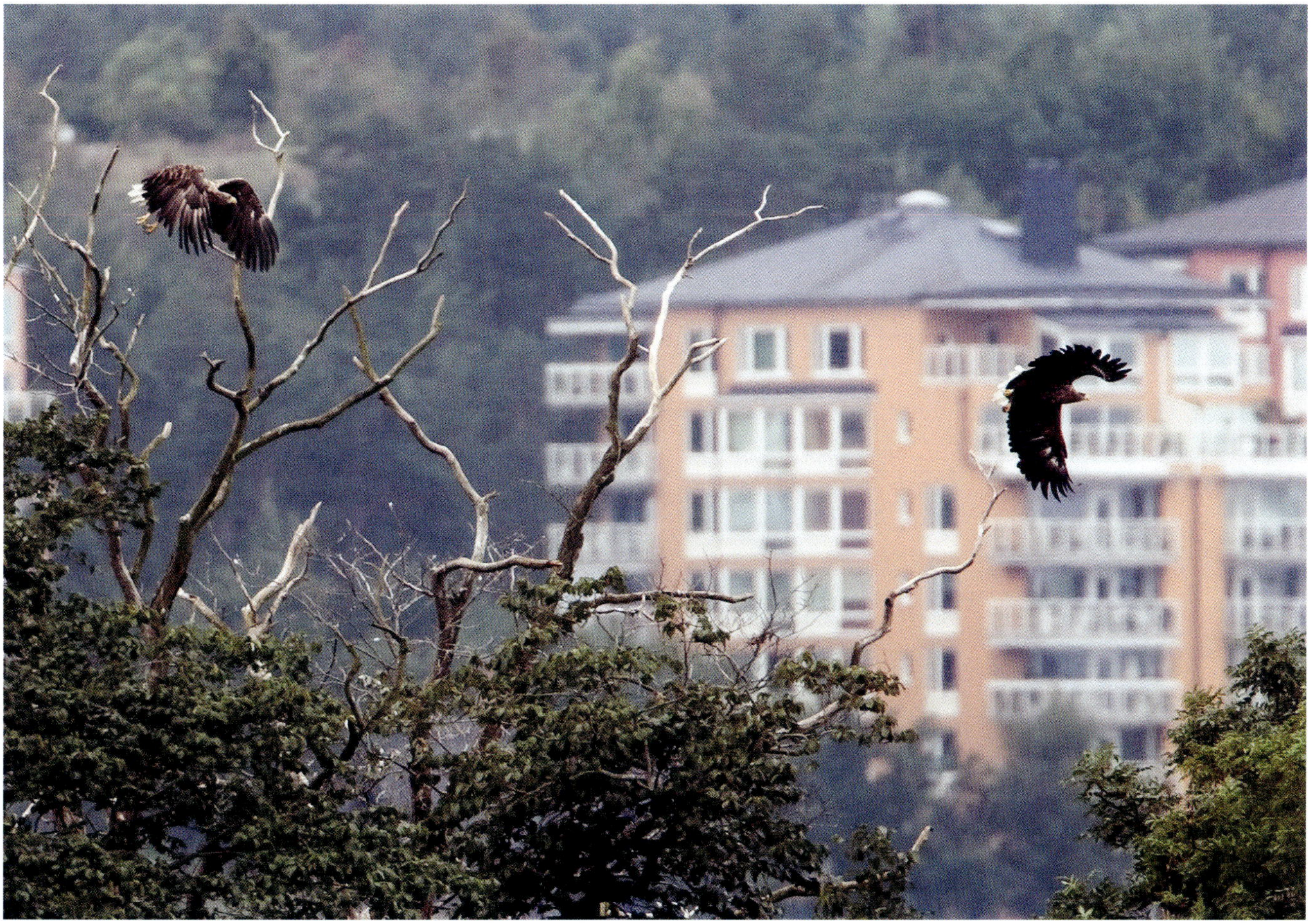

EDDIE GRANLUND

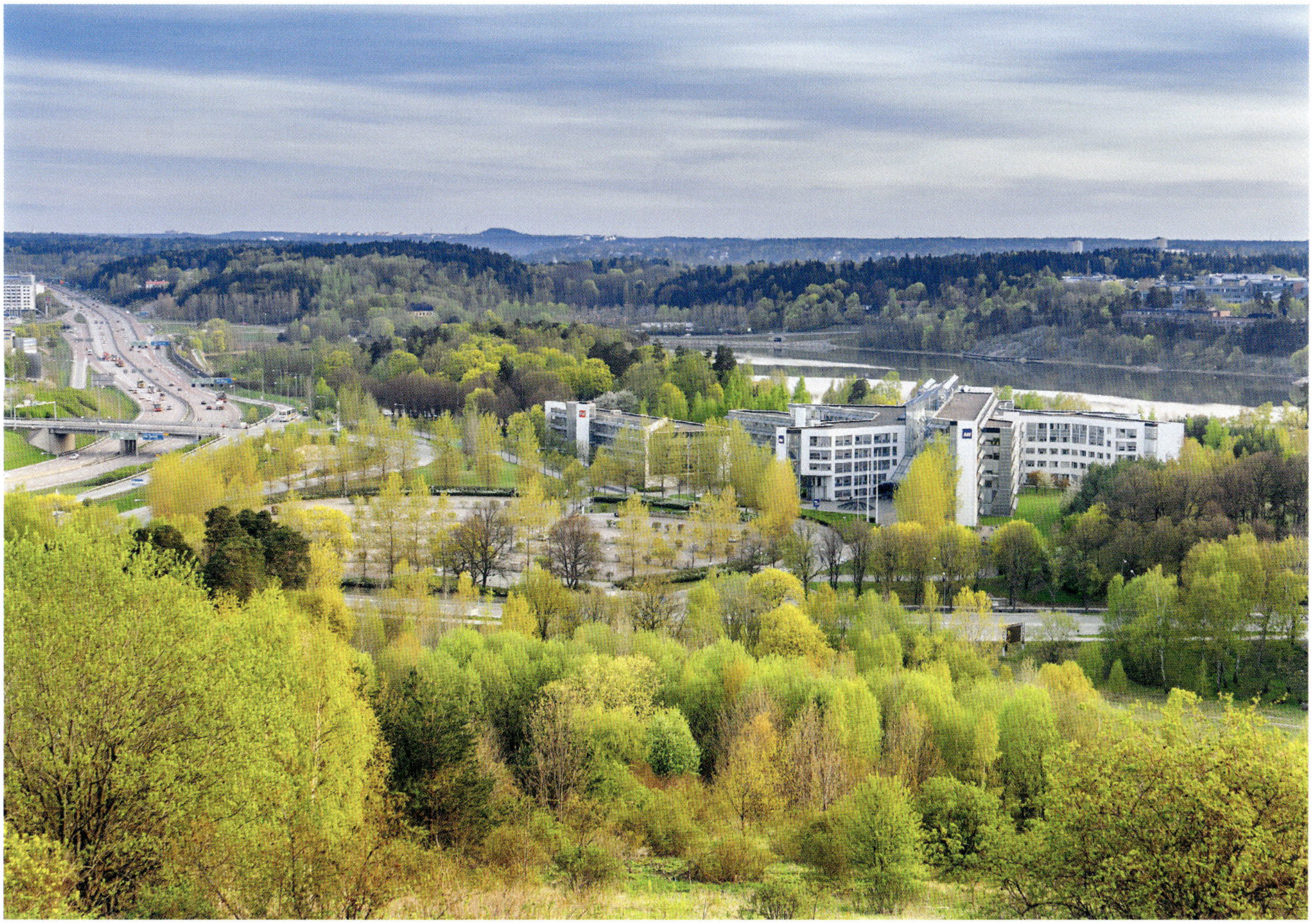

The woods are filled with the "Hoo-hohoho" calls of the tawny owl (*Strix aluco*) in the spring. The landscape has retained its charm and beauty with meadows lined by giant oaks and glimpses of villas.

Furthest out to the east the park meets the archipelago. Two larger islands and one smaller (Fjäderholmarna) form the beginning of the Stockholm archipelago with its 30,000 islands. Fjäderholmarna is home to a large number of seabirds, among them the rare lesser black-backed sea-gull (*Larus fucsus*). Quite often eagles sailing high up in the air can be seen. In the 19th century these islands were outside of Stockholm's control and aquavit could be sold here in large quantities to people who came rowing from the city. What is left today of that trade are a couple of restaurants.

The Royal National City Park is a combination of nature and culture. Buildings, parks, and gardens make up one part of the cultural heritage. Another important part is the maritime heritage—boats (the Wasa ship from 1628, af Chapman, a full-rigger from 1888, a 1915 steam-engine powered ice-breaker and much more), dry-docks, cranes, workshops, etc.—all in the inner harbor of Stockholm.

The legislation to preserve both the rich nature and the cultural heritage in the middle of a metropolitan region had to be innovative and fairly general. Despite giving the area the strictest "national concern" protection, encroachments have not stopped. As previously noted, through history various interests have had an eye for exploiting the park. Existing highways and railways cause noise, disrupt ecological corridors, and damage the heritage landscape. This has not yet been remedied.

The Committee for the Gustavian Park was founded in 1991 by a group of academics, and the Ekoparken Association was founded in 1992 by 22 organizations. A massive development threatened the area but was fought off. Currently, city expansion again encroaches on the park. The Ekoparken Association today includes about 50 organizations. Both the Ekoparken Association and the Gustavian Park committee, together with WWF Sweden, are an official part of the planning process and maintenance of the park and serve in various capacities as watchdogs for the park.

The legislation to preserve both the rich nature and the cultural heritage in the middle of a metropolitan region had to be innovative.

[1] The name was conceived in the report *Ekoparken—A Nature and Culture Reserve in a Big City* (Waldenström, 1990).

[2] Many artists and architects were involved in the design of park and buildings, foremost among them Fredrik Magnus Piper and Louis Jean Desprez.

[3] The name Djurgården (in German, "Tiergarten") is derived from "animal park," the first one having been created in 1579 as a closed in area with elk, reindeer, fallow deer, red deer, and other animals. Later it also for some time housed a lion, baited to fight other animals.

[4] A former marshal in Napoleon's army, by the name of Jean-Baptist Bernadotte, who was invited and came to Sweden in 1810, founding the present royal family.

LEFT: Canoeing on Lake Brunnsviken. Stockholm's city hall is seen in the distance, three kilometers away.

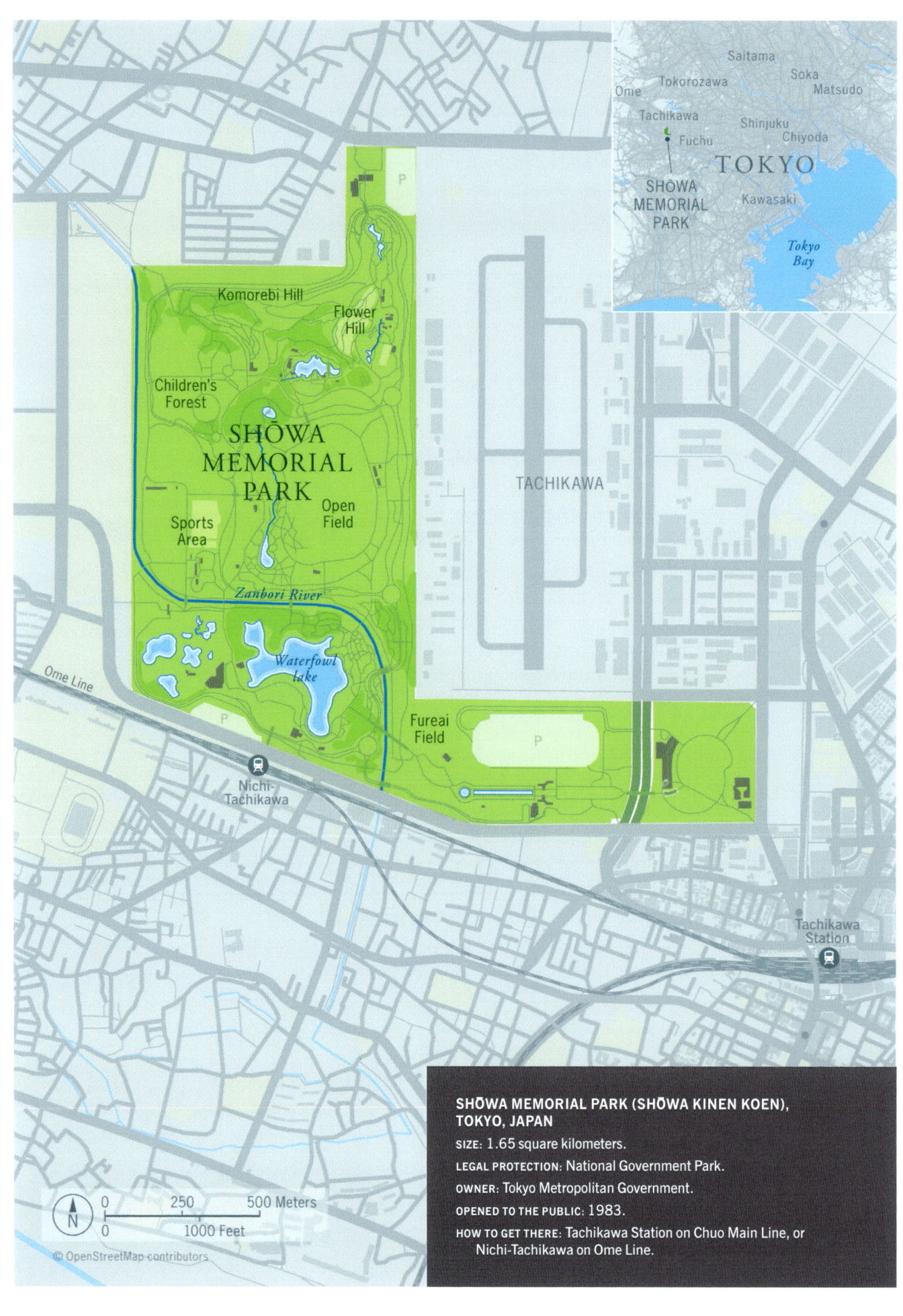

Saitama

Ome Tokorozawa Soka Matsudo

Tachikawa Shinjuku
Fuchu Chiyoda

TOKYO

SHŌWA
MEMORIAL Kawasaki
PARK

Tokyo
Bay

Komorebi Hill

Flower
Hill

Children's
Forest

SHŌWA
MEMORIAL
PARK
Open
Field

Sports
Area

Zanbori River

Waterfowl
lake

TACHIKAWA

Ome Line

P

Fureai
Field

P

Nichi-
Tachikawa

Tachikawa
Station

N

0   250   500 Meters

0   1000 Feet

© OpenStreetMap contributors

**SHŌWA MEMORIAL PARK (SHŌWA KINEN KOEN),
TOKYO, JAPAN**

SIZE: 1.65 square kilometers.

LEGAL PROTECTION: National Government Park.

OWNER: Tokyo Metropolitan Government.

OPENED TO THE PUBLIC: 1983.

HOW TO GET THERE: Tachikawa Station on Chuo Main Line, or
Nichi-Tachikawa on Ome Line.

LIISA EURO WIHMAN
Shōwa Memorial Park, Tokyo, Japan

Shōwa Memorial Park is the largest park in the Tokyo metropolitan area. It consists of over 165 hectares of playgrounds, meadows, forested areas, biking trails, ponds and water features, museums, sports venues, and other recreational facilities. Situated some 30 minutes from central Tokyo by train in the city of Tachikawa, it was created in 1983 to commemorate the 50th anniversary of Emperor Shōwa's reign. Shōwa Memorial Park is classified as a National Government Park.

Tokyo—or Edo, as the city was called before it became the capital of Japan in 1868—has been a densely populated metropolis for hundreds of years. Already by the early 18th century, it was the largest city in the world with an estimated population of 1.3 million.

Despite several severe setbacks in the form of natural and human-inflicted catastrophes, Tokyo has continued to thrive and grow as an urban center and today has 13.3 million inhabitants. However, the city only comprises a part of a larger sprawl of closely interconnected cities and municipalities within the governmental area of Tokyo Metropolis, where around 38 million people live and commute daily for work, education, and entertainment.

With such a long history of dense urban habitation, land is understandably a scarce and valuable commodity. Compared to other major capitals of the world Tokyo trails far behind with its 7 percent land area allocation for public greenspace and parks.

With little or no vacant land available, spaces must usually be converted from one existing function to another. A great historical example of this happened after the fall of the Tokugawa Shogunate and during the following Meiji-restoration in 1868. Large strolling gardens belonging to the luxurious residences of the ousted daimyos (feudal lords) from the Edo period (1603–1867) were converted into public greenspaces. Some of these, like Shinjuku Gyo-en and Koishikawa Korakuen, still provide an escape from hectic city life for Tokyoites.

By the 1930s, most of these gardens had been converted, and new methods for creating green space were needed. After WWII, the focus turned to areas of former military and industrial uses, and/or to landfill and reclamation projects. These new parks are almost always much larger than the parks of the past. Also, they contain new design elements like fountains, flower gardens, and open fields, inspired by city parks in

In the crowded metropolis of Tokyo, where most people live in small apartments with no access to private gardens or even balconies, large urban green spaces are vital for the quality of life.

Europe and North America. However, their overall design often traces back to the strolling gardens of the Edo period, with winding paths tying together the design elements.

Shōwa Memorial Park was one of the former military spaces repurposed as a park—turned into a giant strolling garden that combines both Japanese and Western features. The site had been used as the Tachikawa Airfield, and during WWII it functioned as an army airfield for the aerial defense of Tokyo. After the war, the United States occupied the base and used it for transport flights in both the Korean and Vietnam wars. It was officially closed and formally returned to the Japanese government in 1977.

Despite its concrete-clad military past, the land itself was well-suited for conversion to a park. Located on the vast flatlands of the Musashino plateau, the ground is covered with a thick layer of volcanic loam called "kuroboku." This black soil is rich in humus content and promotes abundant growth and plant diversity, which made the transformation easier than in most reclamation sites.

With this large area of prime, convertible land in their hands, the Tokyo Metropolitan Government had an excellent opportunity to create more green space for citizens. However, it was not their only concern: Tokyo's geographical position at the boundary of three continental plates means that there is a greater than 70 percent possibility of a stronger than magnitude-eight earthquake hitting Tokyo within the next 30 years. This makes detailed planning for an emergency a priority.

When the last major earthquake of that magnitude hit the city in 1923, over 142,000 people perished. Despite vast improvements in building standards and practices, an earthquake of similar magnitude today would still cause major damage for the inhabitants and infrastructure of Tokyo. This is why most parks in Tokyo serve double duty: used primarily as recreational areas, they also function as dedicated emergency evacuation zones, where people can escape the risk of collapsing buildings and possible fires; and where tents can be raised as shelters for the injured and the newly homeless. Called "Disaster Preparedness Parks," they have earthquake-proof toilets, solar-powered lighting for medical and cooking facilities, and are constantly ready to receive millions of people unable to return to their homes after a major earthquake.

In the case of the former Tachikawa Air Base, parts of the area were allocated as a strategic location for integrated emergency services

Families spend whole days here, picnicking and playing games, often blissfully unaware that this 11-hectare field is also used as an emergency evacuation area in case of earthquakes.

TOP RIGHT: Open field with Cosmos.
BOTTOM RIGHT: The Japanese garden.

LIISA WIHMAN

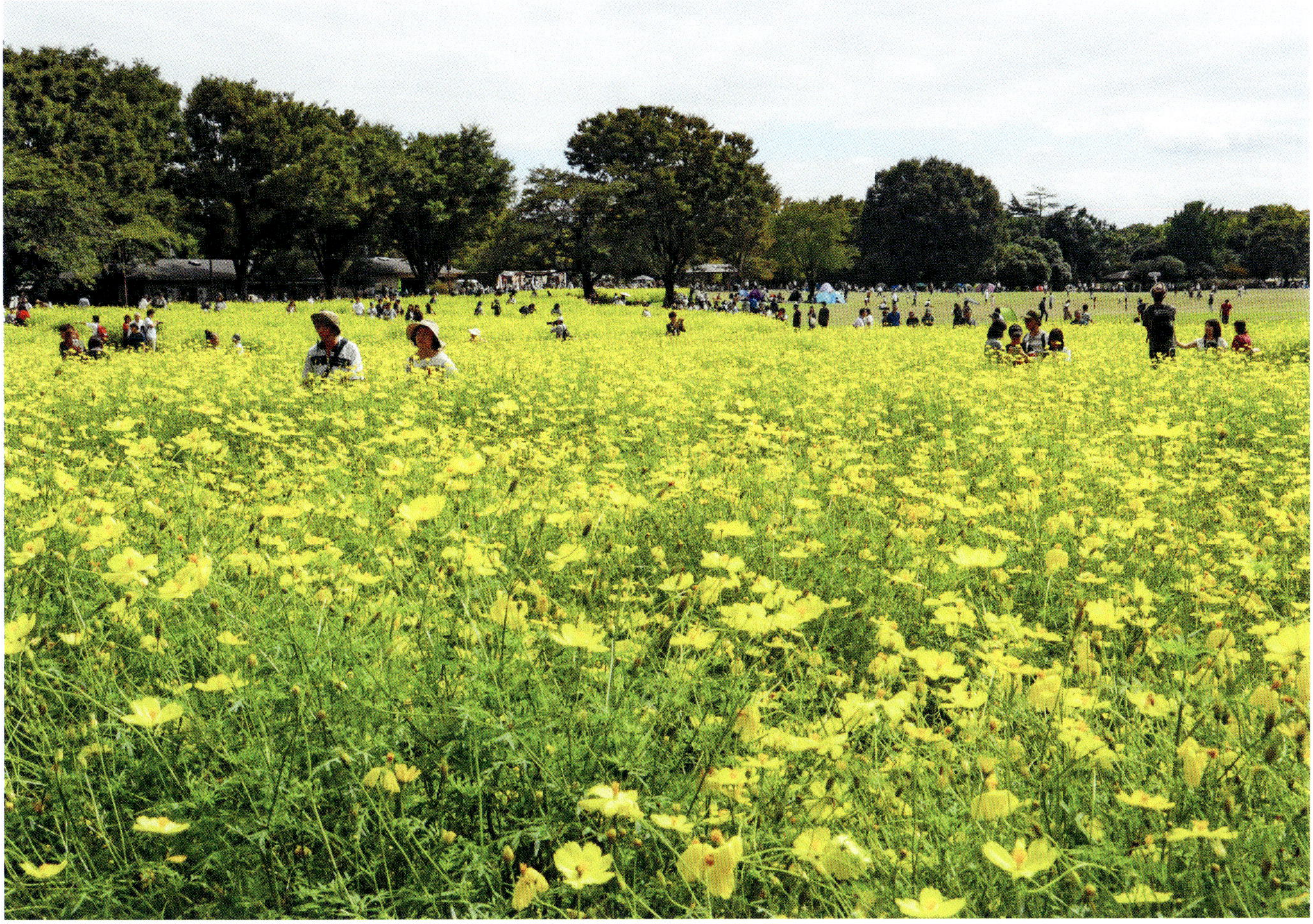

JUMOOBO SHUTTERSTOCK

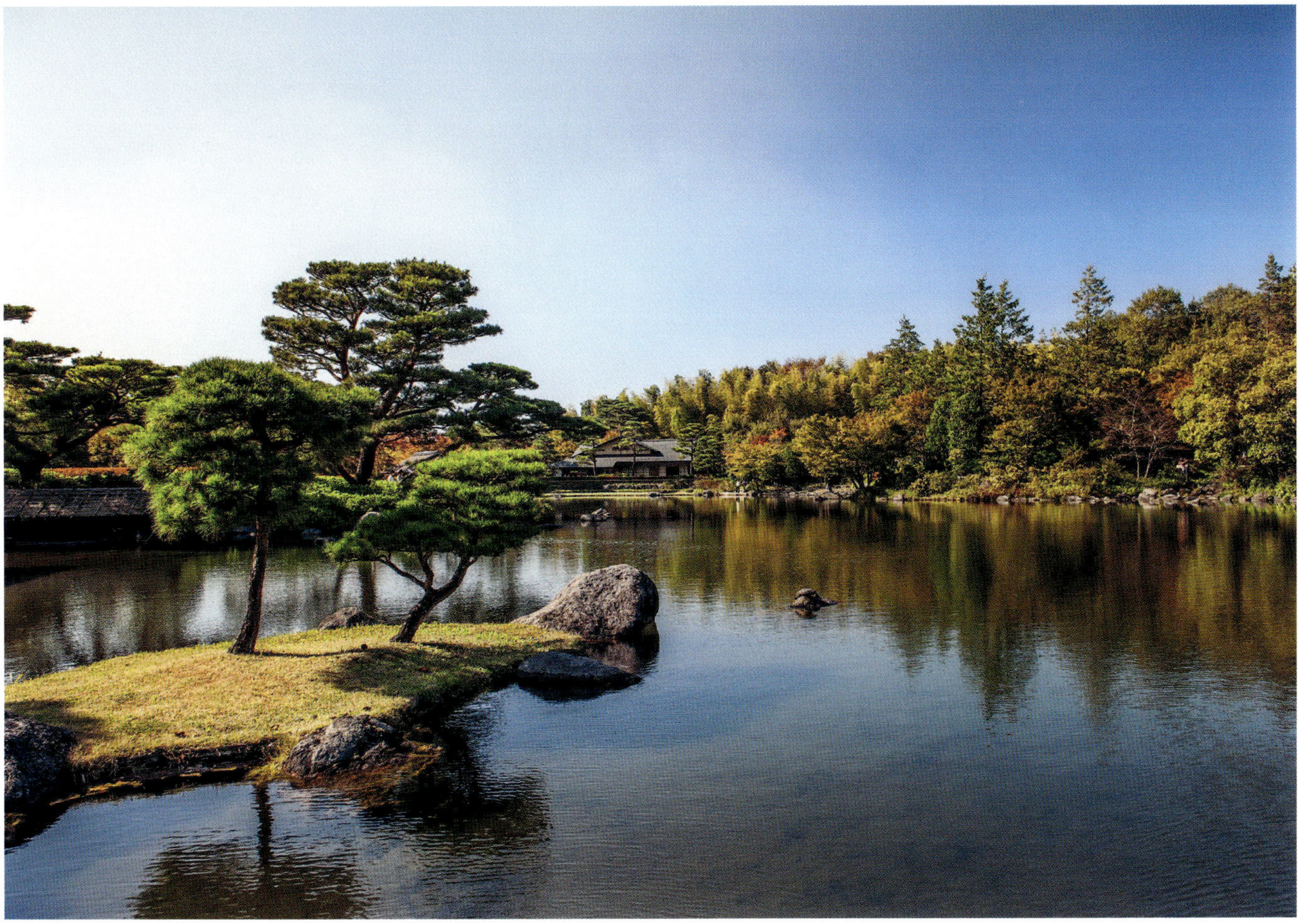

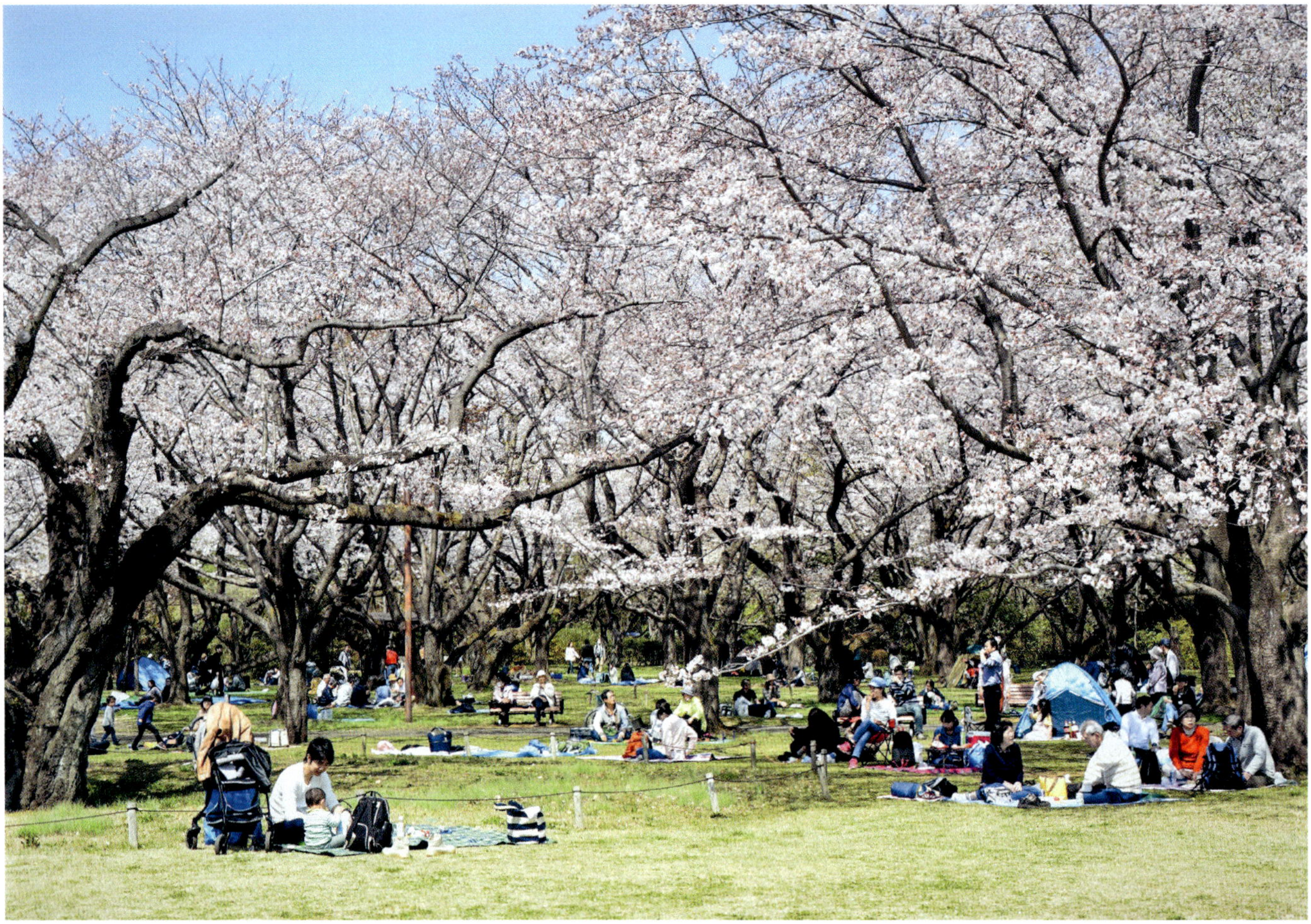
LIISA WIHMAN

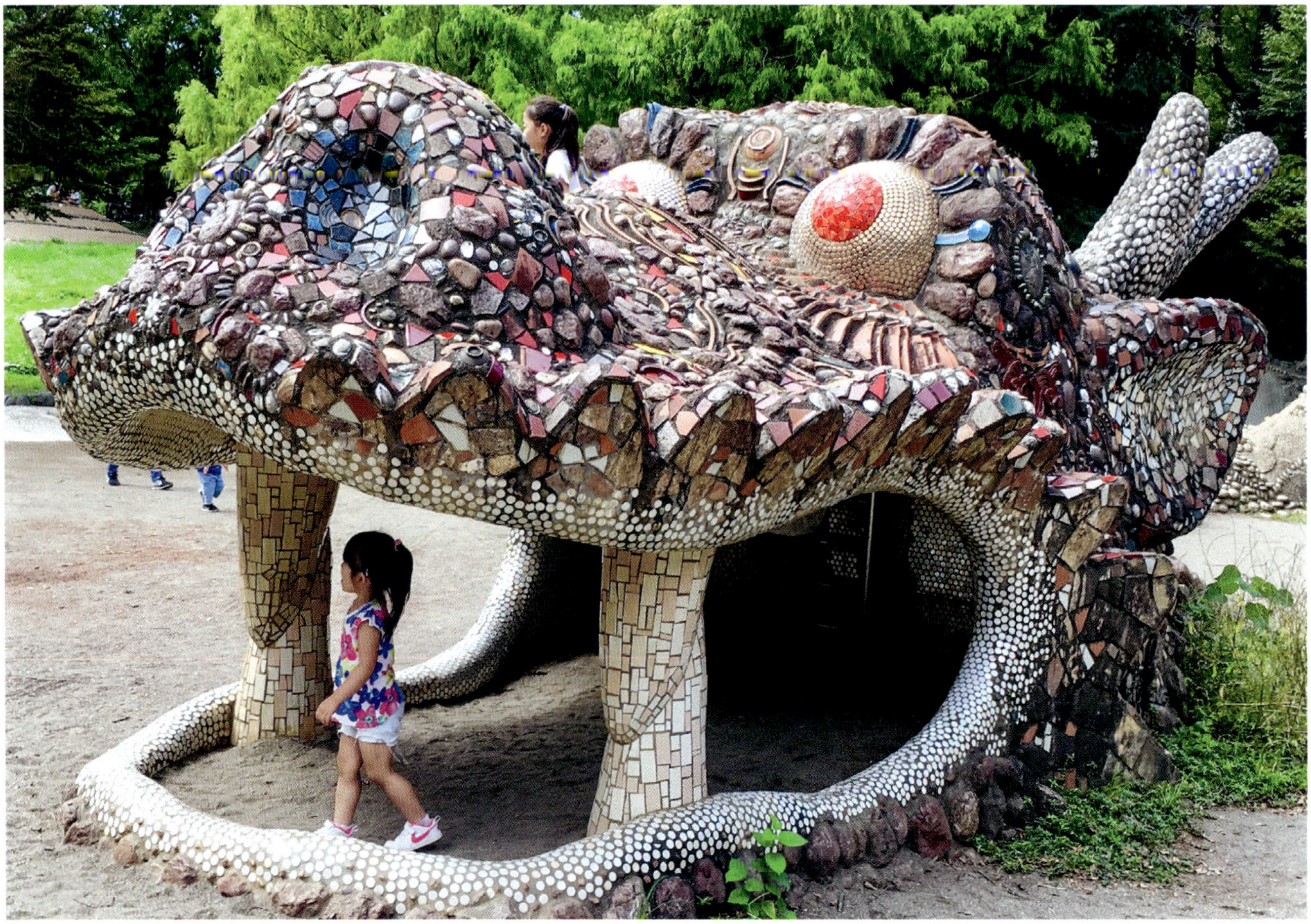
LIISA WIHMAN

including an airport, medical, and stockpiling facilities. The remaining areas were converted into the Shōwa Memorial Park. With incorporated Disaster Preparedness functionality and a large open field the park will also serve as a large-scale evacuation zone.

With 165 hectares of convertible land in their hands, the Tokyo Metropolitan Government decided to create "a natural environment in which to refresh their minds and bodies," together with a hub for culture and arts in the western parts of the metropolis. However, in 1983, at the same time, Tokyo Disneyland opened its gates. Because of budgetary restrictions, the Shōwa Memorial Park entrance needed to be fee-based, and there was a concern that people would not find it worth visiting compared to the colorful amusements of Disneyland. The park plans were ramped up to measure up to this commercial competitor.

The park was divided into zones: culture and exhibition, water, an open field, a sports area with courts for multiple sports, restaurants and cafes, barbeque facilities, children's playground, and forested areas. Tying it all together—like in a giant strolling garden—are walking paths and over 11 kilometers of cycle paths meandering through the landscape.

Outside the fee-based park area, the cultural zone contains the Hana-midori Cultural Center and Shōwa Museum commemorating the life and times of Emperor Shōwa (posthumous name of Emperor Hirohito, who reigned from 1926 until 1989). Inside the fee-based park, water zones include a large pond with rental pedal boats and the Rainbow Pool, which includes wave and adventure pools, offering entertainment for both children and adults.

A vast open field called *Minna no Harappa* ("Everyone's Field") forms the center of the park. In its middle, a century-old large Japanese *Zelkova* (Giant Keyaki) forms a focal point with its vase-formed canopy reaching over 20 meters above the ground. Families spend whole days here, picnicking and playing games, often blissfully unaware that this 11-hectare field is also used as an emergency evacuation area in case of earthquakes. Parts of the *Minna no Harappa* are covered with changing seasonal massed plantings, with blooming poppies in spring, a sunflower maze for children in summer, and huge fields of cosmos in autumn.

There are several themed gardens. In the meticulously trimmed six-hectare Japanese Garden, pines and azaleas mingle with Japanese maples around a central pond, and the National Museum of Bonsai, housed in the park, displays rare bonsai specimens, some of which are

TOP LEFT: Flowering Japanese Zelkova.
BOTTOM LEFT: Dragon Dunes.

hundreds of years old. In the many herb and flower gardens, bulbs and annual plantings change with the seasons. In the Flowering Tree Garden, the plum and peach blossoms are followed by cherries, stretching the flowering season from early January to April. In the Dragonfly Marsh, Wild Grasses trail, and Waterfowl Lake, habitats are formed for native plant species and wildlife, from dragonflies to grasshoppers, and grey herons to spot-billed ducks.

The most artistic expression is found in the "Children's Forest," designed between 1983 and 1987. At this time, new video game consoles and handheld video games were becoming popular, which meant that rather than playing together, children were increasingly engaged in individual indoor play only. When creating this project, Fumiko Takano and Norihiro Kanekiyo wanted to provide a variety of different physical, sensory, and emotional experiences, and so they transformed the children's play area from a traditional playground with conventional equipment into a play environment where children interact with art and with each other. None of the structures was designed for individual play; instead, they were all to be experienced collectively by the children using them simultaneously.

None of the structures were designed for individual play; instead, they were all to be experienced collectively by the children using them simultaneously.

Thus, the Children's Forest was designed to stimulate both the senses and imagination. When entering, the visitor is met by cave-like mosaic grottoes; continuing further, one is enclosed by towering trees and bamboo, only to be confronted by large open areas thereafter. A spiral walkway leads down to a hollow, where water gathers naturally when the water table rises and sinks with the seasons. Close by, tiered pyramids similar to those of the mesoamerican civilizations beg to be explored. In the Dragon Dunes, huge mosaic dragons slither through a giant sandpit, inviting children to climb and scramble over.

Meticulously tended by the Tokyo Metropolitan Government, Shōwa Memorial Park is open to public in all but the most challenging weather conditions. The park is well-visited and gets extremely busy on holidays, when long lines of families with picnic and play gear form at the entrance gates.

In the crowded metropolis of Tokyo, where most people live in small apartments with no access to private gardens or even balconies, large urban greenspaces are vital for the quality of life. In times of catastrophes, they are equipped even to save lives. Like a modern "strolling garden on steroids," the Shōwa Memorial Park offers space for recreation and contemplation; play and exercise; enjoyment for all the senses; and amply fulfills its objective of "providing people with a natural environment in which to refresh their minds and bodies."

CONNAUGHT PLACE

BENGALI MARKET

Ashoka Rd

KG Marg

Tilak Marg

APPU GHAR

RAJPATH AREA

SUNEHRI BAGH ROAD AREA

RABINDRA NAGAR

Dr Zakir Hussain Marg

Subramaniyam Bharti Marg

Prithviraj Road

Mathura Rd

Pragati Maidan

Purana Quila

Zoological Park

Delhi Golf Course

Sundar Nursery

Mahatma Gandhi Marg

Yamuna river

Millennium Indraprastha Park

CULTURAL HERITAGE LANDSCAPE PARK

Safdar Jang tomb

Lodhi gardens

Lodhi Road

Amir Khusro park

Shrine of Saint Hazrat Nizamuddin Auliya

Humayun's Tomb

Hazrat Nizamuddin

Smriti Van (memorial garden)

Barapullah Road

Safdarjung Airport

N

0   0,5   1 Kilometers
0   0,5 Miles

© OpenStreetMap contributors

Old Delhi

NEW DELHI

Airport

CULTURAL HERITAGE LANDSCAPE PARK

CENTRAL DELHI HERITAGE PARK, INDIA

APPROXIMATE SIZE: 5.2 square kilometers.

PROTECTION: Several gardens are World Heritage Sites.

OWNED AND MANAGED BY: Government and private organisations.

HOW TO GET THERE: Metro stations Jor Bagh, Khan Market, JLN Metro Station.

NUPUR PROTHI KHANNA
Central Delhi Heritage Park, India

Delhi is a megacity of 29 million inhabitants dotted by countless historic sites and monuments, now buried under an urban fabric. Consecutive regimes over a millennium have ruled from this land, leaving behind rich archaeological evidence. Delhi today needs a landscape plan that recognizes its many past lives and realizes its importance for the prospect of a healthier life in the city.

As a young landscape architect, I had an opportunity in 1999 to work on the restoration of the Gardens of Humayun's Tomb, a project led by the Aga Khan Trust for Culture (AKTC) and the Archaeological Survey of India.[1] Serving as my guide, mentor, and guru since 1994 was Prof. M.Shaheer, the landscape architect for the Humayun's Tomb and Sundar Nursery projects. The tomb, created during the years 1565–1572, is located to the south of Old Delhi, at that time a barren, dry floodplain that became the necropolis of Sufi saints and Mughal rulers. This area of Delhi contains the elements for creating a cultural heritage park.

Humayun's Tomb precinct covers over 24 hectares. The first example of a Mughal tomb garden on Indian soil, this magnificent structure towers 43 meters over a perfectly laid quadripartite (four part) garden, referred to as "Char Bagh" in local landscape parlance. Its design follows the genre of Persian gardens, popularly referred to as Mughal gardens within the Indian sub-continent. Its horticulture and simple yet innovative use of water create a "paradise on earth." It is noteworthy that Humayun's Tomb Garden went on to inspire the construction of the famous Taj Mahal (1632–1643) within half a century.

Extending our vision beyond this garden shows that it is situated in the midst of a royal necropolis dotted with centuries-old mausoleums and graves, visible today as ruins and rubble dispersed across what would have been an open scrubland. The location of this tomb with the remains of the 2nd Mughal emperor Humayun in a desolate landscape was no accident or coincidence. Witness to nearly seven centuries of tomb building, this land is endowed with spiritual significance by the shrine of a famous Sufi saint, Hazrat Nizamuddin Auliya (13th century).

Humayun's Tomb Gardens and the immediate surroundings have undergone vast changes since the 16th century. Conceived on the banks of the Yamuna river, this link with the water is today entirely obliterated since the river has meandered a few kilometers eastwards. The dense

Delhi today needs a landscape plan that recognizes its many past lives and realizes its importance for the prospect of a healthier life in the city.

Humayun's Tomb (1565–1572), a World Heritage Site since 1993.

planned and unplanned development and subsequent squatter settlements have little connection with the monument and its history. The shrine and the monument coexist but the physical connection is lost. Most monuments in the vicinity have vanished within the built fabric. Some have been fortunate to be part of urban park layouts as a result of colonial and post-independence urban planning. The landscape they inhabit is either completely built over or, where some parts of parkland remain, the space is used as a regular park, entirely erasing their unique history. These tombs stand out, marooned in an unfamiliar setting.

So, how do we connect the stories across roads and buildings—and across time? How do we interpret history through an unbiased design approach addressing different periods and regimes of our chequered past? How can we restore the original scheme without obliterating the additions of subsequent periods? Many of the Islamic tombs were built upon earlier Hindu architecture; and Hindu architecture upon archaeological remains of earlier centuries. From the 16th century until it was claimed by the British, Delhi was a Mughal capital. Earlier capital cities, like Lal Kot, which today lies within Delhi, were Hindu (the story of Delhi is a narrative of cities within a city).

Restoration is full of practical issues. How do we safeguard the historic fabric with an increasing number of visitors? How can the original pathways and water be systematically revived to address present-day

anticipated use? How can we ensure authenticity of historic geometry, materials, and construction? How do we make plant selections when species from the original plant palette are no longer part of catalogs of horticulture nurseries?

Working on this project brought a new awareness to the many layers existing in a cultural landscape. It demonstrated how to weave in a context that has irrevocably changed with time. Here was a potential role for design and interpretation to revive a lost narrative.

The gardens of Humayun's Tomb incorporated monuments that had formed part of the 16th century Mughal Grand Trunk Road[2]. Colonial aspirations transformed this area into a plant nursery, a function that continues to this day. Inaugurated in 2018, the landscape masterplan incorporated the Mughal period historic monuments; usage during the colonial period; and a newly designed park that has already made it to the Times Magazine's list of 100 great places to visit in the world.

The successful completion of this garden restoration work led to another project in its vicinity.

As work at the Humayun's Tomb Gardens was drawing to a close, efforts were made to start work at the adjacent 36-hectare Sundar Nursery (referred to as Azim Bagh in the 16th century). Designed by the same firm that restored the Humayun's Tomb Gardens, the landscape concept acknowledged the historic connection between the two parks. The new

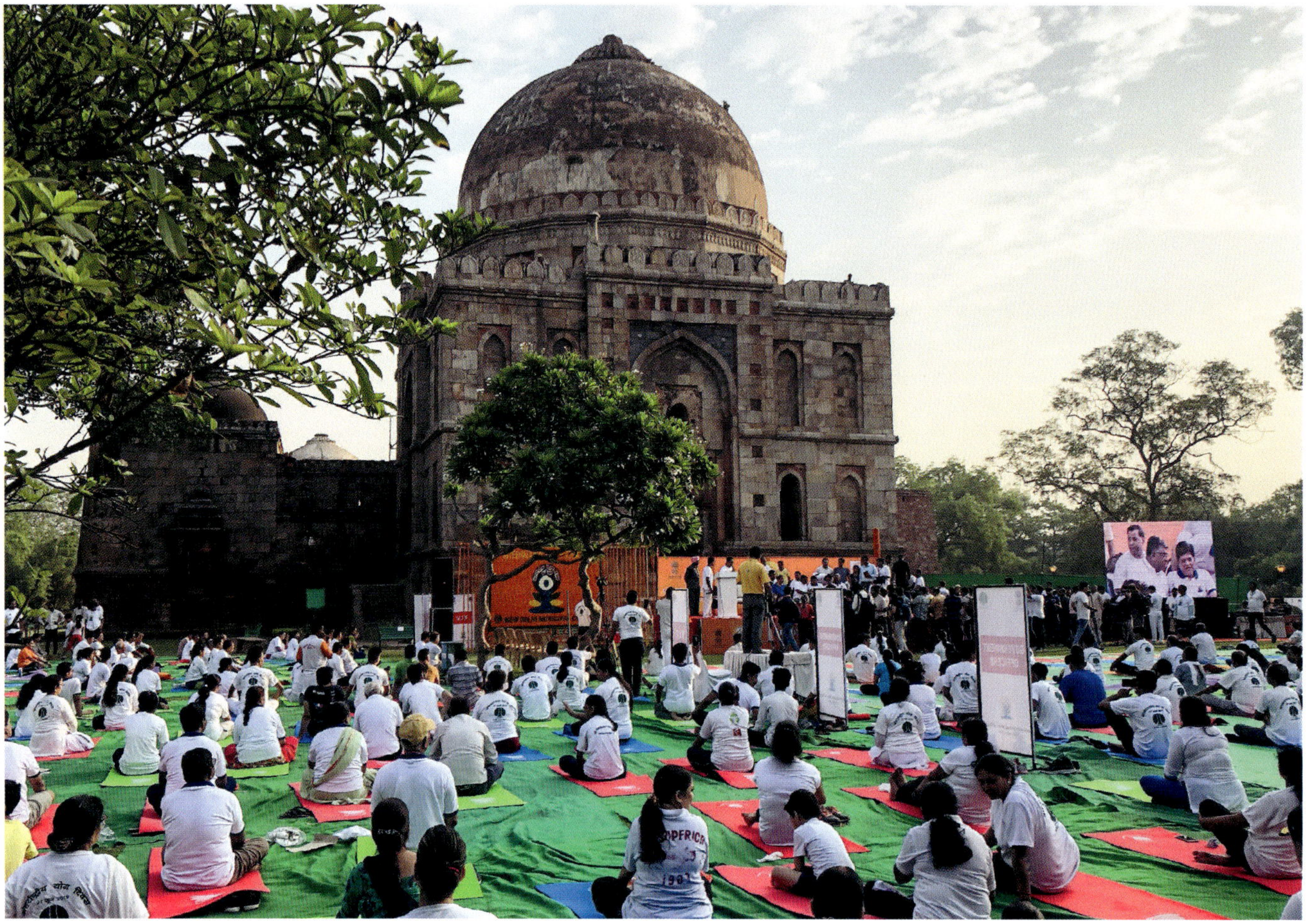

NUPUR PROTHI KHANNA

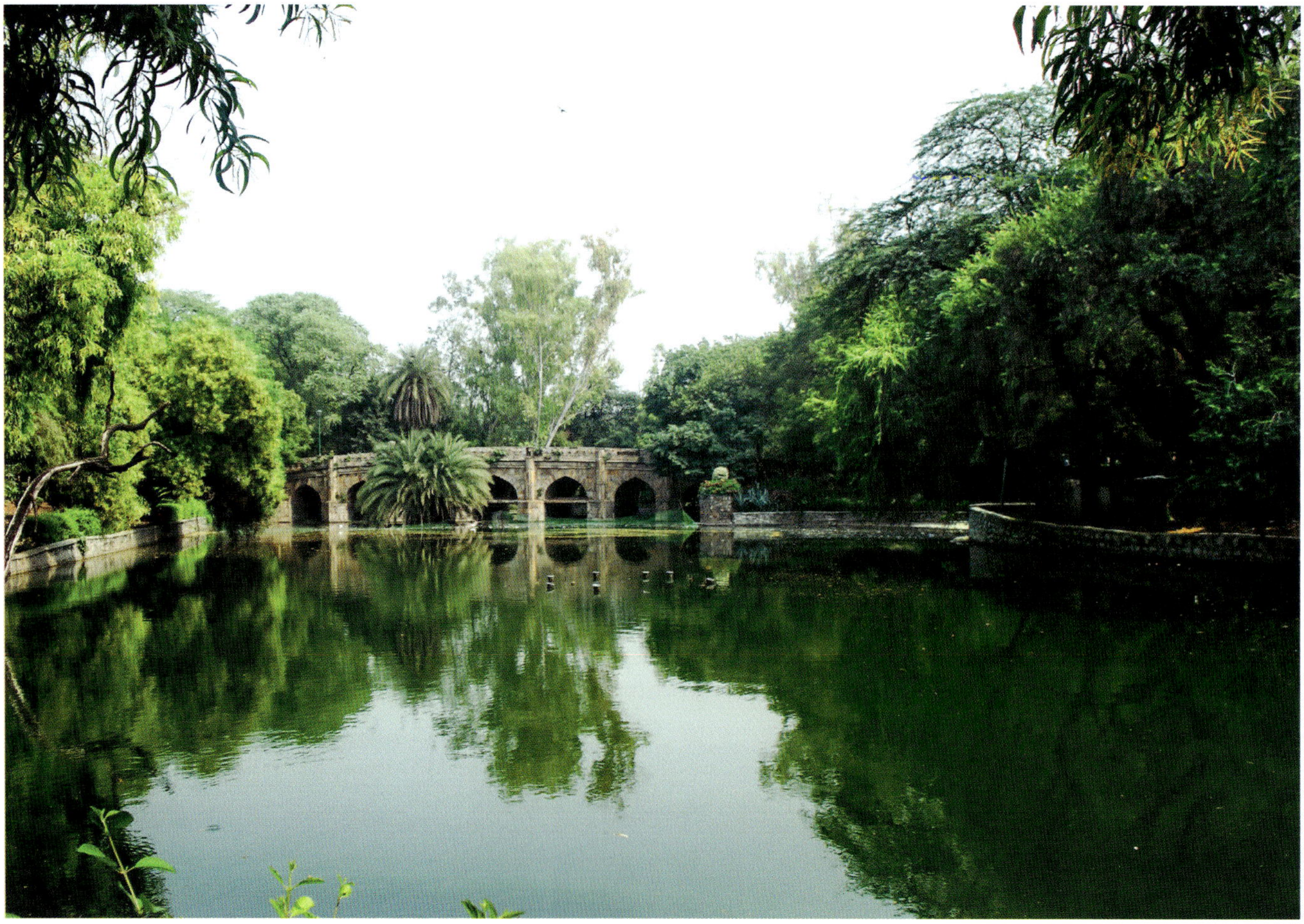

SOWMYA LAKHAMRAJU

design highlighted this by linking the central axis of Sundar Nursery to the layout of Humayun's Tomb.

The Sundar Nursery project celebrates the historicity of place; aesthetics of Mughal landscape design; and showcases that high levels of stone craftsmanship and traditional construction are still prevalent in India. This serves as a repository of past knowledge, a learning platform for the present, and a laboratory for the future through its contextual design vocabulary. The landscape approach ensures uniqueness. It attempts to connect people with their past and highlights the need to consolidate and link open spaces beyond their administrative boundaries.

Moving a step beyond this ensemble it is discernible on the map that Humayun's Tomb, as well as Sundar Nursery, is part of a larger cultural landscape currently fragmented and disconnected due to intense urbanization. Here is an opportunity to scale up this vision for green infrastructure by extending these two garden precincts to conceive a large urban park in central Delhi.

Adjacent to the Humayun's Tomb site and Sundar Nursery there are a host of open spaces, some historic and others that have emerged on the floodplains of the meandering river. These can potentially participate in this green initiative.

To the north of this ensemble is a vast stretch of the City Zoological Park spread over 69 hectares. A haven for flora and fauna, this ecological zone lies in the heart of the city. Northwest of the Zoo is the landscape of the Purana Quila (literally meaning old fort) covering 24 hectares.[3] Usually associated with the 16th century, regular excavations have established a continuous history of occupation on this land since 3–4th century BC.

Humayun's Tomb, Purana Quila, and many other historic fortifications were built along the river edge. Today they sit landlocked as the Yamuna River has changed course. As the river meandered, the flood plain that emerged was used to serve the needs of a bursting city. There appeared a railway line followed soon after by a city Ring Road, irreversibly severing the historic connection with water. As time passed, the flood plains across the railway line, visible from Humayun's Tomb, were used as a landfill site for dumping garbage as this area was then on the outskirts of the city. However, the urban boundary shifted with time and the space was reclaimed and converted into a series of parks. Smriti

Here is an opportunity to scale up this vision for green infrastructure by extending these two garden precincts to conceive a large urban park in central Delhi.

TOP LEFT: Lodhi Garden, incorporating tombs of the Lodhi dynasty (15th–16th century) as "follies" within a 20th century colonial garden. On this occasion being used for International Yoga Day 2019.
BOTTOM LEFT: Lake and old waterworks in Lodhi Garden.

Van (memorial garden), Millennium Park, and Indraprastha Park now add a new layer to the cultural landscape of Delhi.

Moving toward the cityside, westward from the Humayun's Tomb ensemble, a series of open spaces offer a further possibility of expanding a green network within central Delhi: Amir Khusro Park (6,5 hectares); Lodhi Gardens (redesigned as Lady Willingdon Park in 1936 covering 36 hectares); the Delhi Golf Course from the 1930s (89 hectares); and Safdar Jang Tomb (10 hectares).

The cultural landscape of Delhi is speckled with ruins of Lal Kot (11th century) and Mehrauli (13th century), crisscrossed with fortifications of the historic cities of Siri, Tughlaqabad, Jahanpanah (14th century), and Purana Quila (16th century). There are remnants of ancient hydrological structures of Hauz Khas (early 14th century) and Satpula (mid-14th century), and the various "baolis"[4] in an arid landscape. The terrain is home to sacred Sufi saints' shrines, burial grounds of royalty from the Lodhis (mid 15th to early 16th), Sayyids (15th century), Mughals (1526–1857), and Safdar Jang (1754), including the 19th and early 20th century colonial additions and alterations.

Today, the green areas and open spaces that go with the monuments and shrine, discussed in this article, sit disconnected due to roads and housing traversing the landscape. By restoring and connecting them with green infills, dampening traffic, and other such measures, it would be possible to create a large urban cultural heritage park, stretching from the former banks of the Yamuna River up to the Safdar Jung Tomb. It has been over two decades since the work at Humayun's Tomb project began. Delhi comprises many more green areas that should be brought into a green master plan, just to mention a few: the Delhi Ridge; the Central Vista; the many broad, green avenues and the early 19th century layout for New Delhi by Sir Edwin Lutyen. With foresight and planning it will be possible to uncover and present Delhi's rich cultural heritage and at the same time create a healthier city.

[1] The AKTC project is led by Conservation Architect Ratish Nanda, who has headed AKTC in India for two decades. Since Prof. Shaheer's demise in 2015, Yogesh Kapoor has been managing these projects from Shaheer Associates.

[2] The Grand Trunk Road is one of Asia's oldest and longest major roads—founded around 3rd century BC by the Mauryan Empire of ancient India.

[3] Area sizes mentioned in this article are indicative only.

[4] A baoli is a well that has been dug with enormous or many staircases leading down to the water surface at the bottom. Also called stepwell.

RIGHT: Safdar Jang Tomb.

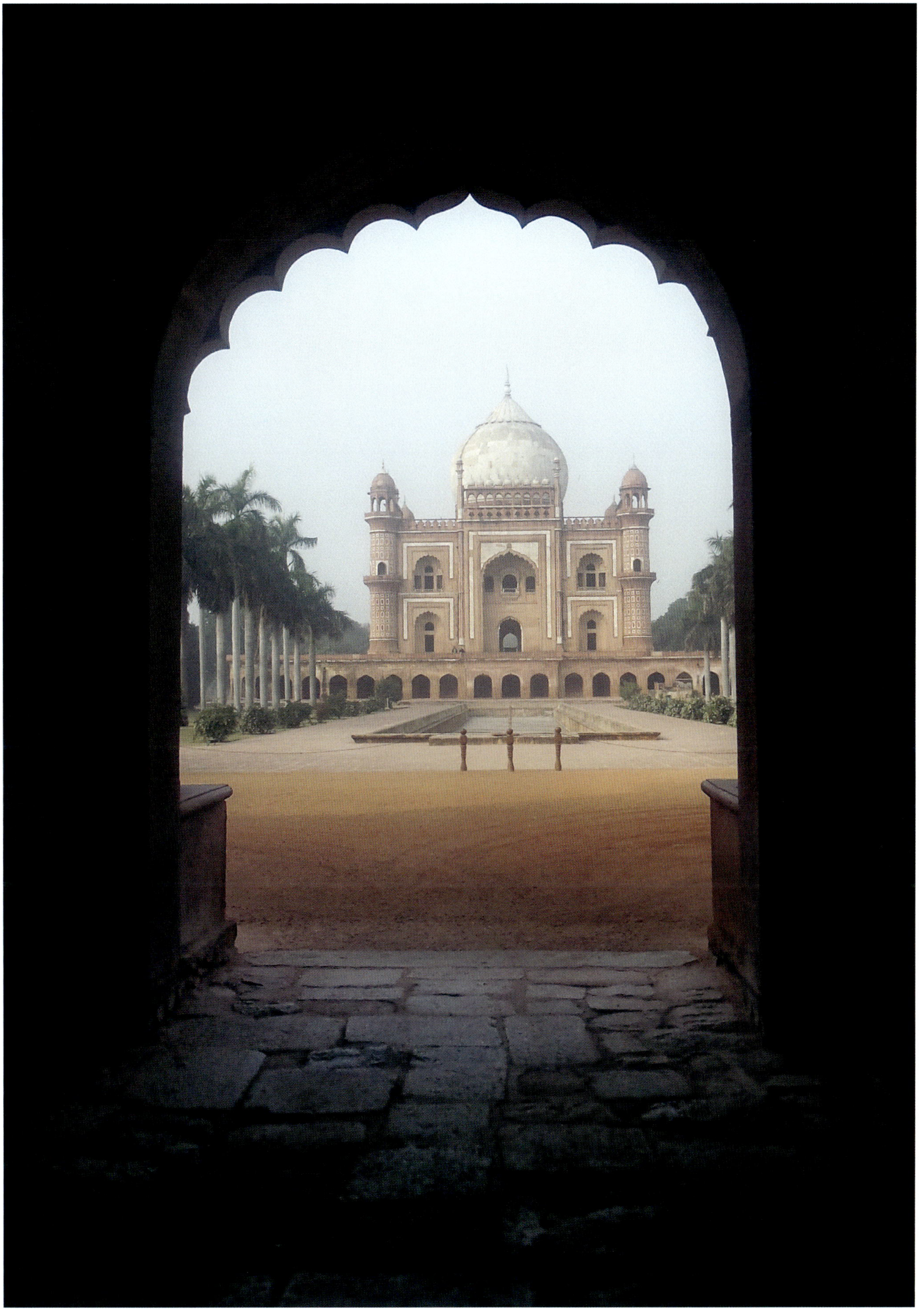

NUPUR PROTHI KHANNA

SUBIACO

Railway Parade

Mueller Park

Hay Street

Bagot Road

Railway Road

Nicholson Road

SHENTON PARK

Thomas Street

Kings Park Road

WEST PERTH

Malcolm Street

Lotterywest Family Area

Rio Tinto Naturescape

Pines Picnic Area

Saw Avenue Picnic Area

May Drive

KINGS PARK

Mounts Bay Road

Mitchell Freeway

Synergy Parkland

Lovekin Drive

Bushland nature trail

State War Memorial

Bodja Gnarning Walk

Broadwalk

Western Australian Botanic Garden

May Drive

Narrows Bridge

Winthrop Avenue

Kwinana Freeway

Forrest Drive

Mounts Bay

SWAN RIVER

Saint George's College

Mounts Bay Road

CRAWLEY

Matilda Bay

Innaloo

Morley

KINGS PARK

Airport

Nedlands

PERTH

Swan River

N

0   250   500 Meters

0   0,3 Mile

© OpenStreetMap contributors

KINGS PARK, PERTH, AUSTRALIA

SIZE: 4.1 square kilometers.

PROTECTION: A-class reserve 1872, extended 1897.

OWNER: The Crown.

MANAGED BY: Botanic Gardens and Parks Authority (BGPA)(https://www.bgpa.wa.gov.au/kings-park).

VOLUNTEER ORGANIZATION: Friends of Kings Park (http://www.friendsofkingspark.com.au).

HOW TO GET THERE: On foot 3 kilometers from central business district, by bus, by car (parking available).

ISAAC MIDDLE
Kings Park, Perth, Australia

What do visitors do when they come to Perth, one of the least-dense, suburbanized cities in the world (Newman 2014)? Indeed, Perth's perceived lack of culture has led it to be called "Dullsville." But Perth has its own special uniqueness that is not to be found in its social and human-constructed environment but rather in its natural environment: the beautiful, if occasionally dangerous, beaches; the iconic, if endangered, Swan River; the immersive, if challenging, bushland. These elements are brought together into a single landscape experience in the city's large urban parks, with no park better achieving this than Kings Park. For visitors to Perth, as well as for its own inhabitants, Kings Park provides excellent possibilities to learn about the Australian landscape, its flora, fauna, and cultural history. Park administration and the community group Friends of Kings Park have made the most of this magnificent resource.

The creation of Kings Park ... was a remarkable act of foresight by Perth's colonizers.

As recounted by Erickson (2009), the creation of Kings Park, which is centered around Mount Eliza and directly overlooks both the Swan River and a then fledgling colonial settlement, was a remarkable act of foresight by Perth's colonizers. The land that Kings Park now occupies was first identified for its public value in 1830 by Perth's Surveyor-General John Septimus Roe—an act for which he is now immortalized with the eponymous Roe Gardens at the southern end of the Botanic Gardens. Under the approaching threat of continued urban encroachment to the west of the central business district, the land on Mount Eliza was officially set aside for public use in 1872. Early on there were several proposals to privatize and develop parts of the park, but they have all been met with significant community resistance, indicating the value of the land to the general public, and thus have been defeated. From its original size of 175 hectares, Kings Park was expanded in 1897 to its current size of 411 hectares.

As it has turned out Kings Park now extends into the very city center. City development eventually surrounded the park and cut it off from other green areas, thus threatening biodiversity in the long term.

The beginning of the park was the creation of a Victorian strolling garden next to the city center, favoring large and contoured open spaces, with the addition of well-maintained plantings and water features. Eventually, the Botanical Garden was created and opened in 1965.

This can be interpreted as a nod to another typical Victorian park element: the promotion of values deemed essential to civic societies such as art, history, science, and the natural world (Taylor 1995). The Victorian heritage is discernible not only in the name of this park but also in the names of two other parks in Perth; Hyde Park and Queens Gardens.

Since 1919 numerous war memorials have been incorporated into this part of the park. There are 50 individual statues and memorials within the park boundaries and an "Honour Avenue," lined with planted trees bearing plaques memorializing individuals who have died in service.

More than two-thirds of the 411 hectares are remnant bushland. The ecological value of this conserved bushland is significant, with Perth being located in one of only 34 global "hotspots" of biodiversity, namely, the South-west Australian Floristic Region (Grose 2011). The bushland in Kings Park is composed primarily of Banksia species, often visually stunning, although a total of 324 individual flora species have been recorded—representing 15 percent of the total flora of the Perth region (the Botanic Gardens contain over 3000 varieties of indigenous flora). Another ecologically significant biotope is limestone heathland on the southern ridge of the park, rising to 65 meters above the Swan River. Such areas are poorly represented in Perth and are the sole habitats to several invertebrate and plant species.

Grose (2011) outlines several risks facing ecological areas in Perth, including the growth of the root-plant pathogen *phytophthora cinnamomi* (commonly known as "dieback"); the changing climate and droughts (predicted to be particularly acute in southern Australia); and of course the ongoing threat of urbanization and fragmentation. As further outlined by Dixon and Moonie (2003), Kings Parks' remnant landscapes—which are completely isolated within the urban landscape from other bushland areas—have undergone significant additional disturbance and degradation. This includes the introduction (both deliberate and accidental) of invasive plant species and the extraction of raw materials. This has not only affected biodiversity but damaged the integrity of the limestone escarpment itself, and led to fears of rocks falling dangerously onto the heavily-used Mounts Bay Road. This initiated a 4-year restoration effort, completed in 2003, comprised of extensive weed removal, slope stabilization, and revegetation (Dixon & Moonie 2003).

A 2014–2019 Management Plan for the entire Kings Park and Botanic Gardens area identifies the practices necessary for maintaining Kings

Perth is located in one of only 34 global "hotspots" of biodiversity.

TOP RIGHT: Perth with Swan River and Kings Park stretching right into the city center.
BOTTOM RIGHT: The Water Garden integrates important historical and cultural educational elements—in this case, the Pioneer Women's Memorial.

BENNY MARTY/SHUTTERSTOCK

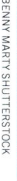

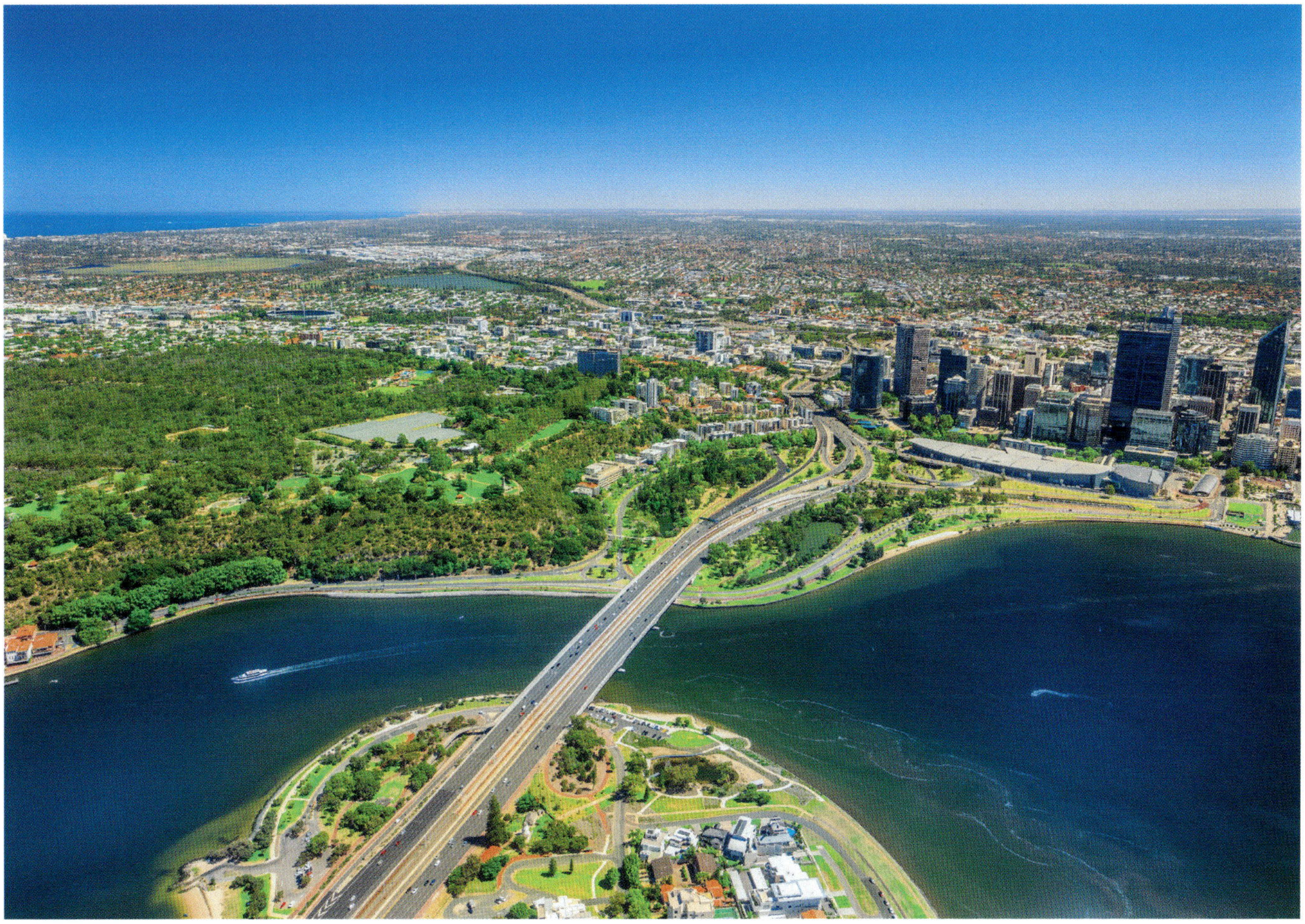

ISAAC MIDDLE

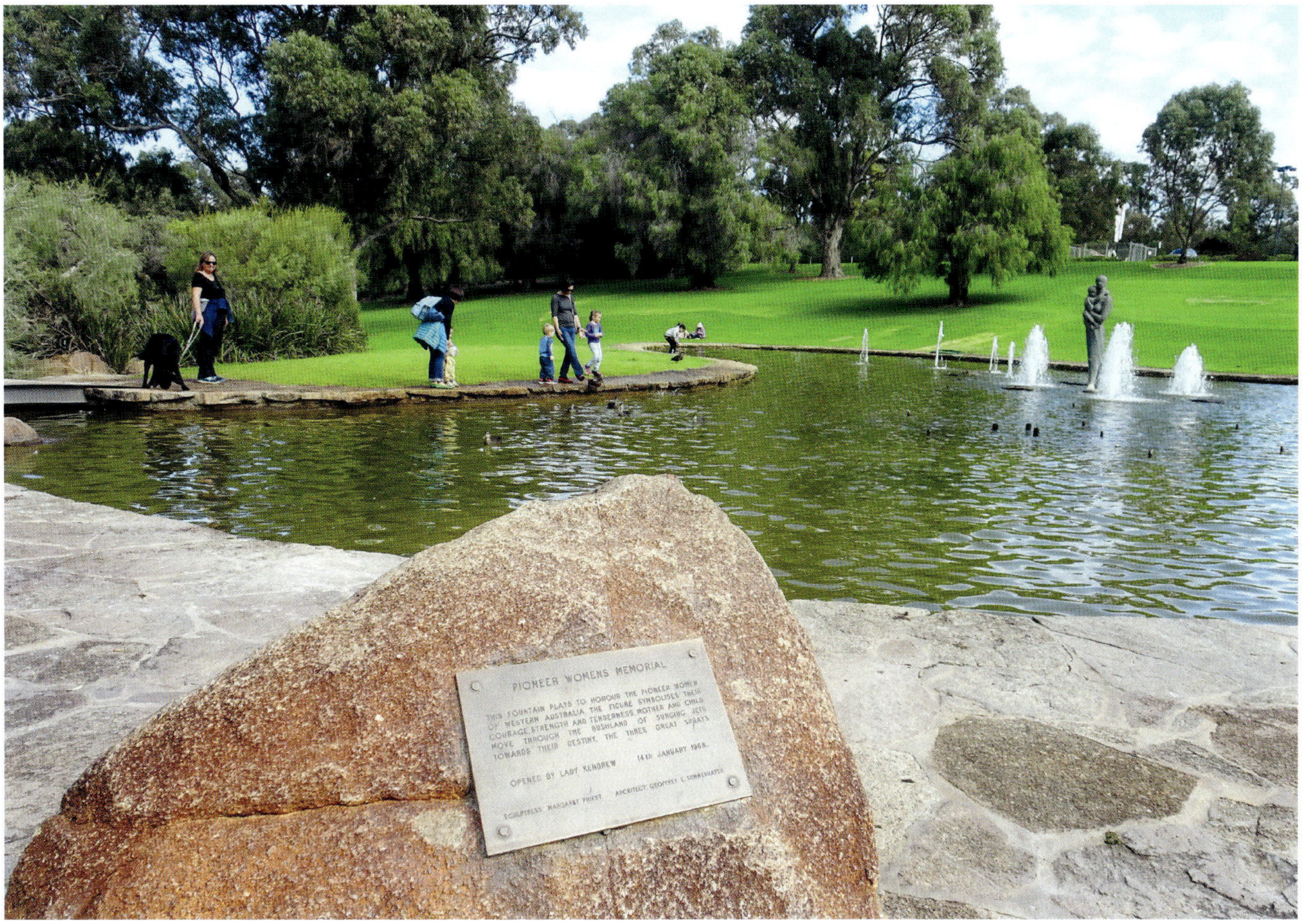

Park as an eco-sanctuary for regionally-significant remnant bushland. The community group Friends of Kings Park is recognized in the plan as providing the governing body with essential assistance in the ongoing management of the park's ecological areas. The group also coordinates and contributes to educational, research, fundraising and publicity activities. At last count, there were over 1,300 paying members; active members were estimated to have contributed over 32,000 hours of volunteer work in 2015/16 (Botanic Gardens and Parks Authority 2016).

The park serves a broader social function: it is a valuable (re)connection with the natural environment. Hence, it helps ameliorate the loss in the understanding of ecological areas by urban residents and it gives the residents a greater understanding of the numerous benefits these areas provide (Miller 2005). Kings Park uniquely and significantly plays this role, being located directly adjacent to high density, inner-urban residential areas with little other access to green space.

Several walks take the visitors on educational tours. One is the Boodja Gnarning Walk, displaying how the Nyoogar people, who lived in this area, used the land for food, tools, medicine, and shelter. The story is told about Waugal, a snake-like rainbow serpent that formed the rivers and lake around which Perth was settled, before retiring to let its body create the Darling Scarp ridge.

While there are many informal bushland paths for the adventurous, they may not be for everyone. For example, several venomous snake species—notably the dugite (*pseudonaja affinis*) and tiger snake (*notechis scutatus*)—are common in this bushland, although they are rarely aggressive unless provoked. Also to be found are several intimidating, yet harmless spider species, such as the Australian golden orb-weaving spider (*nephila edulis*), often to be seen within their impressive webs adjacent to paths. For those more cautious, the Bushland Nature Trail directly west of the Botanic Gardens provides the best entry point to this environment. Another well-maintained path is the Law Walk, which comprises a 2.5 kilometer loop along the southern edge of the park. Complete with explanatory signage, walkers are informed of the local and global significance of the bushland.

Environmental learning is facilitated more actively through a series of formal, hands-on educational programs (Kings Park Education) for school students of all ages. Many utilize the unique interactive environment of the Rio Tinto Naturescape.

LEFT: Baobab tree at Botanical Garden in Kings Park.

Design for All

Cities of the world are about to double in size. In that process, we must not miss the opportunity to build resilient cities, both compact and green. For ages, nature and geology have commanded some respect. Rocky hills, marshes, and lakes have been left untouched. With present day construction skills this is no longer so. Now it depends on us to pay nature due respect, and use our skills with foresight.

JULIA CZERNIAK

Large Parks: Trends (and Possibilities)

The sheer number of large parks being planned, designed, and built around the world today speaks to the important role this beloved landscape type plays, both in our physical environment and in our collective imagination.

What parks are, how they look, and the roles they play in cities have advanced significantly since the time of Frederick Law Olmsted. When, in the late nineteenth century, the preeminent American landscape architect used the word "park," he referred to a large tract of land on which one could produce the effects of a rural landscape for public use and enjoyment. Olmsted distinguished a park from a public square or promenade, which he reserved for more urbanized pleasure. He insisted that his design for New York's Central Park encompass at least five hundred acres (202 hectares), believing this to be the smallest space in which the beauty of nature, together with its healing properties, could be designed in a picturesque geometry. This large size was necessary to fulfill what was then understood as a park's primary role — to provide an image of green and the effects of freshness as an antidote to the industrial city. In accordance with this thinking, Central Park would swell to 843 acres (341 hectares).

Times have changed. Confronting ecologically and culturally disturbed sites such as landfills and deindustrialized parcels requires designers today to challenge the green veneers of the past and advance alternative strategies that ecologically heal the landscape while retelling the complex narrative of a place. Facing the social, ecological, and economic pressures of resilient design, landscape architects have become experts at advancing design approaches that balance fixed elements (such as buildings, circulation patterns, and infrastructure) with those which are flexible, dynamic, and changing (such as wildlife paths, water flows, and seed distribution, as well as resources and client desires). Accommodating the multiple, mutable, and often conflicting constituencies of a park has made eternal optimists out of designers, as regards to the social role of projects: most of us still believe parks to be reliable places for the unfolding of life, human and non-human alike. Furthermore, publics are increasingly engaged in envisioning, operating, and maintaining parks: today, parks are not just *for* us but also *by* us. Designers are increasingly innovating representational strategies, tools, and techniques to enable such participation. Not only does this make parks more useful, it catalyzes local citizens to be stewards of their environment. Together these considerations suggest an evolving shift away from highly tended, singularly controlled, wholly conceived parks to messier, emergent ones, capable of recreating a park's identity in the public imagination.

What follows is a brief survey of contemporary trends in park design. The parks I discuss work to provide services to their urban hosts as well as address significant global challenges, all the while advancing our appreciation of how much design thinking matters. These challenges revolve around water (too much, too little, or in need of cleansing); waste (cleaning it up, or limiting its production); diversity (both biological and social); and pleasure (sensory, didactic, and aesthetic). The sheer number of large parks being planned, designed, and built around the world today speaks to the important role this beloved landscape type plays, both in our physical environment and in our collective imagination.

WATER

The reality of a global water crisis is indisputable. Designers constantly face sites plagued by flooding, drought, and contamination. Organizations such as the World Economic Forum implore us to recognize water's value, use it efficiently, and plan for different futures (World Economic Forum 2020). That the design and planning professions are invoked in this plea is no surprise, as many large parks respond to a diverse set of water issues.

In the northeastern United States the flooding damage caused by Super Storm Sandy in October 2012 prompted the international design competition *Rebuild by Design*.[1] This storm event, and the design research that followed, foregrounded the statistic that 2.5 million people in the New York metropolitan area live in flood zones, many of which are at extreme risk for future damage. In their proposal for the redevelopment of the Meadowlands in New Jersey, the Massachusetts Institute of Technology's Center for Advanced Urbanism, along with Rotterdam-based ZUS landscape architects, used a system of green infrastructure —a mix of existing and proposed wetlands as well as basins, berms, and gates—to protect the site against storm surges. At the center of their proposal is the Meadowpark, composed of 7,800 acres (3,157 hectares) of existing wetland, both tidal and freshwater. The team integrated berms into the wetland system, which allowed them to create a distinction between high and low marshes. Behind the primary protection berm, fresh water marshes and forests play an active role in storm-water management and, equally importantly, provide highly enjoyable places for the public to spend time. This new urban park is not just for water management: it is an infrastructural system around which the community can grow. It is a catalyst that can change people's perception of this place. Historically, the Meadowlands have been perceived as a liability —as a backwater, filled with sewage and waste. Here they are reimagined as an asset: the prized center of an emergent community.

On the western coast of the United States the role of large parks is different. Climate scientists say California faces decades of record drought that will require political will and public investment to address. Designers know that a large park design will also have a positive impact. The 4,000-square-mile (10,360 square kilometer) Los Angeles River Basin is protected by networked flood control channels that carry most runoff water to the ocean, often swiftly and in high volumes. So, in a region that is water hungry and park poor, designers are testing ways to manage what is both a dangerous flood hazard and a precious resource.

The Los Angeles River Revitalization Master Plan, led by a team headed by landscape architect Mia Lehrer, envisions the transformation of a 32-mile (51 kilometer) stretch of concrete-lined river and its adjacent neighborhoods, as both a conduit for the distribution of water and a riparian environment in its own right. This public, green network focuses on water quality and flood protection, while at the same time providing trails and bikeways along the length of the river. One of the more imaginative components of this plan is what the designers call "ecosystem terraces": these stepped landforms slow water, clean it, and (through a system of side channels) collect it underneath the terraces for subsequent reuse.

In China, the water story is radically different. Rapid economic development has taken an environmental toll. As cities grow, parks have begun to play a key role in managing dirty and degraded water. Houtan Park in Shanghai, designed by Kongjian Yu of Turenscape, deploys conventional strategies for cleaning water in innovative and beautiful ways, offering an exemplary approach for large urban parks. This demonstration project is located on a narrow, inaccessible site along the very polluted but nutrient-rich Huangpu River. With the ambition of transforming the site into a living system, the team designed a series of cascades and terraces that both oxygenate the water and reduce suspended sediment. The various species of wetland plants included in the 36-acre (14.5 hectare) demonstration park absorb pollutants from the water,

TOP RIGHT: View of the proposed Meadowpark west of Manhattan, New York.
BOTTOM RIGHT: The New Meadowlands Project, the center of which is the Meadowpark. East of the Hudson River lies Manhattan, in which Central park is depicted.

ZUS

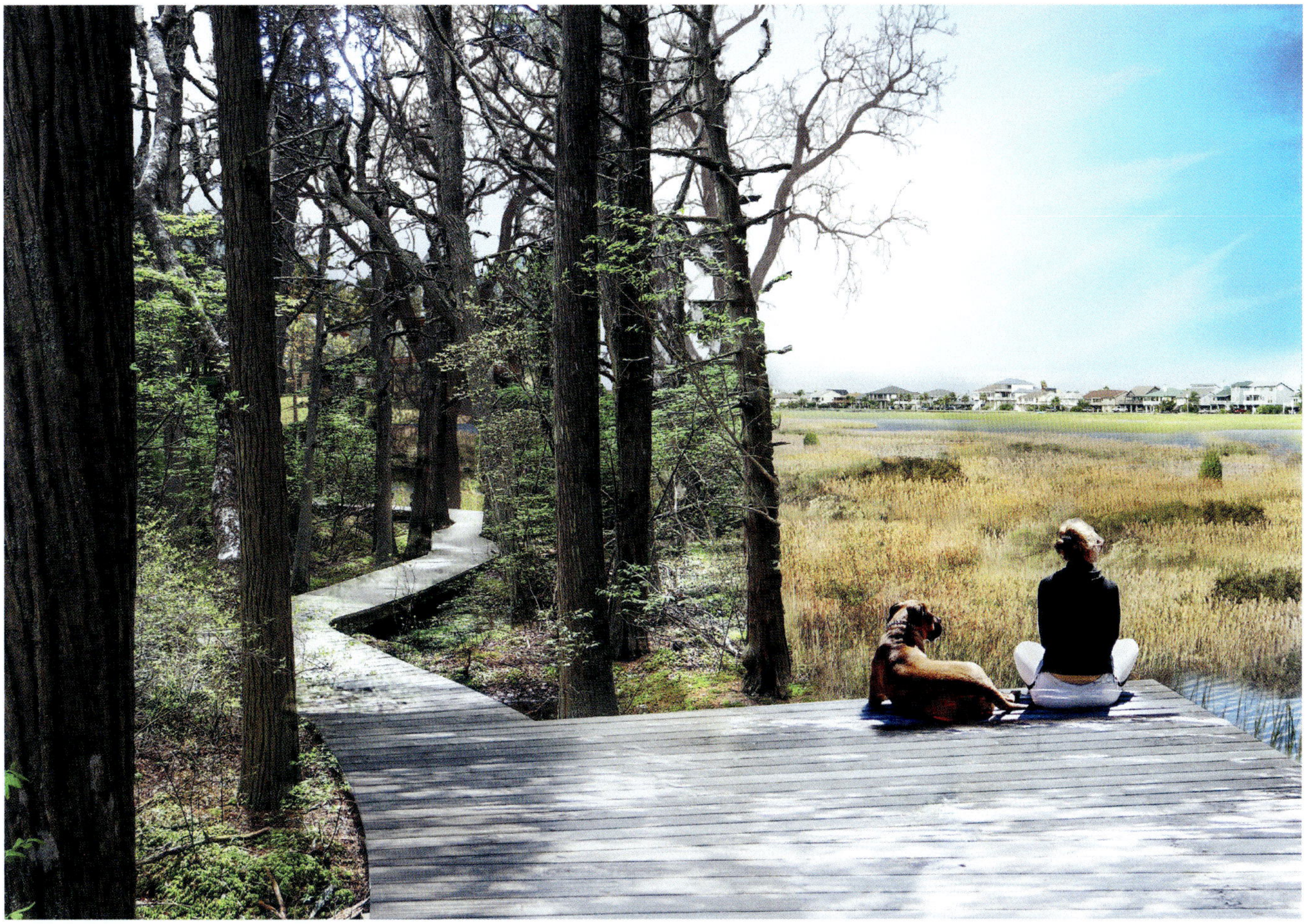

ZUS

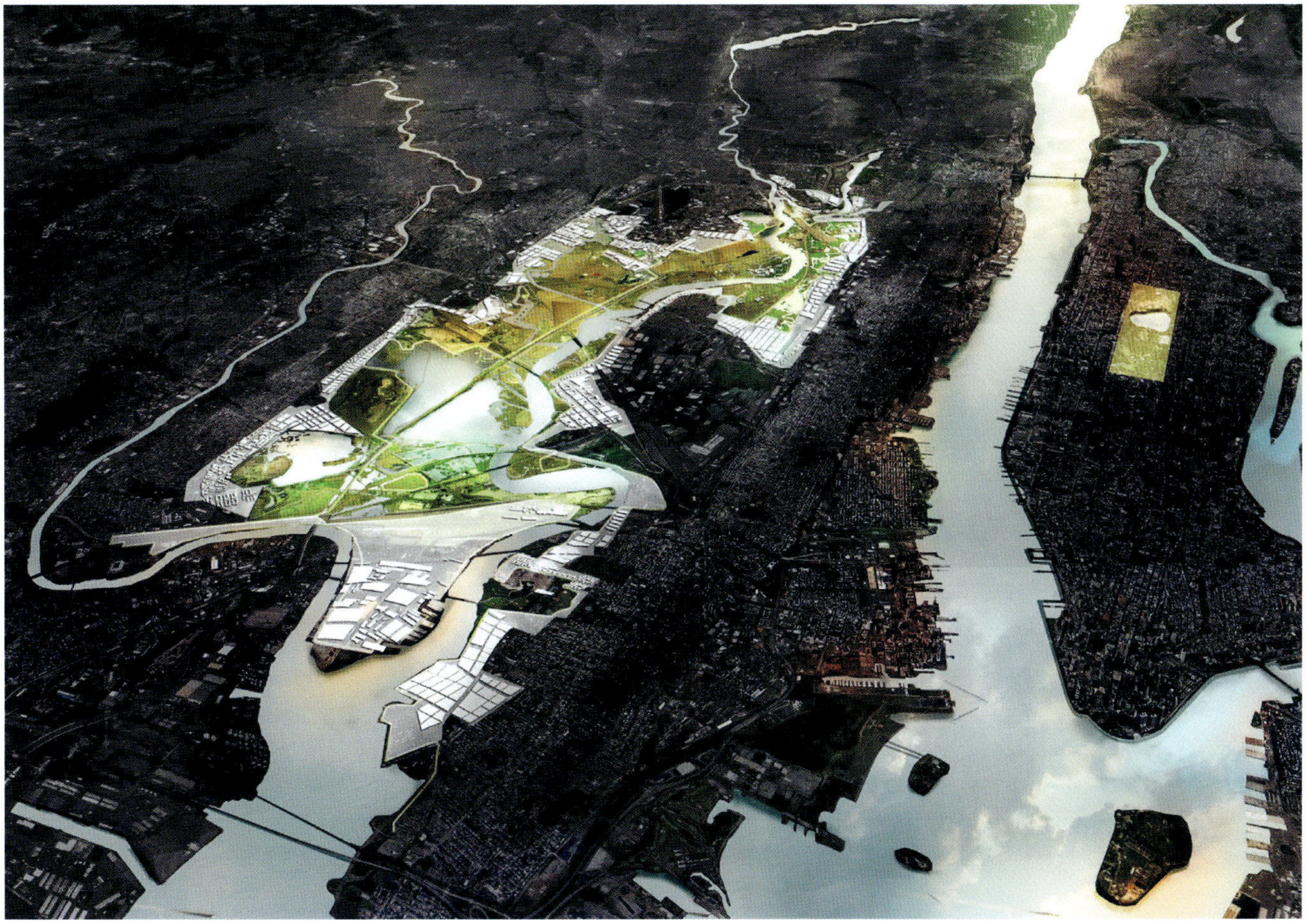

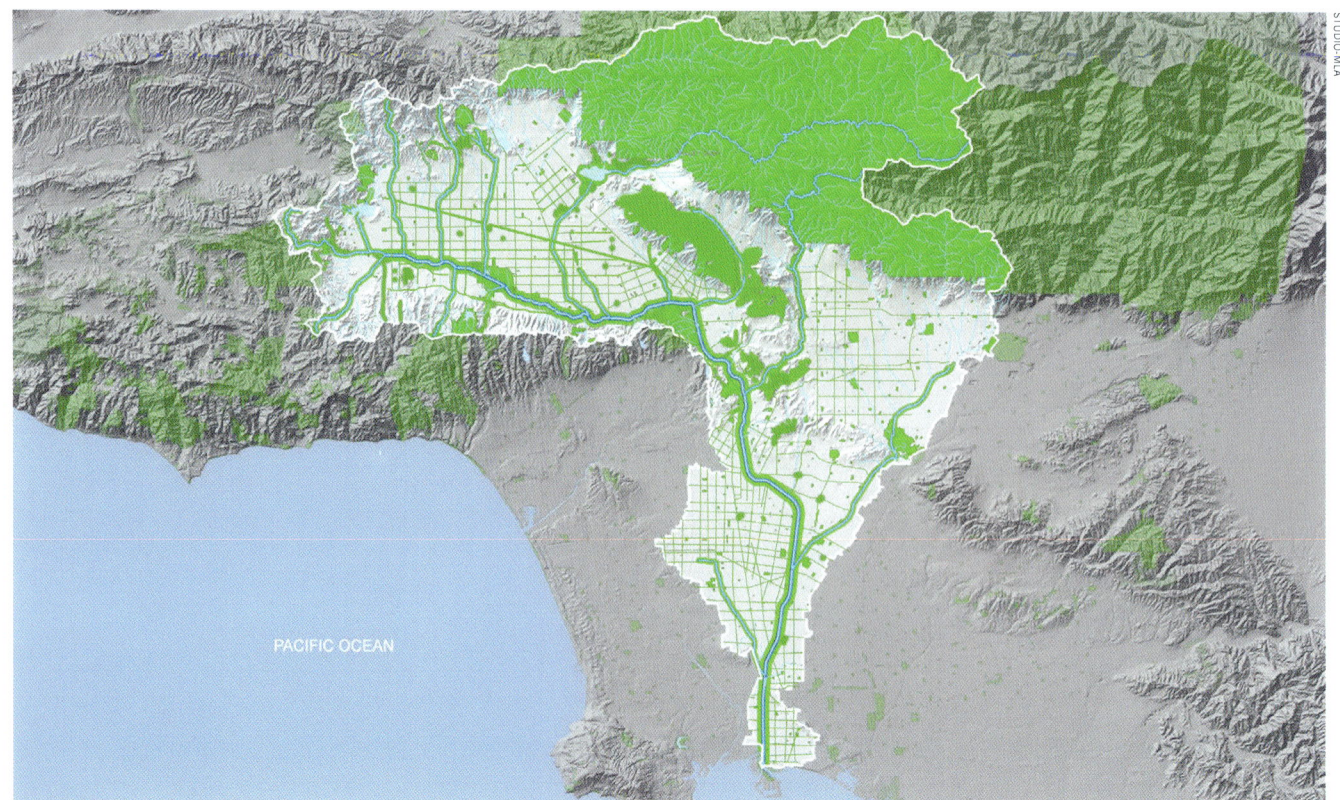

PACIFIC OCEAN

while additional crops such as sunflower and rice filter excess nutrients and enhance the plant palette. Cleaner water is fed back into the river downstream and utilized to create pleasant water features.

In a much larger park—Qunli Stormwater Park—Turenscape gives these strategies a different form. At the center of a new urban district in the Heilongjiang Province of China the firm transforms a wetland that had been cut off from adjacent water systems into a prized public park. Using a cut-and-fill technique, the firm produces a system of ponds and mounds around the circumference of the wetland that collect storm-water, which is cleansed, stored, and then released into the aquifer after filtration. Native wetland grasses and meadows grow in the ponds, and groves of silver birch trees reinforce the park's edge. Pathways, plat-forms, and towers set into this landscape enable interaction between people and the park, and a skywalk affords residents views from above (see separate article).

Juxtaposing this twenty-first century park and its nineteenth-century counterpart, Central Park, two very different visions of nature emerge. In Olmsted's project, nature is carefully constructed—and maintained. More than a century later, in this contemporary Chinese example in particular, but evident as a trend globally, nature is understood as an ongoing process, where landscape is embraced as a messy medium. Not only does the park serve its urban host by cleaning its precious water re-source, it provocatively defines the new aesthetics of urban parks today, lending identity and distinction to the communities that use them.

WASTE

The second trend in contemporary park design involves the rather alchemical transformation of wasted landscapes—more specifically, closed landfills and deindustrialized lands—into public amenities. Remarkably, a World Bank report argues that the growth rate of mu-nicipal solid waste, a by-product of contemporary life, has outstripped that of urbanization (Hoorneg and Bhada-Tata 2012). Park designers have the opportunity not only to provide a public amenity but also to raise awareness about the global waste crisis.

Park Ariel Sharon leverages its siting on a waste landscape in robust and visible ways that other similar projects (such as Field Operations's Freshkills Park) do not. Begun by Peter Latz+Partners, part of this

TOP AND MIDDLE LEFT: View of the proposed "eco-system terraces," before and after. BOTTOM LEFT: The open space strategy for the Los Angeles River Revitalization Master Plan (LARRMP) included the LA River, its tributaries, green streets, and regional and local parks and natural areas.

2,000-acre (809 hectare) park is built atop a 60-meter-high mound called the Hiriya, which covers 25 million tons of waste. Located on the Ayalon Plain outside Tel Aviv, the mound appears as an unconventional topographic feature, encouraging the design team to nickname it "the mystic mountain."

The designers kept the exterior of this mountain intact, masking its true identity, which thanks to a robust investment in recycling, offers us less that of a green image of nature and more of a working landscape. Amid a range of landscapes, recreational opportunities, and open spaces, one receives glimpses of the project's landfill rehabilitation strategies, nestled on the edge of the project's high plateaus. One facility produces bio-gas to create electricity. Another recovers methane gas as a source of revenue. A third facility recycles tires and building materials. Additionally, a garden adjacent to the Visitor's Center, on the eastern plateau, treats sewage with the help of bacteria-flourishing plant roots; the water is subsequently used for irrigation. In these ways, Ariel Sharon Park unapologetically advances recycling and composting strategies, managing waste that would otherwise end up in landfills like the one on which the park rests.

Freshkills Park, in New York City, indexes its waste landscape in subtler but no less important ways. Field Operations's design strategy privileges the park's constructed nature over any overt reference to its landfill past—a past that residents of Staten Island neighborhoods surrounding the vast waste site were more than willing to forget. Although at first glance the competition scheme can be understood as creating a pastoral landscape, the project thoroughly outlines a succes-sional strategy aimed at establishing a landscape ecology. By remediating the soil, stabilizing the slope, removing invasive plants, establishing grassland, introducing woody material, and adaptively managing the site over time, the project is clearly not a naturalism (Czerniak 2007).

Moving on to Germany's Ruhr Valley, Emscher Landscape Park provides an example of a vast design enterprise. Here, seven greenbelts and more than two hundred redevelopment projects have together transformed a wasteland, replete with scars of heavy industry, into a rich landscape mosaic. Plants, animals, and people are returning to the region, amid monuments to the area's industrial heritage reinterpreted by designers. The wild beauty emerging between collieries, blast furnaces, spoils, and other industrial ruins is quite extraordinary.

TOP RIGHT: Ariel Sharon park under construction just outside of Tel Aviv, Israel.
BOTTOM RIGHT: LIFESCAPE, Fresh Kills Landfill to Landscape competition entry.

ARIEL SHARON PARK COMPANY, NIV MOSMAN

JAMES CORNER FIELD OPERATIONS

Gardens within the industrial ruins of Duisburg Nord.

One of the seminal projects that reuses regional artifacts in innovative ways is Landschaftspark Duisburg Nord, designed by Peter Latz+Partners. This well-documented project transforms the former Thyssen Iron & Steel Works into a public park. What is particularly intriguing about this healing landscape is the fresh way in which the site's industrial detritus is imaginatively reclaimed for unconventional uses. The site's gasometers, for instance, are transformed and reprogramed. These tanks, which previously stored blast furnace gas (a highly toxic by-product of pig iron production), have now been filled with more than 5 million gallons of water to create a unique diving center. In addition, the site's former ore bunkers—with their thick and slightly canted masonry walls—now provide challenging surfaces for climbers to test their skill. And bucolic strolling gardens hidden within a former sintering plant radically juxtapose the site's productive and destructive past with its restorative present.

These industrial artifacts are not the only form of leftovers that inspire design innovation. Another form of waste—slag heaps, or artificial mounds formed with a by-product of coal mining—radically transform the territory and index its productive, capitalist past. The topographic shift afforded by one of the largest slag heaps in the Ruhr Region, the Halde Beckstraße, is exaggerated by an art installation called the "Tetrahedron." This three-dimensional, climbable steel pyramid extends from the landform, offering uncommon and breath-taking views of the Ruhr Valley. Such work literally transforms the geomorphology of the region: by reusing, rather than eliminating, the slag heaps, this landscape practice plays a key role in producing new perceptions of—and new identity for—this once blighted territory (see Peter Clark's article).

It is not only the design of parks, however, that matters here; it is also their management. More and more designers are designing with contemporary maintenance practices in mind, in an effort to generate less waste. For example, in the Royal Parks of London—and in particular, Hyde Park—a comprehensive turf management strategy advances three different mowing regimes that require varying levels of care. The tall grass meadow is mowed one time per season; a conservation lawn is mowed six to eight times per year; and a mowed lawn is cut weekly. This approach not only favors a biologically diverse landscape but also reduces energy use. In New York's Central Park, shifts to equipment like mulching mowers save having to gather and bag tons of organic debris that is actually healthy to leave in place. Similarly, in Chicago's Millennium Park the use of "big belly" trash compactors reduce labour and expense by relieving crews from regularly emptying trash. Lastly, in Houston's Discovery Green, the use of on-site generated solar power saves on construction material, in that there is no longer a need to run electrical conduits throughout the site. These seemingly small measures have significantly reduced the amount of waste generated in and by public parks.

DIVERSITY

Noted biologist Edward O. Wilson labels the current rate of species loss "alarming." If human disruption continues at its current rate, he warns, Earth could lose half of its life forms by the year 2100 (Wilson 2003). In this context, the aspiration to design for non-human species and to biologically diversify our environments is one of the most exciting trends in contemporary park design. The landscape architecture practice SCAPE is at the forefront of this trend. Its ongoing work to foster a return of the oyster population to New York Harbor began with their

OYSTER-TECTURE, a vision of oyster reefs in New York harbor.

project Oyster-tecture, which not only nurtures a bivalve community but does so as part of a public landscape and participatory process. The project's major element, a living reef, is constructed from a field of piles connected by a web of "fuzzy" rope, woven by community members and intended to support oyster and mussel growth. By harnessing the biotic processes of oysters, mussels, and eelgrass, the park's emergent soft infrastructure will cleanse millions of gallons of water.[2] Of equal importance, it will establish new relationships between human and non-human species, as well as between New Yorkers and their waterfront, through the hosting of programs and events. In this way the project diversifies not only biologically but socially, as more and more constituents become active stewards of their environment.

SCAPE's Living Breakwaters project expands on these ambitions. Conceived as a living infrastructure to protect Staten Island from storm surges and flooding, the designers shun the conventional sea wall (which divides people and water) and in its place propose a necklace of breakwaters to buffer against wave damage, flooding, and erosion. Micropockets of habitat complexity that they term "reef streets" host a diverse array of fish and shellfish while also effectively protecting the shoreline. The project's design process—like that of Oyster-tecture—is predicted on the belief that involving students at local schools and engaging neighbors in adjacent communities is an essential component of nurturing bio-diversity. Not only do they participate in designing the project and making some of its elements, they will more likely care for the park by understanding the local environment and feeling invested in it. Both projects offer visions for what SCAPE calls "watery" large parks that not only address issues of water quality, rising tides, and community-based development but actively conceive of non-human species as user groups worthy of design's attention.

Hardberger Park, in San Antonio, Texas, exemplifies a land-based park design driven by biodiversity. Located at the convergence of three Texan ecoregions –plains, prairie, and plateau—the park's 300-acre (121 hectare) site affords the project a potentially rich diversity of flora and fauna. Having previously accommodated a dairy farm, however, the landscape "as found" was homogeneous. Surrounded by suburban development, it was also severed in two by a parkway. The designers' – Stephen Stimson Associates in collaboration with Julie Bargman's D.I.R.T. studio—approach to this park aimed to restore the native landscape.

They dedicated only a small portion of its acreage to low-impact recreation (trail systems, playing fields, playgrounds, and picnic areas) and to paying homage to the site's agricultural past; the majority of space was given over to reintroducing biodiversity (preserving heritage oaks, reestablishing woodlands, and reintroducing the endangered oak savanna). To accomplish this end, they followed a unique process: rather than design on paper, the team conducted extensive field work and made many design decisions on site, in order to locate microclimates, maximize habitat, and minimize human occupation.

Thinking about park design in this way—through a genetic rather than a spatial lens—constitutes a brilliant approach to increasing biodiversity. Here, in what the team calls a "cultivated wild" of native plants, the non-human species are those who are free to wander as they please. Facilitating their access is a land-bridge that, once built over the parkway, will serve as a wildlife corridor for the ecoregion's varied species.

PLEASURE

Attention to each of the aforementioned imperatives in park design— valuing water as a precious resource, recognizing waste as a global problem, and engaging with species other than our own—has thankfully not been employed by designers at the expense of the pleasure that well-designed parks give us. Early public parks were often referred to as "pleasure grounds." Here I define "pleasure" quite broadly, as the enjoyment of spaces through the uses they enable, the sensory and aesthetic experiences they provide, and the things they teach us.

The gardens within Hargreaves Associates' Queen Elizabeth Olympic

Park in London provide all three types of pleasure. They provide treasured gathering places, act as cultural artifacts reflecting contemporary aesthetic concerns, and foster appreciation for and knowledge of all things botanical: plant appearance, behaviour, and care, from the economic to the edible. The plant species of the North Park, such as maiden grass, black-eyed susan, aster, and sedum, engage the senses with their rich color and textural differences, which vary from day to night and season to season in a garden comprised of diverse biozones. The gardens of the South Park engage the mind, referencing and teaching us about various epochs in plant collection by British explorers, such as trips to the southern hemisphere indexed by agapanthus, African love grass, and torch lily. The spatial distribution of the gardens in both parks enables a shifting relationship of people to plants. Sometimes plants are the foreground, the object of pleasure and inquiry by users. Sometimes they are spatial background for activities, initially intended to provide the setting to watch the Olympic Games.

Two projects by the Dutch landscape architects West 8 demonstrate the pleasure generated by well-designed parks that give urban residents access to valued assets otherwise invisible, inaccessible, and illegible. The competition site for Yongsan Park, located on a formerly U.S. military base in Seoul, Korea, has been inaccessible to civilians for over one hundred years. The firm's winning scheme reconnects two meaningful features—the mountains to the north and the river to the south—with a spine of, variously, green, open, and public spaces running through a new park. Along this path an artificial lake and enhanced topography together build an illusion of a naturalistic Korean landscape. When built, the park will restore both visual and physical access to the mountains and river to park users and (hopefully) allow these meaningful elements to once again serve as significant symbols of the city itself.

Similarly, West 8's enormously successful Madrid Rio project reclaims the banks of the River Manzanares. By Placing the M30 motorway that rings the city of Madrid within an underground tunnel, the designers were able to link significant urban spaces and to connect people back to this valuable, historic water resource. One of the project's many significant park spaces is the linear green space called Salon de Pinos which is located almost entirely on top of the motorway tunnel, and constituted predominantly by pine trees. The designers describe the pine tree groves, which are able to survive on barren rock in the wild and thus suitable in

TOP RIGHT: Queen Elizabeth Olympic Park, designed by Hargreaves Associates.
BOTTOM RIGHT: Aerial view of Madrid Rio project.

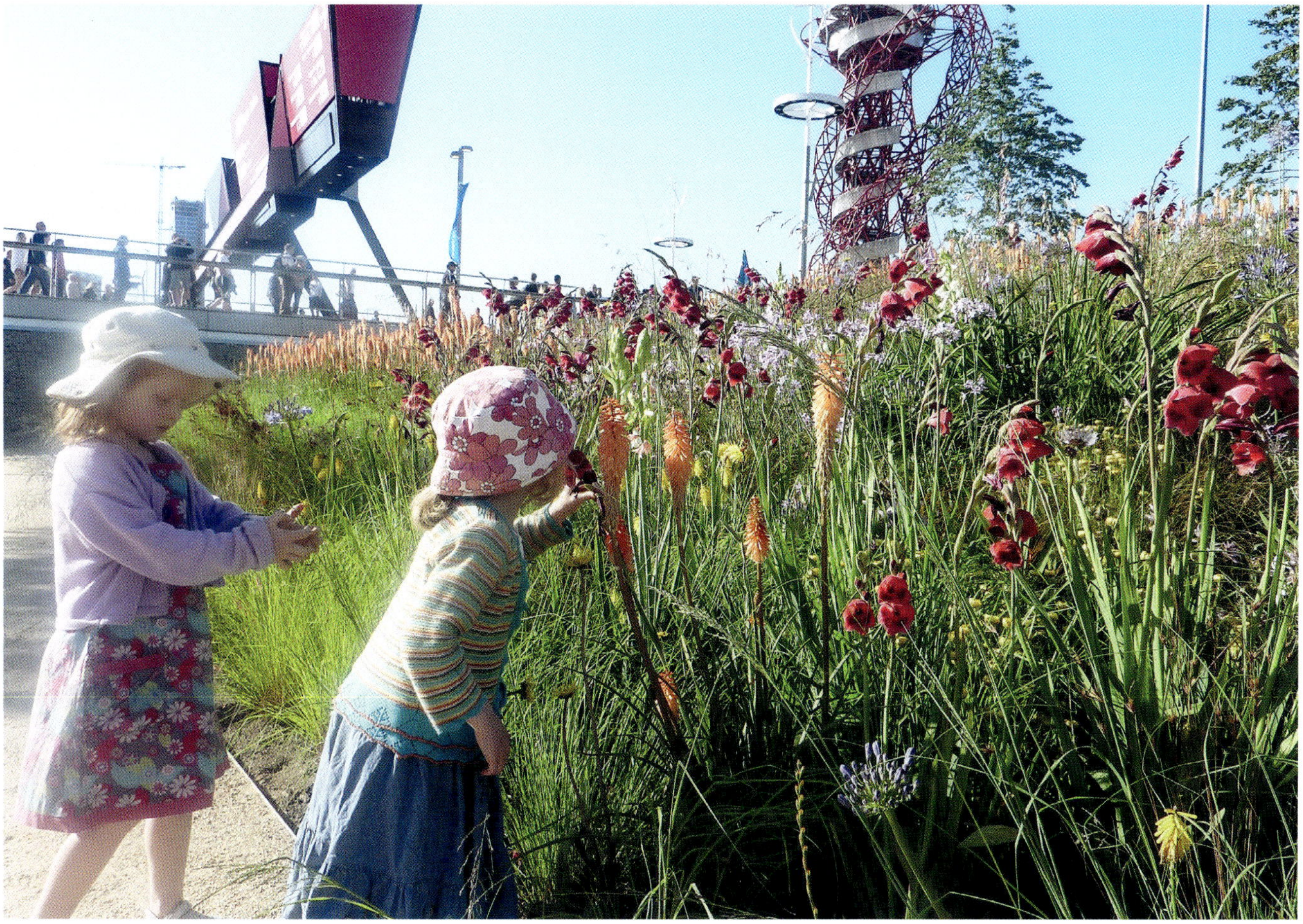
ANDY HARRIS

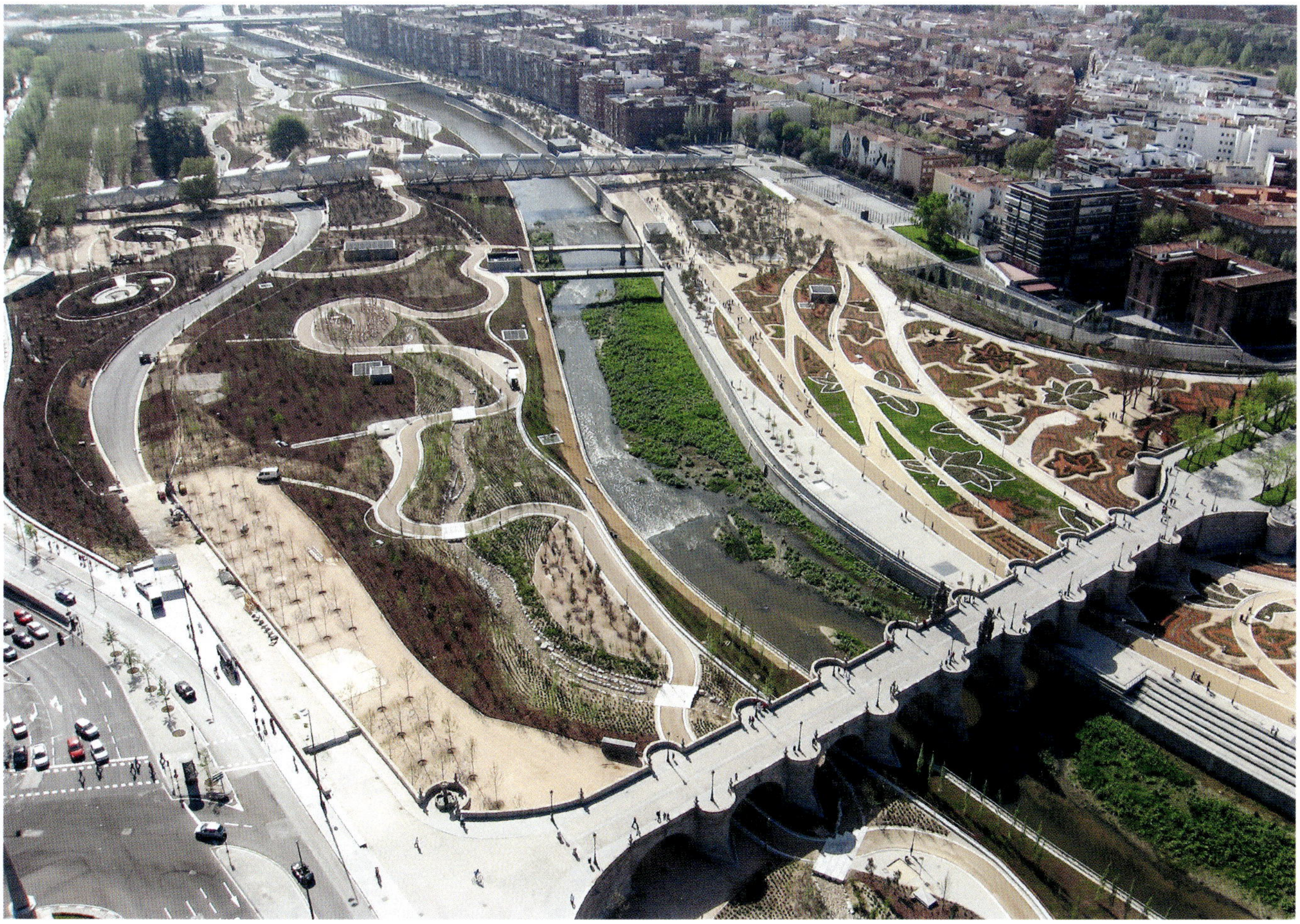
MUNICIPALITY OF MADRID

this context, as planted in a "choreography" that creates a botanical monument. This imbues the reworked site with a memorable identity, while also linking it with the mountain landscape on the city's outskirts. People come to the park for the joy and ease of moving along the river from one place to previously inaccessible places.

POSSIBILITIES

All of the parks discussed demonstrate design vision and talent, strong leadership, and community partnerships, facilitated by abundant resources and solid political will. Their ability to balance the aesthetic, formal, and material preoccupations of our discipline with the equally important scientific and technical requirements is commendable. These accomplishments are, however, the mere beginnings of what is possible within our discipline. Large parks have as yet untapped potential to improve our immediate environs, our cities, and our world at large. The final projects I will mention are not all parks (or if they are, they are small); but each demonstrates strategies that offer provocative possibilities for large parks.

In terms of *water*, what if more of us embraced the use of regionally specific plants, such as those used by Field Operations in Tongva Park in Santa Monica, California? Here the designers advanced drought tolerant plants suitable for a water-hungry environment, while carefully designing for tremendous beauty that cannot be mistaken as natural.

Regarding *waste,* what if more of us employ the three R's of the environment—reduce, reuse, and recycle—in parks specifically, helping ourselves, our community, and the environment by saving money, energy, and natural resources? We can investigate the use of recycled products as building materials, as design practice STOSS did when it used porcelain from recycled toilets in the paving system at Harvard University's campus Plaza.

In terms of *biodiversity,* what if we imagined the homes of non-human species no differently than our own? As lovely as a conventional birdhouse may be, it does not take into account the actual behavioural habits of birds. The experimental "species wall" of Harrison Atelier, which transforms the conventional roadway acoustic barrier into a varied habitat for songbirds, points the way. West 8's pergola in Utrecht's Maxima Park scales this initiative to the realm of a public landscape. The 3.5 kilometer (2.1 mile) long and 6-meter (20 feet) high structure surrounds a green courtyard and marks the transition from the city to the wild. Most importantly, the pergola is intended to attract all kinds of animals and plants, thus creating its own miniature ecosystem for birds, bats, bees, and other fauna. What if we gave circulation for wildlife in our landscapes even more consideration at a different scale, as did the winning entry for ARC Wildlife Crossing Design Competition? Can you imagine if we added a circulation route in Central Park for deer, wild turkeys, hawks, and coyotes (eastern coyotes have recently taken up residence in city parks), building on Olmsted's brilliant nineteenth century strategy of providing separated paths for different users, whether pedestrian, horse, or carriage?

In terms of *pleasure*, what if we aspired to have our projects help urban inhabitants see beyond the natural image of their environments to make visible how a landscape works? SCAPE's Safari 7 does this well. The designers describe a self-guided tour that uncovers the complexity and biodiversity of urban ecosystems along New York City's No. 7 subway line, such as bird life that emerges on top of cut and fill channels, or butterfly colonies that collect in vacant lots. In this spirit, what if we also concerned ourselves with teaching the public about the not-so-nice aspects of our contemporary environments? This is the subject of SCAPE's project Petrochemical America, which shows us the extraordinary degradation that human action has caused and the impact it has had on human lives.

Finally, what if we as designers started to imagine a more inclusive public, as did the Danish architectural firm BIG and landscape architects Topotek in their project Superkilin in Copenhagen which represents resident's cultural interests through diverse symbols that come from over fifty countries.

Large parks have as yet untapped potential to improve our immediate environs, our cities, and our world at large.

What if we were to really value the engagement work designers do with communities, and come to think of parks as not just *for* the community but *by* the community? And what if we continued to work to enhance and advocate the value of open, non-programmed space that we create in large parks—space that enables freedom of assembly and freedom of expression? While not always "pleasurable," these places, from London's Hyde Park to Berlin's Tiergarten, have responded to important needs within human society, and will continue to do so in the future.

Only then will we begin to actualize the great potential of large parks within our cities.

---

[1] The Rebuild by Design competition (2013), led by the (now) Special Envoy for International Water Affairs for the Kingdom of the Netherlands Henk Ovink, was an important initiative of the Hurricane Sandy Rebuilding Task Force and the US Department of Housing and Urban Development.

[2] "Oysters act as a natural filtration system in a body of water, actually pulling particles, toxic chemicals, and other sediment right out of the water. Studies have shown that these natural and efficient water purifiers [each oyster] can filter as much as 50 gallons of water each day." https://www.thoughtco.com/oysters-to-clean-up-polluted-bronx-river-1140682

COPENHAGEN

CHRISTIANSHAVN

VESTERBRO

VALBY

FREDERIKS
BRYGGE

Teglholmen

Valby-
parken

Sluseholmen

Kalveboderne

ØRESTAD

Fasanskov

Kalvebod
Faelled

AMAGER
NATUREPARK

No access

Pinseskoven

Kongelunden

No access
April 1–July 15

KASTRUP

Øresundsmotorvejen, E20

TÅRNBY

AMAGER

Københavns
Lufthavn

VIBERUP

DRAGØR

Sydstranden

ØRESUND

SØVANG

No access

Kongens Lyngby

Hellerup

Nørrebro

COPENHAGEN

Sydhavnen

Kastrup

Salt-
holm

Airport

AMAGER
NATUREPARK

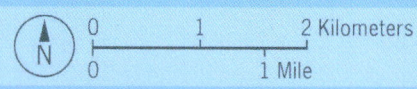

0   1   2 Kilometers
0   1 Mile

N

© OpenStreetMap contributors

AMAGER NATURE PARK

SIZE: 3.5 square kilometers.

OPENED: 1984.

LEGAL PROTECTION: Nature park in 2015.

OWNERS: Municipalities and the Danish state.

HOW TO GET THERE: Vestamager Metro station.

OLE CASPERSEN
Amager Naturepark, Copenhagen, Denmark

Amager Nature Park is a large green area on the outskirts of Copenhagen, capital of Denmark. Originally, Kalvebod and Amager Common were shallow waters but landfill and reclamation have changed it significantly during the last century. The area has been used by the military since the 17th century for practice in shooting with artillery. Over time, the military ceased its activities, and in 1984 the area was opened to the general public. A greater conservation in 1990 enhanced the green and recreational development. The plan process culminated in 2015, when the area was designated as a Nature Park.

Amager Nature Park comprises more than 3,500 hectares. The green area starts less than three kilometers from Rådhuspladsen (the very center of Copenhagen) at Amager Fælled and continues over Kalvebod Fælled and Kongelunden to the south to Sydstranden (South Beach), to end up by the city of Dragør, 12 kilometers from the center of Copenhagen. The area is owned by the municipalities of Copenhagen, Tårnby, and Dragør, and the Danish state.

The shallow area off the western coast of Amager has been the subject of interest since the Swedes used the ice-covered area during the siege of Copenhagen 1658–1660. The military's use of the area started at Amager Fælled in 1680, where the shallow area was used as a shooting area for artillery. At that time, Amager Fælled was considerably smaller than today.

In 1934 a proposal was submitted for the reclamation of Kalvebod Strand. The military was in need of new shooting areas. The proposal was adopted in 1939, and in 1943, during the WWII German occupation, the reclaiming of Kalvebod Common was completed. A total of 2,480 hectares were drained, and Amager's area had become approx. 1/3 larger. For the artillery exercises shooting and observation hills were established. Today, the former shooting and observation hills are the only significant elevated areas in the low-lying landscape.

In 1956 the area was ready for military activities and was used for that purpose, albeit with successively decreasing intensity. During this period awareness of the environment increased, also within the military, and as a result, an area of 374 hectares in the southwest part of Kalveboden was laid out in 1952 as a nature reserve for birds and other animals. In 1983 the military left the place.

B 1872 TÅRNBY CITY AND LOCAL LIBRARY

The dam work has reached 5 kilometers into Kalvebodløbet.

Since then drainage and landfill and, in particular, day-to-day renovation by Copenhagen caused the northern part of Amager Fælled to grow significantly. In the beginning of the 1980s the landfill ended and over time, plant and animal life has evolved, especially around the wetlands. Since 1989 the area has been laid out for recreational purposes. To limit the growth of scrubs and trees, parts of the area are grazed by cattle and sheep.

Amager has traditionally always been relatively undeveloped with regards to woodland and forest, but this was especially pronounced during the Swedish siege of Copenhagen in 1658–1660. Although the stay was short term, almost all forests were eradicated on Amager. Therefore, in 1818, the afforestation of Kongelunden was initiated in the south eastern part of the area. The planting went relatively fast and by 1830, 800,000 trees had been planted. The forest was originally privately owned but is now state-owned. It consists mostly of deciduous forest and it is known for its rich bird life. The forest is widely visited, and when combined with the nearby Pinseskov it forms the largest forest area on Amager.

TOP RIGHT: The combination of grassland areas, scrubs, and forest give the visitors good possibilities for enjoying wildlife experiences in the area, although you are never really far away from urban areas.
BOTTOM RIGHT: Sydstranden (South Beach).

MARTIN RIVERO

OLE HJORTH CASPERSEN

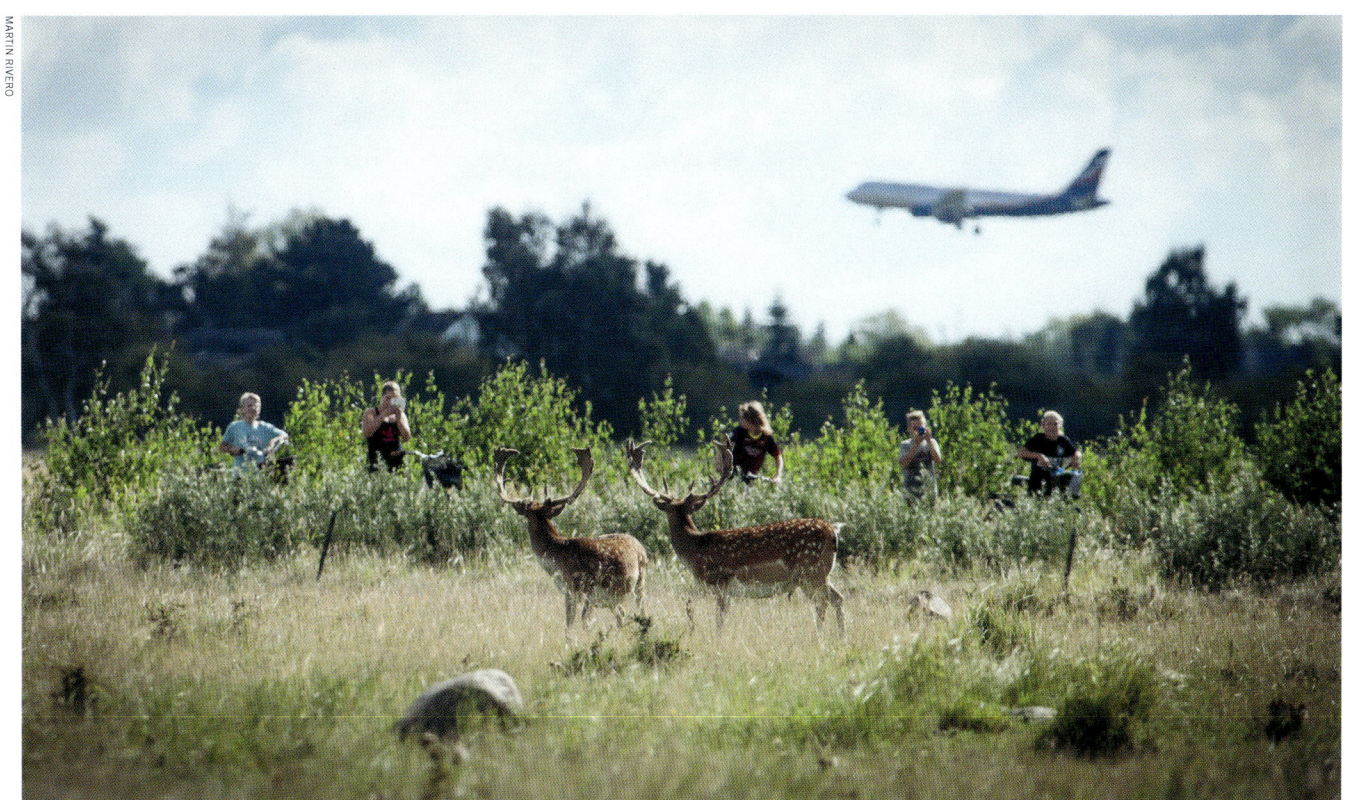

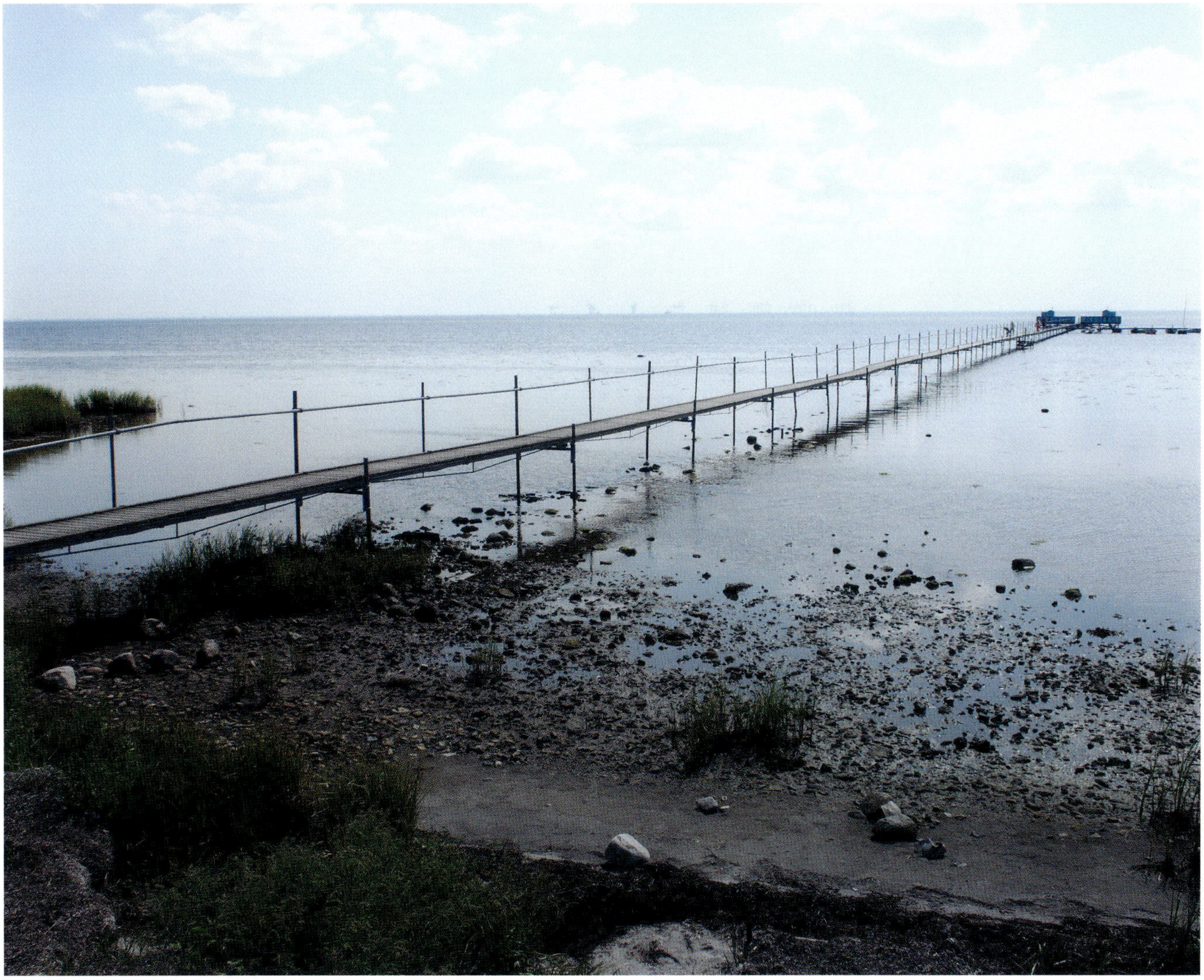

In the southern part of the natural park from Kongelund towards Dragør there is a narrow strip of marshes, which have a high cultural heritage. It is the last original coastline near Copenhagen. The coast consists of reeds, scrubs, and hawthorn which alternate with short sandy beaches and smaller dunes. The hinterland consists of large areas with meadows and dunes, some of which are protected through conservation and Natura 2000 designation. The coast and the shallow water off the coast are widely used by birds as a resting place.

An important decision was taken in 1989 to begin major nature conservation in the area. Conservation was intended to secure and enhance the significant natural history values associated with the area and to provide better recreational opportunities. In 1990, the area was finally cleared of ammunition, and therefore, free access to the entire area was permitted, except for the southern reserve, which is still maintained as a bird sanctuary.

After Vestamager was dammed in 1944, the former seabed and the shallow waters have emerged as a low-lying wetland. More than 500 plant species have been registered in the area. Water content and salinity play an important role in the variation of vegetation, which differs between the salt marshes, commons, reed-beds, and forest areas. Large areas are populated with rugged grass such as *Calamagrótis epigéios*, but there are also several remarkable species, such as blue iris (*Iris spuria)*, orchids, and less eye-catching, but very rare species such as the mushroom Jensen's wax hat (*Neohygrocybe ingrate)*. Several of the rarest species are found on the ancient salt marshes.

As the moisture and the salt content in the soil fell on the sedimentary area, the conditions for woodland improved, and to the south the forest Pinseskoven has grown to be Denmark's largest birch forest, dominated by birch, oak, ashes, and linden. The forest is the only place in Denmark where the rare blue butterfly *Apatura Ilia* still breeds. To the north is the smaller Phasanian forest while the rest of Kalvebod Fælled consists mostly of salt marshes and a few scattered scrubs and woodlands.

Using nature management, in the form of grazing combined with elevated water levels, attempts to keep the vegetation low and create better conditions for the waders. The area is an important breeding area for waders, and there are several species that live in the wildlife reserve around Klydesøen. The birdlife is quite extensive and Kalvebod Fælled is designated as a Natura 2000 and bird protection area. Amager Nature

LEFT: Kongelunden is the largest forest on the island of Amager, hence is often visited by residents of the area.

Park is an important area for resting birds, and especially in fall and spring thousands of migrating geese can be seen. It is not uncommon to spot a sea eagle looking for prey in the area. There are three bird-watching towers on the dyke.

Of Denmark's 14 different amphibians, at least 9 of the species can be found in the Nature Park including the three rarest species. The green-brown toad and beach toad stick to the beach beds. They require saline water to breed. Of larger domestic animals there are horses, cattle, and sheep.

The battle for the large area, however, was not over. New construction was underway in connection with plans for the new town of Ørestad. Plans for housing at Kalvebod Fælled and Amager Fælled have been suggested several times, especially in the 1960s. However, only two larger plans have been realized. The development of 2,500 homes began in 1962, but in 1970 the activities stopped. The second development started in 1991 when a bill was passed on the future development of Ørestad. The new homes in Ørestad are located immediately next to the green area. The construction of Ørestad has removed part of the green area, but at the same time the establishment of the western metro line has significantly improved the accessibility of Kalvebodkilen.

Finally, in Spring 2015 a large part of Amager was designated as a nature park. It was the culmination of a long process. The goal is to protect and convey a variety of landscape and natural qualities and create the framework for a greater experience as you move into the landscape park.

The location of Amager Nature Park is unique. Studies show that people now use nature more often than before, but also spend less time on transport to and from the visited area. This gives Amager Nature Park an advantage over many other green areas because it is close to very densely populated areas.

However, exploiting this potential requires a significant improvement in current access conditions. In 2017, Amager Nature Park received 97 million DKK of funding for outdoor activities from Nordea Foundation, which allows for the start of future development.

Studies show that people now use nature more often than before, but also spend less time on transport to and from the visited area.

TOP RIGHT: Ørestad ends at Vestamager metro station. The significant residential buildings end the district to the south and contrast with the protected area.
BOTTOM RIGHT: The Nature Park consists of a large flat area that comprises meadow, salt marshes, and wetland. Over the years, scrubs and deciduous forest have developed in the area.

OLE HJORTH CASPERSEN

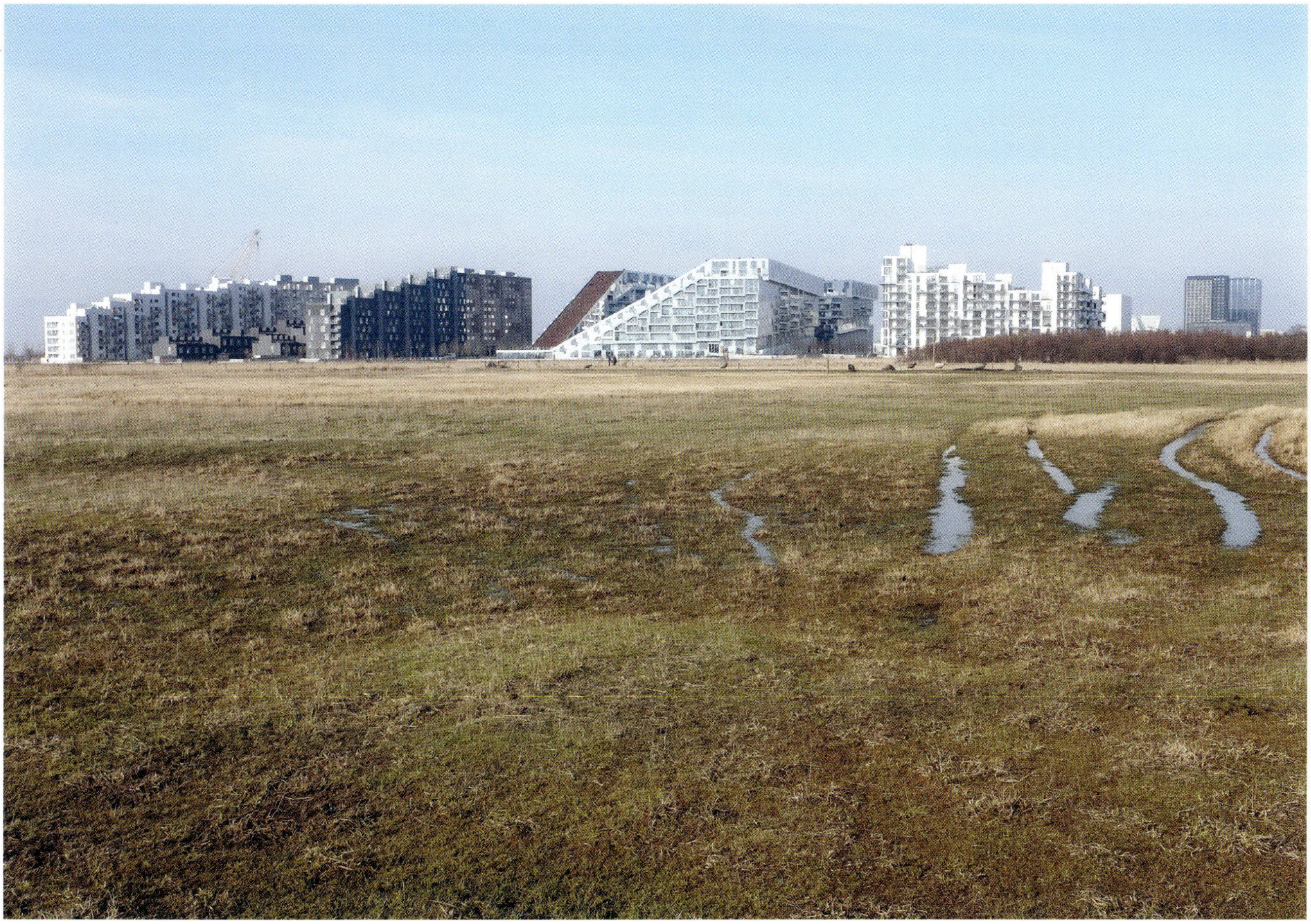

MARTIN RIVERO

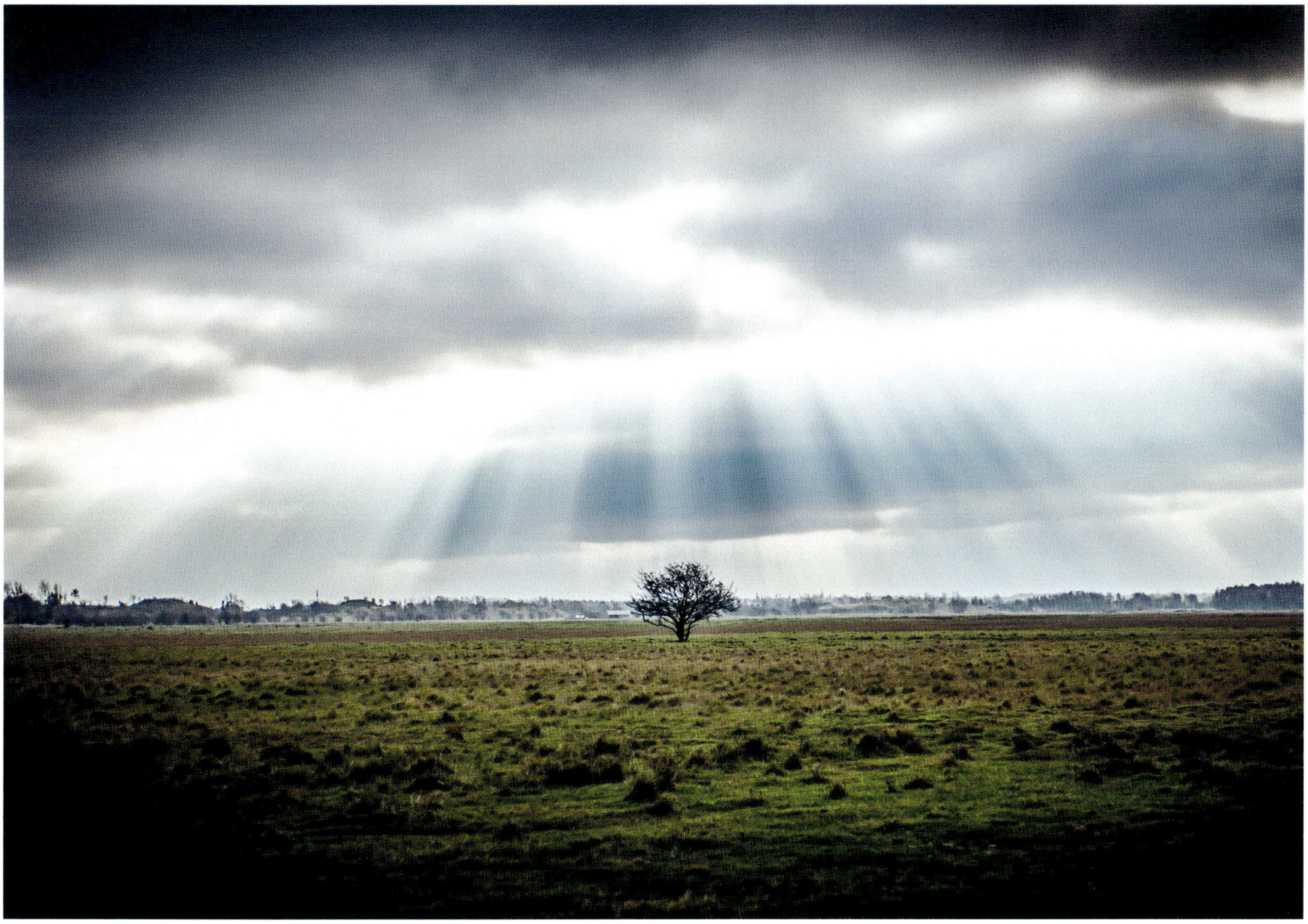

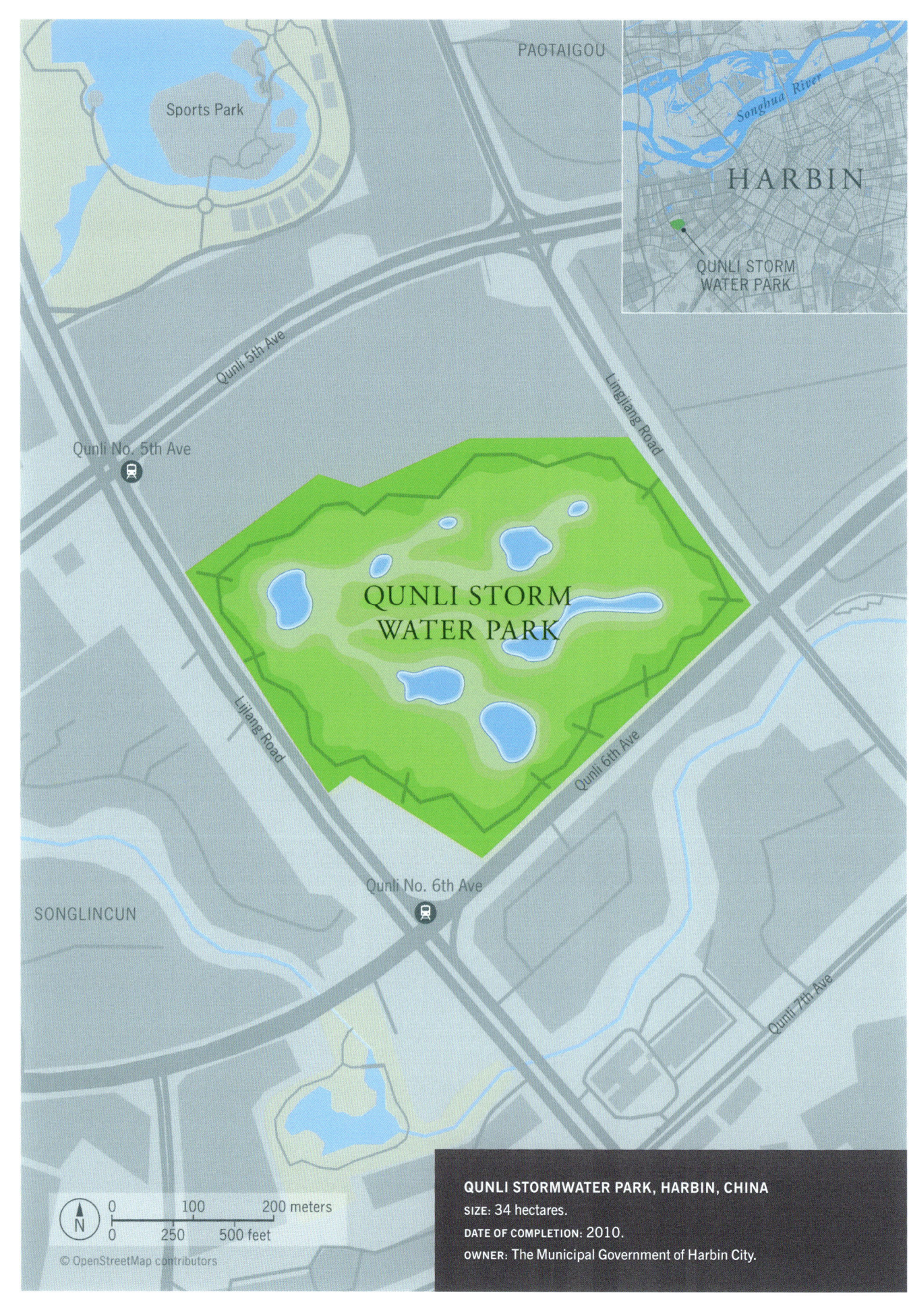

PAOTAIGOU

Sports Park

Songhua River

HARBIN

QUNLI STORM
WATER PARK

Qunli 5th Ave

Lingjiang Road

Qunli No. 5th Ave

QUNLI STORM
WATER PARK

Lijiang Road

Qunli 6th Ave

Qunli No. 6th Ave

SONGLINCUN

Qunli 7th Ave

N

| 0 | | 100 | | 200 meters |
| 0 | 250 | | 500 feet | |

© OpenStreetMap contributors

QUNLI STORMWATER PARK, HARBIN, CHINA
SIZE: 34 hectares.
DATE OF COMPLETION: 2010.
OWNER: The Municipal Government of Harbin City.

KONGJIAN YU

Qunli Stormwater Park, Harbin, China

Planning started in 2006 for a new 2,733-hectare urban district, Qunli New Town, for the eastern outskirts of Harbin in northern China. More than one-third of a million people are expected to live there. While about 16.4 percent of the land was zoned as permeable green space, the majority of the former flat plain would be covered with impermeable concrete. The annual rainfall is 567 millimetres, with June, July, and August accounting for 60 to 70 percent of annual precipitation. Floods and waterlogging have occurred frequently in the past.

In mid-2009 Turenscape landscape architects were commissioned to design a park of 34.2 hectares right in the middle of this new town. The site was a wetland that had been severed from its water sources by roads and dense development and was under threat. Going beyond the original task of just preserving the wetland, which required reconnecting water networks, the landscape architects transformed the area into an urban stormwater park.

Traditionally, in adapting to the monsoon climate, Chinese cities were built with water resiliency by keeping, both inside and outside of the cities, large open spaces for water. Pools and lakes would retain stormwater during the rainy season and become water sources for the dry season.

The pools and lakes at the same time provided other ecosystem services such as the provision of food and vegetables, including Lotus root (*Nelumbo nucifera*), Zizania (*Zizania latifoli*), and many other aquatic crops such as rice. These flood-adaptive water features and green spaces have also become scenic spots and public gathering spaces—parks—that provide spiritual, recreational services and cultural heritage values.

One good example of such a large park is the famous West Lake in Hangzhou, which was originally created almost 2000 years ago to remediate floods, with a water body of nearly 10 square kilometers. Now the 6.5 square kilometer lake plus its surrounding landscaped area has become the most visited park in a Chinese urban area, with annual visitors of up to 100 million.

In almost all traditional Chinese cities and towns, lakes, ponds and the green spaces around them have been present as necessary infrastructure that provides multiple ecosystem services, while keeping the

Traditionally, in adapting to the monsoon climate, Chinese cities were built with water resiliency by keeping, both inside and outside of the cities, large open spaces for water.

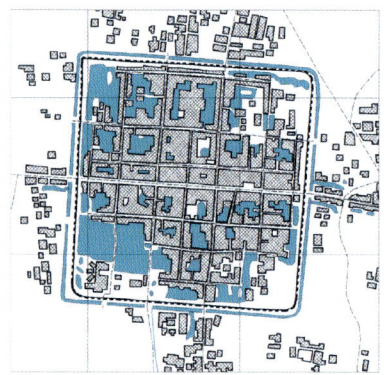 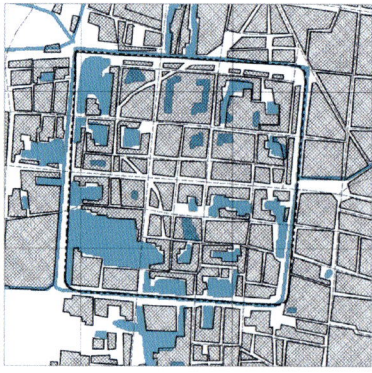 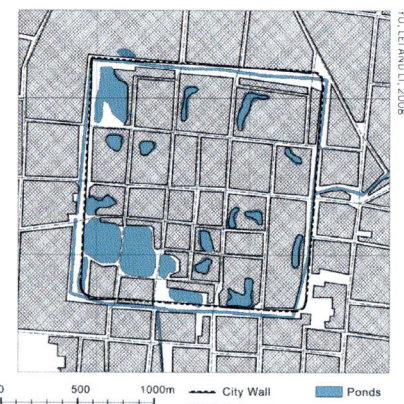

Hezhe City; the disappearance of stormwater remediating ponds and park over the years from 1960, 1983 and 2000 respectively.

settlement water resilient. This infrastructure collects, cleans, and stores stormwater and infiltrates it into the aquifer.

However, in many cities this important infrastructure has been sacrificed for urban development, now causing inundation. One example is the Hezhe City in Shandong Province. The city suffered a serious inundation in 2006.

Urban floods caused by stormwater are becoming an urgent global issue. Along with the expansion of urbanization, and climate change that is said to cause un-predictable precipitation, these floods fundamentally threaten the stability of cities and regions. This is particularly acute in China, where most of the cities are under the climatic influence of monsoons; 70–80 percent of the annual precipitation falls in the summer monsoon season, and in some extreme cases 20 percent of the annual rainfall can occur in a single day. In Beijing, for example, the average annual precipitation is only about 500 millimeters, but on 23 June 2018 the rainfall in the city reached 50–120 millimeters in a single day, and on 12 July 2012 more than 250 millimeters fell in the city and 79 people drowned. Serious urban floods have been hitting major cities in China even after normal rainfall, mainly because of the expansion of impermeable surfaces.

How can a large park be built fast and economically so that it can be managed inexpensively afterwards? For fast urbanizing cities, like those in China, it is not affordable to build large, conventional parks because

TOP RIGHT: Flood remediation space designed and used as a large park in the city, The Yinzhou West Lake in Fuyang City, Anhui Province, China.
BOTTOM RIGHT: Traditional Town Nanshe, in Dong Guan City, Guangdong province, China.

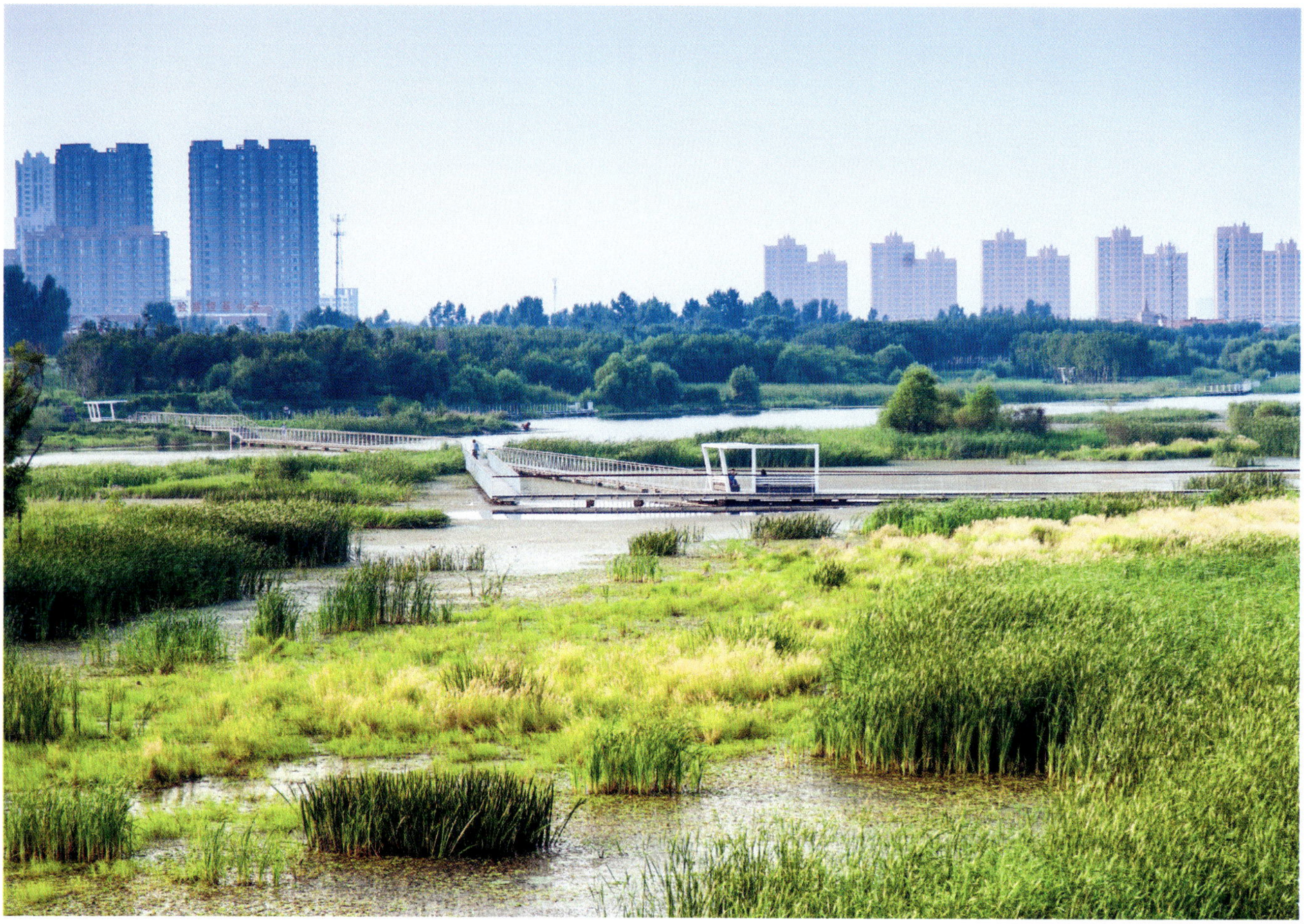

KONGJIAN YU/TURENSCAPE

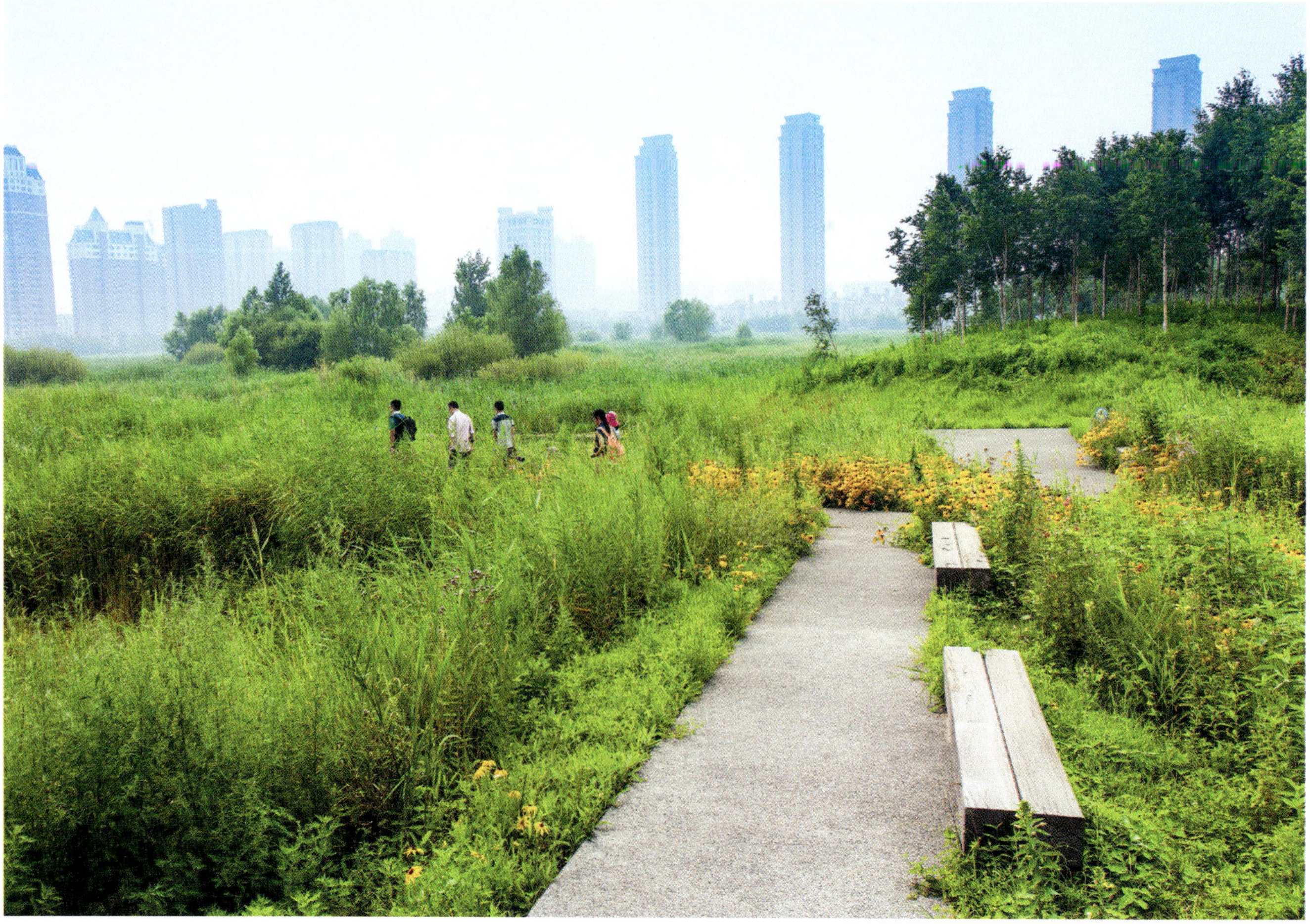

KONGJIAN YU/TURENSCAPE

KONGJIAN YU/TURENSCAPE

Hehze City, 2006.

it takes too long. The challenge in Harbin, therefore, was to build a park
fast and at low cost.

Building such a large park in one year, as required by the ambitious
new mayor, was not an easy job. Based on farming experience, a cut-
and-fill-technique was used. Chinese peasants have transformed large
scale landscapes for millennia using this technique to enable massive
crop planting.

The site was a left-over piece of land in the midst of a densely-built
urban area. The simple cut-and-fill technique was used to create a
necklace of ponds and mounds surrounding the former wetland.
For the peasant, cut-and-fill is one integrated action, not two: farming
earthworks are created on-site, and thus require minimum labour cost
and minimum transport of material. The major core of the wetland was
left untouched for natural evolution and transformation. The pond-
and-mound ring surrounds the periphery of the wetland and creates a
stormwater filtering and cleansing buffer zone for the core wetland.

Stormwater from the newly built urban area is collected into a pipe
around the circumference of the wetland and is then evenly distributed
into the wetland after being filtered and cleansed through the ponds.
Native wetland grasses and meadow plants are grown in ponds of various depths and the natural evolution process is initiated. Groves of native silver birch trees (*Betula*) are grown on the mounds to create a dense

TOP LEFT: Harbin Cultural Center Wetland Park.
BOTTOM LEFT: Qunli Stormwater park.

QUNLI STORMWATER PARK, HARBIN, CHINA

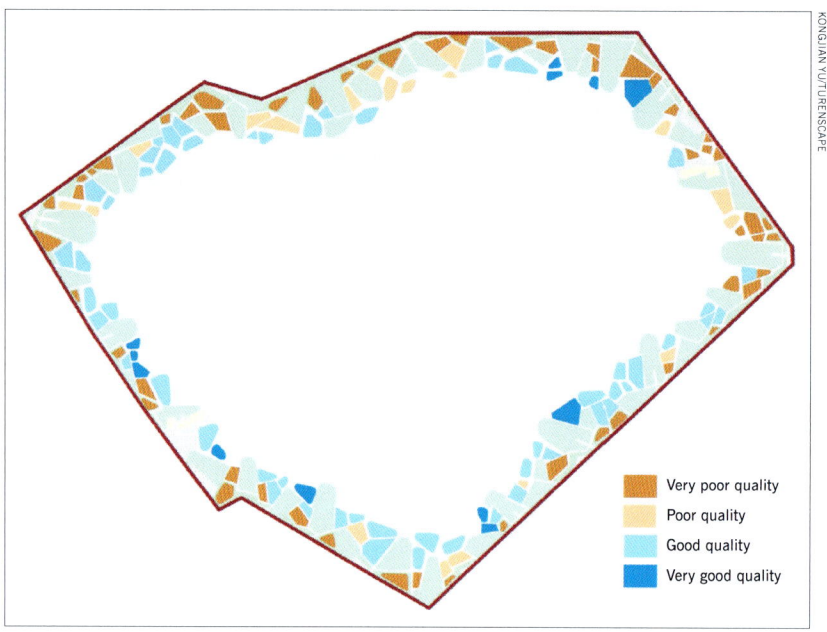

The observed change of the water quality in the bio-swales at the periphery of the park.

Very poor quality
Poor quality
Good quality
Very good quality

KONGJIAN YU/TURENSCAPE

forest setting. This also creates a welcoming landscape filter between nature and the city.

Qunli Park took one year and cost as little as 25 percent of a normal park of the same size.

Following the prevailing convention, people commonly trust engineering solutions to solve the urban flooding problems caused by stormwater by installing larger drainage pipes, more powerful pumps, or stronger dykes. The single-minded engineering approach for stormwater is problematic because of multiple issues that must be taken into consideration. Constructing an underground pipe system that has a large enough capacity to handle the extreme torrential rains, which occur quite often in the monsoon climate, is a huge investment, wasteful, and almost impossible. Such projects also present a management and maintenance burden in the future. Yet it is not simply the flood issues that need solutions.

Over the whole globe in general, and in China in particular, freshwater is in short supply. In Chinese metropolitan areas, the drop in the underground water table has become a serious issue. Of 660 Chinese cities 400 are experiencing some degree of water shortage. Beijing has seen a drop of 1.5 metres on average every year for the past three decades.

RIGHT: Qunli park. The stormwater park has now been listed as a National Urban Wetland Park.

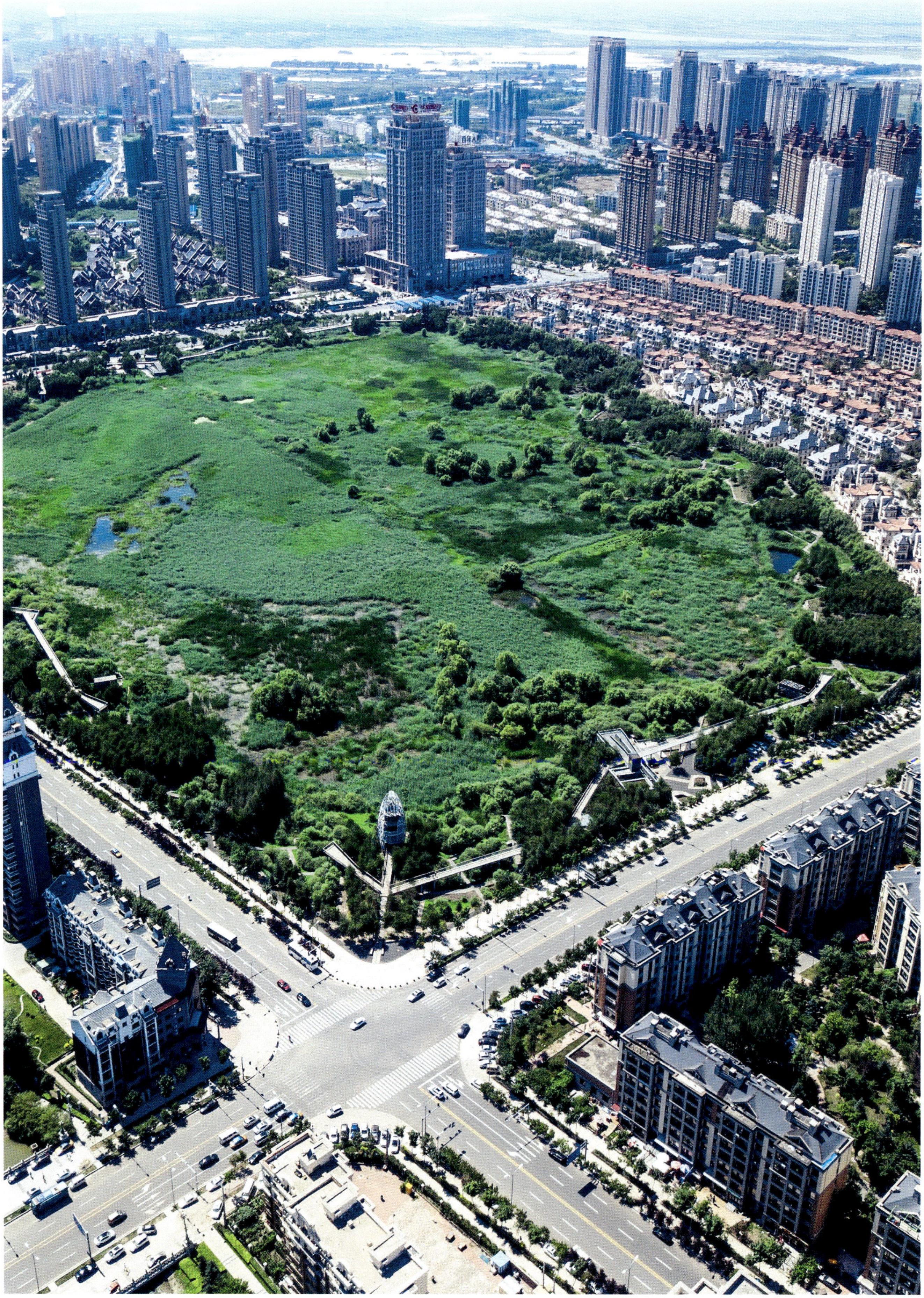

This is mainly due to the overuse of water supplies but also because of almost no aquifer recharge. The lack of recharge is because most of the rainfall has been drained away through pipes or through channelized rivers, instead of percolating naturally into the ground.

Finally, the engineering of stormwater through pipes and channelized rivers leads to the disappearance of surface water features such as brooks, ponds, wetlands, and other water sensitive habitats in the city. In addition, much more irrigation is needed for parks and green spaces in the city.

The Qunli park can retain and filtrate up to 500,000 cubic meters of stormwater, and has successfully solved the stormwater inundation problem for an area of 3 square kilometers (10 times the area of the park). If a city can allocate 10 percent of its total area as green sponge areas, it can solve the stormwater problem.

At the same time, the water quality of the wetland has been dramatically improved. A natural sanctuary has been created that provides diverse habitats for native plant species. The number of native species increased significantly over the years after the park was built.

The same concept has been used in many other cities in China, as well as in other countries. Most notable in China are the Harbin Cultural Center Wetland Park (118 hectares), the Minghu Wetland Park in Liupanshui City (120 hectares), and the Dong'an Wetland Park, Sanya City (67 hectares), while in Russia are the Kaban Lakes in Kazan City (186 hectares).

If a city can allocate 10 percent of its total area as green sponge areas, it can solve the stormwater problem.

LEFT: Qunli Stormwater Park, Harbin, China.

Nairobi, Kenya, seen from Nairobi National Park.

ADAM VAN NIEUWENHUIZEN
Urban Parks in Africa

Africa has a rich tapestry of cultures, landscapes, and urban development spread over the vast continent in landscapes ranging from deserts, high mountains, tropical forests, savannahs, and more. Where parks are concerned, the continent is better known for the massive natural conservation areas in central and southern Africa than for its urban parks. These large natural areas, like the Serengeti and Kruger National Park, are generally not accessible to local urban dwellers with limited disposable income. In recent years there seems to have been a resurgence in the development of new urban parks in some African cities, providing public spaces that are relevant to their urban, ecological, and cultural contexts.

The colonial past casts a shadow over all restoration ambitions. In several notable instances, the city park in colonized Africa was a leafy, flowery, palmy political display of power. These parks, now in a bad state, were often created by colonial powers in the past as a triumphal urban landscape celebrating a successful occupation but, at the same time, establishing dominion over nature. Military style, geometric control of the landscape, flagpoles, spiked guns, and cannonball pyramids set the tone, along with flowerbeds and always, whatever the climate, impossibly meticulously-clipped green lawns. "Exotic" parts have also existed in the form of a testbed for experimental, commercial plants and trees. When restoring such parks, should their former grandeur be strived for? Or should there be a completely new type of African urban park?

A common thread in the last 50 years has been rapid urbanization with infrastructure and services not designed for the increased urban population. The role of parks in cities is, however, increasingly being appreciated as offering opportunities that are more than simply aesthetic but rather also provide space for passive or active recreation and ecosystem services. Where the pressures on public funding greatly outstrip the availability of budgets, they need to provide the city and residents with a value that can be tangibly appreciated, such as improving the health and well-being of urban ecosystems and the residents.

The high density of rapid unplanned urbanization in relatively small areas and poor economic conditions mean space, especially quiet safe places, for play, relaxation, and recreation, is rare. Children play in the streets amid bustling noisy urban life. The areas available for the establishment of new parks are often areas not suitable for human habitation, i.e., along wetlands and in floodplains. City officials are starting to realize the importance of ecosystem services provided by these floodplains and river systems and that leaving them isolated leads to unsafe and undervalued places. People generally do not really value undeveloped urban open-spaces as positive environments and often use them as public toilets, informal settlements, and dumping areas. Developing open spaces into urban parks can help to change this mindset and make people see the values of urban green spaces.

In this chapter, we illustrate a selection of urban parks to give an idea of urban park developments in Africa, without the pretense to give a complete and balanced view.

The parks presented in this chapter are not very large. There are large urban parks in some cities, but they should more adequately be called peri-urban parks. Well-known examples are Nairobi National Park and Table Mountain National Park in Cape Town.

AL-AZHAR PARK

As part of a development project for a deprived area in Cairo, a 30-hectare park was created in 1999–2005. The project, financed by the Aga Khan Trust, aimed at developing a low-income section of Cairo and included vocational training, renovating residential buildings, health care, microfinance to help small businesses start-up—and a large park! The construction of the park and the restoration of cultural monuments were meant to be catalysts for social and economic development and the overall improvement of the quality of life in the district.

The park was created on a garbage dump. The project included the excavation and extensive restoration of the 12th century Ayyubid wall. Cairo is void of green areas, so this park is a big addition. The park helps ensure that the three water reservoirs in the park are kept clean. The park serves many purposes: a place of pride for inhabitants in a deprived area, a recreation area, a meeting place, and provision of clean water. Some two million people visit the park per year, and special festivals can attract up to 40,000 people.

This is just one example of the Aga Kahn Trust's involvement in creating or restoring urban parks in Africa and elsewhere. Other examples in Africa are the Mali National Park in Bamako (see below) and Nairobi National Park in Kenya and in India a park in Dehli (see separate article).

TOP RIGHT: View from terraced gardens in front of the Union Buildings in Pretoria.
BOTTOM RIGHT: A view from Al-Azhar park toward the Citadel with the Muhammad Ali mosque in Cairo, Egypt.

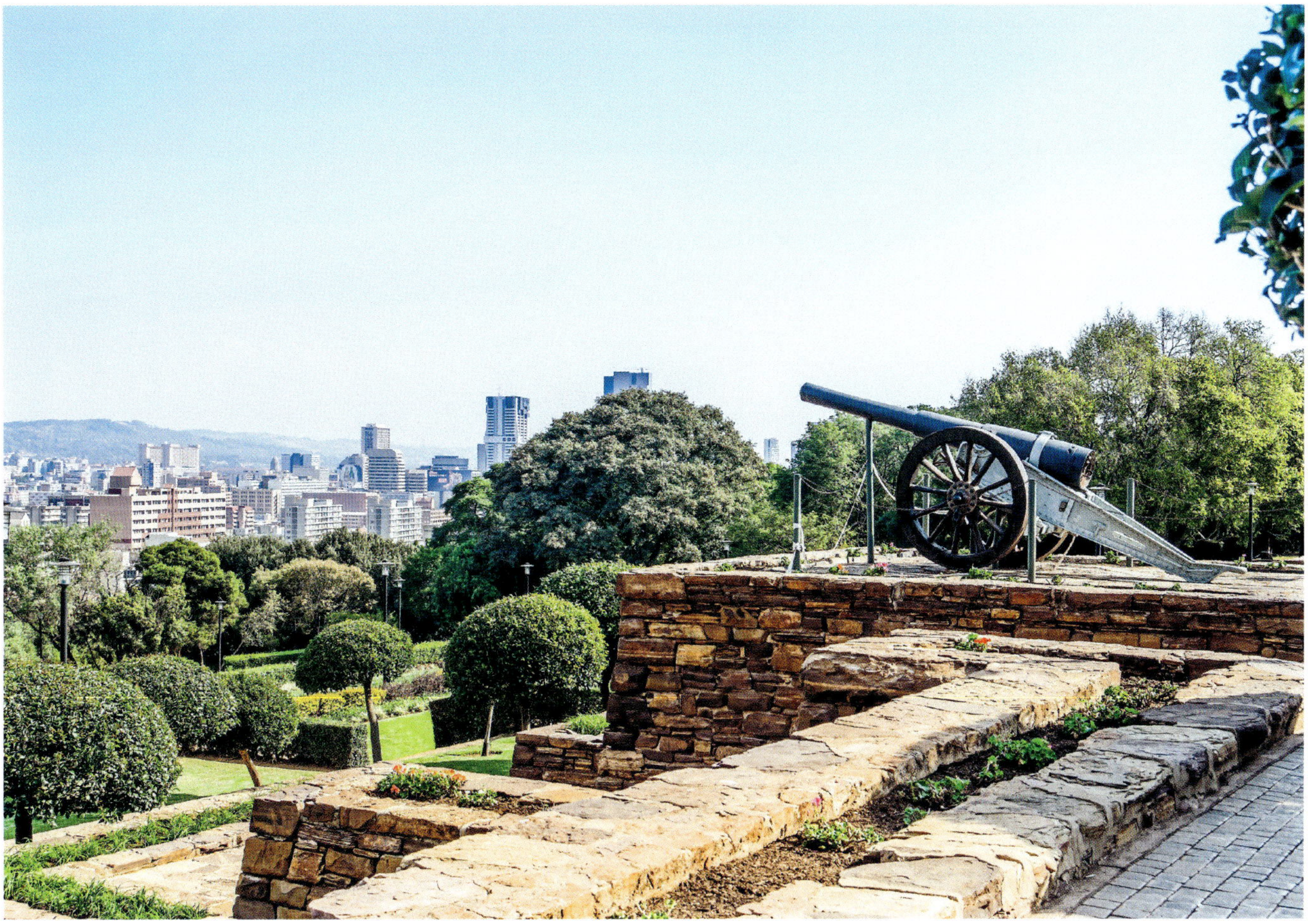

LARANIK SHUTTERSTOCK

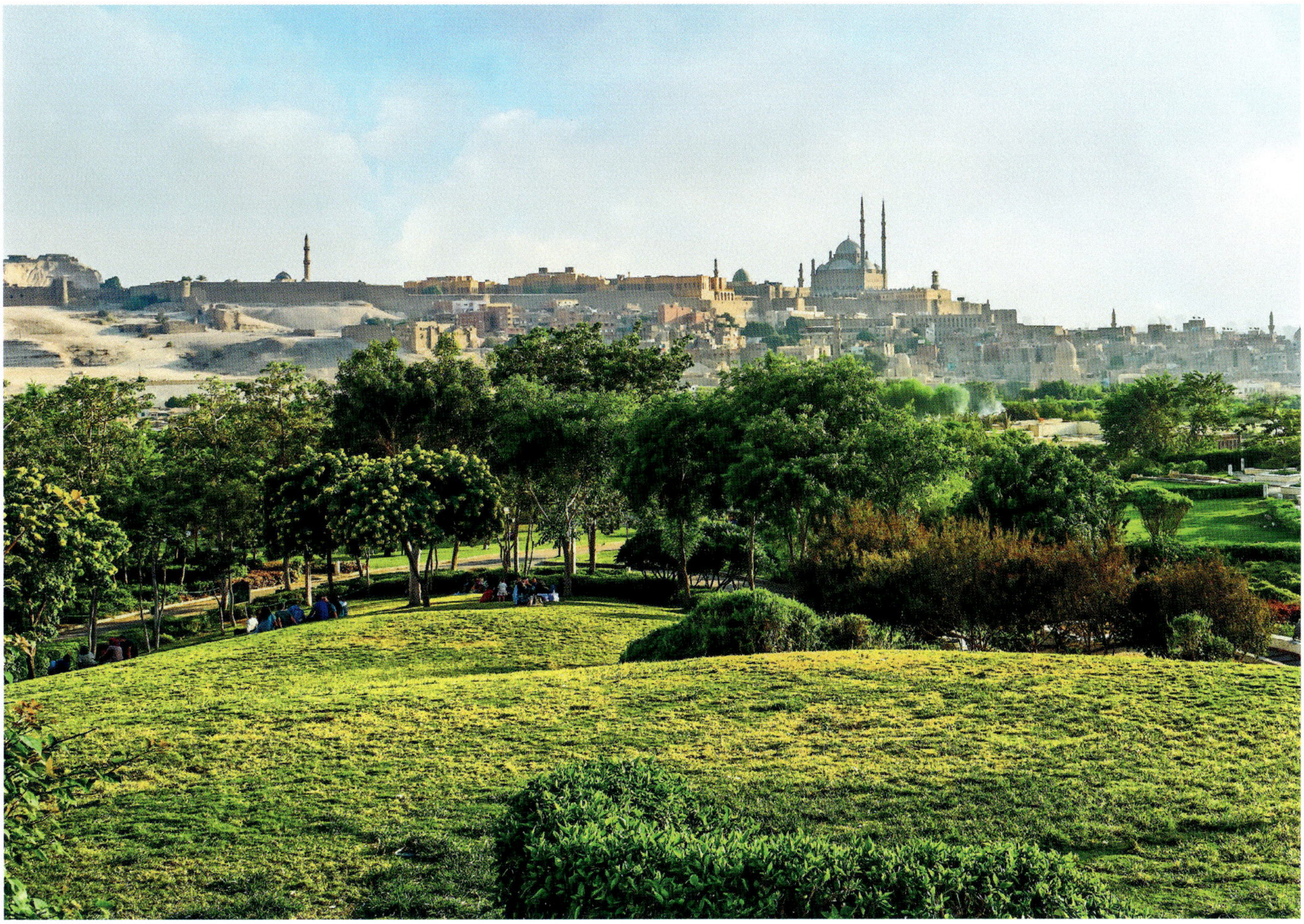

EFESENKO SHUTTERSTOCK

EARTHWORKS LANDSCAPE ARCHITECTS

BRUCE WHITEHOUSE

KING'S BEACH PARK

King's Beach Park is situated in Port Elizabeth in the Eastern Cape Province of South Africa. It is a 1.6 kilometer long golden sandy beach, bordered by vast lawns, suitable for picnic and play. During the apartheid years in South Africa, segregation along racial lines was most visible in public places. In 1987 Allan Hendrickse, a colored politician, famously swam at King's Beach with around 150 members of his Labour Party in defiance of the Separate Amenities Act. The beach was then a whites-only beach. In the years after apartheid, the park became derelict and poorly used. Recently a redevelopment was undertaken by the Mandela Bay Development Agency, aiming to change it into a place of social integration where the current generation of South Africans could play together and knit a new society. The first step was to improve visual permeability and legibility so as to showcase the park and its amenities, thereby improving its use and sense of safety.

Southern Africa is experiencing devastating periods of drought, arguably due to the effects of climate change. The park could therefore not be irrigated using potable water. A valuable water source was discovered across the road where an apartment building was pumping shallow fresh groundwater out of its basement into the stormwater system and out to sea. This water was redirected into the park and became a playful stream flowing into a lake. Today the entire park is irrigated from this source.

The design concept was to integrate play into the whole park rather than confining it to zoned play-areas. This way parents and children can play together, inspire creative play, and develop family relationships. The aim was also to encourage lots of children to play together, for instance, the slides are wide, accommodating several children at a time. Similarly, the skate park was designed for skaters and BMX riders of all levels and ages and designed to be an inclusive open area. The park was separated from the surrounding urban fabric by berms and planting.

A series of painted low walls (depicting a fable of the king who lost his crown) surrounds the play stream, which carries the water from the spring (stormwater inlet) to the lake. The play elements in the park are a mix of bespoke custom-made elements and standard play park items, open lawns, wide walkways for skating and cycling, adventure play areas, and swimming pools.

Metal cut-outs of birds were inserted into the concrete surface to add variety, interest, and playfulness to the walkways. The landscape

In recent years there seems to have been a resurgence in the development of new urban parks in some African cities, providing public spaces that are relevant to their urban, ecological, and cultural contexts.

TOP RIGHT: King's Beach.
BOTTOM RIGHT: Picnic in Mali National Park in Bamako.

architects developed the idea of a crown, which changes into birds that fly away. This refers to letting go of the ego as you enter the park and setting yourself free. The inserts are repeated in different places in the park; sometimes they simply cross the path and give the impression of a shadow of a massive albatross, or flock of small arctic terns.

PARC NATIONAL DU MALI

Until recently, if you were looking for a place to go in Bamako, the capital of Mali, that might approximate the Western notion of a park, you were basically out of luck. The city has its share of plazas, squares, and monuments, but none intended for people actually to spend time in, unless you count the street vendors who have gradually colonized most of Bamako's public spaces. In late 2010, however, a new space opened up next to the National Museum. It is simply called le Parc National du Mali, and it offers Bamako residents new possibilities for leisure.

The aim was also to encourage lots of children to play together, for instance, the slides are wide, accommodating several children at a time.

The new park, which sits on 17 hectares of land, was funded by the Aga Khan Trust for Culture and is run in partnership with the Malian government. The site used to be a seldom-visited arboretum. It is still home to thousands of trees, including many unusual and rare species, which now share space with walkways, gardens, and recreational equipment.

The park includes a fitness trail, a bike path, a gym, several eateries ranging from basic to upscale, a childcare center, and three separate playgrounds for children.

Access to the park isn't free, however. It costs the equivalent of US$ 1 for a Malian adult, $ 0.60 for a Malian child. This fact does limit the number and range of people who can visit, yet on average the park receives 500 visitors per day. It has been a big success overall, and for good reason: there is literally no other place in town where children can play on swings, families can picnic in the grass, or couples can relax together in a safe, pleasant natural setting.

Young people, just hanging out in the park, behave in ways they aren't allowed to elsewhere. You don't see kissing—this is Mali, after all, where public displays of affection range from low-key to non-existent. But you do see men and women holding hands and generally being close to one another. It is hard to do this outside the park without attracting unwanted attention from relatives and neighborhood gossips. Inside the park, however, there seems to be an assumption of some degree of privacy.

The Parc National du Mali has quickly become a favorite place in Bamako. It makes for a welcome getaway from the city's noise, traffic, and pollution.

CHIVEVE RIVER PARK: BEYOND FLOOD CONTROL

An ongoing project at the time of writing this article is the Chiveve River Park in Beira, Mozambique. The City of Beira was developed on the banks of the Chiveve River estuary near the mouth of the mighty Pungwe river in Northern Mozambique. The Chiveve is a small river that empties itself at the mouth of the Pungwe River. It stretches 5 kilometers through the historic city center, from the coast up to its source at Goto wetland and informal settlement. The creation of the Chiveve River Park was initiated by an engineering project to deal with flood control in relation to climate change funded by the German Kreditanstalt für Wiederafbau (KFW) and World Bank in 2017. The park encompasses the whole of the Chiveve river basin from its source to the sea.

Beira is a city that is extremely vulnerable to intense tropical storms and tidal flux. When Beira's harbor was built the river mouth outlet to the sea was channelled through a culvert in the harbor wall. Water would become trapped in the river, causing flooding. There are areas in the city that flood regularly when extreme rainfall coincides with high tides. During the engineering works it was recognized that the floodplain of the river could provide an opportunity for urban public space in the form of a park. The overall objective was to improve the functioning of the ecosystem, while also providing a place that would be welcoming, accessible, and provide a sense of nature.

A serious problem is that plastic bottles, bags, and all other rubble generated in the city have been dumped onto the river banks to be swept away with the next tide. The local population is not aware of the dangers of plastic, which were introduced relatively recently. Before plastic, glass, and tin came into wide-spread usage, all containers were organic and completely biodegradable: when the banana leaf served as the food container, the leaf was simply dumped on the ground and became compost. It is only through totally banning plastic, as instituted by Kenya, Rwanda, Mauritania, and Morocco, or by extensive educational programs that this culture will change.

An important design principle has been to respond to the surrounding local character rather than import design influences from elsewhere in the world. The aim has been to build local pride in the park, for instance, the layout of the Botanical garden is inspired by allotments surrounding rural villages. A total of 2,250 trees will be planted in the park and 162,000 shrubs and groundcovers with the majority bearing edible fruit, seeds, or having medicinal value. Walkways and boardwalks are inspired by the patterns derived from lowland rivers and streams in more natural

The colonial past casts a shadow over all restoration ambitions.

environments outside the City. Access by bridges, pathways, and boardwalks link the park to the surrounding urban area. A pedestrian and cycle route network provide connection from the harbor through to the wetland at the source of the river and into the mangrove forests. The use of local timber will hopefully encourage people to see the value of what they have rather than export it.

Culture and heritage are important elements in the park. There is a wonderful old historic building, Casa dos Bicos, an example of the strange and fantastic modernist architecture that occurred in Mozambique in the 1960s. Next to it is the Casa del Cultura that serves as the art and cultural center with an amphitheater.

The wetland at the source of the river is surrounded by the Goto Informal Settlement. People fish in the river, grow vegetables, and harvest reeds, but due to organic waste the water quality is extremely poor. Children play in the river and on its banks among piles of rubbish. The environmental focus for this area is to protect the already strangled wetland from further urban encroachment while seeking to improve the water quality of the river downstream through ecological water treatment. A series of formal safe play-areas is envisaged, to be built by trained local craftsmen.

THE NEW URBAN AGENDA

The developments described in this article harmonize well with the United Nations Habitat New Urban Agenda (NUA), adopted in 2016, and also with the United Nations Global Sustainability goals (SDGs), adopted the year before. SDG 11 states the mission to "make cities and human settlements inclusive, safe, resilient, and sustainable." This includes the aim to "by 2030, provide universal access to safe, inclusive and accessible, green and public spaces, in particular for women and children, older persons and persons with disabilities [which] stresses the need for access to green."

The related examples of urban park initiatives in Africa contribute towards this goal. Much more, though, is to be done. One other fine example of recreating an urban park is that of Agodi Gardens in Ibadan, Nigeria. That story is related in a separate article.

ABOWE: The lower part of the Chiveve River Park near the Ocean.

RIGHT: Children's playground today.

BELOW: Renovated river bank.

BOTTOM RIGHT: Wetland at the source of river Chiveve.

Nairobi National Park borders the city to the south.

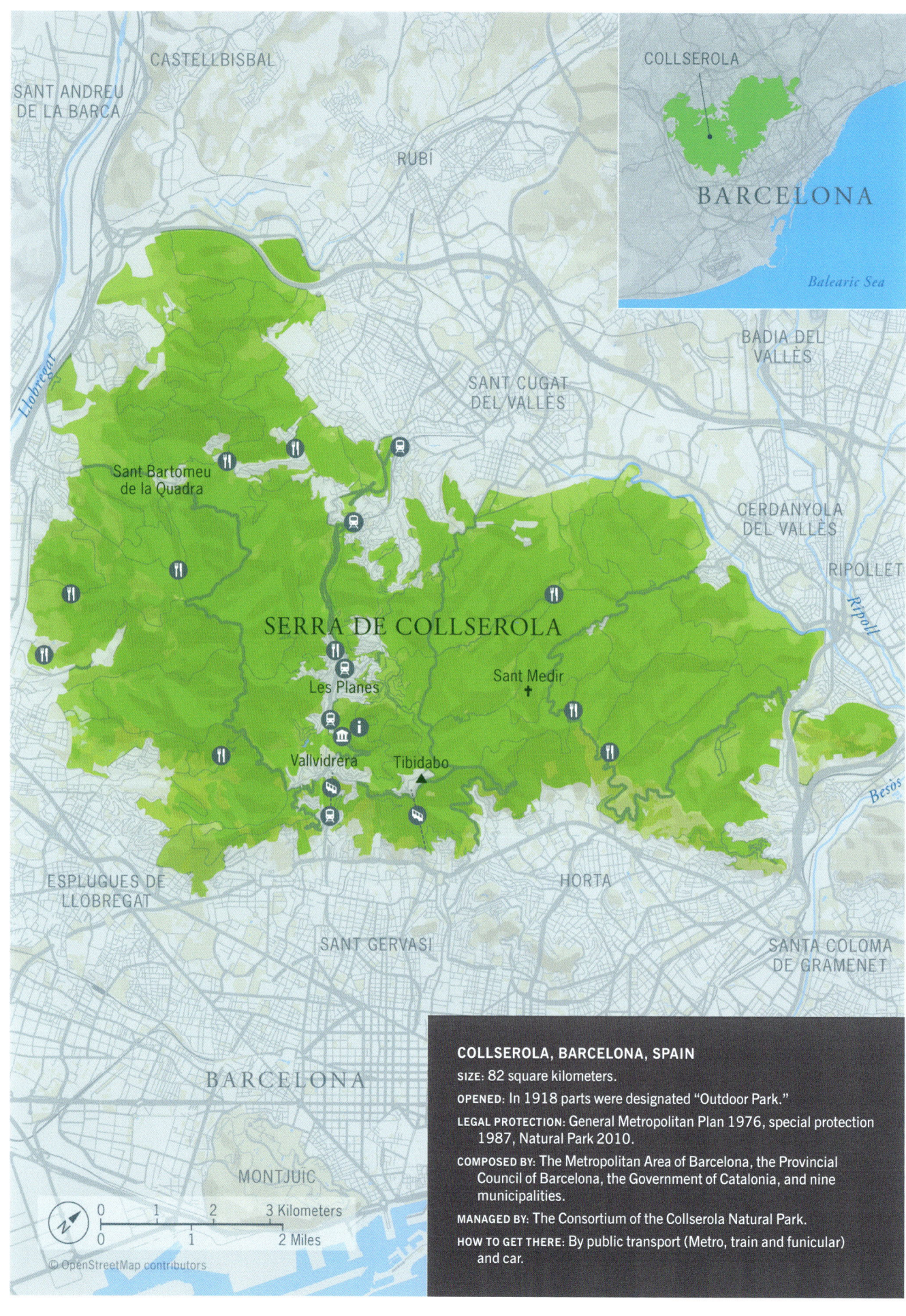

CASTELLBISBAL

SANT ANDREU
DE LA BARCA

RUBÍ

COLLSEROLA

BARCELONA

Balearic Sea

BADIA DEL
VALLÈS

SANT CUGAT
DEL VALLÈS

CERDANYOLA
DEL VALLÈS

RIPOLLET

Sant Bartomeu
de la Quadra

Llobregat

Ripoll

SERRA DE COLLSEROLA

Sant Medir

Les Planes

Vallvidrera

Tibidabo

Besòs

ESPLUGUES DE
LLOBREGAT

HORTA

SANTA COLOMA
DE GRAMENET

SANT GERVASI

BARCELONA

MONTJUIC

0 1 2 3 Kilometers
0   1   2 Miles

© OpenStreetMap contributors

COLLSEROLA, BARCELONA, SPAIN

SIZE: 82 square kilometers.

OPENED: In 1918 parts were designated "Outdoor Park."

LEGAL PROTECTION: General Metropolitan Plan 1976, special protection 1987, Natural Park 2010.

COMPOSED BY: The Metropolitan Area of Barcelona, the Provincial Council of Barcelona, the Government of Catalonia, and nine municipalities.

MANAGED BY: The Consortium of the Collserola Natural Park.

HOW TO GET THERE: By public transport (Metro, train and funicular) and car.

MARIÀ MARTÍ VIUDES AND JOSEP MASCARÓ CATALÀ
Collserola, Barcelona, Spain

The typical image of the city of Barcelona, in which you can see a mountain background or skyline beyond the city's limits, illustrates the context of Collserola Natural Park. At 8,200 hectares. it constitutes one of the largest peri-urban parks in Europe.

Collserola Natural Park is a huge, green, mostly forested "island" in the midst of the Barcelona Metropolitan Area and is surrounded by a dense urban fabric occupied by 3.5 million inhabitants.

The Collserola mountain range is 17 kilometers long and 6 kilometers wide and forms part of the Serralada Litoral, the mountain chain that runs parallel to the coast. The park is delimited by two rivers that cut through this chain: the Besòs river to the northeast and the Llobregat river to the southwest. The maximum altitude is the top of Tibidabo (512 m), where an iconic temple is located.

The massif is located also between two plains: one to the southeast of the city of Barcelona and another to the northwest of the Vallès depression. The massif rises abruptly from the sunny slopes of Barcelona and descends gently towards the Vallès plain (the shady side) forming several valleys there with gentle slopes. These different physical and climatic conditions give rise to a multiplicity of landscapes.

How is it possible that this extraordinary natural area has been preserved in a territory where such intense development has taken place?

The Collserola mountains underwent intense agricultural development during the seventeenth and eighteenth centuries. The medieval farmhouses were enlarged and renovated, and several are preserved largely intact to this day. In 1860, with the appearance of the phylloxera disease in France, vine cultivation was extensively undertaken in Collserola, and as a result forests were reduced. Later, they recovered when phylloxera finally reached Collserola and a period of forestry took place. Pine trees were planted during the first half of the 20th century, until logging was subsequently abandoned due to lack of profitability, and consequently the forests fell into neglect. This allowed for a recovery of forest species such as holm oak and other oak trees, which had hitherto been neglected due to their slow growth. The forest we have today is a product of the interaction between humans and nature.

The enjoyment of nature was discovered during the second half of the 19th century, in the pursuit of a healthy outdoor life, and so in the

FREDERICH BALLELL

Summer festival in the Tibidabo forest with song by the choir L'Orfeó Català, early 20th century.

mid-nineteenth century Collserola began to be used for leisure. The first summer homes of Barcelona residents were built in the high areas (Vallvidrera) and new communication routes were opened up. Collserola ceased to be a barrier and became a meeting place and a destination for excursions.

The beginning of the 20th century produced a qualitative and quantitative change in the use of the mountain range: In 1901 the Tibidabo amusement park was inaugurated, located at the highest point of the mountain range and connected to the city by a spectacular funicular. In 1917 the railroad crossed the mountain range. Emblematic projects also appeared in the area e.g. the Grand Hotel Restaurant Rabassada (1899).

In fact, the existence of a rich historical and architectural heritage is one of the most outstanding characteristics of Collserola.

In 1918 the architect, urban planner, and gardener, Nicolau Maria Rubió Tudurí, implemented a project categorizing free spaces in the city of Barcelona. This scheme assigned Vallvidrera, Tibidabo, and Sant Medir to the category of outdoor parks. This was the first attempt by the city to protect the mountain and Collserola.

Throughout the 20th century, urban development moved closer to the mountain, and so the 1953 County Plan defined the forests as reserves destined for public use. Together with the geomorphology of the mountain range, this contributed to the conservation of Collserola's natural spaces. This was the second step in the protection of Collserola.

TOP RIGHT: Temple on Tibidabo.
BOTTOM RIGHT: Today's funicular leading up to the amusement park.

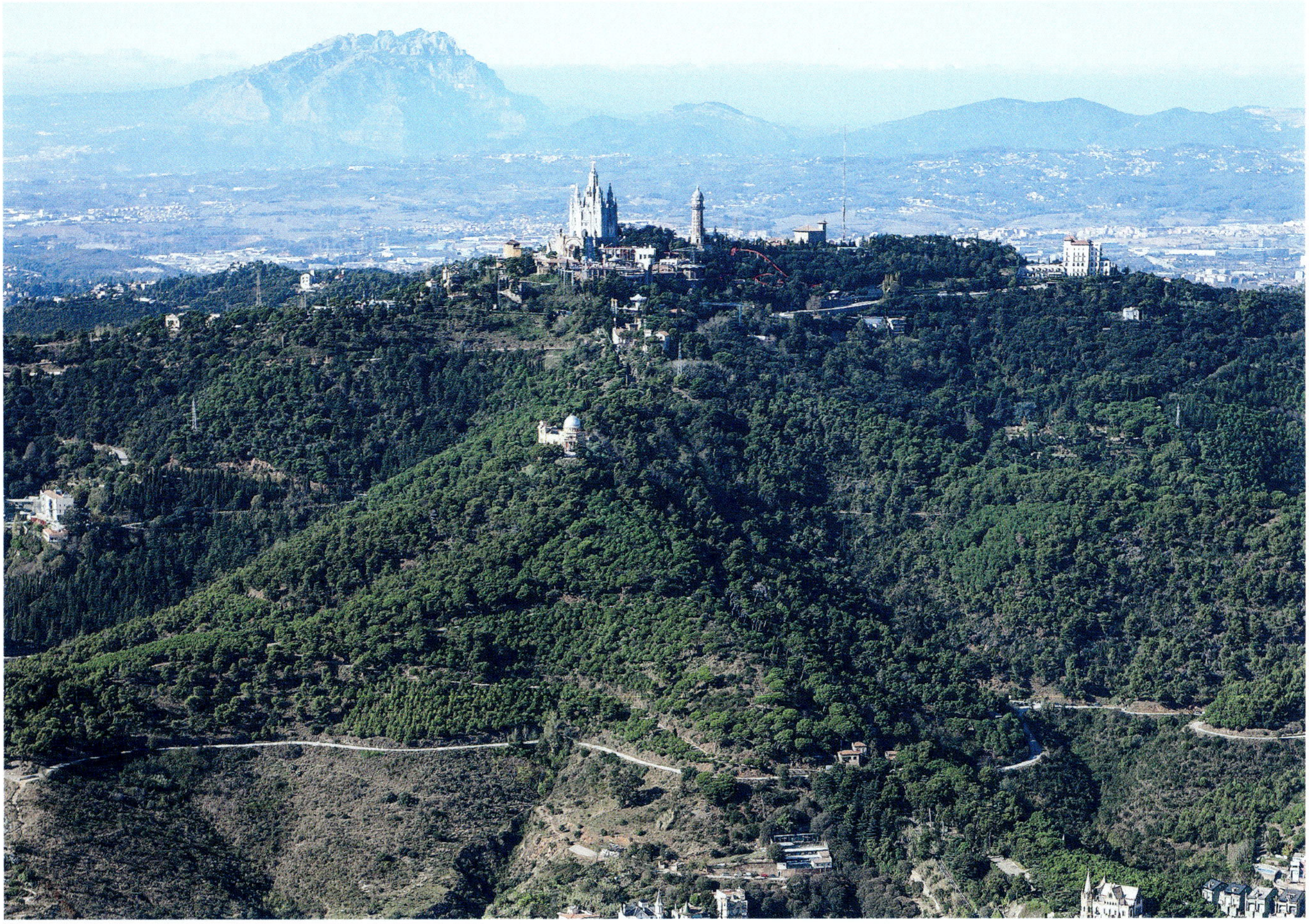

BIMSA

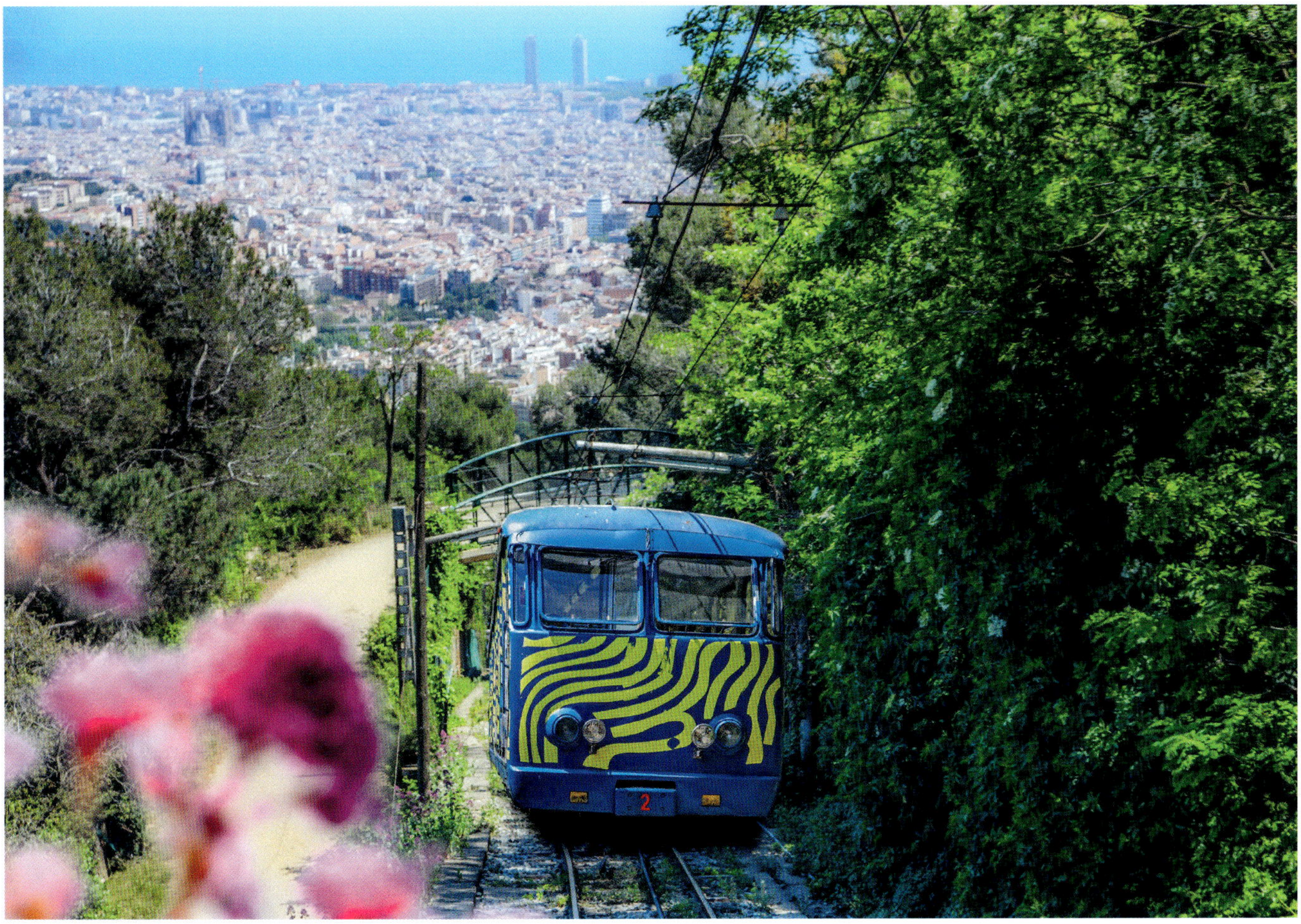

PARC D'ATRACCIONS TIBIDABO S.A.

View over Barcelona from Collserola Natural Park

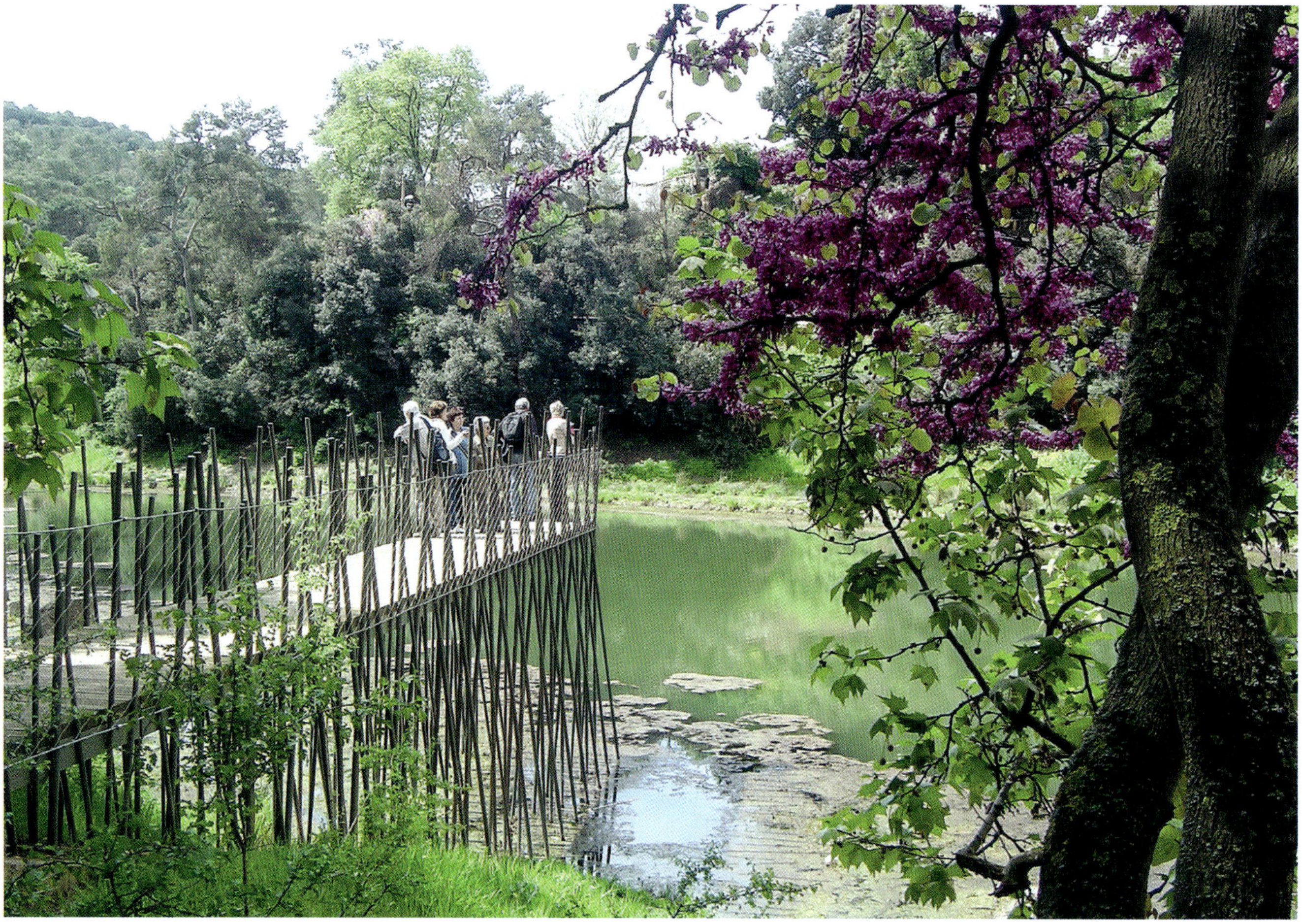

JOSEP MASCARÓ/CPNSC

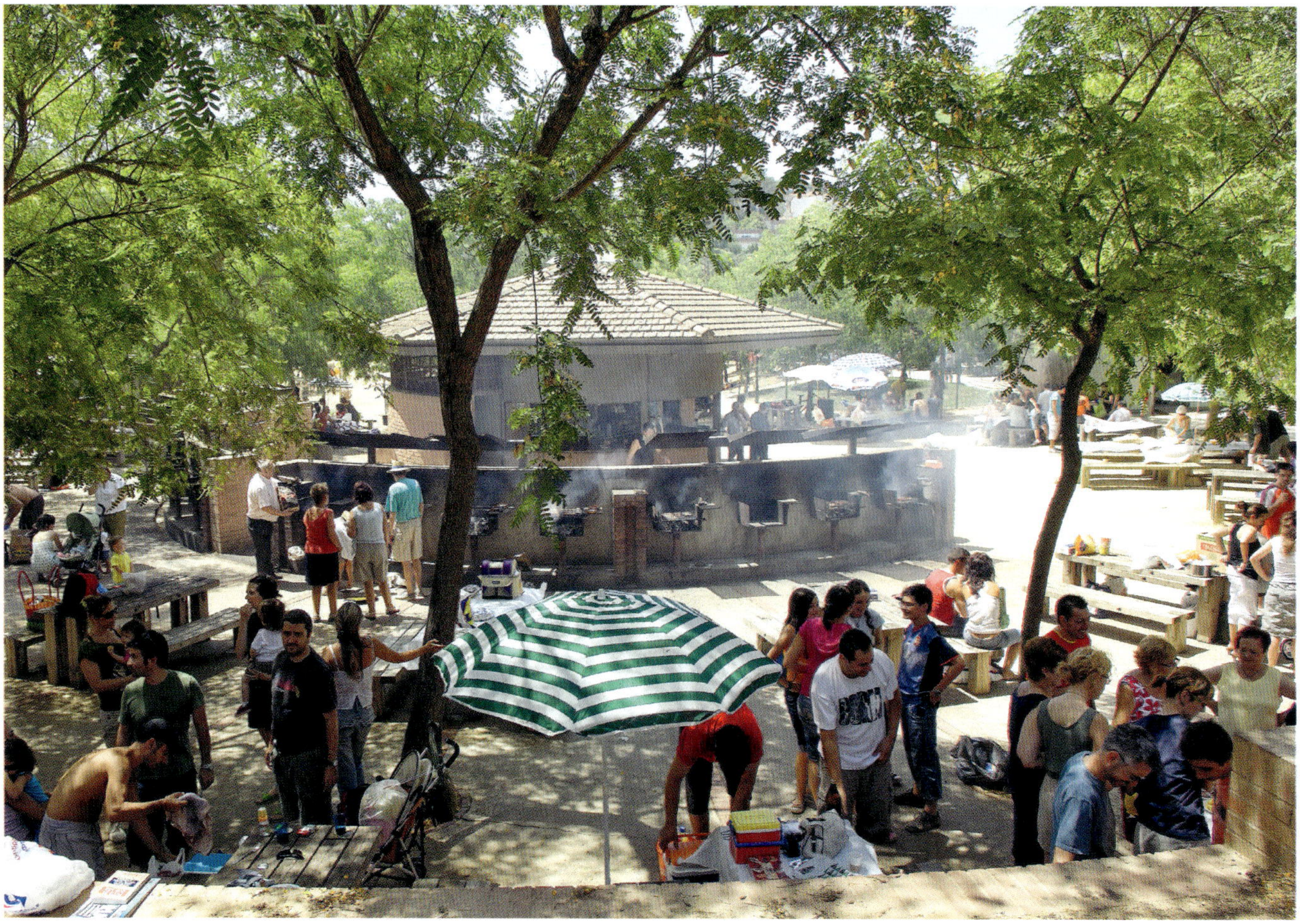

ROBERT PEÑA /CPNSC

The third step is a series of plans. A period of accelerated and disorderly expansion of the metropolis ended during the 1970s. In a decade, one million people from other regions of Spain had immigrated into Catalonia. More than half of the population of Catalonia now lived on less than 2 percent of its area.

To impose order on the existing mess and to improve the social and environmental quality of the cities, Barcelona undertook city planning, which involved a Special Plan for the Regulation and Protection of the Collserola Park in 1987, the definitive step to guarantee the protection of the mountain range. In 2006, Collserola Park was included in the Natura 2000 Network and, in 2010, the Government of Catalonia declared Collserola a Natural Park. A new Special Protection Plan is underway in the year 2020.

In 1987 a special body for the management and guardianship of the park was created. It has evolved into the Collserola Consortium, which includes the Metropolitan Area and the nine municipalities with territory in the park. Today it manages a budget of around six million Euros and a workforce of 70 people.

The 1987 plan pursued three objectives: to preserve the natural resources and the ecological balance, to protect and restore the historical heritage, and at the same time develop a park for nature-based leisure activities.

The current landscape of Collserola is the result of an evolution in which human activity has had a great influence. It presents a great diversity of landscapes. The vegetation is a mosaic of forest, scrub, shrub and brushwood, as well as grasslands and crops. The scarcest environment is the riparian forest, associated with the few permanent watercourses.

Today the mountain range is an extensive pine forest dotted with areas of holm oak and regular oak forests. The reality is that in the more shaded and humid areas there is a strong, dynamic, natural selection that is gradually leading to the disappearance of the pine trees and their replacement by oaks.

A very important phenomenon that shapes the landscape is that of forest fires. Two explosive elements coincide during the summer: high temperatures and minimum rainfall. Consequently, a large part of the management of the natural space is focused on the prevention and extinction of forest fires. When a fire destroys a forest, new vegetation appears that is called scrub, a shrubby formation, between two and four

The mosaic landscapes where the remains of old crops are mixed with grasslands, clear shrub lands, rambles, narrow valley bottoms, and the forest edge, are where we find the highest wildlife diversity.

TOP LEFT: Here aquatic invertebrates can be found.
BOTTOM LEFT: Leisure area with barbecues.

meters in height, formed mainly by sprouting plants. This scrub forms the embryo of the succession towards the future holm oak and oak forest.

The dry grasslands are maintained as meadows and do not evolve into scrub, or later towards brushwood, as long as they are repeatedly affected by fires or used as pasture for livestock.

The multitude of environments supports a rich and diverse fauna.

Mammals as emblematic as the wild boar, but also other less known ones such as the common genet inhabit the forests. The goshawk, a noteworthy bird of prey, occupies the quietest stands of the park. Various species of small wild birds such as tits, the short-toed tree creeper, the blackbird, woodpeckers, and wood pigeons are to be found. The dense bush formations are preferred by warblers, which are very loud but often difficult to see.

At the beginning of autumn, we can also observe the two-tailed pasha, a large butterfly which is known locally as the strawberry tree butterfly, since this tree provides its essential sustenance.

The most peripheral parts of the park are much less forested and dominated by crops as well as meadows and scrub areas. The wildlife which is present here differs greatly from that of more humid habitats. The most characteristic birds of prey are the buzzard, and we may also see the majestic short-toed eagle flying above, a species which, although it breeds in the woods, visits open spaces in search of the snakes which constitute its main prey.

The mosaic landscapes where the remains of old crops are mixed with grasslands, clear shrub lands, brambles, narrow valley bottoms, and the forest edge, are where we find the highest wildlife diversity. If we had to choose a representative species of these Mediterranean mosaics we would certainly opt for the badger, a species which is in a slightly fragile situation, but which still builds its imposing setts[1] in earthy slopes, always hidden in brushwood and forest environments.

In the open dry grasslands, we find grasshoppers and bugs of high scientific and conservation interest.

The aquatic environments, formed by small, typically Mediterranean water courses, host native fish species, for example the red-tailed barbel. In some stretches of less frequented streams of the park, communities of aquatic invertebrates that are indicators of an unexpected, ecological rarity, have been found. Small lakes, ponds, and reservoirs host amphibians such as the green frog and the tree frog, whilst springs in forests and small torrents in the valleys are the preferred environments for salamanders.

The fauna of Collserola faces various threats. There is loss, fragmen-

Currently, mass public visits to parks and nature are a big concern in all of Europe.

tation and isolation of habitats due to exploitation, as well as roadkill. Also, the high intensity of visitor use generates disturbances and puts the most sensitive species at risk. The release of unwanted domestic pets into the natural environment also has a high impact on wildlife. Problems for native amphibians generated by exotic fish and aquatic turtles are well known. However, less well known is the impact caused by the recent release of Vietnamese pot-bellied pigs on the local population of wild boars, which still maintain a high degree of genetic purity.

The management of a peri-urban natural park like Collserola is complicated since with so many people and urban areas—even within the park—this necessarily implies intense interventions in the landscape, with roads and railway lines, electric power lines, pipelines for water, gas, etc., as well as illegal occupation, dumping, spills, etc., alongside growing public use.

Currently, mass public visits to parks and nature are a big concern in all of Europe. Since 2010 the park administration is working with citizens and various organizations, in order to reverse misuse that damages the park. At present 5 million people a year visit the park.

A network of tracks and trails has been established to host authorized collective activities. Starting from a network of 702 kilometers this has now been reduced to 324 kilometers for activities on foot and 250 kilometers for cycling. This is intended to create spaces that are free from intense public use.

In short, Collserola is a great challenge, requiring the commitment of the local administrations to guarantee the existence of and access to an extraordinary natural space for a great number of people. We believe that it contributes to the quality of life of its citizens and that it makes the city more friendly, resilient, and sustainable.

[1] A sett is a badger's den, a network of tunnels and numerous entrances.

Wild boar grazing with a view of Barcelona.

REFERENCES

WHY CITIES NEED LARGE PARKS

Angel, S., Parent, J., Civco, D.L. and Blei, A.M. (2011). *Making Room for A Planet of Cities.* Cambridge, MA: Lincoln Institute of Land Policy.

Boone, C. G., Buckley, G. L., Grove, J. M. and Sister, C. (2009). Parks and People: An Environmental Justice Inquiry in Baltimore, Maryland. *Annals of the Association of American Geographers.* 99(4), pp. 767–787.

Czerniak, J. and Hargreaves, G. (eds.) (2007) *Large Parks.* Princeton Architectural Press: New York.

Elmqvist, T., Seta, H., Handel, S.N., van der Ploeg, S., Aronson, J., Blignaut, J.N., Gómez-Baggethun, E., Nowak, D.J., Kronenberg. J. and de Groot, R. (2015). Benefits of restoring ecosystem services in urban areas, *Current Opinion in Environmental Sustainability,* 14:101–108

Fragklas, M., Güneralp, B., Seto, K.C. and Goodness, J. (2013). A Synthesis of Global Urbanization Projections. In: *Urbanization, Biodiversity and Ecosystem Services: Challenges and Opportunities, A Global Assessment,* eds. Elmqvist et al. Heidelberg: Springer.

Kabisch, N., Storbach, M., Haase, D. and Kronenberg, J. (2016). Urban green space availability in European cities. *Ecological Indicators* (open access).

Li, X., Zhou, Y., Eom, J., Yu, S., & Asrar, G. R. (2019). Projecting global urban area growth through 2100 based on historical time series data and future Shared Socioeconomic Pathways. *Earth's Future,* 7,351–362.

Schantz, P. (2006). The Formation of National Urban Parks: A Nordic Contribution to Sustainable Development? In: Clark, P. (Ed). *The European City and Green Space; London, Stockholm, Helsinki and S:t Petersburg, 1850–2000.* Aldershot, England: Ashgate, pp. 159–174.

Statens offentliga utredningar (SOU 1996:38), *Nationalstadsparker.* Stockholm: Fritzes.

Steenbergen, C. and Reh, W. (2011). *Metropolitan Landscape Architecture—Urban Parks and Landscapes.* Thoth publishers: Bussum, the Netherlnds

United Nations (2018). *The World's Cities.* Available at: https://read.un-ilibrary.org/human-settlements-and-urban-issues/the-world-s-citiesin-2018_c93f4dc6-en#page1

Waldenström, H. (1990). *Ekoparken—natur-och kulturpark i storstad.* (Ekoparken – A Nature and Culture Reserve in a Big City) Ståthållareämbetet.

VALUES OF LARGE-VERSUS-SMALL URBAN GREENSPACES AND THEIR ARRANGEMENT

Bennett, A. F. (2003). *Linkages in the Landscape: The Role of Corridors and Connectivity in Wildlife Conservation.* Gland, Switzerland and Cambridge, UK: IUCN, The World Conservation Union.

Benton-Short, L. and Short, J. R. (2008). *Cities and Nature.* London: Routledge.

Chang, C. R., Li, M. H. and Chang, S. D. (2007). A preliminary study of the local cool-island intensity of Taipei city parks. *Landscape and Urban Planning,* 80, 386–395.

Davis, A., Hunt, W., Traver, R. and Clar, M. (2009). Bioretention technology overview of current practice and future needs. *Journal of Environmental Engineering,* 135, 109–117.

Forman, R. T. T. (1995). *Land Mosaics: Ecology of Landscapes and Regions.* New York: Cambridge University Press.

Forman, R. T. T. (2004). *Mosaico territorial para la region metropolitana de Barcelona.* Barcelona: Editorial Gustavo Gili.

Forman, R. T. T. (2008). *Urban Regions: Ecology and Planning Beyond the City.* New York: Cambridge University Press.

Forman, R. T. T. (2014). *Urban Ecology: Science of Cities.* New York: Cambridge University Press.

Forman, R. T. T. (2019). *Towns, Ecology, and the Land.* New York: Cambridge University Press.

Forman, R. T. T. and Wu, J. (2016a). Where are the best places for the next billion people? Think globally, plan regionally. In: D. Geneletti (Ed.) *Handbook on Biodiversity and Ecosystem Services in Impact Assessment.* Cheltenham, UK: Edward Elgar Publishers, pp. 453–473.

Forman, R. T. T. and Wu, J. (2016b). Where to put the next billion people. *Nature,* 537, pp. 608–611.

Forman, R. T. T., Galli, A. E. and Leck, C. F. (1976). Forest size and avian diversity in New Jersey woodlots with some land-use implications. *Auk,* 93, 356–364.

Gilbert, O. L. (1991). *The Ecology of Urban Habitats.* London: Chapman & Hall.

Giuliano, W. M. (2005). Lepidoptera—habitat relationships in urban parks. *Urban Ecosystems,* 7, pp. 361–370.

Giusti de Perez, R. C. and Perez, R. A. (2008). *Analyzing Urban Poverty: GIS for the Developing World.* Redlands, California: ESRI Press.

Houck, M. C. and Cody, M. J. (2000). *Wild in the City: A Guide to Portland's Natural Areas.* Portland, Oregon: Oregon Historical Society Press.

Hurley, S. E. and Forman, R. T. T. (2011). Stormwater ponds and biofilters for large urban sites: modeled arrangements that achieve the phosphorus reduction target for Boston's Charles River, USA. *Ecological Engineering,* 37, pp. 850–863.

Keenan, J. M. and Weisz, C. (2017). *Blue Dunes: Climate Change by Design.* New York: Columbia University Press.

Konijnendijk, C. C., Nilsson, K., Randrup, T. B. and Schipperijn, J., eds. (2005). *Urban Forests and Trees.* New York: Springer.

Kowarik, I. and Korner, S., eds. (2005). *Wild Urban Woodlands: New Perspectives for Urban Forestry.* New York: Springer.

Loeb, R. E. (2006). A comparative flora of large urban parks: intraurban and inter-urban similarity in the megalopolis of the northeastern United States. *Journal of the Torrey Botanical Society,* 133, pp. 601–625.

McPherson, E. G. (1994). Benefits and costs of tree planting and care in Chicago. In: E. G. McPherson, D. J. Nowak and R. A. Rowntree (eds.). *Chicago's Urban Forest Ecosystem: Results of the Chicago Urban Forest Climate Project.* General Technical Report NE-186. Radnor, Pennsylvania: U.S. Department of Agriculture, Forest Service, pp. 115–134.

New York Times. (February 16, 1997), p. 1.

Opdam, P. (1991). Metapopulation theory and habitat fragmentation: a review of Holarctic breeding bird studies. *Landscape Ecology,* 5, pp. 93–106.

Ozawa, C. P. (2004). *The Portland Edge: Challenges and Successes in Growing Communities.* Washington, D.C.: Island Press.

Pezzoli, K. (1998). *Human Settlements and Planning for Ecological Sustainability.* Cambridge, Massachusetts: MIT Press.

Simberloff, D. and Abele, L. (1976). Island biogeography theory and conservation practice. *Science,* 191, pp. 285–286.

Spirn, A. W. (1984). *The Granite Garden: Urban Nature and Human Design.* New York: Basic Books.

Todd, D. K. and Mays, L. W. (2005). *Groundwater Hydrology.* New York: John Wiley.

UN Population Division. (2017). *World Urbanization Prospects: The 2017 Revision.* New York: United Nations.

von Stulpnagel, A., Horbert, A. and Sukopp, H. (1990). The importance of vegetation for the urban climate. In: H. Sukopp, S. Hejny and I. Kowarik (eds.), *Urban Ecology*. The Hague, Netherlands: SPB Academic Publishing, pp. 175–193.

Wang, H. and Tassinary, L.G. (2019). Effects of greenspace morphology on mortality at the neighborhood level: a cross-sectional ecological study. *Lancet & Planetary Health*, 3, pp. 460–468.

Welty, C. (2009). The urban water budget. In *The Water Environment of Cities*, ed. L. A. Baker. New York: Springer, pp. 17–28.

LARGE URBAN PARKS AND URBAN GREEN SPACE: A HISTORICAL PERSPECTIVE

Baycan-Levent, T. et al. (2009). A multi-criteria evaluation of green spaces in European Cities, *European Urban and Regional Studies*, 16.

Brantz, D. (2017). The urban politics of nature … Berlin 1800–2014, in Clark et al., (eds.).

Brantz, D. and Dümpelmann, S. (eds.) (2011). *Greening the City: Urban Landscapes in the Twentieth Century*. London: University of Virginia Press.

Clark, P. (2017). Urban green space in a globalising world, in Clark, P., Niemi, M. and Nolin, C. (eds.).

Clark, P. (ed.) (2006). *The European City and Green Space*. London, Stockholm, Helsinki and St Petersburg: Aldershot.

Clark, P., Nolin, C. and Niemi, M. (eds.) (2017). *Green Landscapes in the European City*, London: Routledge.

Clark, P. and Menjot, D. (2019). *Subaltern City? Alternative and Peripheral Urban Spaces in the Pre-Modern Period, (13th to 18th centuries)*. Serie: Studies in European Urban History (1100–1800), vol. 46.

Conway, H. (1991). *People's Parks: The Design and Development of Victorian Parks in Britain*. Cambridge University Press.

Czerniak, J. (2007). Legibility and resilience, in Czerniak and Hargreaves.

Czerniak, J. and Hargreaves, G. eds. (2007). *Large Parks*. Princeton University Press: Princeton, New York.

Duxbury, G. (2016), article in *Guardian*, 29th September.

Elliot, P. A. (2016), *British Urban Trees. A Social and Cultural History*. Winwick: White Horse Press.

Garside, P.L. (2006). Politics, ideology and the issue of open space in London, 1939–2000, in Clark, P. (ed.).

Hamadeh, S. (2007). Public spaces and the garden culture of Istanbul in the 18th century, in Aksan, V. H. and Goffman, D. (eds.) *The Early Modern Ottomans*. Cambridge University Press.

Hannikainen, M. O. (2014). *The Greening of London*. Helsinki: Taylor & Francis.

Hannikainen, M. O. (2017). London's green spaces in the late 20th century, in Clark et al., (eds.).

Hargreaves, G. (2007). Large parks: A Designer's Perspective, in Czerniak and Hargreaves (eds.).

Mack, J. and Parscher, J.S. (2017), The right to the garden: allotments and the politics of urban green space in Sweden, in Clark et al., (eds.).

Ojala, A. et al. (2017). Impacts of residential infilling on private gardens, in Clark et al., (eds.).

Orvell, M. and Meikle, J.L (eds.) (2009). *Public Space and the Ideology of Place in American Culture*. New York: Rodopi.

Owen, J. (2014), article in *Financial Times*, 1st August.

Pinol, J.L. (2017). Vegetation and green spaces in Paris, in Clark et al., (eds.).

Reeder, D. (2006). London and green space 1850–2000, in Clark (ed.).

Schantz, P. (2006). The formation of National Urban Parks, in Clark (ed.).

Taylor, D.E. (2009). *The Environment and the People in American Cities 1600s–1900s*. London: Duke University Press.

von Hoffman, A. (1988). Of Greater Lasting Consequence, Frederick Law Olmsted and the Fate of Franklin Park, Boston. *Journal of the Society of Architectural Historians*, 47.

https://en.wikipedia.org/wiki/Amsterdamse_Bos

http://whc.unesco.org/en/list/881;

https://en.wikipedia.org/wiki/Temple_of_Heaven

https://en.wikipedia.org/wiki/Sanjay_Gandhi_National_Park

http://www.fieldoperations.net/project-details/project/freshkills-park.html

URBAN PARK TRADITIONS IN IRAN AND TEHRAN

John-Alder, K. (2016). Paradise reconsidered: the early design history of Pardisan Park in Tehran, *Contemporary Urban Landscapes of the Middle East*, ed. M. Gharipour. London: Routledge,.

McHarg, L. (1975). *Pardisan: Plan for an environmental park in Tehran*, The Mandata Collaborative/Wallace, McHarg, Roberts and Todd, Philadelphia, Pennsylvania.

Farahani, M.L., Motamed, B. and Jamei, E. (2016). Persian gardens: Meanings, symbolism, and design, *Landscape online*, vol. 46.

Mashayekhi, A. (2018). The 1968 Tehran master plan and the politics of planning development in Iran (1945–1979), *Planning Perspectives (online)*, May 24.

Mirsadeghi, P. (2016). Tehran and the lost nature, in *Urban Change in Iran*, F. F. Arefian and S. H. I. Moeini (Eds.), Springer.

Malekzadeh, R. (1992). *The history of old gardens and park design*, Municipality of Tehran, Tehran Parks and Green Spaces organization.

Shobeiri S. (2018). The Role of Large Parks in Human-Nature Interaction and Socio-Cultural Sustainability in Cities—Context: Tehran and the River-Valleys, *Spaces and Flows: An International Journal of Urban and ExtraUrban Studies*, 9(2), pp. 45–78.

Sedaghatfar, J. (1971). *Entopia*. Kent State University, School of Architecture.

PHOENIX PARK, DUBLIN, IRELAND

International charters respected in the maintenance and development of Phoenix park:

Recommendation concerning safeguarding of the Beauty and Character of Landscape and Sites UNESCO, 1962

Convention for the Protection of the Architectural Heritage of Europe, 1985 (Ireland is a signatory)

The Athens Charter for the Restoration of Historic Monuments (1931)

The Venice Charter (1964)

The Burra Charter (1979, revised 1999)

The Florence Charter on Historic Gardens (1982)

Charter for the Protection and Management of the Archaeological Heritage, The Valetta Convention (1990)

Principles for the Recording of Monuments, Groups of Buildings and Sites (1996)

International Cultural Tourism Charter (1999)

PAVLOVSKY PARK, ST PETERSBURG, RUSSIA

Emelyanova, A. (2017). *Analysis of the "White Birch Area" in Pavlovsky Park*. Master's Thesis. St. Petersburg State Forest Technical University, St. Petersburg (in Russian).

Ignatieva, M. (2005). Music for the eyes: The historical restoration of the White Birch Area of Pavlovsky Park in St. Petersburg, Russia. *Ecological Restoration* 23:83–88.

Ignatieva, M., Melnichuk I., Cherdantseva O. & Lukmazova E. (2015). History and restoration of the St. Petersburg Summer Garden: Returning to the roots. In: *Garden History* 43: 199–217.

Hayden, P. (2005). *Russian Parks and Gardens*. Frances Lincoln Ltd.

Ilyinskaya, N. (1993). *Reconstruction of Historic Landscape Architecture Objects*, St Petersburg: Stroiisdat (in Russian).

Kuchumov, A. (1980). Pavlovsk. Lenisdat, Leningrad (in Russian).

BIRKENHEAD PARK, WIRRAL, UNITED KINGDOM

Davey, E (2008) *Birkenhead, A History*, Phillimore, West Sussex. (now The History Press Ltd.)

Davey, E. (2010), 'A Complete and Constant Superintendence': The Cheshire Parks and Gardens of Edward Kemp (1817–1891, *Cheshire History*, Cheshire Local History Association (50), pp. 71–99.

Davey, E. (2018). Edward Kemp (1817–1891): 'an able and useful man', in *Garden History* 46: supplement 1, Autumn 2018.

Lee, R. (2013). *The People's Garden? A History of Crime and Policing in Birkenhead Park* (Vol. 1). Liverpool.

Lee, R. (2018). The Challenge of Managing the First Publicly funded Park: Edward Kemp as the 'fixed' Superintendent of Birkenhead Park, 1843–1891, in *Garden History* 46: supplement 1, Autumn 2018.

CENTRAL PARK, NEW YORK, UNITED STATES

Miller, S.C. (2003), *Central Park, An American Masterpiece*. New York: Abrams.

Miller. S.C, (2009). *Seeing Central Park*. New York: Abrams.

Miller, S.C. (2011), *Strawberry Fields: Central Park's Memorial to John Lennon*. New York: Abrams.

Rosenzweig, R. and Blackmar, E. (1992). *The Park and the People*. Cornell University Press.

CAN NATURE REALLY AFFECT OUR HEALTH?

Améen, L. (1915). Trettio år. *Svenska Turist-föreningens årsskrift*, s. IX–XIV. Stockholm: Svenska Turistföreningen.

van den Berg, A., Joye, Y., de Vries, S. (2019). Health Benefits of Nature: An Introduction. In *Environmental Psychology*, 2 ed. (eds. Steg, L, de Groot, J.I.M.) Chichester: John Wiley & Sons.

van den Bosch, M., Bird, W. (2018). *Oxford Textbook of Nature and Public Health: The role of nature in improving the health of a population*. Oxford: Oxford University Press.

Bratman, G.N., Hamilton, J.P., Hahn, K.S., Daily, G.C., Gross, J.J. (2015). Nature experience reduces rumination and subgenual prefrontal cortex activation. *Proceedings of the National Academy of Sciences of the United States of America* 112(28):8567–72.

Ceci, R., Hassmén, P. (1991). Self-monitored exercise at three different RPE intensities in treadmill vs field running. *Med Sci Sports Exerc.* 6:732–8.

Engemann, K., Pedersen, CB., Arge, L., Tsirogiannis, C., Mortensen, PB., Svenning, JC. (2019). Residential green space in childhood is associated with lower risk of psychiatric disorders from adolescence into adulthood. *Proc Natl Acad Sci USA.* pii: 201807504.

Hohwü Christensen, E. (1945). Friluftslivets fysiologi och hygien. In: Ivar Holmquist & Carl Nordenson (eds.). *På skidor*. Stockholm: Föreningen för skidlöpningens och friluftslivets främjande i Sverige. pp. 18–26.

Maas, J., Verheij, RA., de Vries, S., Spreeuwenberg, P., Schellevis, FG., Groenewegen, PP. (2009). Morbidity is related to a green living environment. *J Epidemiol Community Health* 63(12):967–73.

Markevych, I., Schoierer, J., Hartig, T., Chudnovsky, A., Hystad, P., Dzhambov, AM., de Vries, S., Triguero-Mas, M., Brauer, M., Nieuwenhuijsen, MJ., Lupp, G., Richardson, EA., Astell-Burt, T., Dimitrova, D., Feng, X., Sadeh, M., Standl, M., Heinrich, J., Fuertes, E. (2017). Exploring pathways linking greenspace to health: Theoretical and methodological guidance. *Environ Res.* 158:301–317

Mitchell, R. and Popham, F. (2008). Effect of exposure to natural environment on health inequalities: an observational population study. *Lancet* 372(9650):1655–60.

Nawrath, M., Kowarik, I. & Fischer, LK. (2019). The influence of green streets on cycling behaviour in European cities. *Landscape and Urban Planning* 190, 103598

Schantz, P. (2003). Fysisk aktivitet och hälsa kräver goda miljöer. Vilka är de och hur skapas de? (Physical activity and health require good environments. Which are they and how are they created?) *Svensk Idrottsforskning* 1:6–11.

Ulrich, RS. (1984). View through a window may influence recovery from surgery. *Science* 224: 420–421.

Ulrich, RS., Simons, RF., Losito, BD., Fiorito, E., Miles, MA., Zelson, M. (1991). Stress recovery during exposure to natural and urban environment. *J Environ Psychol* 11:201–230.

de Vries, S., Verheij, R.A., Groenewegen, P., Spreeuwenberg, P. (2003). Natural environments—Healthy environments? An exploratory analysis of the relationship between greenspace and health. *Environment and Planning A* 35(10):1717–1731

de Vries, S., Claßen, T., Eigenheer-Hug, S-M., Korpela, K., Maas, J., Mitchell, R., Schantz, P. (2011). Contributions of Natural Environments to Physical Activity. Theory and Evidence Base. In *Forests, Trees and Human Health,* Eds. K. Nilsson, M. Sangster, C. Gallis, T. Hartig, , S. de Vries, K. Seeland, & J. Schipperijn, Berlin: Springer Verlag, pp. 205–243.

Wahlgren L. and Schantz P. (2012). Exploring bikeability in a metropolitan setting: stimulating and hindering factors in commuting route environments. *BMC Public Health* 12:168.

Wahlgren L. and Schantz P. (2014). Exploring Bikeability in a Suburban Metropolitan Area using the Active Commuting Route Environment Scale (ACRES). *Int J Environ Res Public Health* 11 (8) 8276–8300.

WHO (1946). *Preamble to the constitution of the World Health Organization as adopted by the International Health Conference, New York, 1946*. Entered into force on 7 April 1948. New York: WHO, pp. 19–22.

WHO Europe (2016). *Urban green spaces and health. A review of evidence.* Copenhagen: WHO Europe.

A GREEN INFRASTRUCTURE PLAN FOR GOTHENBURG, SWEDEN

Göteborg (2014). *Grönstrategi för en tät och grön stad* (Green strategy for a green and dense city). Park och naturnämnden, 2014-02-10.

TIJUCA NATIONAL PARK, RIO DE JANEIRO, BRAZIL

Coelho-Netto, A.L., Avelar, A.S., Fernandes, M.C., Lacerda W.A. (2007). Landslide susceptibility in a mountainous geoecosystem, Tijuca Massif, Rio de Janeiro: The role of morphometric subdivision of the terrain. *Geomorphology* 87 (3),120–131.

Coimbra, M.S., Martinelli, G., Menezes, P.C. (2001). *Parque Nacional da Tijuca: 140 anos da reconstrução de uma floresta*. Rio de Janeiro: Editora Ouro sobre Azul.

Herzog, C. (2013). *Cities for All: (re)learning to live with Nature*. Rio de Janeiro: Mauad X.

Herzog, C.P. and Finotti, R. (2013). Local Assessment of Rio de Janeiro City: Two Case Studies of Urbanization Trends and Ecological Impacts. In: T. Elmqvist, et al. (eds.). *Urbanization, Biodiversity and Ecosystem Services: Challenges and Opportunities, A Global Assessment*). Heidelberg: Springer open.

Silva, R. F. (2015). *O Rio antes do Rio*. Rio de Janeiro: Editora Babilônia Cultural.

Vieira, A.C.P. (2001). *Lazer e Cultura na Floresta da Tijuca: história, arte, religião, fauna, flora e literatura*. Rio de Janeiro: Makron Books.

CHAPULTEPEC FOREST, MEXICO CITY, MEXICO

Fideicomiso Probosque de Chapultepec (2011). *Bosque de Chapultepec*, Mexico City.

WHY LARGE PARKS MATTER FOR BIODIVERSITY AND ECOSYSTEM SERVICES

Aguilera, G., Ekroos, J., Persson, A. S., Pettersson, L. B. and Öckinger, E. (2018). Intensive management reduces butterfly diversity over time in urban green spaces. *Urban Ecosystems.*

Aronson, M. F. J., Lepczyk, C. A., Evans, K. L., Goddard, M. A., Lerman, S. B., MacIvor, J. S., Nilon, C. H. and T. Vargo (2017). Biodiversity in the city: key challenges for urban green space management. *Frontiers in Ecology and the Environment* 15:189–196.

Baldock, K. C. R., Goddard, M. A., Hicks, D. M., Kunin, W. E., Mitschunas, N., Osgathorpe, L. M., Potts, S. G., Robertson, K. M., Scott, A. V., Stone, G. N., Vaughan, I. P. and Memmott J. (2015). Where is the UK's pollinator biodiversity? The importance of urban areas for flower-visiting insects.

Proceedings of the Royal Society B: Biological Sciences 282.

Boverket (2007). *Bostadsnära natur—Inspiration & vägledning.*

Butt, N., Shanahan, D. F., Shumway, N., Bekessy, S. A., Fuller, R. A., Watson, J. E. M., Maggini, R. and Hole, D. G. (2018). Opportunities for biodiversity conservation as cities adapt to climate change. *Geo: Geography and Environment* 5:e00052.

CBD (Convention on Biological Diversity) (2012). *Cities and Biodiversity Outlook.* CBD, Montreal.

Chong, K. Y., Teo, S., Kurukulasuriya, B., Chung, Y. F., Rajathurai, S. and Tan, H. T. W. (2014). Not all green is as good: Different effects of the natural and cultivated components of urban vegetation on bird and butterfly diversity. *Biological Conservation* 171:299–309.

EEA (2015). Exploring nature-based solutions. *The role of green infrastructure in mitigating the impacts of weather- and climate change-related natural hazards*. EEA Technical report No 12/2015. ISSN 1725-2237. European Environment Agency, Luxembourg.

Elmqvist, T., Setälä, H., Handel, S. N., van der Ploeg, S., Aronson, J., Blignaut, J. N., Gómez-Baggethun, E., Nowak, D. J., Kronenberg, J. and de Groot, R. (2015). Benefits of restoring ecosystem services in urban areas. *Current Opinion in Environmental Sustainability* 14:101–108.

Foley, J. A., Defries, R., Asner, G. P., Barford, C., Bonan, G., Carpenter, S. R., Chapin, F. S., Coe, M. T., Daily, G. C., Gibbs, H. K., Helkowski, J. H., Holloway, T., Howard, E. A., Kucharik, C. J., Monfreda, C., Patz, J. A., Prentice, I. C., Ramankutty, N. and P. K. Snyder (2005). Global consequences of land use. *Science* 309;570–574.

Geschke, A., S. James, Bennett, A. F. and D. G. Nimmo (2018). Compact cities or sprawling suburbs? Optimising the distribution of people in cities to maximise species diversity. *Journal of Applied Ecology* 55:2320–2331.

Gómez-Baggethun, E., and D. N. Barton (2013). Classifying and valuing ecosystem services for urban planning. *Ecological Economics* 86:235–245.

Haaland, C., and C. K. van den Bosch (2015). Challenges and strategies for urban green-space planning in cities undergoing densification: A review. *Urban Forestry &*

Urban Greening 14:760–771.

Hedblom, M. and Söderström, B. (2008). Woodlands across Swedish urban gradients: Status, structure and management implications. *Landscape and Urban Planning* 84:62–73.

Ignatieva, M., Eriksson, F., Eriksson, T., Berg, P. and M. Hedblom (2017). The lawn as a social and cultural phenomenon in Sweden. *Urban Forestry & Urban Greening* 21:213–223.

Ignatieva, M., and M. Hedblom (2018). An alternative urban green carpet. *Science* 362:148–149.

Ives, C. D., Lentini, P. E., Threlfall, C. G., Ikin, K., Shanahan, D. F., Garrard, G. E., Bekessy, S. A., Fuller, R. A., Mumaw, L., Rayner, L., Rowe, R., Valentine, L. E. and Kendal, D. (2016). Cities are hotspots for threatened species. *Global Ecology and Biogeography* 25:117–126.

Lin, B. B., and Fuller, R. A. (2013). FORUM: Sharing or sparing? How should we grow the world's cities? *Journal of Applied Ecology* 50:1161–1168.

McDonald, R. I., Beatley, T. and Elmqvist, T. (2018). The green soul of the concrete jungle: the urban century, the urban psychological penalty, and the role of nature. *Sustainable Earth* 1:3.

Narango, D. L., Tallamy, D. W. and P. P. Marra (2018). Nonnative plants reduce population growth of an insectivorous bird. *Proceedings of the National Academy of Sciences* 115:11549.

Shanahan, D. F., Lin, B. B., Bush, R., Gaston, K. J., Dean, J. H., Barber, E. and R. A. Fuller (2015). Toward improved public health outcomes from urban nature. *American Journal of Public Health* 105:470–477.

Soga, M., and Gaston, K. J. (2018). Shifting baseline syndrome: causes, consequences, and implications. *Frontiers in Ecology and the Environment* 16:222–230.

Soga, M., Yamaura, Y., Koike, S. and Gaston, K. J. (2014). Land sharing vs. land sparing: does the compact city reconcile urban development and biodiversity conservation? *Journal of Applied Ecology* 51:1378–1386.

Stott, I., Soga, M., Inger, R. and Gaston, K. J. (2015). Land sparing is crucial for urban ecosystem services. *Frontiers in Ecology and the Environment* 13:387–393.

Sushinsky, J. R., Rhodes, J. R., Possingham, H. P., Gill, T. K. and Fuller, R. A. (2013).

How should we grow cities to minimize their biodiversity impacts? *Global Change Biology* 19:401–410.

Sushinsky, J. R., Rhodes, J. R., Shanahan, D. F., Possingham, H. P. and R. A. Fuller (2017). Maintaining experiences of nature as a city grows. *Ecology and Society* 22.

The Royal Society (2014). *Resilience to extreme weather.* The Royal Society Science Policy Centre report 02/14. 02/14, The Royal Society Science Policy Centre, London SW1Y 5AG.

WHO (World Health Organsation) (2016). *Urban green spaces and health.* WHO Regional Office for Europe, Copenhagen.

KARAWATHA FOREST, BRISBANE, AUSTRALIA

Bond, A.R. and Jones, D.N. 2008. Temporal trends in use of fauna–friendly underpasses and overpasses. *Wildlife Research*, 35(2), pp.103–112.

Pell, S. and Jones, D. 2015. Are wildlife overpasses of conservation value for birds? A study in Australian sub–tropical forest, with wider implications. *Biological Conservation*, 184, pp. 300–309.

McGregor, M., Matthews, K. and Jones, D. 2017. Vegetated Fauna Overpass Disguises Road Presence and Facilitates Permeability for Forest Microbats in Brisbane, Australia. *Frontiers in Ecology and Evolution*, 5, p.153.

McGregor, M.E., Wilson, S.K. and Jones, D.N. 2015. Vegetated fauna overpass enhances habitat connectivity for forest dwelling herpetofauna, *Global Ecology and Conservation*, 4, pp. 221–231.

Taylor, B. and Goldingay, R. 2012. Restoring Connectivity in Landscapes Fragmented by Major Roads: A Case Study Using Wooden Poles as "Stepping Stones" for Gliding Mammals. *Restoration Ecology*, 20(6), pp.671–678.

SANJAY GANDHI NATIONAL PARK, MUMBAI, INDIA

Monga, S. (2000). *City Forest: Mumbai's National Park.* India Book House.

THE PARKS OF NEW ORLEANS, UNITED STATES

Barry, J. (1997). *Rising Tide: The Great Mississippi Flood and How it Changed America.* Simon and Schuster.

Goodyear, Sarah (2014). *Living with Water to Stop Flooding in New Orleans, How Do You Keep a Sinking City Afloat?* Available at: https://nextcity.org/

Waggonner and Ball architects (2013). *Greater New Orleans Urban Water Plan.*

INCLUSIVE LARGE PARKS FOR EVERYONE

Carson, R. (1962). *Silent Spring.* Boston: Houghton Mifflin Company.

O'Donnell, P. (2016). Enabling Access to public spaces to advance economic, environmental and social benefits, in *Culture Urban Future, Global Report on the Culture for Sustainable Urban Development.* UNESCO.

UNESCO (2011). *Recommendations on the Historic Urban Landscape (HUL).*

United Nations, World Commission on Environment and Development (WCED) (1987). *Our Common Future (Brundtland Report).* Oxford University Press.

United Nations (2015). *Transforming our world: the 2030 Agenda for Sustainable Development.* Resolution adopted by the General Assembly on 25 September 2015.

United Nation Habitat (2016), *The New Urban Agenda.*

Wilson, E.O. (1984). *Biophilia.* Boston: Harvard University Press.

World Health Organization (WHO) (2017), *Urban green spaces: a brief for action.*

AMSTERDAMSE BOS, THE NETHERLANDS

Dupont, S., van der Werf, J. and Heeren, J. (2019). *Amsterdamse Bos—A biography of an urban forest.* Bussum: THOTH Publishers.

SILESIA PARK, CHORZÓW, POLAND

Fortuna-Antoszkiewicz, B., Łukaszkiewicz J., Wiśniewski P. (2013–2014). *The Gen. Jerzy Ziętek Voivodship Park of Culture and Recreation in Chorzów. General Inventory of Park Stand (Part 1) and Studies and Analyzes (Part 2).* Manuscript: Warsaw, SGGW; Silesia Park, Chorzów.

Fortuna-Antoszkiewicz B., Łukaszkiewicz J., Wiśniewski P. (2016). *The Voivodship Park of Culture and Recreation in Chorzów (Silesia Park)—history and spatial composition.* Urbanity and Architecture Files, Krakow Section of Polish Academy of Sciences, VOL. XLIV, pp. 203–213.

Fortuna-Antoszkiewicz B., Łukaszkiewicz J., Wiśniewski P. (2017). *The transformation of vegetation's composition 60 years after the establishment of Silesia Park.* Urbanity and Architecture Files, Krakow Section of Polish Academy of Sciences, XLV, pp. 193–215.

Knobelsdorf W. ed. (1972). *Oaza pod rudym obłokiem: Śląski Park Kultury i Wypoczynku.* Editorial "Śląsk" (Wydawnictwo "Śląsk"), Katowice.

Łukaszkiewicz J., Fortuna-Antoszkiewicz B. (2017). *Silesia Park in Chorzów / Poland— the successful re-naturalization of industrial landscape after 60-years.* Miškininkystė ir Kraštotvarka, Forestry and Landscape Management 2017 1 (12). Lithuania, pp. 25–34.

Niemirski W., Słotwiński K. (eds.) (1963). *Śląski Park Kultury.* Editorial "Śląsk" (Wydawnictwo "Śląsk"), Katowice.

ALBERT PARK, MELBOURNE, AUSTRALIA

Barnard, J. and Keating, J. (1996). *People's Playground. A History of Albert Park.* Chandos.

Parks Victoria (2019). *Albert Park Masterplan.*

Willis, D. (2019). *A Guide to Historic St Kilda,* St Kilda Press.

LONDON: A NATIONAL PARK CITY, UNITED KINGDOM

Loudon, J. C. (1829). *Hints for Breathing Places for the Metropolis. A Landscape Architecture London Plan.*

London National Park City (2020) Available at: http://www.nationalparkcity.london Accessed 19 April 2020.

Urban Good CIC (2017), *London National Park City Map.*

THE CITY MEETS NATURE

Backéus, I., Emanuelsson,U. & Petersson, M. (2016). *The Rural Landscapes of Northeast Asia,* Fri Tanke.

Emanuelsson, U. (2009), *The rural landscapes of Europe, How man has shaped European nature,* Stockholm: Formas.

Environmental Protection Agency (Naturvårdsverket), Sweden. Long Time Planning for Sustainable Landscapes. Available at: https://www.naturvardsverket.se/Documents/publikationer6400/978-91-620-8807-1.pdf?pid=22473

PERI-URBAN PARKS

Fedenatur, (2004). *The place of peri-urban natural spaces for a sustainable city.* Report to the European Commission, DG Environment. January.

Sundseth, K. & Raeymaekers, G., (2006). Biodiversity and Natura 2000 in urban areas. A review of issues and experiences of nature in cities across Europe. Ecosystems Ltd, November.

LOSINY OSTROV, MOSCOW, RUSSIA

Arshinova S.N. (2009). Geoinformational Methods in Ecodiagnostical Studies for the Losiny Ostrov National Park Territory. In *Cartography in Central and Eastern Europe. Lecture Notes in Geoinformation and Cartography.* Gartner G. and Ortag F. eds. Berlin, Heidelberg: Springer. pp. 193–198.

Kavtaradze, D.N. (2018). Urbanization of Biosphere: from Mega- to Ecopolises. In: V. I. Vasenev et al. (eds.). *Megacities 2050: Environmental Consequences of Urbanization.* Springer Geography.

Voronin, F.N. and Kiseleva, V.V. eds. (2014). *Scientific works on the Losiny Ostrov National Park.* Moscow: Typography ABT Group (in Russian).

BOGOTÁ'S ENVIRONMENTAL CIRCUIT, COLOMBIA

Territorial Planning Plan of Bogotá (POT), 2019.

FLORENCE AND FARO, TWO CITIES AND THEIR CONVENTIONS: PARKS AND EVERYDAY LIFE IN MAJOR CITIES

Council of Europe, (2009). *Heritage and Beyond,* Strasbourg: Council of Europe. (Available online).

Olwig, K. R. 2019. *The Meanings of Landscape: Essays on Place, Space, Environment and Justice.* Routledge.

UNESCO, (2011). Historic Urban Landscape declaration (HUL).

THE ROYAL NATIONAL CITY PARK, STOCKHOLM, SWEDEN

Brusewitz, G. and Ekman, H. (1995). Ekoparken, Djurgården–Haga–Ulriksdal, Stockholm: Wahlström & Widstrand (in Swedish).

Hammarström, T. and Utgren, L. (2004), Ekoparken—the Royal Parks of Djurgården and Haga. Örebro: Gullers.

Holm, L. and Schantz, P. (2002). *Nationalstadsparken—ett experiment i hållbar utveckling. Studier av värdefrågor, lagtillämpning och utvecklingslinjer.* (National City Park—a sustainable development experiment. Values, Law Application and Future.) Formas (in Swedish).

Sporrong, U. (ed.) (2018). Platsens kulturella betydelse. Juridiken kring nationalstadsparken. Konferenser (96). Stockholm: Kungl. Vitterhetsakademien (in Swedish).

Waldenström, H. (1990). *Ekoparken—ett sammanhängande natur- och kulturlandskap.* Ståthållareämbetet (in Swedish).

CENTRAL DELHI HERITAGE PARK, INDIA

Khanna, N. P. (2015). Open Cities, Closed Spaces: Envisioning Inclusive Landscape Design. Delhi: *Architecture Design,* 32(9): 54–58.

Krishen, P. (2006). *Trees of Delhi: A Field Guide,* NTACH & Delhi Tourism. Delhi: Dorling Kindersley Publishing.

Nanda, R. (2016). Working in Paradise with the Master, Mohammad Shaheer. *Manzar,* 7 (33): 28–37.

Sharma, Y.D. (1964). *Delhi and its Neighbourhood.* New Delhi: Organizing Committee, XXVI International Congress of Orientalists.

Wescoat Jr., J. L. (2012). The Changing Cultural Space of Mughal Gardens, *A Companion to Asian Art and Architecture,* Brown, D.S. and R. M., Hutto (eds.). New Jersey: John Wiley & sons, pp. 201–229.

KINGS PARK, PERTH, AUSTRALIA

Botanic Gardens and Parks Authority (2016). *Botanic Gardens and Parks Authority 2015–2016 Annual Report.*

Dixon, B., & Moonie, P. (2003). Ecological restoration of a cliff face in Kings Park and Botanic Gardens, Perth, Western Australia. *Botanic Gardens Conservation International,* 4(1).

Erickson, D. (2009). *A Joy Forever: The Story of Kings Park and Botanic Garden.* Botanic Gardens and Parks Authority.

Grose, M. (2011). Considering Ecological Imperatives in Public Open Space in a Global Hotspot of Biodiversity. *Landscape Review,* 13(2): 6–25.

Miller, J. R. (2005). Biodiversity conservation and the extinction of experience. *Trends in Ecology and Evolution,* 20(8): 430–434.

Newman, P. (2014). Density, the Sustainability Multiplier: Some Myths and Truths with Application to Perth, Australia. *Sustainability,* 6(9): 6467–6487.

Stratton, J., & Trainer, A. (2016). Nothing happens here: Songs about Perth. *Thesis Eleven,* 135(1): 34–50.

Taylor, H. A. (1995). Urban Public Parks, 1840–1900: Design and Meaning. *Garden History,* 23(2): 201–221.

LARGE PARKS: TRENDS (AND POSSIBILITIES)

Czerniak, J. (2007), Legibility and Resilience, in *Large Parks.* J. Czerniak and G. Hargreaves (eds.). New York: Princeton Architectual Press.

Hoorneg, D. and Bhada-Tata, P, (2012). *What a Waste: A Global Review of Solid Waste Management,* World Bank.

Wilson, E., O., 2003. *The Future of Life,* New York: Vintage Press.

World Economic Forum (2020). *The Global Risks Report 2020.*

AMAGER NATUREPARK, COPENHAGEN, DENMARK

Caspersen O.H, Olafsson A.S (2010). Recreational mapping and planning for enlargement of the green structure in Greater Copenhagen. *Urban Forestry & Planning* 9(2): 100–112.

Forsvarministeriet & Miljøministeriet (1974): *Forsvarets arealer.* En beretning fra Forsvarministeriets Naturfredningsudvalg.

Naturstyrelsen (2014). *Amager—Naturparkplan 2015–2020.* Naturstyrelsen Hovedstaden, Tårnby Kommune, Dragør Kommune, og By og Havn.

QUNLI STORMWATER PARK, HARBIN, CHINA

Liu Jie, 2017, *Empirical Research of Designed Ecologies Based on Three Urban Wetland Parks,* Ph.D. Dissertation at Peking University

Yu Kongjian, Zhang Lei and Li Dihua, 2008, Living with Water: Flood Adaptive Landscapes in the Yellow River Basin of China, *Journal on Landscape Architecture, Autumn 2008,* pp. 6–17.

AUTHORS

LANE ADDONIZIO is Vice President for Planning, Design & Construction and is responsible for managing the physical planning effort and restoration program for Central Park. Her background is city planning, public space improvement and historic preservation.

CENTRAL PARK CONSERVANCY

HOOSHMAND ALIZADEH is Associate Professor of Urban Design, University of Kurdistan in Iran, has served as an academic member of the Department of Urban Planning and Design and is now at the Institute for Urban and Regional Research in the Austrian Academy of Sciences, Vienna.

SARA CEDAR MILLER is a Historian Emerita and has for many years been historian and photographer for Central Park Conservancy. She has written several books about Central Park.

HANNAH MRAKOVCIC

KAITLYN CHOW, Master of Environmental Science, is a writer and project coordinator for Parks Canada, working on the establishment of Rouge National Urban Park. After a youth spent in the Rocky Mountains of western Canada, she learned first-hand the value of urban parks after moving to Toronto.

PETER CLARK is Emeritus Professor of European Urban History at the University of Helsinki. From 1985 to 1999 he was director of the Centre for Urban History at Leicester University UK. He is one of the co-founders of the European Association for Urban History and has published extensively on European and global urban development.

CATA PORTIN

JULIA CZERNIAK is Professor and Associate Dean at Syracuse University in the United States. Educated both as an architect and landscape architect, her work—which moves between design and writing—draws on the intersection of these disciplines. She is co-editor of the book Large Parks (2007) and editor of Downsview Park Toronto (2002).

SYRACUSE UNIVERSITY

ELIZABETH DAVEY, M.A., is a retired teacher and lecturer with a particular interest in landscape history. For over a decade she has made a special study of the history of Birkenhead Park and of the life and work of the Park's superintendent, the prolific Victorian landscape architect, Edward Kemp. She has published a History of Birkenhead.

PAM SULLIVAN

PHYLLIS ELLIN is an architectural historian who specializes in the World Heritage program for the U.S. National Park Service. She is now retired from the Service but is still working as a contractor for them.

BARBARA GORDON

URBAN EMANUELSSON is an ecologist and Professor at the Swedish University of Agricultural Sciences, where he started up the Center for Biodiversity in 1995 and led it until 2008. Major works are books on the shaping of European and North-East Asian landscapes.

PRIVATE SOURCE

GRAHAM FAIRCLOUGH worked in the UK government agency English Heritage until 2012 and is presently a research member of the McCord Centre for Landscape at Newcastle University (UK) and co-editor of the journal *Landscapes*. He coordinated the EU CHeriScape network (2012–2020) and is currently a member of the EU HERILAND project.

PRIVATE SOURCE

C.CAMBON

RICHARD T. T. FORMAN, Professor Emeritus at Harvard University, is often considered a "father" of landscape ecology and road ecology. He has helped spearhead urban ecology, and recently pioneered town ecology. His primary scholarly interest links science with spatial patterns to interweave nature and people on the land.

CARINA GRAN

PER HALLÉN is a senior lecturer at the Department of Economy and Society, Unit for Economic History in Gothenburg. One of his research themes is urban economy and history with a special interest in the economic aspect of urban planning and its effect on green space.

PIOTR WIŚNIEWSKI

BEATA FORTUNA-ANTOSZKIEWICZ, PhD, is Assistant Professor of Landscape Architecture at Warsaw University of Life Sciences (WULS-SGGW), Poland, and a member of the Association of Polish Architects (SARP) and Polish Dendrology Society (PTD).
She is professionally involved in conservation and protection of historical parks, as well as shaping and care of urban greenery.

ODETTE VAN DIJK

JAN HEEREN is educated as a political scientist. Since 2012 he has been working as a senior policy advisor in the Amsterdam Forest, part of the Sports and Recreation Department of the City of Amsterdam.

GAUTAM S PATEL

DEBI GOENKA is an Executive Trustee of Conservation Action Trust, an NGO working with the environment. He has all his life been active in environmental protection in India. He has served on various Committees appointed by the Bombay High Court and the Government. He stresses the role of forests in protecting the water security of India.

REBECA DIOS

CECILIA HERZOG is an Urban Landscape Planner, and Professor at Pontifical Catholic University of Rio de Janeiro, Department of Architecture and Urbanism. She also works as a consultant for Global Environmental Facility in Brazil on the III Sector Dialogue on Nature-based Solutions between EU and Brazil.

OFFICE OF PUBLIC WORKS

MARGARET GORMLEY is the Chief Park Superintendent with the Office of Public Works in Ireland where she is responsible for historic parks and gardens conservation, restoration, management, and presentation, including the Phoenix Park.

PRIVATE SOURCE

OLE HJORTH CASPERSEN, University of Copenhagen, researches ecological geography with emphasis on environmental issues, land use, planning, landscape dynamics, and present and historical agroecosystems. Much of his research uses GIS and aims at development of methods for planning and management of recreational landscape experiences.

KYOUMARS HABIBI is Associate Professor of urban planning at the University of Kurdistan, Iran. He holds a PhD from the University of Tehran, Iran, on urban planning and GIS. He has been the head of the Urban Planning and Design Department 2017–2019. His research mainly focuses on urban planning, urban regeneration, GIS, and urban tourism.

PRIVATE SOURCE

FRAN HORSLEY has worked in park management in the Victorian public sector for almost 30 years. She has a particular passion for the importance of urban parks as intrinsic to the health of communities and in shaping healthy, livable cities. She is currently leading strategic planning for metropolitan Melbourne.

VERA MIRONOVA

MARIA IGNATIEVA is Professor of Landscape Architecture at the School of Design, University of Western Australia. She has been working in academia in Russia, USA, Sweden, New Zealand, and Australia, and is an Honorary Doctor of St. Petersburg State Forest Technical University.

MARIÀ MARTÍ/CPNSC

JOSEP MASCARÓ CATALÀ is an architect and landscape architect educated at the University of Barcelona. He has worked in architecture and landscaping in Latin America and Spain. From 1988 to 2018, he was the chief architect of projects and works in the Collserola Park and participated in the drafting of the "Special Protection Plan for Collserola Park."

PRIVATE SOURCE

DMITRY KAVTARADZE is an urban ecologist and a professor at the Faculty of Biology, Moscow State University. He has been fully immersed in eco-city design—Ecopolis —near Moscow and is especially engaged in creating education simulation games for sustainable development, both for urban planners and children.

PRIVATE SOURCE

MEL MCGREGOR, PhD, is a wildlife ecologist who has worked to develop our understanding of how habitat connectivity is impeded by linear infrastructure and how this can be overcome.

SREEJITH

NUPUR PROTHI KHANNA has degrees in physical planning and landscape architecture from SPA, Delhi, India; and heritage conservation studies from York University, U.K. Founder of a research based design practice, Beyond Built, she has focused on public spaces across India. She is currently based in Stockholm. She has held positions in ICOMOS, IFLA/ISOLA, and is now a board memeber of ICOMOS.

CURTIN UNIVERSITY

ISAAC MIDDLE, PhD, Curtin University, Perth, Western Australia, focuses on the planning and design of public and private green spaces—including neighborhood parks, sports grounds, coastal and river reserves, and backyards—with the aim of getting more people outdoors, active, and in contact with nature.

MARÍA CLAUDIA LÓPEZ was Secretary of Culture, Recreation, and Parks in the city of Bogotá, Colombia, until December 2019, and was engaged in the master planning of the city.

LILIA HAUA MIGUEL has been the CEO of the Pro Bosque de Chapultepec Trust Fund for 13 years, working with fund-raising for remodeling, rehabilitation and preservation of the Chapultepec Forest in Mexico City. She is a member of the International Chamber of Commerce in Mexico, acting as Vice President of its Environment Commission.

PIOTR WIŚNIEWSKI

JAN ŁUKASZKIEWICZ, PhD, is an Assistant Professor at the Landscape Architecture Department, Warsaw University of Life Sciences (SGGW), Poland, and a member of the Association of Polish Architects (SARP); Polish Dendrology Society (PTD); and Polish Association of Green Roofs (PSDZ). The maintenance of trees and urban greenery is a major professional interest, in both writing and design.

YUHINA MONGA

SUNJOY MONGA is a full-time travel and wildlife author and photographer in India. He organizes programs on Indian wildlife, environment, and bird-watching for schools, corporations, and overseas agencies. He was awarded the BBC-British Gas International Wildlife Photographer of the Year prize, and has made TV series on nature.

EARTHWORKS LANDSCAPE ARCHITECTS

ADAM VAN NIEUWENHUIZEN is the founder and managing director of Earthworks Landscape Architects, which focuses on upgrading transformed urban and rural parks in Africa into sustainable, inspiring, and functional places. He is a part-time lecturer at the University of Cape Town's Department of Landscape Architecture.

SAHAR COSTON-HARDY

ERIC TAMULONIS is a landscape architect who leads the consulting practice, PARRICUS design, focused largely on planning and designing major public parks. A board member of World Urban Parks and City Parks Alliance, he believes parks are among the great cultural achievements of our society.

JIM DONOVAN

PATRICIA M. O'DONNELL is a preservation landscape architect and urban planner. She founded the firm Heritage Landscapes LLC in the USA in 1987. She has led preservation and revitalization projects for more than 200 public park landscapes and is deeply committed to inclusive, just, and flourishing public parks for uplifting everyday urban life.

EARTHWORK LANDSCAPE ARCHITECTS

TESSA TOERIEN practices landscape architecture at the Earthworks Landscape, alongside Adam and a small band of other designers. With a degree in architecture and a master's degree in landscape architecture, her interest lies in delivering contextually relevant solutions which are environmentally and socially responsible, and fun. She has worked on local and international public space projects, parks, and playgrounds.

PRIVATE SOURCE

TERESA PASTOR was the Coordinator of Fedenatur for 15 years and, after its integration with Europarc Federation, she is currently in charge of the peri-urban parks dossier. She holds a PhD in Biology from the University of Barcelona.

PROVINCIAL PARKS REGION ANTWERP

PETER VERDYCK has a PhD in Biology and as a director of the Provincial Parks of the Antwerp Region he is responsible for the management of four green areas: the parks Rivierenhof, Vrieselhof, and Hof Van Leysen, and the natural heathland area Kesselse Heide.

PRIVATE SOURCE

ANNA SOFIE PERSSON, PhD, works at the Center for Environment and Climate Research, Lund University, Sweden. Her research concerns biodiversity and ecosystem services in farmland landscapes and urban areas, with a particular focus on insect pollinators.

JOSEP MASCARÓ CPNSC

MARIÀ MARTÍ VIUDES has a PhD in biology from the Universitat Autònoma de Barcelona. He has managed the Collserola Natural Park since 1993 and served as Secretary General of FEDENATUR—European Federation of Periurban Parks (1997–2017)—and is currently a member of the Council of the Europarc Federation.

PETER SCHANTZ, Dr Med Sc, is professor in human biology, including the multidisciplinary field of human movement, health, and environment, at the Swedish School of Sport and Health Sciences, GIH, in Stockholm, Sweden.

PRIVATE SOURCE

LIISA WIHMAN has a Masters degree in Arts, with a major in Art History. She has been working as a freelance writer and art exhibition coordinator in several countries, recently in Tokyo. Gardens and garden design is one of her major interests.

TJARK WILKE

TORSTEN WILKE, Dipl.-Ing. Landscape Planning, University of Hannover, has since 2010 been working as coordinator for strategic spatial development of the urban green-blue infrastructure for the city of Leipzig.

JOAN BURTON WHENT

STEPHEN WILKINSON has been a chartered town planner for over 30 years and is a former President of the Royal Town Planning Institute. Until 2019 he was Head of Planning and Strategic Partnerships at the Lee Valley Regional Park Authority.

JAN ŁUKASZKIEWICZ

PIOTR WIŚNIEWSKI is a graduate in Landscape Architecture from Warsaw University of Life Sciences (SGGW), Poland. He is professionally associated with the issues of conservation, care, and protection of urban greenery, and is a member of the Association of Polish Architects (SARP) and Polish Dendrology Society (PTD).

JUDY LING WONG is a Founder of the National Park City Foundation. She is a painter, poet, and environmental activist, best known as the Honorary President of the charity Black Environment Network. Judy is a vision caster and a major voice on policy towards social inclusion.

KONGJIAN YU/TURENSCAPE

KONGJIAN YU is a professor and Dean of the Graduate School of Landscape Architecture at Peking University; founder of Turenscape landscape architects; and lead designer of Qunli Stormwater Park and other similar parks.

ACKNOWLEDGMENTS

I would like to offer my warmest thanks to Taco I. Matthews for her ever impressive illustrations, and to Richard Murray for the effective photographs.

Richard T. T. Forman

This 'survey' would not have been possible without the thoughtful suggestions of over fifty of my landscape colleagues (including designers, academics, practitioners, and theorists) from around the world. I thank each of them for their generous sharing, which enabled this crowd-sourced research. This work was first presented as a keynote lecture in the Large Parks in Large Cities conference at the Museum of Natural History, Stockholm, Sweden, in September 2015.

Julia Czerniak

Special thanks to: Natalia Danilina, Director of the Russian
NGO Zapovedniks; Grigori Eremkin, biologist of Moscow State University; and Victor Solodushkin, Head f the Children's Ecological and Educational Center of the Federal State National Park Losiny Ostrov.

Dmitry Kavtaradze

I want to thank Jukka-Pekka Flander at the Ministry for Environment in Finland, who supplied valuable information about the Finnish National Urban Parks.

Richard Murray

Anthony Wain provided substantial input to the article on African Urban parks.
The text on Parc National du Mali is from https://bridgesfrombamako.com/2011/10/07/bamakos-parc-national/. By courtesy of Bruce Whitehouse.
I express my thanks to both of them.

Adam van Nieuwenhuizen

Permission has been granted to include the SDG icons in this book. The content of this publication has not been approved by the United Nations and does not reflect the views of the United Nations or its officials or Member States. Information about the SDGs can be found at the website https://www.un.org/sustainabledevelopment/

INDEX

PARK PEOPLE

Editor: Richard Murray
Cover: Central Park, photo by Sara Cedar Miller,
    courtesy Central Park Conservancy
Maps: Martin Thelander
Graphic design: Göran Dyhlén, Judith Wernholm
Language review and proof reading:
    Frances Boylston and Hannah Clarkson